TRANSFER RNA
in
PROTEIN SYNTHESIS

Edited by

Dolph L. Hatfield, B.A., M.A., Ph.D.
Research Biologist
Laboratory of Experimental Carcinogenesis
National Cancer Institute
National Institutes of Health
Bethesda, Maryland

Byeong J. Lee, B.A., M.A., Ph.D.
Assistant Professor
The Institute for Molecular Biology and Genetics
Seoul National University
Seoul, Korea

Robert M. Pirtle, B.S., Ph.D.
Associate Professor
Department of Biochemistry
University of North Texas/Texas College
of Osteopathic Medicine
Denton, Texas

CRC Press
Boca Raton   Ann Arbor   London   Tokyo

Library of Congress Cataloging-in-Publication Data

Transfer RNA in protein synthesis / editors, Dolph L. Hatfield, Byeong J. Lee, Robert M. Pirtle.
    p. cm.
    Includes bibliographical references and index.
    ISBN 0-8493-5698-9
    1. Transfer RNA. 2. Proteins—Synthesis. I. Hatfield, Dolph L.
II. Lee, Byeong J. III. Pirtle, Robert M.
    [DNLM: 1. Codon—physiology. 2. Proteins—biosynthesis. 3. RNA, Transfer—physiology. QU 58 T77112]
QP263.5T73T72   1992
574.87'3283—dc20
DNLM/DLC
for Library of Congress   92-4137
                                                                             CIP

    This book represents information obtained from authentic and highly regarded sources. Reprinted material is quoted with permission, and sources are indicated. A wide variety of references are listed. Every reasonable effort has been made to give reliable data and information, but the author and the publisher cannot assume responsibility for the validity of all materials or for the consequences of their use.

    Neither this book nor any part may be reproduced or transmitted in any form or by any means, electronic or mechanical, including photocopying, microfilming, and recording, or by any information storage and retrieval system, without permission in writing from the publisher.

    Direct all inquiries to CRC Press, Inc., 2000 Corporate Blvd., N. W., Boca Raton, Florida, 33431.

© 1992 by CRC Press, Inc.

International Standard Book Number 0-8493-5698-9

Library of Congress Card Number 92-4137
Printed in the United States 1 2 3 4 5 6 7 8 9 0
Printed on acid-free paper

DEDICATION

The editors take great pleasure in dedicating this book to their loving wives, Mary J. Wilson-Hatfield, Hwan Yeon Lee, and Irma Lindsey Pirtle, and to their loving children, Hugh H. Hatfield and his wife Cherryl, Sandra L. Hatfield, Michéle J. Hatfield, and Hye Jin Lee and Hye Seo Lee.

EDITORS

Dolph L. Hatfield, Ph.D. is a research biologist at the Laboratory of Experimental Carcinogenesis, National Cancer Institute, National Institutes of Health, Bethesda, Maryland.

Dr. Hatfield received his B.A., M.A., and Ph.D. degrees in 1958, 1960, and 1962, respectively, in the Genetics Foundation at the University of Texas, Austin, Texas. After postdoctoral training in the laboratories of Drs. James B. Wyngaarden, Duke University Medical Center, Durham, NC, Marshall Nirenberg, National Institutes of Health, Bethesda, MD, and Jacques Monod, Pasteur Institute, Paris; he joined the National Cancer Institute, National Institutes of Health, where he has served as a research biologist since 1967. He was awarded a Robert A. Welsh Foundation Scholarship for studies at the University of Texas, a PHS Postdoctoral Fellowship for studies at Duke University, an American Cancer Society Postdoctoral Fellowship for studies at NIH, and a National Cancer Institute Postdoctoral Fellowship for studies at the Pasteur Institute.

Dr. Hatfield has been a member of the American Society for Biochemistry and Molecular Biology since 1971. His research interests are in the occurrence and role of nonsense suppressor tRNAs and in the mechanism of expression of tRNAs in mammalian cells and in the role that specific hypomodified bases in tRNAs, which occur in cancer tissue, may have in the malignancy process.

Byeong J. Lee, Ph.D. is an assistant professor in the Institute for Molecular Biology and Genetics and is also associated with the Department of Microbiology at Seoul National University, Seoul, Korea.

Dr. Lee received his B.A., M.A., and Ph.D. degrees in 1978, 1982, and 1986, respectively, in the Department of Microbiology, at Seoul National University. He was a visiting fellow from 1986 to 1989 and then a visiting associate from 1989 to 1991 in the Laboratory of Experimental Carcinogenesis, National Cancer Institute, National Institutes of Health, Bethesda, MD.

Dr. Lee is a member of the American Society of Biochemistry and Molecular Biology, the Korean Society of Molecular Biology, and the Microbiological Society of Korea. His research interests are in the organization, expression, and function of eukaryotic tRNA genes.

Robert M. Pirtle, Ph.D. is Associate Professor of Biochemistry at the University of North Texas in Denton, and is also associated with the Department of Biochemistry and Molecular Biology at the Texas College of Osteopathic Medicine in Fort Worth.

Dr. Pirtle graduated in 1967 with a B.S. in Applied Psychology from the Georgia Institute of Technology in Atlanta, where he was supported by a General Motors Scholarship. After a year of graduate study in psychology at the University of Louisville in Kentucky, he entered the doctoral program in biochemistry in the University of Louisville School of Medicine, and received a Ph.D. in Biochemistry in 1973. His postdoctoral training in nucleic acids biochemistry and molecular biology was done at the State University of New York at Stony Brook, where he had an NIH postdoctoral fellowship. From 1980 to 1987, he was an Assistant Professor of Biochemistry at the University of North Texas/Texas College of Osteopathic Medicine. There, in 1987, he became an Associate Professor.

Dr. Pirtle is a member of the American Society of Biochemistry and Molecular Biology, American Society for Microbiology, American Chemical Society, American Association for the Advancement of Science, and the honorary society Sigma Xi. In addition to teaching biochemistry and molecular biology to graduate, undergraduate, and medical students, he has been the recipient of research grants from the National Institutes of Health, the National Science Foundation, the Texas Advanced Research Program, The Robert A. Welch Foundation, the Research Corporation, and the Samuel Roberts Noble Foundation. His current major research interests relate to the structure and chromosomal organization of human transfer RNA genes, the regulation of tRNA gene expression, and the posttranscriptional modification and processing of tRNA.

CONTRIBUTORS

John Abbotts, Ph.D.
Research Associate
Laboratory of Biochemistry
National Cancer Institute
National Institutes of Health
Bethesda, Maryland

Glenn R. Björk, Ph.D.
Professor
Department of Microbiology
University of Umeå
Umeå, Sweden

Conal J. Burgess, B.A.
Research Student
Department of Genetics
Trinity College
Dublin, Ireland

In Soon Choi, B.A., M.A., Ph.D.
Visiting Fellow
Laboratory of Experimental
  Carcinogenesis
National Cancer Institute
National Institutes of Health
Bethesda, Maryland

Elizabeth Cowe, B.Sc.
Research Student
Department of Genetics
Trinity College
Dublin, Ireland

James F. Curran, Ph.D.
Assistant Professor
Department of Biology
Wake Forest University
Winston-Salem, North Carolina

André Dietrich, D.Sci.
Directeur de Recherches
Institut de Biologie Moléculaire
  des Plantes
Université Louis Pasteur
Strasbourg, France

Laurence Maréchal Drouard, D.Sci.
Chargé de Recherches
Institut de Biologie Moléculaire
  des Plantes
Université Louis Pasteur
Strasbourg, France

George A. Gutman, Ph.D.
Professor
Department of Microbiology and
  Molecular Genetics
University of California, Irvine
Irvine, California

Dolph L. Hatfield, B.A., M.A., Ph.D.
Research Biologist
Laboratory of Experimental
  Carcinogenesis
National Cancer Institute
National Institutes of Health
Bethesda, Maryland

G. Wesley Hatfield, Ph.D.
Professor
Department of Microbiology and
  Molecular Genetics
University of California, Irvine
Irvine, California

Toshimichi Ikemura, D.Sci.
Professor
Department of Population Genetics
National Institute of Genetics
Mishima, Japan

Jae-Eon Jung, B.A., M.A., Ph.D.
Visiting Fellow
Laboratory of Experimental
  Carcinogenesis
National Cancer Institute
National Institutes of Health
Bethesda, Maryland

Ulf Lagerkvist, Ph.D.
Professor
Department of Medical
 Biochemistry
Gothenburg University
Göteborgs, Sweden

Byeong J. Lee, B.A., M.A., Ph.D.
Assistant Professor
The Institute for Molecular
 Biology and Genetics
Seoul National University
Seoul, Korea

Andrew T. Lloyd, B.A., Ph.D.
Research Fellow
Department of Genetics
Trinity College
Dublin, Ireland

Kevin J. Mitchell, B.A.
Research Student
Department of Genetics
Trinity College
Dublin, Ireland

Susumo Ohno, D.V.M., Ph.D., D.Sci.
Chair
Department of Theoretical Biology
Beckman Research Institute of
 The City of Hope
Duarte, California

Leo Pallanck, B.S.
Graduate Student
Department of Developmental
 Biology and Cancer
Albert Einstein College of
 Medicine
Bronx, New York

Jack Parker, Ph.D.
Professor
Department of Microbiology
Southern Illinois University
Carbondale, Illinois

Irma L. Pirtle, B.S., Ph.D.
Research Assistant Professor
Department of Biochemistry
University of North Texas/
 Texas College of Osteopathic
 Medicine
Denton, Texas

Robert M. Pirtle, B.S., Ph.D.
Associate Professor
Department of Biochemistry
University of North Texas/
 Texas College of Osteopathic
 Medicine
Denton, Texas

LaDonne Schulman, Ph.D.
Professor
Department of Developmental
 Biology and Cancer
Albert Einstein College of
 Medicine
Bronx, New York

Paul M. Sharp, B.S., Ph.D.
Lecturer
Department of Genetics
Trinity College
Dublin, Ireland

Jacques-H. Weil, D.Sci.
Professor
Institut de Biologie Moléculaire
 des Plantes
Université Louis Pasteur
Strasbourg, France

Samuel H. Wilson, M.D.
Chief, NAE
Laboratory of Biochemistry
National Cancer Institute
National Institutes of Health
Bethesda, Maryland

Michael Yarus, Ph.D.
Professor
Department of Molecular, Cellular,
 and Developmental Biology
University of Colorado
Boulder, Colorado

TABLE OF CONTENTS

Chapter 1
tRNA in the Molecular Biology of Retroviruses 1
Samuel H. Wilson and John Abbotts

Chapter 2
The Role of Modified Nucleosides in tRNA Interactions 23
Glenn R. Björk

Chapter 3
Correlation Between Codon Usage and tRNA Content in
Microorganisms 87
Toshimichi Ikemura

Chapter 4
Aminoacyl-tRNA (Anticodon): Codon Adaptation in
Higher Eucaryotes 113
Dolph L. Hatfield, Jae-E. Jung, Byeong J. Lee, and In S. Choi

Chapter 5
Adaptation of tRNA Population to Codon Usage in
Cellular Organelles 125
Laurence Maréchal-Drouard, André Dietrich, and Jacques-H. Weil

Chapter 6
Differential tRNA Gene Expression in Eukaryotes 141
Robert Pirtle and Irma Pirtle

Chapter 7
Codon Pair Utilization Bias in Bacteria, Yeast, and Mammals 157
G. W. Hatfield and George A. Gutman

Chapter 8
Variations in Reading the Genetic Code 191
Jack Parker

Chapter 9
Selenocysteine, a New Addition to the Universal Genetic Code 269
Dolph L. Hatfield, In S. Choi, Byeong J. Lee, and Jae E. Jung

Chapter 10
tRNA Discrimination in Aminoacylation 279
Leo Pallanck and LaDonne Schulman

Chapter 11
The Translational Context Effect 319
Michael Yarus and James Curran

Chapter 12
Codon Discrimination in Translation 367
Ulf Lagerkvist

Chapter 13
Universal Rule of TA/CG Deficiency and TG/CT Excess 381
Susumu Ohno

Chapter 14
Selective Use of Termination Codons and Variations in Codon Choice ..397
**Paul M. Sharp, Conal J. Burgess, Andrew T. Lloyd,
and Kevin J. Mitchell**

Index 427

Chapter 1

tRNA IN THE MOLECULAR BIOLOGY OF RETROVIRUSES

Samuel H. Wilson and John Abbotts

TABLE OF CONTENTS

I. Introduction: Retroviruses and Their Life Cycle 2

II. Historical Evidence for tRNA Involvement 5

III. tRNA in HIV-1 Life Cycle 7

IV. HIV-1 RT and $tRNA_3^{Lys}$ Interaction 8

V. Retroviral Nucleocapsid Protein Acts as a Replication-Accessory Factor 13

VI. Conclusions 17

Acknowledgments 17

References 18

I. INTRODUCTION: RETROVIRUSES AND THEIR LIFE CYCLE

Eighty years ago, Peyton Rous was the first to identify a filterable agent that caused tumors in chickens, and it is now known that this agent, Rous sarcoma virus (RSV), is a retrovirus, converting its RNA genome to DNA during its life cycle. Viruses of this type were neglected for several decades, until strains that caused tumors in mammals were isolated. These agents were identified as retroviruses in the late 1960s and 1970s with the discovery of reverse transcriptases and proviruses transmitted in the germ line.[1,2] More recently, retroviruses have gained further attention as the presumed causative agents of T cell leukemia and acquired immune deficiency syndrome (AIDS). Indeed, major sections of this chapter will focus on tRNA involvement in the life cycle of the AIDS virus, human immunodeficiency virus, type 1 (HIV-1).

Figure 1 is a diagram depicting the structural and biochemical features of a typical retroviral particle. The particle contains a homodimer of the RNA genome and several copies of the viral enzyme reverse transcriptase (RT). The viral RNA contains a capped nucleotide on its 5' end and a poly A tail on its 3' end. Viral RNA and RT are surrounded by a protein core, which in turn is surrounded by an envelope consisting of viral proteins and membrane from the previous host cell. Retroviral particles attach to cells through envelope proteins which recognize normal cell surface proteins; entry is believed to proceed by receptor-mediated endocytosis. Once inside cells, the viral particle uncoats to release an enzymatically active nucleoprotein complex that carries out reverse transcription, replication, and eventually integration of product double-stranded DNA into the host genome. When integrated into the host chromosome, viral DNA enters a latent phase, reminiscent of phases in the phage λ life cycle.[2]

Figure 2 illustrates the organization of the viral genome in its integrated, proviral state. The provirus contains two long terminal repeat sequences (LTR). The major genes are termed *gag*, *pol*, and *env*. The gene *gag* (group specific antigen) codes for proteins which make up the viral core. *Pol* codes for the reverse transcriptase and integrase proteins. *Env* codes for envelope proteins. Viral proteins are expressed as long precursors that are processed by a protease, which depending on the virus may be expressed from either the *gag* gene or the *pol* gene, or in a separate reading frame.[4] Different viral strains may contain additional genes, which code for oncogenic products or proteins that regulate viral expression, such as HIV-1 *tat* or human T-cell lymphotropic virus, type 1 (HTLV-1) *tax*. The provirus remains latent in the integrated state until stimuli induce viral expression by mechanisms that are not well-characterized. At this point, the cellular machinery carries out viral transcription and translation. With the synthesis of viral RNA and protein, assembled particles exit by budding through the cell membrane, which contributes to the viral envelope.

Figure 3 shows steps in retroviral replication, in which genomic RNA is converted into a double-stranded DNA form capable of integration into the host genomic DNA. The primer for DNA synthesis is a specific tRNA, which

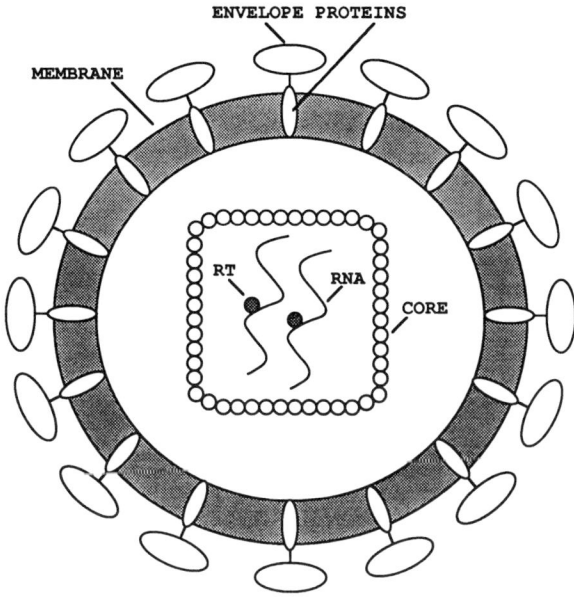

FIGURE 1. Retrovirus particle. The retroviral particle contains, among other components, the RNA viral genome and reverse transcriptase (RT), which are surrounded by a protein core. The core in turn is surrounded by the viral particle, containing lipid membrane from the host cell and viral envelope proteins.[3]

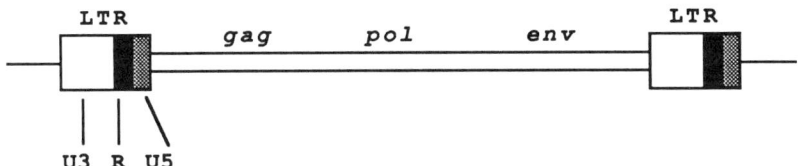

FIGURE 2. Organization of the retroviral genome. The genome is indicated in the proviral form, double-stranded DNA integrated into host DNA. The components of the long terminal repeat (LTR) from the 5' end, are U3, R, and U5, described in Figure 3. The order of the *gag*, *pol*, and *env* genes are indicated. Details are contained in the text.[2]

is derived from the host cell and packaged into virions. The nucleotide sequence at the 3' end of the appropriate primer tRNA hybridizes to a complementary 18-19 residue viral sequence, near the 5' end of the RNA genome. DNA synthesis using RNA as template is carried out by the viral reverse transcriptase. The initial product of elongation of the tRNA is "minus strong stop DNA," the result of DNA synthesis to the 5' end of the genome (Figure 3, step 1). The product molecule (minus viral strand) is then transferred to the 3' end of genomic RNA, annealing to a terminal repeat therein (step 2). Experiments with heterozygous virus particles indicate that this first strand "jump" is intermolecular.[5] The viral reverse transcriptase also possesses RNase

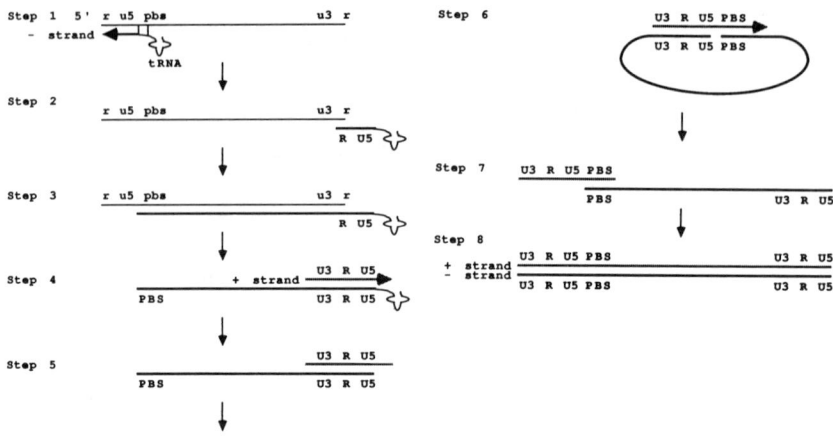

FIGURE 3. Model for retroviral replication. RNA is indicated by thin lines and lower-case designations for sequence regions. DNA is indicated by thick lines and upper-case designations for sequence regions; the minus strand is solid, the plus strand is shaded. The repeated sequence r is found at both ends of the viral RNA; sequences u5 and u3 are associated with the 5' and 3' r regions, respectively; pbs is the primer binding site to which the tRNA primer anneals. Viral RNA also contains a 3' poly A tail which is not replicated and is omitted from the figure for simplicity. Regions shown are not necessarily to scale; for example, pbs typically is 18 nucleotides and is located ~150 nucleotides from the 5' end. Details are described in the text.[5,6]

H activity, which degrades the RNA strand of DNA:RNA hybrids. This RNase H activity may facilitate transfer of the minus strand by removing genomic RNA, thus eliminating the need for the minus strand to melt before subsequent annealing on the 3' end.

The minus strand is then elongated to positions opposite the 5' region of genomic RNA (Figure 3, step 3). Elongation during this step stops at about the 5' region of the primer binding site (PBS).[7] This is consistent with a template whose r and u5 region had been removed by RNase H activity prior to the minus strand jump. During elongation in step 3, RNase H activity degrades viral RNA hybridized to DNA, and in the process produces a specific fragment, designated as the polypurine tract, just 5' of the u3 sequence. This tract serves as primer for the initiation of DNA-directed DNA synthesis of the plus strand, which is also carried out by reverse transcriptase (step 4). Plus strand synthesis proceeds into the primer binding site region of the tRNA. In Moloney murine leukemia virus (M-MuLV), 18 nucleotides of the primer tRNA are copied, which corresponds to the region of tRNA which hybridizes to viral RNA. Termination of synthesis at this point appears to be precise;[8] the next position on the primer tRNA is a modified base, which may signal a halt to synthesis.[6] RNase H activity of reverse transcriptase then removes the tRNA (step 5). Experiments with model substrates and reverse transcriptase of avian retrovirus indicate that the RNase H activity removes the tRNA primer precisely at the RNA-DNA junction, leaving the tRNA intact. RNase H cleavage occurs after synthesis of 18 nucleotides complementary to the tRNA.[9]

These 18 nucleotides facilitate a second strand transfer; this is a shift of the plus DNA strand to anneal to the PBS region of the DNA minus strand. This transfer is known to be intramolecular and may proceed by a circular intermediate (Figure 3, steps 6 and 7).[5] Synthesis of both strands is then completed to produce double-stranded DNA; synthesis may proceed from the circular intermediate, which would require strand-displacement activity of the reverse transcriptase. This double-stranded product is the precursor for integration into host DNA.

In the scheme shown in Figure 3, tRNA is part of the newly-synthesized minus strand during steps 1 through 4; further discussion will focus on phenomena relevant to these steps. We note that host tRNAs play other important roles in the molecular biology of retroviruses through their involvement in ribosomal frameshifting and in-frame suppression during translation of the viral *gag-pol* precursor protein. This topic has been extensively discussed in recent reviews[68,69] and in Chapter 9 of this book, and therefore will not be covered here.

II. HISTORICAL EVIDENCE FOR tRNA INVOLVEMENT

Work with RNA tumor viruses in t he 1960s identified 70S RNA as the major nucleic acid entity in virions; subsequent work demonstrated that this species is a dimer of the 35S genomic RNA.[10] In the 1970s, small RNA molecules (4S to 9S) were found in retroviral virions, in addition to the high molecular weight genomic RNA.[11] Of these smaller species, the most abundant component was 4S RNA. It was also observed that if 70S RNA was heat-denatured, one could identify 35S genomic RNA, but a 4S RNA species was also released. Thus, two different populations of 4S RNA were identified in retroviral particles, designated "free" and "70S-associated." It was subsequently learned that the major component of both populations of 4S RNA was tRNA.[10]

Dahlberg et al.[12] identified a 4S RNA species which was hydrogen-bonded to the 35S genomic RNA of Rous sarcoma virus (RSV); the 4S RNA served as a primer for *in vitro* DNA synthesis with the 35S RNA and purified RT. Subsequent research identified the 4S RNA in RSV as tRNA^{Trp},[13] and the primer for M-MuLV as tRNA^{Pro}.[14] It was then observed that RT from avian myeloblastosis virus (AMV) would carry out DNA synthesis with a tRNA^{Trp} primer on RSV genomic RNA or with a tRNA^{Pro} primer on M-MuLV genomic RNA.[15] Thus, the appropriate tRNA primer for a given retrovirus appeared to correlate with the presence of complementary sequences on the genomic RNA to which the 3' end of tRNA anneals.[16]

In addition to the respective tRNA primers, however, virions were found to contain other tRNAs as part of the "free" 4S RNA population. The tRNAs in retrovirus particles are derived from host cells, yet their composition represents a concentration in a subset of the cellular tRNA population, rather than a random mixture of the cellular tRNA. Moreover, tRNA populations are specific for categories of virus.[10] For example, a comparison of 4S RNA in virions of RSV and Rous associated virus (RAV) revealed that RAV contained much higher levels of tRNAs for Arg, Asn, Asp, and Glu. Both viruses contained a population of tRNAs clearly different from those of infected or uninfected chicken cells.[17] No function has yet been identified for these nonpriming tRNAs.

Particles of some retroviruses may contain 10 or more species of tRNA. In contrast, virions of mouse mammary tumor virus (MMTV) contain only two major tRNA species. One species is identified as $tRNA_{1,2}^{Lys}$, and the other as $tRNA_{3}^{Lys}$. In MMTV virions, these species are present in a ratio of approximately 2:1, but $tRNA_{3}^{Lys}$ is the primer for viral replication and is found more frequently than $tRNA_{1,2}^{Lys}$ in association with viral RNA.[18] The next section describes the evidence that $tRNA_{3}^{Lys}$ is also the primer for HIV.

To summarize, only a few tRNA primers have been identified to date for retroviruses: $tRNA^{Trp}$ for avian viruses, $tRNA^{Pro}$ for murine leukemia viruses and most other mammalian viruses, $tRNA_{1,2}^{Lys}$ for Visna virus,[19] $tRNA_{3}^{Lys}$ for MMTV, HIV-1 and -2, and for some other mammalian viruses. These tRNAs share an unusual feature in the pseudouridine loop. The consensus sequence among all tRNAs in this loop is G-U-Ψ-C. This same region is G-Ψ-Ψ-C for $tRNA^{Trp}$ and $tRNA^{Pro}$, and G-Tm-Ψ-C for the HIV-1 primer $tRNA_{3}^{Lys}$, where Tm is 2′-O-methyl ribosylthymine.[18] For these tRNA primers, the 3′-terminal 18 nucleotides are complementary to the corresponding viral RNA. The Ψ in the pseudouridine loop is a few nucleotides 5′ of the primer sequence. Thus, this region is not involved directly in priming, but it remains to be seen whether it might have some regulatory role in retroviral replication, such as conferring a strong stop site.

In avian cells, $tRNA^{Trp}$ represents only about 2% of total cellular tRNA, but this species, which is the primer for avian sarcoma virus (ASV) replication, makes up 30 to 50% of the tRNA in ASV particles. Studies with mutant virus which lacked detectable RT activity indicated that these strains showed no selectivity in tRNAs in viral particles; that is, tRNAs in these particles were not enriched in $tRNA^{Trp}$.[20] In contrast, a mutant which failed to package viral RNA, inserting cellular mRNA into its viral particles, was found to contain tRNAs of the same composition as wild-type particles. These workers also showed that *in vitro* binding of avian cellular 4S RNA to AMV RT revealed a selected population of tRNAs identical to the tRNA population in ASV particles.

Similarly, Levin et al.[21] developed a system in which virus maturation is uncoupled from synthesis of viral RNA by actinomycin D treatment of cells chronically infected with MuLV. Although this treatment leads to a rapid

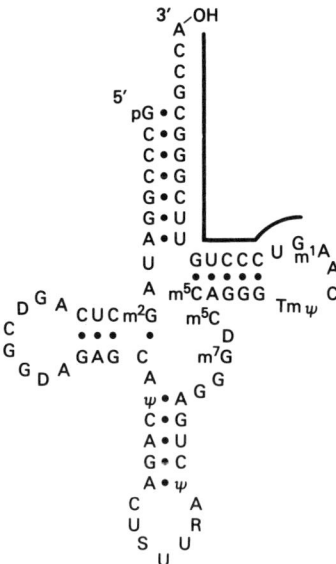

FIGURE 4. Nucleotide sequence of rabbit liver tRNA$_3^{Lys}$ presented in cloverleaf form. Modified nucleotides are: D, dihydrouridine; Ψ, pseudouridine; S, 5-(methoxycarbonyl-methy)-2-thiouridine; R, N-[N-(9-β-D-ribofuranosyl-2 methylthiopurin-6-yl)carbamoyl] threonine; Tm, 2′-O-methyl ribosylthymine; at other positions, m indicates a methylated G or A, with the position of methylation indicated. The solid line indicates 18 nucleotides at the 3′ end of the tRNA complementary to the HIV-1 primer binding site.[26]

cessation of cellular and viral RNA synthesis, virus particle assembly and release continue. These particles have morphology characteristic of MuLV and contain functional RT, and *gag* and *env* proteins.[22] The particles, however, lack viral 70S genomic RNA and are noninfectious. This lack of genomic RNA has no impact on the selective incorporation of a subset of host cell tRNAs into the virion, including the primer tRNA^{Pro}.[23] These results [20,23] indicate that the RT and perhaps other virion protein components are the major determinant of the tRNA composition in viral particles; the genomic RNA has little role in tRNA selectivity.

III. tRNA IN HIV-1 LIFE CYCLE

tRNA$_3^{Lys}$ was assigned as the probable primer for HIV-1 replication on the basis of nucleic acid sequencing results by Wain-Hobson et al.,[24] Ratner et al.,[25] and Raba et al.[26] The sequence of rabbit liver tRNA$_3^{Lys}$ and its posttranscriptional modifications are illustrated in Figure 4. The 3′ 18 residues of the sequence are base-pair complementary to a sequence in the 5′ region of HIV-1 genomic RNA considered to be the PBS. Furthermore, Darlix and associates[27] have shown that tRNA$_3^{Lys}$ can act as a functional primer for synthesis of minus strand strong stop DNA by the HIV-1 RT starting at the normal *in vivo* primer binding site on HIV-1 genomic RNA. This *in vitro* DNA synthesis reaction required the HIV nucleocapsid protein, as well as the RT and tRNA$_3^{Lys}$. Kleiman et al.[28] reported that HIV-1 virions contain a subset of five to ten species of cellular tRNA, depending upon the cell type analyzed. Several species were tightly associated with the genomic RNA, and these presumably included tRNA$_3^{Lys}$, based upon analogy with other retroviral systems, i.e., tight association reflects annealing between the tRNA and the primer binding site. These tightly associated tRNA species were not present in mutant virions lacking *pol* and *gag* gene products, confirming a requirement for these proteins in tRNA selection in the HIV-1 system.

IV. HIV-1 RT AND tRNA$_3^{Lys}$ INTERACTION

Much of the recent interest in tRNA retrovirology stems from the role of human tRNA$_3^{Lys}$ in HIV-1 replication. Due to the intense investigation of the HIV-1 RT, workers in this field recently have learned a great deal about the interactions of this RT with its cognate replication tRNA, and we will focus attention in this review on the HIV-1 system.

By analogy with the AMV system we can expect that the HIV RT has a role in two of the key retroviral metabolic events involving the primer, tRNA$_3^{Lys}$: (1) the enzyme participates in the selection process by which tRNA$_3^{Lys}$ is included in the subset cellular tRNA molecules packed into the virion, and (2) the enzyme participates in the positioning or annealing of tRNA$_3^{Lys}$ onto the PBS of the genomic RNA. It seems implicit for both of these events that the RT has the capacity to specifically bind to tRNA$_3^{Lys}$. Indeed, Barat et al.[29] and Sallafranque-Andreola et al.[30] demonstrated that recombinant RTs prepared from *E. coli* and yeast, respectively, are capable of forming complexes with tRNA$_3^{Lys}$, and the stability of these complexes was greater than complexes formed with at least one reference, nonprimer tRNA, suggesting a degree of specificity for tRNA$_3^{Lys}$ binding. Therefore, the first expectation mentioned above appears to be borne out. These experiments will be summarized in the following paragraphs. There is evidence that the second expectation may be true as well. Preliminary RNase sensitivity experiments by Sarih-Cottin et al.[31] indicate that the normally annealed acceptor stem of tRNA$_3^{Lys}$ is melted in the complex of RT-tRNA$_3^{Lys}$. Since the acceptor stem and the T stem must be melted in order for tRNA$_3^{Lys}$ to hybridize with the PBS, this melting effect of RT on the acceptor stem clearly points to an active role by the enzyme in the primer annealing process. Barat et al.,[27] using an *in vitro* RNA synthesis system that depends upon annealing of tRNA$_3^{Lys}$ to the PBS in a fragment of HIV genomic RNA, found that the RT stimulates annealing of tRNA$_3^{Lys}$ to the PBS. Therefore, the RT appears to be an active participant in tRNA$_3^{Lys}$ assembly into a productive DNA synthesis complex.

Indeed, considerable evidence has emerged during the last two years documenting binding between the RT and tRNAs or tRNA analogues. The first evidence came from detailed steady-state kinetic studies of the DNA polymerase activity of the virion-derived, immunoaffinity-purified p66/p51 heterodimer. This purified enzyme was studied with the synthetic template·primer system of poly r(A)·oligo d(T), and the results were consistent with a kinetic mechanism in which the enzyme free in solution combines first with the template·primer complex and then combines with the incoming dNTP substrate, i.e., dTTP. These studies also revealed that the primer molecule oligo d(T) could combine with the free enzyme and thereby inhibit the reaction (Figure 1c in reference 32). Nevertheless, the enzyme's affinity for template·primer appeared to be much greater than the affinity for the free primer. These ideas on the capacity of free RT to bind a primer analogue were subsequently confirmed in a study of the effect of heterologous primer analogue molecules on the DNA polymerase activity.[33] The synthetic oligonucleotide primer analogue $d(C)_{28}$ was found to inhibit the polymerase activity in the poly r(A)·oligo d(T)-directed dTMP incorporation system, and the inhibition by $d(C)_{28}$ was competitive with the template·primer, indicating that $d(C)_{28}$ could bind to the free enzyme and remove it from the reaction scheme. The $d(C)_{28}$ K_i value was 560 nM. Interestingly, the phosphorothioate derivative of $d(C)_{28}$, i.e., $Sd(C)_{28}$, also acted as a competitive inhibitor of the poly r(A)·oligo d(T)-directed dTMP incorporation reaction, but the K_i value was only about 3 nM, indicating relatively tight binding between the free enzyme and the sulfur-substituted primer analogue. In addition to confirming the RT plus primer analogue binding reaction, these results suggested the important idea that increased hydrophobic binding characteristics in the polynucleotide, i.e., sulfur vs. oxygen at nonbridging groups, could increase binding affinity to the enzyme (for discussion see Reference 33).

Subsequently, the Litvaks and their associates[30] used glycerol gradient centrifugation to observe complex formation between a yeast-derived recombinant p66/p51 heterodimer and a mixture of bovine $tRNA_{1,2}^{Lys}$ and $tRNA_{3}^{Lys}$; bovine $tRNA_{3}^{Lys}$, and human $tRNA_{3}^{Lys}$, incidentally, have the same nucleotide sequence. Under the incubation and centrifugation conditions used, a reference nonprimer tRNA, bovine $tRNA^{Trp}$, failed to bind. These workers also showed that $tRNA^{Lys}$ is able to protect the RT against heat inactivation at 45°C, whereas the control $tRNA^{Trp}$ is not. This effect between the enzyme and $tRNA^{Lys}$ is presumably secondary to complex formation, and similarly, $tRNA^{Lys}$ partially protects the enzyme from trypsin inactivation, whereas a reference nonprimer tRNA, bovine $tRNA^{Val}$, does not.

In a parallel, but independent study, Darlix and associates[29] used a Northwestern blotting approach to demonstrate binding between *E. coli*-derived recombinant p66/p51 RT and purified bovine $tRNA_{3}^{Lys}$. This binding reaction was not inhibited by a 20-fold excess of reference tRNAs, bovine $tRNA^{Pro}$ or $tRNA^{Trp}$, indicating a significant measure of RT specificity for $tRNA_{3}^{Lys}$ vs. the two reference tRNAs. These studies were extended by use of a gel mobility shift assay in which the stoichiometry between RT heterodimer and bovine $tRNA_{3}^{Lys}$ for maximal binding appeared to be 1:1. Bovine reference tRNAs, tRNA^{Pro} and tRNA^{Trp}, failed to compete with tRNA$_3^{Lys}$ binding at a 100-fold molar excess over tRNA$_3^{Lys}$, suggesting much greater binding free energy for the RT-tRNA$_3^{Lys}$ complex than for corresponding complexes with the reference tRNAs. These workers analyzed domains of tRNA$_3^{Lys}$ responsible for contacts with the RT by an elegant chemical crosslinking approach involving transdiamine-dichloro-platinum. This agent forms coordinate complexes between purine residues in the tRNA and Cys or Met residues in the protein. Crosslinked tRNA$_3^{Lys}$-RT complexes were formed and then digested with T1 RNase. The tRNA oligonucleotide(s) remaining bound to the enzyme by virtue of crosslinking were then isolated and sequenced. The sequence corresponded to a 12-residue nucleotide which is the predicted T1 digestion product from the anticodon domain of tRNA$_3^{Lys}$. The results suggested close contact between the anticodon loop of the tRNA and the primer binding site of RT. This general picture is reminiscent of a key role of the anticodon domain in the tRNA-synthetase recognition,[34-37] and is, of course, reminiscent of specific contacts deduced from x-ray diffraction patterns of co-crystals of tRNA^{Gln} and tRNA^{Asp} and their respective tRNA synthetases.[38,39] The anticodon loop and stem of tRNA$_3^{Lys}$ contains the sequence 5'-CAGACUSUURAΨCUG-3', where S is 5-(methoxycarbonyl-methyl)-2-thiouridine and R is N-[N-(9-β-D-ribofuranosyl-2 methylthiopurin-6-yl)carbamoyl] threonine. These two hypermodified residues of U and A are attractive candidates for contributing at least a portion of the specificity of tRNA$_3^{Lys}$-RT binding.

The picture surrounding the question of specificity of RT-primer binding became more complex when a report by Bordier et al.[40] showed that the recombinant HIV RT is able to bind several nonprimer tRNAs, such as bovine tRNA$_{1,2}^{Lys}$, bovine tRNA^{Phe}, and bovine tRNA^{Trp}. It is intriguing that these nonprimer tRNAs and the tRNA$_3^{Lys}$ share a common feature of having a highly modified anticodon loop. Several other nonprimer tRNAs such as bovine tRNA^{Val}, bovine tRNA^{Pro}, and bovine tRNA^{Gly} failed to bind and do not have similarly modified anticodon loops. The Litvak group went on to show that the tRNA binding activity of their tRNA$_3^{Lys}$ preparation could be traced to RNase T1 digestion products corresponding to the anticodon loop in the tRNA. For example, an isolated 12-residue T1 digestion product was able to block RT DNA polymerase activity by virtue of binding to the enzyme and competing with the substrate template·primer, activated DNA.[40] Taken together, these results implicated the anticodon loop of tRNA$_3^{Lys}$ in RT binding and also indicated that the binding mechanism is not specific for tRNA$_3^{Lys}$ and does not require the full tRNA molecule.

Barat et al.[27] found that tRNA$_3^{Lys}$ synthesized *in vitro* with SP6 RNA polymerase is capable of binding to recombinant HIV RT. This binding reaction, however, appeared to be less avid than binding between RT and the natural bovine tRNA$_3^{Lys}$ used earlier. That is, competition of gel retardation complex was found for nonprimer tRNA^{Pro} in the case of the synthetic tRNA$_3^{Lys}$, whereas no such competition was seen for natural tRNA$_3^{Lys}$. Thus, the implication is that posttranscriptional modifications in tRNA$_3^{Lys}$ (anticodon loop)

TABLE 1
Summary of HIV-1 p66 Homodimer RT Binding as Revealed by Competition Experiments[a]

Competitor[b]	Concentration[c] required for 50% competition [M]	K_D [M][d]
5'p-d(T)$_8$	—	2.5×10^{-6}
5'p-d(T)$_{16}$	—	0.5×10^{-6}
5'OH-d(C)$_{19-24}$	200×10^{-9}	98×10^{-9}
Rabbit tRNA$_3^{Lys}$	80×10^{-9}	30×10^{-9}
Rabbit tRNA$_{1,2}^{Lys}$	120×10^{-9}	52×10^{-9}
E. coli tRNA$_1^{Val}$	80×10^{-9}	30×10^{-9}
E. coli tRNA^{Phe}	80×10^{-9}	30×10^{-9}
E. coli tRNA^{fMet}	80×10^{-9}	30×10^{-9}

[a] Adapted from Reference 42.
[b] Purified oligodeoxynucleotide primers or natural tRNAs.
[c] 5'-[^{32}P]-d(T)$_8$ probe is at 1.25 µM, equal to ~one-half K_D.
[d] Values for 5'p-d(T)$_8$ and 5'p-d(T)$_{16}$ were measured directly. Other values were calculated from competition experiments at 25°C.

contribute to the binding free energy, but are not required for binding. The idea of tRNA specificity imparted by posttranscriptional modification is well known, for example, from the work on the translational specificity of tRNA^{Ile}.[41] Taken together, the results from the Darlix[27,29] and Litvak [30,40] groups indicate that two recombinant p66/p51 heterodimer RTs are capable of binding both the replicative primer tRNA$_3^{Lys}$ and a variety of nonreplicative bovine tRNAs, such as tRNA^{Pro}, tRNA^{Trp}, and tRNA$_{1,2}^{Lys}$. A degree of binding specificity is observed, however, since some bovine tRNAs either do not bind or bind only weakly. Both groups implicated the tRNA$_3^{Lys}$ anticodon domain in making important contacts with the enzyme.

Thermodynamic parameters for RT-tRNA$_3^{Lys}$ binding have recently been obtained by Sobol et al.[42] These workers used photochemical cross-linking with a pulse of UV light to trap complexes formed between RT and the synthetic primer analogue oligo d(T). In this way, concentrations of bound and free oligo d(T) could be measured under equilibrium binding conditions, and RT binding affinity for other primers, such as tRNA$_3^{Lys}$, could be readily determined by competition assays. The experiments were conducted with *E. coli*-derived recombinant p66 homodimer RT. Results from this work are summarized in Table 1. The K_D for RT plus oligo d(T)$_{16}$ binding was about 500 nM, whereas the K_D for oligo d(T)$_8$ binding was about 2500 nM. This difference in affinity may reflect the size of the binding groove on the surface of the enzyme. Rabbit tRNA$_3^{Lys}$ was able to bind to the enzyme and compete for oligo d(T) binding. This competition indicated a K_D for RT-tRNA$_3^{Lys}$ binding of 30 nM, or a free energy change of -11.1 kcal/mol at 25°C. Similar K_D values for

RT-tRNA binding were observed with the reference "nonpriming" rabbit tRNA$_3^{Lys}$ and with three reference tRNAs from *E. coli*, tRNA^{Val}, tRNA^{Phe}, and tRNA^{Met}. Therefore, this enzyme fails to exhibit specificity among the tRNAs tested, but does nevertheless exhibit avid binding to the tRNA$_3^{Lys}$.

The results to date, however, clearly indicate that HIV-1 RT binding to template·primer complexes is far more avid than binding to free primer. For example, Huber et al.[43] made use of presteady state experiments to measure the K_D of an *E. coli* recombinant p66/p51 heterodimer preparation of RT to poly r(A)·oligo d(T). The K_D value was found to be ~3 nM, which is two orders of magnitude lower than K_A values for binding to free oligo d(T) (Table 1).

Binding between another *E. coli* recombinant HIV-1 RT p66/p51 heterodimer and synthetic heteropolymer oligodeoxynucleotide template·primer analogues also have been studied by Müller et al.[44] These workers made use of fluorescent probes incorporated into the oligonucleotide primer to obtain binding parameters under equilibrium conditions. A K_D value of 3 nM for the probe-containing template·primer was found, and this value then was dissected to rate constants of k_{on} 10^7 M^{-1}sec^{-1} and k_{off} 0.04 sec^{-1}. Competition experiments revealed that the K_D for poly r(A)·oligo d(T) binding was 0.35 nM, which again is much lower than K_D values for RT binding to free primer.

The results showing binding between the enzyme and free primer (tRNA$_3^{Lys}$ and d(C)$_{19-24}$) are in-line with the minimal reaction scheme for DNA synthesis proposed earlier from kinetic studies, in that the free enzyme is able to bind primer in the absence of dNTP and template; the K_D values for binding are in the range of those obtained by the kinetic approach. It is interesting to consider that the free energy of tRNA$_3^{Lys}$ binding in 10 mM NaCl, 6 mM MgCl$_2$ at pH 8.2 is equal to –11.1 kcal/mol. This free energy is somewhat greater than the free energy of binding usually observed with sequence-nonspecific nucleic acid binding proteins, which are generally on the order of –7 to –9.5 kcal/mol.[45,46] As already noted by Litvak,[30,40] LeGrice,[47] and Darlix[27,29] and their coworkers, an improved understanding of this relatively high affinity protein-nucleic acid interaction between RT and tRNA$_3^{Lys}$ may lead to an approach for identifying a class of specific inhibitors of the enzyme.

The ionic requirement for the interaction between oligo d(T)$_8$ and RT was examined by Sobol et al.[42] who conducted the binding reaction in the presence of increasing concentrations of NaCl. Although binding is inhibited by NaCl, a plot of the data according to the procedure of Record et al.[48] revealed a modest effect of increasing ionic strength on K_A.[42] Thus, the relationship between log K_A and log [NaCl] was described by a linear plot with slope equal to about –1, and substantial binding remained at the reference NaCl concentration of 1 *M* where ionic interactions are expected to be minimal. Thus, oligo d(T)$_8$ binding involves as little as one charge-charge interaction between the protein and DNA lattice, with the majority (~80%) of the binding energy being contributed by nonelectrostatic forces.

Sobol et al.[42] extended these studies to the localization of the primer binding site within the primary sequence of RT. Controlled proteolysis with V8 protease of a mixture of 5′[^{32}P]d(T)$_{16}$-crosslinked RT and uncrosslinked RT resulted in the initial formation of a ~40 kDa labeled complex (p40), followed by the formation of a stable ~16 kDa labeled complex (p16), as determined by SDS-PAGE and autoradiography. The stability of this RT fragment to further proteolysis was not surprising, as Stammers and co-workers had reported that a 30 kDa tryptic fragment of RT is resistant to further hydrolysis for several hours.[49]

The localization of the 5′[^{32}P]d(T)$_{16}$ binding site in p66 was conducted by a combination of Western blot analysis and protein microsequencing of the V8 protease fragments.[42] The p40 complex was probably comprised of both DNA (5′[^{32}P]pd(T)$_{16}$; ~4 kDa) and peptide (~36 kDa), and Western blot analysis with a monocolonal antibody (mAb19) to the N-terminal region of RT indicated that the complex contained the N-terminal region of RT. Further, microsequence analysis of p40 and a corresponding unlabeled Coomassie blue stained peptide of ~36,000 Da revealed that both had the N-terminal sequence PISPIETVPVKLKPG. This N-terminal sequence is identical to that of the parent p66 RT. These data, therefore, initially assigned the binding site to amino acids number 1 to ~300 and confirmed that the labeled p40 complex had one molecule of d(T)$_{16}$. This initial localization of the primer binding site to the N-terminal region of p66 seemed consistent with reports in which the DNA polymerase function of HIV RT was suggested to be in the N-terminal region (amino acids number 1–440), while the RNase H domain was in the C-terminal region.[50-52]

Continued V8 proteolysis of p40 resulted in the relatively stable ^{32}P-labeled p16 complex, the generation of which was proportional to the hydrolysis of ^{32}P-labeled p40.[42] Western blot analysis indicated that ^{32}P-p16 is not recognized by mAb19, suggesting that the fragment no longer contained the amino-terminal region of p66. Sequence analysis established that the p16 complex (12 kDa protein plus 4 kDa oligonucleotide) contained the N-terminal amino acid sequence IGQHRTKIEELRQHL and corresponded to amino acids 195 to approximately 300 of the primary sequence of HIV-1 RT. Sequence analysis of an unlabeled 12 kDa peptide produced in the same V8 protease digestion revealed that it had, as expected, the same N-terminal sequence as labeled p16. These findings on localization of the oligo d(T) binding region to the central portion of the 560 residue RT should be useful in further studies to characterize interactions involved in RT-primer binding.

V. RETROVIRAL NUCLEOCAPSID PROTEIN ACTS AS A REPLICATION-ACCESSORY FACTOR

Miller et al.[53] identified a "zinc finger" domain in the transcription factor IIIA (TFIIIA) of *Xenopus*, and they proposed a mechanism for binding to nucleic acids through this domain. These workers found that TFIIIA binds 7–11 zinc atoms, and its deduced amino acid sequence shows nine tandem repeating units of the sequence Cys-X$_{2-5}$-Cys-X$_{12}$-His-X$_{3-4}$-His, where X is any amino acid. The observation that TFIIIA is inactive in the absence of zinc further led to the proposal that the repeat motif Cys and His residues form coordination complexes with zinc, creating finger regions that bind in grooves of DNA.[53] Berg [54] proposed a structure for these finger domains, which later was generally supported by two-dimensional NMR studies of a single zinc finger peptide.[55] This structure consists of a two-stranded antiparallel β sheet that includes the two cysteine residues, a turn, and then a helix that includes the two histidine residues.[56]

Following the proposal of the zinc finger domain in TFIIIA, Berg [57] conducted a data bank search and identified groups of proteins with similar sequences. One such group consisted of the nucleocapsid (NC) proteins of retroviruses, which contain sequences of the form $Cys-X_2-Cys-X_4-His-X_4-Cys$ as a putative metal-binding domain. Retroviral proteins of one class, typified by Moloney MuLV, contain one copy of this sequence. Proteins of a second class, including the HIV-1 NC, contain two such sequences separated by 5 to 11 amino acid residues.[57]

The NC entities are low molecular weight, basic proteins which can bind nucleic acid; in virions, they are associated with genomic RNA. NC proteins are produced from the *gag* gene of retroviruses, which encodes proteins necessary for viral particle assembly and encapsidation of the viral RNA genome. Under current models of viral assembly, *gag* expresses a protein precursor which assembles under the host cell membrane,[58] and which is then proteolytically cleaved to produce several smaller proteins, including NC, found in the mature virion. Figure 5 indicates the precursor *gag* (Pr^{gag}) proteins and the arrangement of smaller proteins for three different retroviruses. In addition to NC, the other proteins indicated are matrix (MA), a hydrophobic protein associated with cell membranes, which becomes part of the viral envelope, and capsid (CA), the major capsid structural protein. In RSV, the protease (PR) is also translated as part of the *gag* precursor, whereas, in the other viruses, PR is translated in a different reading frame.

Studies with synthetic peptides containing viral NC sequences have indicated their ability to bind metal ions in reduced form.[56,62,63] There is one report that NC from AMV failed to bind zinc,[64] but this may have been due to oxidation of cysteine residues and formation of disulfide bonds.[56] NMR methods were recently applied to examine the structure of the HIV-1 NC and two synthetic peptides representing the putative zinc-binding domains of this protein. Summers et al.[63] examined an 18-residue peptide containing the sequence of the first metal-binding domain of HIV. This peptide binds one equivalent of zinc and is stabilized by at least seven internal hydrogen bonds within the 14-residue zinc-binding domain. These workers report differences from the TFIIIA zinc finger; because of tight turns in the four-residue regions between the metal-binding residues, they describe the peptide as a "knuckle" rather than a finger,[63] and characterize it as a "retroviral-type" zinc finger, rather than the "classical" zinc finger of TFIIIA.[65]

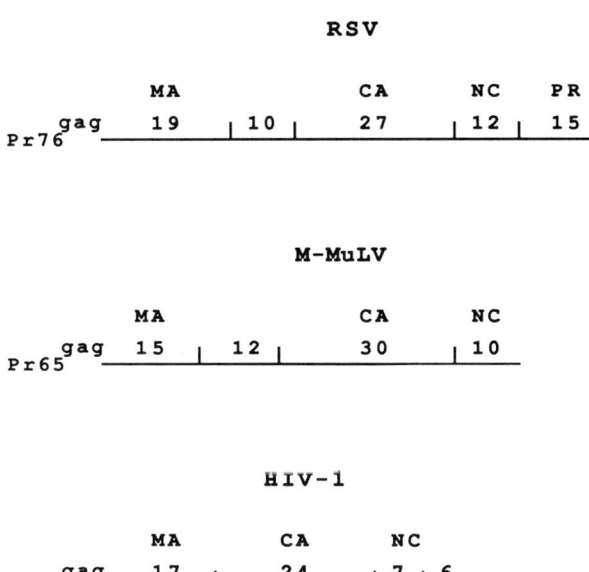

FIGURE 5. Arrangement of *gag* proteins. The relationship of *gag* proteins is shown for Rous sarcoma virus (RSV), Moloney murine leukemia virus (M-MuLV), and human immunodeficiency virus (HIV). Pr represents a precursor protein which is subsequently processed; the amino-terminus is on the left. MA is matrix protein, CA is capsid protein, NC is nucleocapsid protein, and PR is protease. In RSV, PR is translated in-frame with the other *gag* proteins; in the other viruses, PR is translated in a different reading frame. Numbers indicate the apparent molecular weights of proteins in kDa. Functions have not been identified for the RSV p10, M-MuLV p12, or HIV-1 p6 proteins.[4,59-61]

Subsequent analysis was conducted of the full NC protein, isolated directly from infectious HIV-1 particles, and a synthetic peptide representing the second finger region.[65] The full protein, comprising 55 amino acids, contains 2 putative zinc-binding regions, separated by 7 amino acids. The NC bound two mole equivalents of zinc tightly, and zinc binding was accompanied by NMR spectral shifts attributed to histidines at residues 23 and 44. The same shifts were seen with zinc binding of the individual synthetic peptides, designated F1 (residues 13–30), and F2 (residues 34–51). Two-dimensional NMR spectra of the F1 and F2 peptides were similar, suggesting similar structures; one difference was that sharp signals were observed for F2 only after cooling the sample below 5°C, suggesting that the zinc-binding domain for this peptide is less stable than for the F1 structure. With the full NC, the two-dimensional NMR spectrum represents a superposition of the spectra of the two peptides, and sharp signals are seen for both zinc-binding regions at temperatures up to 30°C. These results suggest that the second domain has an increased stability when located in the full protein. Preliminary data from these workers also indicated that zinc copurified with mature HIV-1 particles, suggesting that zinc may be associated with NC in viral particles.[65]

The ability of NC proteins to bind single-stranded DNA suggested an involvement in the packaging of the viral genome, and experiments with mutants provide further support for this function. Gorelick et al.[61] produced NC mutants of HIV-1, where either one or two Cys residues in a metal-binding domain were replaced with Ser, and one mutant where the first metal-binding domain was deleted. These mutants produced viral particles and RT activity at levels comparable to the wild-type virion. However, mutant viral particles contained 2 to 20% genomic RNA as compared to wild-type, and the mutants were less infective than wild-type by factors of at least 10^5.[61]

Similar experiments were performed with the NC of M-MuLV, which contains a single metal binding domain of sequence: Cys-Ala-Tyr-Cys-Lys-Glu-Lys-Gly-His-Trp-Ala-Lys-Asp-Cys. Changing one or more Cys residues to Ser reduced viral infectivity below detectable levels. Changing Tyr or Trp to Ser also destroyed infectivity.[66] Mutant virus particles lacked detectable levels of viral RNA, but did contain cellular RNA. The suggestion is that the metal-binding domain is necessary for specific packaging of the retroviral genome.[66] Méric and Goff[60] also produced mutants with the M-MuLV finger domain, but they left the metal-complexing residues unaltered, and mutated either Tyr, Gly, or Trp. All these mutants showed reduced infectivity, and contained from 1 to 20% viral RNA as compared to wild-type. One mutant, which had a Tyr to Ser change, contained 20% viral RNA as compared to wild-type, but was unable to replicate, since no viral cDNA could be detected with this mutant. This suggested an additional replication defect distinct from RNA packaging for NC.[60]

An additional function for NC is further indicated by *in vitro* experiments which support a role for this protein in tRNA priming. Prats et al.[67] found that NC of RSV promoted annealing of the correct primer tRNA^{Trp} to RSV RNA. Similarly, NC of M-MuLV promoted annealing of the correct primer tRNA^{Pro} to M-MuLV RNA. In each case, tRNA annealing was not seen with viral RNA where the primer binding site was deleted. Each NC protein also promoted dimerization of its respective viral RNA. These workers also examined an NC mutant of RSV with a two amino acid insertion in the first finger domain. This mutant is partially able to dimerize and encapsidate viral RNA, but the virions are not infectious. Virions with this mutant were found to be unable to support cDNA synthesis, and tRNA^{Trp} was identified in mutant virions but was not associated with viral RNA, in contrast to wild-type. These workers attributed this defect to a failure of the mutant NC to position tRNA primer on the PBS. Annealing activity of NC protein may also facilitate the first strand transfer during retroviral replication (see Figure 3). Full length viral DNA is synthesized in RSV virions only in the presence of NC, while RSV cDNA of 101 nucleotides (the distance between pbs and the 5' end of the genome) accumulates in the absence of NC.[67]

Experiments have also been conducted *in vitro* with HIV-1 proteins.[29] The RT and NC each showed the ability to bind to HIV-1 viral RNA and the replication primer tRNA$_3^{Lys}$; the HIV RT was able to recognize tRNA$_3^{Lys}$ even in the presence of a 100-fold excess of competing tRNA^{Trp} or tRNA^{Pro}. RT alone, however, was not able to initiate DNA synthesis in the presence of tRNA$_3^{Lys}$ and complementary viral RNA; and synthesis was only observed with the addition of HIV-1 NC. In the latter system, NC of RSV could substitute for HIV-1 NC. These experiments support the idea that NC is required for efficient annealing of the tRNA primer to viral RNA.[29] In this system the NC protein exhibited strong annealing activity for the tRNA$_3^{Lys}$-PBS interaction.

VI. CONCLUSIONS

We have focused in this chapter on the roles of host tRNA molecules in the replication cycle of the retroviral genome. As tRNA plays an essential role in genome replication, one would expect that tRNA may be exploited to exert control over the viral replication process. Similarly, tRNAs could be exploited for regulatory roles in viral protein production through their requirement in protein synthesis translational mechanisms[68,69] (see Chapter 9 for review). Experiments on expression of modified tRNAs in cells and of tRNA-targeted inhibitors should be informative toward evaluating these interesting possibilities.

One would expect that further investigations of the biochemistry of tRNA involvement in retroviral replication will follow at least two themes. The first line of investigation will be further elucidation of the tRNA binding domains of RT and other proteins, such as nucleocapsid protein. Research in this area will also investigate mechanisms by which such proteins are able to select tRNAs for packaging into virions, and to distinguish the correct species for priming replication. With the identification of nucleocapsid protein as a replication accessory factor, a second theme of research is likely to be the reconstitution and investigation of model systems which more closely approximate *in vivo* replication.

ACKNOWLEDGMENTS

We gratefully acknowledge Dr. Martha Delahunty, National Institute of Child Health and Human Development, for sharing her knowledge of the retroviral literature. This work was done while John Abbotts held a National Research Council-NIH Research Associateship.

REFERENCES

1. **Bishop, J. M.**, Retroviruses, *Ann. Rev. Biochem.*, 47, 35, 1978.
2. **Varmus, H.**, Retroviruses, *Science*, 240, 1427, 1988.
3. **Gallo, R. C.**, The first human retrovirus, *Sci. Am.*, 225(6), 88, 1986.
4. **Wong-Staal, F. and Gallo, R. C.**, Human T-lymphotropic retroviruses, *Nature (London)*, 317, 395, 1985.
5. **Panganiban, A. T. and Fiore, D.**, Ordered interstrand and intrastrand DNA transfer during reverse transcription, *Science*, 241, 1064, 1988.
6. **Goff, S. P.**, Retroviral reverse transcriptase: synthesis, structure, and function, *J. Acq. Imm. Def. Syn.*, 3, 817, 1990.
7. **Gilboa, E., Mitra, S. W., Goff, S., and Baltimore, D.**, A detailed model of reverse transcription and tests of crucial aspects, *Cell*, 18, 93, 1979.
8. **Roth, M. J., Schwartzberg, P. L., and Goff, S. P.**, Structure of the termini of DNA intermediates in the integration of retroviral DNA: Dependence on IN function and terminal DNA sequence, *Cell*, 58, 47, 1989.
9. **Omer, C. A. and Faras, A. J.**, Mechanism of release of the avian retrovirus tRNA^{Trp} primer molecule from viral DNA by ribonuclease H during reverse transcription, *Cell*, 30, 797, 1982.
10. **Waters, L. C. and Mullin, B. C.**, Transfer RNA in RNA tumor viruses, *Prog. Nucleic Acid Res. Mol. Biol.*, 20, 131, 1977.
11. **Sawyer, R. C. and Dahlberg, J. E.**, Small RNAs of Rous sarcoma virus: Characterization by two-dimensional polyacrylamide gel electrophoresis and fingerprint analysis, *J. Virol.*, 12, 1226, 1973.
12. **Dahlberg, J. E., Sawyer, R. C., Taylor, J. M., Faras, A. J., Levinson, W. E., Goodman, H. M., and Bishop, J. M.**, Transcription of DNA from the 70S RNA of Rous sarcoma virus. I. Identification of a specific 4S RNA which serves as primer, *J. Virol.*, 13, 1126, 1974.
13. **Harada, F., Sawyer, R. C., and Dahlberg, J. E.**, A primer ribonucleic acid for initiation of in vitro Rous sarcoma virus deoxyribonucleic acid synthesis, *J. Biol. Chem.*, 250, 3487, 1975.
14. **Peters, G., Harada, F., Dahlberg, J. E., Panet, A., Haseltine, W. A., and Baltimore, D.**, Low-molecular-weight RNAs of Moloney murine leukemia virus: identification of the primer for RNA-directed DNA synthesis, *J. Virol.*, 21, 1031, 1977.
15. **Haseltine, W. A., Panet, A., Smoler, D., Baltimore, D., Peters, G., Harada, F., and Dahlberg, J. E.**, Interaction of tryptophan tRNA and avian myeloblastosis virus reverse transcriptase: Further characterization of the binding reaction, *Biochemistry*, 16, 3625, 1977.
16. **Panet, A., Haseltine, W. A., Baltimore, D., Peters, G., Harada, F., and Dahlberg, J. E.**, Specific binding of tryptophan transfer RNA to avian myeloblastosis virus RNA-dependent DNA polymerase (reverse transcriptase), *Proc. Natl. Acad. Sci. U.S.A.*, 72, 2535, 1975.
17. **Wang, S., Kothari, R. M., Taylor, M., and Hung, P.**, Transfer RNA activities of Rous sarcoma and Rous associated viruses, *Nature New Biol.*, 242, 133, 1973.
18. **Peters, G. and Glover, C.**, tRNAs and priming of RNA-directed DNA synthesis in mouse mammary tumor virus, *J. Virol.*, 35, 31, 1980.
19. **Sonigo, P., Alizon, M., Staskus, K., Klatzmann, D., Cole, S., Danos, O., Retzel, E., Tiollais, P., Haase, A., and Wain-Hobson, S.**, Nucleotide sequence of the Visna lentivirus: relationship to the AIDS virus, *Cell*, 42, 369, 1985.
20. **Peters, G. G. and Hu, J.**, Reverse transcriptase as the major determinant for selective packaging of tRNAs into avian sarcoma virus particles, *J. Virol.*, 36, 692, 1980.
21. **Levin, J. G., Grimsley, P. M., Ramseur, J. M., and Berezeky, I. K.**, Deficiency of 60-70S RNA in murine leukemia virus particles assembled in cells treated with actinomycin D, *J. Virol.*, 14, 152, 1974.

22. **Levin, J. G. and Rosenak, M. J.,** Synthesis of murine leukemia virus proteins associated with virions assembled in actinomycin D-treated cells: evidence for persistence of viral messenger RNA, *Proc. Natl. Acad. Sci. U.S.A.,* 73, 1154, 1976.
23. **Levin, J. G. and Seidman, J. G.,** Selective packaging of host tRNAs by murine leukemia virus particles does not require genomic RNA, *J. Virol.,* 29, 328, 1979.
24. **Wain-Hobson, S., Sonigo, P., Danos, O., Cole, S., and Alizon, M.,** Nucleotide sequence of the AIDS virus, LAV, *Cell,* 40, 9, 1985.
25. **Ratner, L., Haseltine, W., Patarca, R., Livak, K. J., Starcich, B., Josephs, S. F., Doran, E. R., Rafalski, J. A., Whitehorn, E. A., Baumeister, K., Ivanoff, L., Petteway, S. R., Pearson, M. L., Lautenberger, J. A., Papas, T. S., Ghrayeb, J., Chang, N. T., Gallo, R. C., and Wong-Staal, F.,** Complete nucleotide sequence of the AIDS virus, HTLV-III, *Nature (London),* 313, 277, 1985.
26. **Raba, M., Limburg, K., Burghagen, M., Katze, J. R., Simsek, M., Heckman, J. E., RajBhandary, U. L., and Gross, H. J.,** Nucleotide sequence of three isoaccepting lysine tRNAs from rabbit liver and SV40-transformed mouse fibroblasts, *Eur. J. Biochem.,* 97, 305, 1979.
27. **Barat, C., LeGrice, S. F. J., and Darlix, J.-L.,** Interaction of HIV-1 reverse transcriptase with a synthetic form of its replication primer, tRNA$_3^{Lys}$, *Nucleic Acids Res.,* 19, 751, 1991.
28. **Kleiman, L., Caudry, S., Boulerice, F., Wainberg, M. A., and Parniak, M. A.,** Incorporation of tRNA into normal and mutant HIV-1, *Biochem. Biophys. Res. Comm.,* 174, 1272, 1991.
29. **Barat, C., Lullien, V., Schatz, O., Keith, G., Nugeyre, M. T., Grüninger-Leitch, F., Barré-Sinoussi, F., LeGrice, S. F. J., and Darlix, J.-L.,** HIV-1 reverse transcriptase specifically interacts with the anticodon domain of its cognate primer tRNA, *EMBO J.,* 8, 3279, 1989.
30. **Sallafranque-Andreola, M.-L., Robert, D., Barr, P. J., Fournier, M., Litvak, S., Sarih-Cottin, L., and Tarrago-Litvak, L.,** Human immunodeficiency virus reverse transcriptase expressed in transformed yeast cells, *Eur. J. Biochem.,* 184, 367, 1989.
31. **Sarih-Cottin, L., Sallafranque-Andreola, M.-L., Robert, D., Tarrago-Litvak, L., and Litvak, S.,** personal communication. 14th International tRNA Workshop, (Abstract), Rydzyma, Poland, May 4–9, 1991.
32. **Majumdar, C., Abbotts, J., Broder, S., and Wilson, S. H.,** Studies on the mechanism of human immunodeficiency virus reverse transcriptase: steady-state kinetics, processivity and polynucleotide inhibition, *J. Biol. Chem.,* 263, 15657, 1988.
33. **Majumdar, C., Stein, C. A., Cohen, J. S., Broder, S., and Wilson, S. H.,** Stepwise mechanism of HIV reverse transcriptase: primer function of phosphorothioate oligodeoxynucleotide, *Biochemistry,* 28, 1340, 1989.
34. **Schulman, L. H. and Pelka, H.,** In vitro conversion of a methionine to a glutamine acceptor tRNA, *Biochemistry,* 24, 7309, 1985.
35. **Schulman, L. H. and Pelka, H.,** Anticodon switching changes the identity of methionine and valine transfer RNAs, *Science,* 242, 765, 1988.
36. **Schulman, L. H. and Abelson, J.,** Recent excitement in understanding transfer RNA identity, *Science,* 240, 1591, 1988.
37. **Sampson, J. R., DiRenzo, A. B., Behlen, L. S., and Uhlenbeck, O. C.,** Nucleotides in yeast tRNA^{Phe} required for the specific recognition by its congate synthetase, *Science,* 243, 1363, 1989.
38. **Rould, M. A., Perona, J., Söll, D., and Steitz, T.,** Structure of *E. coli* glutaminyl-tRNA synthetase complexed with tRNA^{Gln} and ATP at 2.8 Å resolution, *Science,* 246, 1135, 1989.
39. **Ruff, M., Krishnaswamy, S., Boeglin, M., Poterszman, A., Mitschler, A., Podjarny, A., Rees, B., Thierry, C., and Moras, D.,** Class II aminoacyl transfer RNA synthetases: crystal structure of yeast aspartyl-tRNA synthetase complexed with tRNA^{Asp}, *Science,* 252, 1682, 1991.

40. **Bordier, B., Tarrago-Litvak, L., Sallafranque-Andreola, M.-L., Robert, D., Tharaud, D., Fournier, M., Barr, P. J., Litvak, S., and Sarih-Cottin, L.,** Inhibition of the p66/p51 form of human immunodeficiency virus reverse transcriptase by tRNA^{Lys}, *Nucleic Acids Res.,* 18, 429, 1990.
41. **Muramatsu, T., Nishikawa, K., Nemoto, F., Kuchino, Y., Nishimura, S., Miyazawa, T., and Yokoyama, S.,** Codon and amino-acid specificities of a transfer RNA are both converted by a single post-transcriptional modification, *Nature (London),* 336, 179, 1991.
42. **Sobol, R. W., Suhadolnik, R. J., Kumar, A., Lee, B. J., Hatfield, D. L., and Wilson, S. H.,** Localization of a polynucleotide binding region in the HIV-1 reverse transcriptase: implications for primer binding, *Biochemistry,* 30, 10623, 1991.
43. **Huber, H. E., McCoy, J. M., Seehra, J. S., and Richardson, C. C.,** Human immunodeficiency virus 1 reverse transcriptase: template binding, processivity, strand displacement synthesis, and template switching, *J. Biol. Chem.,* 264, 4669, 1989.
44. **Müller, B., Restle, T., Reinstein, J., and Goody, R. S.,** Interaction of fluorescently labeled dideoxynucleotides with HIV-1 reverse transcriptase, *Biochemistry,* 30, 3709, 1991.
45. **Nadler, S. G., Merrill, B. M., Roberts, W. J., Keating, K. M., Lisbin, M. J., Barnett, S. F., Wilson, S. H., and Williams, K. R.,** Interactions of the A1 heterogeneous nuclear ribonucleoprotein and its proteolytic derivative, UP1, with RNA and DNA: evidence for multiple RNA binding domains and salt-dependent binding mode transitions, *Biochemistry,* 30, 2968, 1991.
46. **Wu, C.-W. and Singer, P. T.,** The mechanism of RNA polymerase, in *The Eukaryotic Nucleus: Molecular Biochemistry and Macromolecular Assemblies,* Strauss, P. R. and Wilson, S. H., Eds., Telford Press, Caldwell, N. J., 1990, chap. 15.
47. **LeGrice, S. F. J., Schatz, O., and Darlix, J.-L.,** Ribonuclease H and primer tRNA binding activities of HIV reverse transcriptase as therapeutic targets, in *Advances in Molecular Biology and Targeted Treatment for AIDS,* Kumar, A., Ed., Plenum Press, New York, 1991, 55.
48. **Record, M. T., Lohman, T. M., and de Haseth, P.,** Ion effects on ligand-nucleic acid interactions, *J. Mol. Biol.,* 107, 145, 1976.
49. **Lowe, D. M., Aitken, A., Bradley, C., Darby, G. K., Larder, B. A., Powell, K. L., Purifoy, D. J. M., Tisdale, M., and Stammers, D. K.,** HIV-1 reverse transcriptase: crystallization and analysis of domain structure by limited proteolysis, *Biochemistry,* 27, 8884, 1988.
50. **Johnson, M. S., McClure, M. A., Feng, D.-F., Gray J., and Doolittle, R. F.,** Computer analysis of retroviral *pol* genes: Assignment of enzymatic functions to specific sequences and homologies with nonviral enzymes, *Proc. Natl. Acad. Sci. U.S.A.,* 83, 7648, 1986.
51. **Larder, B. A., Purifoy, D. J. M., Powell, K. L, and Darby, G.,** Site-specific mutagenesis of AIDS virus reverse transcriptase, *Nature (London),* 327, 716, 1987.
52. **Hansen, J., Schulze, T., Mellert W., and Moelling, K.,** Identification and characterization of HIV-specific RNase H by monoclonal antibody, *EMBO J.,* 7, 239, 1988.
53. **Miller, J., McLachlan, A. D., and Klug, A.,** Repetitive zinc-binding domains in the protein transcription factor IIIA from *Xenopus* oocytes, *EMBO J.,* 4, 1609, 1985.
54. **Berg, J. M.,** Proposed structure for the zinc-binding domains from transcription factor IIIA and related proteins, *Proc. Natl. Acad. Sci. U.S.A.,* 85, 99, 1988.
55. **Lee, M. S., Gippert, G. P., Soman, K. V., Case, D. A., and Wright, P. E.,** Three-dimensional solution structure of a single zinc finger DNA-binding domain, *Science,* 245, 635, 1989.
56. **Berg, J. M.,** Zinc fingers and other metal-binding domains, *J. Biol. Chem.,* 265, 6513, 1990.
57. **Berg, J. M.,** Potential metal-binding domains in nucleic acid binding proteins, *Science,* 232, 485, 1986.
58. **Bolognesi, D. P., Montelaro, R. C., Frank, H., and Schäfer, W.,** Assembly of type C oncornaviruses: a model, *Science,* 199, 183, 1978.

59. **Leis, J., Baltimore, D., Bishop, J. M., Coffin, J., Fleissner, E., Goff, S. P., Oroszlan, S., Robinson, H., Skalka, A. M., Temin, H. M., and Vogt, V.**, Standardized and simplified nomenclature for proteins common to all retroviruses, *J. Virol.*, 62, 1808, 1988.
60. **Méric, C. and Goff, S. P.**, Characterization of Moloney murine leukemia virus mutants with single-amino-acid substitutions in the Cys-His box of the nucleocapsid protein, *J. Virol.*, 63, 1558, 1989.
61. **Gorelick, R. J., Nigida, S. M., Bess, J. W., Arthur, L. O., Henderson, L. E., and Rein, A.**, Noninfectious human immunodeficiency virus type 1 mutants deficient in genomic RNA, *J. Virol.*, 64, 3207, 1990.
62. **Green, L. M. and Berg, J. M.**, A retroviral Cys-Xaa$_2$-Cys-Xaa$_4$-His-Xaa$_4$-Cys peptide binds metal ions: spectroscopic studies and a proposed three-dimensional structure, *Proc. Natl. Acad. Sci. U.S.A.*, 86, 4047, 1989.
63. **Summers, M. F., South, T. L., Kim, B., and Hare, D. R.**, High-resolution structure of an HIV zinc fingerlike domain via a new NMR-based distance geometry approach, *Biochemistry*, 29, 329, 1990.
64. **Jentoft, J. E., Smith, L. M., Fu, X., Johnson, M., and Leis, J.**, Conserved cysteine and histidine residues of the avian myeloblastosis virus nucleocapsid protein are essential for viral replication but are not "zinc-binding fingers," *Proc. Natl. Acad. Sci. U.S.A.*, 85, 7094, 1988.
65. **South, T. L., Blake, P. R., Sowder, R. C., Arthur, L. O., Henderson, L. E., and Summers, M. F.**, The nucleocapsid protein isolated from HIV-1 particles binds zinc and forms retroviral-type zinc fingers, *Biochemistry*, 29, 7786, 1990.
66. **Gorelick, R. J., Henderson, L. E., Hanser, J. P., and Rein, A.**, Point mutants of Moloney murine leukemia virus that fail to package viral RNA: evidence for specific RNA recognition by a "zinc finger-like" protein sequence, *Proc. Natl. Acad. Sci. U.S.A.*, 85, 8420, 1988.
67. **Prats, A. C., Sarih, L., Gabus, C., Litvak, S., Keith, G., and Darlix, J. -L.**, Small finger protein of avian and murine retroviruses has nucleic acid annealing activity and positions the replication primer tRNA onto genomic RNA, *EMBO J.*, 7, 1777, 1988.
68. **Jacks, T.**, Translational suppression in gene expression in retroviruses and retrotransposons, *Curr. Top. Microbiol. Immunol.*, 157, 93, 1990.
69. **Hatfield, D. L., Levin, J. G., Rein, A., and Oroszlan, S.**, Translational suppression in retroviral gene expression, *Adv. Virus Res.*, 41, 193, 1992.

Chapter 2

THE ROLE OF MODIFIED NUCLEOSIDES IN tRNA INTERACTIONS

Glenn R. Björk

TABLE OF CONTENTS

I. Introduction 24

II. Role of Modified Nucleosides in tRNA Interactions with Modifying Enzymes 26

III. Role of Modified Nucleosides in tRNA Interactions with Aminoacyl-tRNA Ligases 27
    A. General Modification Deficient tRNAs 27
    B. Role of Modified Nucleosides in Specific Positions 28
        1. Position 8 30
        2. Position 10 30
        3. Position 26 31
        4. Position 32 31
        5. Position 34 (The Wobble Position) 32
        6. Position 35 (Middle Base of Anticodon) 32
        7. Position 37 (3' of the Anticodon) 32
        8. Positions 38 (Anticodon Loop), 39, and 40 (Anticodon Stem) 35
        9. Position 46 35
        10. Position 48 35
        11. Position 54 36
        12. Position 55 36
    C. Conclusions 36

IV. Role of Modified Nucleosides in tRNA Interactions with Translation Factors 36

V. Role of Modified Nucleosides in tRNA Interactions not Apparently Involved in Translation 37

VI. Role of Modified Nucleosides in Anticodon:Codon Interactions 39
    A. General Considerations 39

B. Efficiency of Translation 41
     1. Position 32 42
     2. Position 34 43
     3. Position 35 45
     4. Position 37 46
        a. i^6A Derivative 46
        b. t^6A37 Derivatives 47
        c. Y Base 48
     5. Positions 38, 39, and 40
       (*hisT* Mediated Ψ Formation) 49
     6. Position 54 (Ribothymidine; m^5U54) 49
  C. Codon Context Sensitivity 51
     1. Position 34 52
     2. Position 37 52
  D. Fidelity 54
     1. Missense Errors 54
        a. Position 34 54
        b. Position 37 56
        c. Position 38, 39, and 40
           (*hisT* Mediated Ψ Formation) 58
        d. Position 54 (m^5U54) 59
     2. Frameshifting Errors 59
  E. Codon Choice 62

VII. Summary and Outlook 65

Acknowledgments 68

References 68

I. INTRODUCTION

Transfer RNA interacts with many different molecules in the cell. During the translation process tRNA interacts with aminoacyl-tRNA ligases, initiation factors, elongation factor Tu (EF-Tu), elongation factor G (EF-G), mRNA, ribosomes, and peptidylhydrolase. tRNAs also interact with several ribosomal proteins as well as with different parts of the 16S and the 23S rRNA. Furthermore, tRNA is also involved in such diverse processes as synthesis of chlorophyll, heme and vitamin B12,[1] cell division,[2-5] cell wall biosynthesis,[6] proteolytic degradation,[7,8] and priming reverse RNA-DNA synthesis during replication of retroviruses.[9] Thus, tRNA interacts with many different molecules in the cell and this may be one reason why this fascinating molecule has the highest density of modified nucleosides among all RNA molecules in the cell. In fact, the presence of so many different modified nucleosides in the tRNA introduces local chemical microenvironments in the molecule that may be recognition sites for proteins or nucleic acids and may change the conformation of the tRNA. This review will cover the role of modified nucleosides in all of these different interactions with emphasis on their role in the anticodon:codon interaction.

Transfer RNA from all organisms contains modified nucleosides, which are derivatives of the four normal nucleosides: adenosine (A)*, guanosine (G), uridine (U), and cytidine (C). More than 50 different modified nucleosides have been characterized.[10] All tRNA modifications occur posttranscriptionally, i.e., concomitant with the processing of the primary transcript. Thus, they are an integral part of tRNA maturation and consequently, the corresponding enzymes are tRNA biosynthetic enzymes. Only two modified nucleosides, queuosine (Q) and inosine (I), are synthesized before they are incorporated into tRNA.[11-13] Some of the modified nucleosides are formed by a simple addition of a methyl group to the base or to the 2′ hydroxyl group of the ribose moiety, while others are formed by a complex set of reactions, which result in hypermodified nucleosides. The tRNA modifying enzymes are position-specific. Hence, more than one enzyme is required for the formation of the same nucleoside at different positions in the tRNA. Furthermore, several enzymes are involved in the synthesis of each of the hypermodified nucleosides. For example, synthesis of the 29 different modified nucleosides present in *Salmonella typhimurium*[14] requires about 45 different modifying enzymes corresponding to as much as 1% of the genetic information of this bacterium.[15] These enzymes seem to be synthesized in small amounts and, therefore, the energetic load for the cell is not too extensive. There are 79 tRNA genes (including many duplicates) for 46 tRNA species present in *E. coli*.[16] If one assumes an average tRNA gene to be 150 nucleotides (the mature tRNAs are 75–95 bp in length), the genetic information devoted to the synthesis of the primary transcripts for tRNA is about 12 kb, which would represent about 0.25% of the genetic information available in eubacteria. Thus, at least four times more genetic information is devoted to tRNA modification than to the synthesis of the primary tRNA transcripts. The presence of modified nucleosides in tRNA from different organisms, the genetics and regulatory features for the tRNA modifying enzymes, and the physiological consequences of modification deficiency have recently been reviewed.[15,17-19]

---

* Abbreviations used: Structures of different modified nucleosides as well as their abbreviations can be found in Nishimura.[20] An index and an exponent indicate the number and the position of the substitution, respectively, e.g., 6-dimethyladenosine is abbreviated m^6_2A. k-,m-,c-,n-, o-,t-,i-, and s- are abbreviations of lysine, methyl, carbon, amino, oxy, threonine, isopentenyl, and thio groups, respectively. Other abbreviations: Ψ, pseudouridine; I, inosine; yW, nucleoside of Y base; oyW, nucleoside of peroxy Y base; and Q, queuosine. An enzyme catalyzing the formation of m^5U at position 54 in the tRNA is denoted tRNA(m^5U54)methyltransferase and likewise for other tRNA modifying enzymes. Methionyl-tRNA ligase is denoted MetRS and likewise for other aminoacyl-tRNA-ligases.

II. ROLE OF MODIFIED NUCLEOSIDES IN tRNA INTERACTIONS WITH MODIFYING ENZYMES

Transfer RNA modifying enzymes are highly specific. Sequences close to the target as well as structural features far from the target nucleotide are important determinants for this high specificity.[21] It is also well established that the maturation of tRNA is a sequential process during which tRNA modifying enzymes act. In eukaryotes, the localization of the enzymes either in the nucleus or in the cytoplasm is one parameter in this sequential maturation. Other features are also important, e.g., ribose methylation seems to be a very late event, even in prokaryotes, since such modification requires an almost mature tRNA as the substrate.[22,23] So far, a mutation in a structural gene for one tRNA modifying enzyme seems only to influence the formation of the corresponding modified nucleoside; for instance, the deficiency caused by a genetically defective tRNA modifying enzyme does not prevent other modification reactions from occurring. However, the biosynthesis of 2-methylthio-N^6-(4-hydroxy)isopentenyladenosine (ms^{2}io^{6}A37), which is present in position 37 in tRNAs that read codons starting with U (except tRNA$_{I,V}^{Ser}$), is an example of how one modification is a determinant for another tRNA modifying enzyme in the pathway. The postulated biosynthetic pathway and the genes involved (the genes *miaB/C* and *E* are tentative) are as follows:

$$A37 \xrightarrow{miaA} i^6A37 \xrightarrow{miaB} s^2i^6A37 \xrightarrow{miaC} $$
$$ms^2i^6A37 \xrightarrow{miaE} ms^2io^6A37$$

(The *miaD* gene product may be involved in a demodification step or in the regulation of the synthesis of the *miaA* gene product[24].) In *E. coli*, the last hydroxylation reaction does not occur but this biosynthetic step is present in the closely related organism *S. typhimurium*.[25] The postulated biosynthetic pathway is derived from precursor analyses of tRNA from methionine or cysteine starved *E. coli* cells,[26] as well as from genetic analysis.[27,28] A mutation in the *miaA* gene results in accumulation of only an unmodified A37 in the tRNA but not ms^{2}A37, which would occur if the *miaB* gene product did not require the isopentenyl modification. Furthermore, overexpression of tRNA^{Tyr} upon induction of a phage Φ80Su3 lysogen in *E. coli* results in tRNA^{Tyr} containing either A37, i^{6}A37, or ms^{2}i^{6}A37.[29] Again, no ms^{2}A37-containing tRNA^{Tyr} species is formed under these conditions. Thus, the *MiaB* enzyme seems to have a strict requirement for the i^{6}A37 derivative. This conclusion is also consistent with some early experiments on the biosynthesis of ms^{2}i^{6}A.[30] The starvation of *E. coli* (*rel⁻*,*met⁻*,*cys⁻*) for methionine results in accumulation of a precursor to ms^{2}i^{6}A, which was suggested to be s^{2}i^{6}A37, whereas cysteine or iron starved *E. coli* accumulates i^{6}A.[26,31-33] Cysteine or iron starvation in *S. typhimurium* also results in the accumulation of i^{6}A37 and only a small proportion (5%–12%) is in the hydroxylated form (io^{6}A37).[34]

A mutation likely to be in the *miaB* gene of *S. typhimurium* resulted in tRNA containing i^6A37 and no io^6A37 was present.[35] Thus, the hydroxylation reaction in *S. typhimurium* requires the ms^2 group. Therefore, the modifying enzymes involved in the synthesis of ms^2io^6A act strictly sequentially and some enzymes are dependent on modifications in other positions of the nucleoside.

4-Thiouridine is present at position 8 of many tRNA species including $tRNA_f^{Met}$ from *E. coli*. Modification of s^4U8 or UV crosslinking to C13 strongly influences the activity of the tRNA(Gm18)methyltransferase from *Thermus thermophilus* suggesting that s^4U or U8 is involved in the recognition of the tRNA(Gm18)methyltransferase.[36,37] However, the involvement of the modification of s^4U8 *per se* was not addressed.

III. ROLE OF MODIFIED NUCLEOSIDES IN tRNA INTERACTIONS WITH AMINOACYL-tRNA LIGASES

A. GENERAL MODIFICATION DEFICIENT tRNAs

General methyl-deficient tRNA can be obtained from a *relA⁻, met⁻ E. coli* strain following starvation for methonine.[38] Unfractionated, methyl-deficient tRNA accepts several amino acids to the same extent as normal tRNAs.[39-42] Methyl-deficient tRNA^{Phe}, which normally contains the three methylated nucleosides ms^2i^6A37, m^7G47, and m^5U54, accepts Phe to the same extent as normal methylated tRNA^{Phe},[43] although conflicting results have been reported.[44] In a similar way, t^6A37 deficient tRNA accumulates during threonine starvation of a *relA⁻, thr⁻* strain. Such threonine deficient tRNAs have the same K_m for 12 different aminoacyl tRNA-ligases as compared to normal tRNA.[45] Sulfur-deficient tRNA^{Gln} has normal amino acid acceptance.[46] Thus, modified nucleosides were observed early on not to influence greatly the aminoacylation reaction for several tRNAs. Completely unmodified $tRNA_1^{Tyr}$, $tRNA_2^{Gly}$ suA36, and $tRNA_3^{Thr}$ have been synthesized *in vitro* using purified RNA polymerase.[47] $tRNA_1^{Tyr}$ normally contains eight modified nucleosides (s^4U8 and 9, Gm18, Q34, ms^2i^6A37, Ψ39 and 55, and m^5U54), $tRNA_2^{Gly}$ (suA36) contains four modified nucleosides (U*34, A*37, m^5U54 and Ψ55), and $tRNA_3^{Thr}$ contains seven modified nucleosides (D17, 18 and 21, mt^6A37, m^7G47, m^5U54, and Ψ55). The extent of aminoacylation of these three tRNA species is unaffected by the lack of modified nucleosides (except Ψ55 in all three tRNAs and Ψ39 in $tRNA_1^{Tyr}$ due to their synthesis during the aminoacylation reaction). Therefore, no gross effect on aminoacylation can be attributed to the modified nucleosides normally present in these three tRNAs. Using runoff *in vitro* transcription of a tRNA gene under the control of a promoter recognized by the T7 RNA polymerase, completely unmodified tRNA can be synthesized in quantities suitable for various physical and biological studies.[48] Comparisons of the kinetics of the aminoacylation reactions with cognate and noncognate aminoacyl-tRNA-ligases are shown in Table 1. Since all unmodified tRNAs, except one (*E. coli* $tRNA_1^{Ile}$, see discussion in Section II.B.7), are able to accept the cognate amino acid, the modified nucleosides are not a prerequisite for most aminoacylation reactions *in vitro*. However, in all cases but one (*E. coli* tRNA^{Asp}), the cognate interactions with the unmodified species has, albeit small, kinetic characteristics different from that of the fully modified species. This suggests that the modified nucleosides moderately influence either, directly or indirectly, (through conformational changes) the recognition by the cognate aminoacyl-tRNA-ligases. This effect can be enhanced by changing the physical conditions of the reaction. High Mg^{2+} concentration maximizes the velocity of the transfer of Phe whereas a low Mg^{2+} concentration in the presence of 1mM spermine favors accuracy of the aminoacylation reaction. At the high accuracy conditions, the unmodified tRNA^{Phe} shows a considerable decrease in the relative V$_{max}$ (and consequently a decreased specificity constant), whereas the K$_m$ is not altered compared to native tRNA (Table 1). At a high Mg^{2+} concentration, the unmodified tRNA^{Phe} folds normally whereas at a low Mg^{2+} concentration, the unmodified species do not adopt the native conformation as shown by NMR.[49] Unmodified yeast tRNA^{Asp} also has a changed conformation as analyzed by Pb^{2+} cleavage and by chemical modification.[50] The presence of the modified nucleosides stabilizes the biologically active conformation of the tRNA, whereas the unmodified tRNA^{Asp} is disrupted in the D- and T-loop tertiary interaction. Such unmodified tRNA is mischarged by ArgRS with considerable efficiency.[51] The major effect is in the rates of mischarging (k$_{cat}$), whereas the K$_m$ for unmodified tRNA^{Asp} in the noncognate interaction increases only twofold. To ensure that the tRNA is functionally efficient, there are some identity determinants of which specific nucleotides, like those in the anticodon region, play an important role.[52] Equally important are the antideterminants, which discriminate between cognate and noncognate tRNAs. Perret et al.[51] suggest that in the case of yeast tRNA^{Asp} the modified nucleosides act as antideterminants. However, as seen in Table 1, this does not seem to be a general mechanism since unmodified *E. coli* tRNA specific for methionine and valine, and yeast tRNA^{Phe} are not mischarged.

B. ROLE OF MODIFIED NUCLEOSIDES IN SPECIFIC POSITIONS

The role of a specific modified nucleoside in the interaction with aminoacyl-tRNA-ligases has been studied using tRNA species either lacking a modified nucleoside or containing a derivative of a specific modified nucleoside. Such tRNAs have been produced by specific chemical treatments, by overproduction of tRNA in such a way that a specific modifying enzyme is unable to properly modify the tRNA, or through the isolation of mutants defective in the synthesis of a specific modified nucleoside. Furthermore, chemical modification of a specific modified nucleoside has also been used to study the possible interactions between a modified nucleoside and the tRNA.

TABLE 1
Kinetic Parameters for Aminoacylation of Unmodified tRNAs Relative to those of the Native tRNA

tRNA	K_m	V	V/K_m	Conditions	Misacylation	Ref.
Ec Ala	0.88	0.67	0.71	10 mM Mg$^+$		53b
Ec Asp	0.97	1.0	1.0	10 mM Mg^{2+}		53
Ec Gln	0.29	0.087	0.26			54
Ec His	1.1	0.77	0.67	10 mM Mg^{2+}		55
Ec Ile1			3×10^{-4}			56
Ec Met(m)	1.8	0.86	0.46	8mM Mg$^+$	None observed for ArgRS, GlnRS, GluRS, LysRS, IleRS, PheRS and ValRS	57
Ec Thr3	3.3	1.2	0.36	8mM Mg^{2+}		58
Ec Val1	2.0	0.75	0.38	11 mM Mg$^+$	None observed for MetRS, GlnRS, GluRS, LysRS, IleRS and PheRS	57
Ec Val1	1.0	0.94	0.94	15 mM Mg$^+$		57b
Ec Val1	0.72	0.43	0.63	10 mM Mg$^+$		57c
Ec Ser1	2.5	0.76	0.31	10 mM Mg$^+$	None observed with TyrRS	57d
Ec Tyr	1.3	0.97	0.71	10 mM Mg$^+$	None observed with SerRS	57d
Yeast Asp	0.64	0.58^a	0.91^a	15 mM Mg$^+$	Misacylation observed with ArgRS but not with HisRS, PheRS or ValRS	51
Yeast Phe	4.0	0.81	0.21	Buffert A; maximizes the activity of PheRS; 15 mM Mg$^+$	None observed with TyrRS	48

TABLE 1 (continued)
Kinetic Parameters for Aminoacylation of Unmodified tRNAs Relative to those of the Native tRNA

tRNA	K_m	V	V/K_m	Conditions	Misacylation	Ref.
Yeast Phe	3.8	0.22	0.058	Buffert B; high accuracy buffert; 2mM Mg$^+$;1 mM Spermidine		
Mycoplasma Gly	n.d.	0.90	n.d			59

[a] This value is relative k_{kat} and not V_{max} and thus the relative specificity constant in this case is k_{cat}/K_m.

1. Position 8

Cyanogen bromide reacts specifically and quantitatively with s^{4}U8 and other sulfur-containing modified nucleosides.[60] From these and other chemical modification studies, s^{4}U8 was shown to have no effect on amino acid acceptance by the tRNA.[37,61-67] Furthermore, cyanoethylation of s^{4}U8 or crosslinking of s^{4}U8 to C13 by UV irradiation do not affect the aminoacylation reaction.[68-70] Using NMR analysis of 5-fluorouracil-substituted tRNA^{Val}, the Michael adduct between ValRS and 5-fluorouracil at position 8 of tRNA^{Val} that has been proposed to be a common intermediate in the aminoacylation reaction[71] is not formed.[72] Thus, these data suggest that s^{4}U8 does not play any important role in the aminoacyl-tRNA recognition process.

2. Position 10

Yeast PheRS aminoacylates *E. coli* tRNA^{Phe} less efficiently than yeast or wheat germ tRNA^{Phe}.[73] The former tRNA^{Phe}, unlike the latter two, lacks m^{2}G10. Using a purified tRNA(m^{2}G10)methyltransferase from yeast, the G10 was methylated in *E. coli* tRNA^{Phe}. In the heterologous interaction with yeast PheRS, the presence of the m^{2}G10 modification in *E. coli* tRNA^{Phe} increases the V_{max} 10-fold whereas the K_m decreases slightly. In fact, the increased V_{max} for the m^{2}G10 containing *E. coli* tRNA^{Phe} becomes similar to the V_{max} for the yeast tRNA^{Phe}. Using the homologous *E. coli* PheRS, the methylation of G10 in the normally m^{2}G10 deficient *E. coli* tRNA^{Phe} decreases the V_{max} 10-fold without changing the K_m. Thus, the same modification in different tRNAs can either enhance or retard a reaction. Although it is clear from the data of Roe et al.[73] that the introduction of a methyl group in position 10 of *E. coli* tRNA^{Phe} changes the aminoacylation kinetics, the precise interpretation of these data with respect to the recognition characteristics of yeast PheRS is unclear. The introduction of the methyl group might change the structure of *E. coli* tRNA^{Phe} to which the *S. cerevisiae* and the *E. coli* PheRS respond. Note, however, that position 10 is not one of the five bases important in the recognition by yeast PheRS.[74] Clearly, the presence of modified nucleosides influences the structure of tRNA and these results might merely reflect this fact. The heterologous system (yeast PheRS/*E. coli* tRNA^{Phe}) used by Roe et al.[73] might be especially sensitive to a slight conformational change due to the introduction of a methyl group in position 10 of *E. coli* tRNA^{Phe}. In fact, Peterkofsky[75] observed that the yeast LeuRS showed a much more stringent requirement for methyl groups in the interaction with *E. coli* tRNA than with the homologous enzyme.

3. Position 26

An *S. cerevisiae* mutant exists that is unable to synthesize m$^2_2$G26 in its tRNA.[76] The rate of serylation of m$^2_2$G26 lacking tRNA^{Ser} is the same as that of wild-type tRNA^{Ser}, but the extent is decreased by 10%.[77] Since all cytoplasmic species of yeast tRNA^{Ser} contain m$^2_2$G26,[78] this methylated nucleoside has no direct effect on the aminoacylation reaction.

4. Position 32

Very little information of the functional significance of a modification at position 32 is available, but chemical removal of the sulfer group of s^{2}C32 in tRNA^{Arg} does not influence the cognate aminoacylation reaction.[79]

5. Position 34 (the wobble position)

The wobble base is part of the recognition site for the ArgRS.[80] Still, chemical modification of I34 with acrylonitrile does not influence the interactions between this enzyme and the tRNA$^{Arg}_1$.[81] Transfer RNA$^{Met}_m$ contains the modified nucleoside ac^{4}C34 in its wobble position. Treatment with sodium bisulfite removes the acetyl group and results in a tRNA$^{Met}_m$ having an unmodified C34. Although the wobble base of tRNA^{Met} is part of the major identity element for the MetRS, the acetyl group of ac^{4}C34 is not of critical importance in this aspect.[82]

Two isoleucine isoaccepting tRNAs, tRNA$^{Ile}_1$ and tRNA$^{Ile}_2$ (an isoacceptor present in minor levels), are present in *E. coli*. tRNA$^{Ile}_1$ contains G34 and tRNA$^{Ile}_2$ contains the modified nucleoside lysidine (k^{2}C34 or L34) in their wobble position. Lysidine, a derivative of cytidine, is synthesized by the addition of the amino acid lysine (K) at position 2 of cytosine.[83] Even though, the chemical structures of G34 (present in tRNA$^{Ile}_1$) and k^{2}C34 (present in tRNA$^{Ile}_2$) are quite different, IleRS recognizes both species. By exchanging the k^{2}C34 of the tRNA$^{Ile}_2$ with C34 *in vitro*, the anticodon of this tRNA^{Ile} is converted to CAU, which is the anticodon of tRNA^{Met}. Such a mutated tRNA^{Ile} with anticodon CAU is efficiently misacylated with Met.[84] Schulman and collaborators[82,85] have established that C34 or ac^{4}C34 is a positive identity determinant for the MetRS. Thus, the tRNA^{Ile} modification of C34 to k^{2}C34 acts as an antideterminant towards MetRS thereby preventing a lethal misacylation and as a positive identity element for IleRS. In addition, variants of tRNA$^{Ile}_1$ with A34, U34, or C34 in place of G34 is not charged with isoleucine.[86] Thus, the IleRS from *E. coli* must recognize structural elements common to the chemical different nucleosides G34 and k^{2}C34 as the identity elements.

Chemical modification by cyanogen bromide of mnm^{5}s^{2}U34 present in tRNA^{Gln}, tRNA^{Glu}, and tRNA^{Lys} significantly reduces the amino acid acceptance activity of these tRNAs.[46,61,87,88] These results indicate a direct interaction between mnm^{5}s^{2}U34 and the corresponding aminoacyl-tRNA ligases although some conflicting results exist.[89] Furthermore, the GlnRS forms an H bond with mnm^{5}s^{2}U34.[90] Thus, the corresponding aminoacyl-tRNA ligase may make physical contact with the wobble nucleoside mnm^{5}s^{2}U34, which is consistent with the X-ray analysis of the GlnRS-tRNA^{Gln} complex.[91] In fact, the 2-thio group of mnm^{5}s^{2}U may be bound in the same way as the 2-keto group of C34, present in one of the two glutomine isoacceptors.[91a]

The hypermodified nucleoside Q or its glycosyl derivatives (manQ and galQ) are present in position 34 of tRNAs specific for Asn, His, Asp, and Tyr from eubacteria and eukaryotes except for yeast. The presence of manQ34 in the tRNA^{Asp} from mammals compared to the unmodified G34 results in a higher V_{max} and a lower K_m[92] (Table 2). tRNA^{Tyr} from *E. coli* has Q34 as the wobble nucleoside. The mutant, *tgt*, contains an unmodified G34 instead of the normal modified Q34 at the wobble position of this tRNA species. tRNA^{Tyr} from such a mutant has a small but according to Noguchi et al.[93] a definite difference in K_m (Table 2). However, histidine tRNAs from rabbit reticulocytes with or without Q34 have the same K_m.[94] Thus, the hypermodified nucleoside Q34 is of importance in some aminoacylation reaction of tRNAs but with no effect in others.

6. Position 35 (middle base of anticodon)

Native yeast tRNA^{Tyr} has pseudouridine (Ψ35) as the middle nucleoside of the anticodon. A change from Ψ35 to U35 at this position increases the K_m twofold for the cognate TyrRS at a high Mg^{2+} concentration and greater than twofold at low Mg^{2+} concentration.[95] Results from other chemical modifications at position 35 suggest that specific hydrogen bonds are formed between TyrRS and the nitrogen hydrogens of positions 1 and 3 of Ψ35.[95] Thus, Ψ35 might be directly involved in the recognition process between the tRNA and TyrRS.

7. Position 37 (3' of the anticodon)

Bisulfite treatment of *E. coli* tRNA^{Phe} results in the chemical modification of only the side chain of ms^{2}i^{6}A37. Such treatment does not influence the aminoacylation kinetics of this tRNA.[96] Neither iodination now desulfurization by Raney nickel of ms^{2}i^{6}A37 influence the aminoacylation of tRNA^{Phe}, tRNA^{Tyr}, or tRNA^{Ser}.[97,98] Infection by a defective lambda phage, which harbors the gene for the suppressor derivative of tRNA^{Tyr}, results in an overproduction of this tRNA. The enzymes involved in the synthesis of ms^{2}i^{6}A37 are unable to properly synthesize this nucleoside when the primary transcript is overproduced. This results in the appearance of two undermodified tRNA^{Tyr} species, which have either i^{6}A37 or unmodified A37 in place of the fully modified derivative.[29] Both undermodified species showed the same rate of acylation at low tRNA concentrations. In *E. coli* overproduced tRNA^{Phe} completely devoid of ms^{2}i^{6}A37 shows the same kinetics of aminoacylation as the ms^{2}i^{6}A-containing species[99] (Table 2). If *E. coli* is grown in the presence of iron-binding proteins or under iron starvation, the tRNAs possess the undermodified derivative i^{6}A.[31-33] Such undermodified tRNA^{Phe} and tRNA^{Trp} have the same kinetics of aminoacylation as that of fully modified tRNAs.[100] Note that different degrees of modification of A in position 37 (A37, i^{6}A37, and ms^{2}i^{6}A3) do not affect the aminoacylation reaction. Starvation of a *Lactobacillus* strain for mevalonic acid (a precursor to the isopentenyl (i^6) side chain of i^{6}A37) results in a tRNA lacking i^{6}A37. Such a tRNA accepts several amino acids to the same extent as the i^{6}A-containing tRNAs.[101] Thus, all results strongly suggest that the ms^{2}i^{6}A37 does not interfere with the aminoacylation reaction. Note, however, no systematic analysis has been performed, which addresses the question whether or not ms^{2}io^{6}A37 acts as an antideterminant for noncognate aminoacyl-tRNA ligases.

Yeast tRNA^{Phe} contains the hypermodified Y base in position 37. This base can specifically be removed by mild acid treatment, which results in a tRNA having only a ribose group at position 37. This tRNA^{Phe}, which lacks the Y base, has the same K_m as Y-containing tRNA and has 40% of the normal V_{max} in the aminoacylation reaction.[102] Since the tRNA is totally lacking a base in position 37, these minor effects suggest that the modification *per se* might not influence the aminoacylation reaction.

Completely unmodified tRNA$_1^{Ile}$ has a much lower amino accepting activity than the native tRNA$_1^{Ile}$ (Table 1) suggesting that the native modification may be an identity element for tRNA$_1^{Ile}$. Base substitutions at position 37 of tRNA$_1^{Ile}$ are critical for the aminoacylation reaction.[56] Therefore, t^{6}A37, which is normally present in native tRNA$_1^{Ile}$, may be of utmost importance in the aminoacylation reaction of tRNA$_1^{Ile}$. However, tRNA^{Ile}, which partially lacks t^{6}A37, is also normally aminoacylated.[45] One explanation to these apparently conflicting results may be that Miller et al.[45] used a mixture of modified and unmodified tRNA$_1^{Ile}$ in their determination of the K_m, whereas Yokoyama[56] used a pure tRNA species. Obviously, more experiments must be done to establish the role of t^{6}A in the recognition process of IleRS. Unmodified tRNA$_3^{Thr}$, which normally contains mt^{6}A37, shows almost the same kinetic characteristics as native tRNA$_3^{Thr}$[58] (Table 1). Thus, t^{6}A37 is not an identity element for TheRS. Still, it is not unreasonable that the IleRS and ThrRS have different identity elements.

Nucleosides G34, U35, and C36 in the anticodon loop and the discriminator nucleoside G73, but not G37, are important for the aspartylation of yeast tRNA^{Asp}.[103] Completely unmodified yeast tRNA^{Asp} is mischarged by argRS.[51] Two Ψs and m^{1}G37 were suggested to be important as antideterminants. Synthesis *in vitro* of the two Ψs does not prevent mischarging by argRS

TABLE 2
Kinetic Parameters for Aminoacylation of tRNAs Lacking a Specific Modified Nucleoside Relative to Wild-Type tRNA

tRNA	K_m	V	V/K_m	Conditions	Ref.
Ec. Arg s^2C32^-	0.94	0.94	1.0		79
Ec. Met ac^4C34^-	1.0	1.0	1.0		82
Ec. Phe $ms^2i^6A37^-$	1.0	1.1	0.93	10mM Mg^{2+}	99
Ec. Tyr $Q34^-$	1.3	0.89	0.68		93
Ec. Tyr i^6A37 instead of ms^2i^6 A37	n.d.	Same rate	n.d.		29
Ec. Tyr $ms^2i^637^-$	n.d.	Same rate	n.d.		29
Ec. Tyr m^5U54^-	1.2	1.0	0.83	20 mM Mg^{2+}	118
Ec. Tyr s^4U8^-	0.93	1.0	0.93		62
St. His Ψ^-	0.8	0.88	1.1		108
Yeast Tyr $\Psi35^-$	1.8	1.0	0.56	15 mM Mg^{2+}	95
Yeast Tyr $\Psi35^-$	3.3	0.8	0.74	1.1 mM Mg^{2+}; 1mM Spermine	95
Yeast Phe $Y35^a$	1.0	0.4	0.4	15 mM Mg^{2+}	102
LM. Asp $Q34^-$	1.8	0.76	0.43		92
Rabbit His Q^-	0.7	1.0	1.4		94

[a] $tRNA^{Phe}_Y$– obtained after mild acid treatment has only a ribose, i.e., no base, instead of the Y base.

suggesting that m^1G37 may be the modified nucleoside acting as an antideterminant for the yeast $tRNA^{Asp}$.[104]

In summary, some modified nucleosides (ms^2i^6A and its derivatives) present in position 37 are not involved in the cognate interaction with the aminoacyl-tRNA ligases. Position 37 is not an identity element for many aminoacyl-tRNA ligases.[52] Thus, this lack of an effect of modification at this position in the cognate interaction is not surprising. However, in one case, $tRNA_1^{Ile}$, position 37 is likely an identity element and consequently t^6A37 may be of importance in the interaction with IleRS, although not with ThrRS. Furthermore, the m^1G37 modification of the yeast $tRNA^{Asp}$ may be an antideterminant.

8. Positions 38 (anticodon loop), 39, and 40 (anticodon stem)

Mutations in the *hisT* gene of *S. typhimurium* result in a lack of Ψ in positions 38, 39, or 40 in several tRNA species.[105] Such undermodified $tRNA^{His}$, $tRNA^{Leu}$, and $tRNA^{Lys}$ have the same charging levels *in vivo* as native tRNA counterparts.[106,107] Furthermore, the kinetics of acylation of the Ψ-deficient $tRNA^{His}$ from *S. typhimurium* is identical to that of the wild-type $tRNA^{His}$.[108]

9. Position 46

Chemical modification of m^7G46 in $tRNA^{Phe}$ does not influence the phenylation reaction.[109] In accordance with this observation, Sampson et al.[74] have shown that position 47 is not part of the recognition site for PheRS. However, enzymatic introduction of m^7G46 in $tRNA_f^{Met}$ from *Bacillus subtilis*, using a purified tRNA (m^7G46) methyltransferase from *E. coli*, slightly changed the kinetics of $tRNA^{Phe}$ acylation.[110] Thus, the presence of m^7G might slightly change the structure of $tRNA^{Phe}$, which thus indirectly influences the conformation of the identity elements of PheRS.

10. Position 48

Derivatives of acp^3U48 have been prepared using different chemical agents.[111-113] In all but two cases,[114,115] the chemically modified tRNA is able to normally accept the amino acid as the native tRNA does. The kinetic parameters, K_m and V_{max}, were not affected by introducing an acetyl group and was only slightly affected when more bulkier groups were introduced.[113] Thus, the acp^3U48 modification probably does not participate directly in the recognition process.

11. Position 54

Transfer RNA from a *trmA* mutant of *E. coli* lacks m^5U54 in its tRNA.[116] The initial rate of aminoacylation of m^5U lacking $tRNA^{Arg}$, $tRNA^{Met}$, and $tRNA^{Phe}$ under different ionic conditions is identical to that of the m^5U-containing tRNA.[117] A $tRNA^{Tyr}$ with either a U54 or an m^5U54 modification has the same K_m and V_{max} with respect to charging.[118] However, the presence of m^5U54 as well as $\Psi55$ in yeast $tRNA^{Ala}$ enhances the recognition between the tRNA and the cognate AlaRS.[119] *Thermus thermophilus* is an extreme thermophilic bacteria. tRNAs from these bacteria grown at a high temperature contain s^2m^5U54. The extent of thiolation increases with increasing temperature, and increases the stability of the tRNA. However, the presence of the s^2 group has no influence on amino acid acceptance.[120]

12. Position 55

Siddiqui and Ofengand[68] modified $\Psi 55$ with acrylonitrile. Although the modification probably acts specifically on $\Psi 55$, it was not directly demonstrated. Nevertheless, the modification inactivated the ability of $tRNA_f^{Met}$ to be charged. Since $\Psi 55$ has not been implicated as an identity element for MetRS, these results are not easily reconciled with the current models for MetRS recognition.[58] However, the acrylonitrile treatment might completely disrupt the D- and T-loop interactions resulting in a conformation not recognized by MetRS. As previously stated for position 54, $\Psi 55$ may also enhance the recognition process between yeast $tRNA^{Ala}$ and the cognate synthetase.

C. CONCLUSIONS

Clearly, nucleotides in the anticodon are important identity determinants for many aminoacyl-tRNA ligases.[52] Therefore, the involvement of k^2C34 in the identity of $tRNA^{Ile}_{minor}$, mnm^5s^2U34 in the identity of $tRNA^{Glu}$, $tRNA^{Gln}$, and $tRNA^{Lys}$, $\Psi 35$ in the identity of yeast $tRNA^{Tyr}$, and Q34 in the identity of $tRNA^{Asp}$ and $tRNA^{Tyr}$ is not surprising. However, the lack of any effect on the aminoacylation reaction of ac^4C34 in $tRNA^{Met}$ and I34 in $tRNA^{Arg}$ is more difficult to reconcile with these data. Modifications in position 37 may, e.g., t^6A37, m^1G37, or may not, e.g., $ms^2i(o)^6A37$, yW37, be involved as determinants or antideterminants in the recognition process. The lack of influence in other positions, e.g., in position 8 in the case of s^4U8, is reasonable since these positions have so far not been shown to be directly involved in the recognition process. However, these studies also clearly show that when many modified nucleosides are absent, conformational changes may influence the kinetic parameters. In the case of yeast $tRNA^{Asp}$, lack of numerous modified bases strongly influences the misacylation activities. This suggests that the conformational changes brought about by the presence of modified nucleosides are indirectly involved as identity determinants. However, in two cases (possibly m^1G37 in yeast $tRNA^{Asp}$ and k^2C34 in $tRNA_2^{Ile}$) the modified nucleoside(s) act(s) as (an) antideterminant(s).

IV. ROLE OF MODIFIED NUCLEOSIDES IN tRNA INTERACTIONS WITH TRANSLATION FACTORS

A unique feature of the cytoplasmic initiator methionine tRNAs from plants and fungi is the presence of a modified nucleoside at position 64.[121] In *S. cerevisiae*, this modified nucleoside is a 2'-ribosylated adenosine whereas wheat germ contains a 2'-ribosylated guanosine.[122] The specific removal of these modifications by periodate oxidation allowed the initiator tRNA$_i^{Met}$ from yeast and wheat germ to read internally located AUG codons. However, this lack of modification did not prevent the initiator tRNA$_i^{Met}$ from participating in the initiation process.[121] Position 64 is part of the minimal structure required for elongator tRNAs to interact with the bacterial elongation factor Tu(EF-Tu).[123] Eukaryotic aminoacyl-tRNAs bind efficiently to bacterial EF-Tu and the mechanism of binding to eukaryotic EF-1a-GTP and to prokaryotic EF-Tu-GTP may be similar.[121] In fact, yeast initiator tRNA$_i^{Met}$ lacking the modification at position 64 binds to EF-Tu-GTP more efficiently than the native initiator tRNA$_i^{Met}$. Thus, the modification in position 64 of eukaryotic cytoplasmic initiator tRNA$_i^{Met}$ may prevent the binding of tRNA$_i^{Met}$ to elongation factors and consequently, prevents it from participating in the elongation cycle. Elongation factor EF-Tu-GTP binds to the tRNA at the aminoacyl end and stem and at the T-stem, i.e., on the same side of the L-shape on which position 64 is exposed.[124] Thus, the anticodon stem and loop are not covered by EF-Tu. Furthermore, EF-Tu does not cover the TΨC-region of loop IV. Consequently, lack of m^{5}U54 (rT54) has no effect on the formation of the tRNA^{Phe}-EF-Tu-GTP and tRNA^{Lys}-EF-Tu-GTP ternary complexes.[125]

In *Streptococcus faecalis,* initiation of protein synthesis is normally mediated by formylated Met-tRNA$_f^{Met}$. However, when grown in the absence of folic acid the initiation of protein synthesis occurs with unformylated Met-tRNA$_f^{Met}$ since tetrahydrofolate is the cofactor for both the formylation of initiator tRNA and the synthesis of m^{5}U in G$^+$ bacteria (unlike other eubacteria and eukaryotes that use AdoMet as the cofactor in the synthesis of m^{5}U). Consequently, tRNA from folate starved cells is deficient in m^{5}U54.[126-129] Unformylated Met-tRNA$_f^{Met}$ lacking m^{5}U54 binds to ribosomes in response to initiation factors and poly(A,G,U). However, the unformylated Met-tRNA$_f^{Met}$ containing m^{5}U54 does not.[127] Thus, tRNA$_f^{Met}$ initiates protein synthesis independent of formylation in the absence of m^{5}U54 modification. Although not shown, these experiments suggest that the m^{5}U54 and formylation interfere in binding to some initiation factors. Formylation as well as m^{5}U54 are known to induce conformational changes in tRNA$_f^{Met}$.[127,130] Interestingly, initiation of protein synthesis in eukaryotic systems utilizes an unformylated initiator tRNA^{Met} that has A54 instead of m^{5}U54.

V. ROLE OF MODIFIED NUCLEOSIDES IN tRNA INTERACTIONS NOT APPARENTLY INVOLVED IN TRANSLATION

tRNA is also involved in cellular metabolism not apparently involved in translation. Mutations in bacterial tRNA genes [(tRNA$_{minor}^{Arg}$ *(dnaY)*, tRNA$_1^{Ser}$*(divE),* and tRNA$_2^{Ser}$*(supH/ftsM/supU)*] have been shown to affect DNA replication and/or cell division.[2-5] Modified nucleosides *per se* have so far not been implicated in the bacterial cell cycle. However, modification of tRNA may be directly involved in the control of the mammalian cell cycle. In such cells there are three tRNA^{Lys} species with different primary sequences.[131] One of them, tRNA$_2^{Lys}$ may occur in an undermodified form, tRNA$_4^{Lys}$. The latter appears in dividing or proliferating tissue and is almost completely absent in nondividing or nonproliferating cells.[132,133] The amount of tRNA$_4^{Lys}$ is correlated to the rate of cellular proliferation and this lysine tRNA becomes the predominant species in leukemia cells grown in tissue culture. Maturation of tRNA$_4^{Lys}$ involves several intermediates, two of which are tRNA$_5^{Lys}$ and tRNA$_6^{Lys}$.[134] The cell cycle of mammalian cells comprises four successive phases. Following M phase, the cells divide and enter the G1 phase, in which the cells prepare themselves for the S phase. DNA is synthesized in the latter phase. The level of tRNA$_4^{Lys}$ decreases before the cells are arrested in the G1 phase, and returns to a high level before the cells resume growth.[135] Therefore, it seems as if tRNA$_4^{Lys}$ exerts its effect early in the G1 phase. Several temperature-sensitive mutants blocked in the G1 phase were screened for changed levels of tRNA$_4^{Lys}$.[134] One mutant, ts-694, showed a decrease in the level of tRNA$_5^{Lys}$ and an increase of tRNA$_4^{Lys}$ at the restrictive temperature, but a reverse in levels at the permissive temperature. An enzyme extract from the mutant is defective in this first step of modification at the nonpermissive temperature, but not at the permissive temperature. This observation suggests that the defect at high temperature is at the first modification reaction. These results suggest that specific tRNA modifications may be necessary for the commitment of cell division.

Less conclusive data for a relationship between cell cycle and tRNA modifications exist for *S. pombe*. The antisuppressor mutation *sin3*, which influences the level of mcm^{5}s^{2}U in tRNA, leads to an increase in cell length.[136] Furthermore, allosuppressor mutations that also modulate the efficiency of nonsense suppressor tRNAs are allelic with the cell cycle control gene *cdc25*.[137] Preliminary data on altered tRNA modification patterns in allosuppressor strains have been obtained.[138] Finally, the *cdr* mutant that has in addition a changed cellular division response to nitrogen depletion is also an allosuppressor.[139]

Mutations in the *miaA* gene, which is the structural gene for the tRNA (i^{6}A37) synthetase,[140] result in an unmodified A37 in tRNAs that read codons starting with U. Furthermore, mutations in the *miaA* gene also result in a mutator phenotype.[141] The presence of *miaA* mutations raises the frequency of spontaneous mutations 5 to 29-fold depending on the lesions monitored. The induced mutations are preferably GC to TA transversions.[24] Limitation of iron results in i^{6}A37 instead of ms^{2}i^{6}A$^{31-33}$ and also increases the spontaneous mutation frequency.[140] The ms^{2}i^{6}A modification has been suggested to act as a physiological switch that modulates spontaneous mutation frequency.[141]

tRNA is also involved in several cellular reactions that are not directly connected to translation. It is directly involved in the conjugation of ubiquitin and in protein degradation.[7,8,142] The involvement of modified nucleosides in these reactions is unknown. A specific tRNA^{Gly} is involved in cell wall biosynthesis in *Streptococcus epidermidis*.[6,143] Interestingly, this tRNA^{Gly} is unable to bind to ribosomes and it is unusually devoid of modified nucleosides. Also, the major tRNA^{Gly} from *E. coli* donates glycine to the cell wall lipopolysaccharide.[144] δ-Aminolevulinic acid (δ-ALA) is a precursor in the biosynthesis of chlorophyll, heme, and vitamin B_{12}. Its synthesis occurs via at least two biosynthetic pathways, of which one involves the conversion of glutamate to δ-ALA. Glutamate is first activated at the α-carboxyl group by ligation to tRNA^{Glu} and this reaction is catalyzed by the GluRS. The activated Glu is then reduced and catalyzed by an amino transferase reaction to form δ-ALA[1]. The wobble base in this tRNA is mnm^{5}s^{2}U34. Two tRNA^{Glu} species differing in the extent of modification of mnm^{5}s^{2}U34 participate equally well in the synthesis of δ-ALA.[145a] Thus far, no evidence has been obtained for any physiological role for mnm^{5}s^{2}U34 in the control of δ-ALA synthesis.

VI. ROLE OF MODIFIED NUCLEOSIDES IN ANTICODON:CODON INTERACTIONS

A. GENERAL CONSIDERATIONS

tRNA from all organisms are frequently modified in the anticodon stem and loop. Position 34 (the wobble base) and position 37 (the 3' base next to the anticodon) are frequently modified with a great variety of modified nucleosides. In fact, no other position in tRNA has such a large variety of modified nucleosides.[15] As pointed out by Nishimura,[10] a correlation may be made between the identity of the modified nucleoside present, e.g., in position 37, and the coding capacity of that tRNA. Thus, tRNAs that read codons starting with U usually have a hydrophobic nucleoside, such as i^{6}A37 or its derivatives, whereas tRNAs that read codons starting with A always have a hydrophilic nucleoside, such as t^{6}A37 or derivatives thereof. Codons that start with C or G are read by tRNAs that have an unmodified purine in position 37 or, more frequently, methylated derivatives like m^{1}G37, m^{1}I37, m^{2}A37, or m^{6}A37. Since an A-U base pair is energetically weaker than a G-C base pair, the A-U base pair in the first position of the codon may require a stabilizing hypermodified nucleoside next to it to enhance the fidelity of the interaction. Interestingly, tRNAs from all organisms (eukaryotes, eubacteria, and archaebacteria) that recognize codons starting with A contain t^{6}A37 or its derivatives. Such an evolutionary conservation of i^{6}A37 derivatives is not seen with tRNAs that recognize codons starting with U. This group of tRNAs from eukaryotes and eubacteria also contains other hydrophobic nucleosides such as yW (nucleoside of base Y), whereas such tRNAs from archaebacteria have m^{1}G37. This evolutionary conservation of certain modified nucleosides and the regularities towards codon reading capacities of tRNAs suggest a common function irrespective of the origin of the tRNA.[146] As pointed out by Yarus,[147] the coding information of the tRNA may in fact extend beyond the anticodon towards the 3' side of the tRNA as far as to position 44. This area includes positions 38, 39, and 40, that are frequently modified to Ψ in many tRNA species and position 37, that, as previously stated, is frequently modified.[15,17,146,148,149]

Several hypermodified nucleosides are also present in position 34, the wobble base.[15,17,146,148,149] The recognition capacity of the modified nucleoside inosine (I34) was integrated into the wobble hypothesis.[150] Inosine (I), which is a derivative of A, is suggested to pair not only with U but also with A and C. However, it was soon realized that other modified nucleosides in this position violated Crick's wobble hypothesis. Nishimura and collaborators[151,152] discovered that tRNA^{Val} containing the wobble nucleoside cmo^{5}U34 not only paired with A and G according to the wobble hypothesis but also with U ["Nishimura rule" as stated by Ninio[153]]. In fact, organisms may be classified into three groups depending on whether cmo^{5}U34/mcmo^{5}U34, mo^{5}U34, or I acts as the wobble base in tRNAs specific for tRNA^{Val}, tRNA^{Ser}, tRNA^{Pro}, tRNA^{Thr}, or tRNA^{Ala}.[10] Moreover, the base pairing capacity of the xm^{5}s^{2}U34 derivatives (x can be, e.g., H-, CH$_3$NH$_2$-groups, etc.) are also in violation of the wobble hypothesis. These modified uridines are restricted in their coding capacity and preferentially pair with A, but less well with G contrary to what would be predicted from the wobble hypothesis. Also in this case it is likely that the function of these modified nucleosides is the same irrespective of the origin of the tRNA, since such xm^{5}s^{2}U34 derivatives are present in tRNAs from both eubacteria and eukaryotes. A structural explanation for the sharp contrast between the codon recognition patterns of the two kinds of modified uridines, xo^{5}U34 or xm^{5}s^{2}U34, has been elucidated.[154] Also, other modified nucleosides, e.g., ac^{4}C34, Q34, in position 34 have been shown to influence the codon recognition pattern. The pattern of modification at position 34 may be related to the identity of the middle base of the anticodon.[149] For instance, the modified nucleosides xm^{5}s^{2}U34 and Q34 derivatives are present in tRNAs having a U as middle base (position 35), whereas the modified nucleosides Gm34, Cm34, I34, and m^{5}C34 are present in tRNAs having an A at position 35. Such a modification pattern may be related to the stability of the anticodon:codon complex, and if so, the function of the modified nucleosides may be to modulate this interaction.[149]

Several kinds of model experiments have been performed to elucidate the function of the modified nucleosides. Formation of complexes of polynucleotides containing modified nucleosides have been studied.[155] Poly(m^{2}A), poly(Ψ), and poly(mo^{5}U) form a copolymer with the cognate polymer, whereas poly(m^{6}A), poly(i^{6}A), and poly(s^{2}U) do not, suggesting that m^{6}A, i^{6}A, and s^{2}U do not form a Watson-Crick base pair. A Hoogsten base pair and not a Watson-Crick base pair was suggested for the the m^{2}A-U base pair.[156] Furthermore, the presence of a methyl group in position 1 of guanine destabilizes the double-helix structure.[157] Therefore, these tRNA modifications frequently present in position 37, m^{2}A, m^{6}A, i^{6}A, and m^{1}G prevent potential base pairing with the mRNA. However, it must be remembered that the milieu in the anticodon region induces a restriction in the conformation of these modified nucleosides, which is not present in a copolymer.

tRNA with complementary anticodons can form dimers,[158] which was utilized by Grosjean and collaborators as a useful representation of an anticodon:codon interaction[149,159,160] Detailed analysis of different tRNA-tRNA complexes revealed three major elements, which stabilize the interaction: (a) a decreased flexibility of the anticodon induced by the anticodon stem, (b) the influence of bases outside the anticodon ("the dangling end effect"), and (c) the presence of modified nucleosides in the anticodon loop, which have a major stabilizing effect due to an enhanced stacking interaction. These three parameters vary in importance in the different tRNA combinations. In tRNA dimers having two A-U base pairs, the major stabilizing effect comes from stacking interactions whereas dimer formation between GC-rich anticodons are more dependent on the standard entropy term. The most surprising observation from these studies is that the stability of the dimer is not dependent on the GC content of the anticodon.[160] The ms^2i^6 modification in position 37 has a strong stabilizing effect mainly due to stacking interaction of the modified adenine to the anticodon triplet. Interestingly, most if not all of this effect was attributed to the ms^2 modification.[161] Furthermore, t^6A37 and yW37 also impose a stabilizing effect on the anticodon-anticodon interaction.[159,162] For wobble bases, both stacking interactions and flexibility constraints are responsible for the stabilization effect observed for tRNA dimers.[163] These results are consistent with the "extended anticodon hypothesis",[147] which is based on the correlation between sequences of tRNAs and their translational efficiency *in vivo*. Thus, in the anticodon:codon interactions, we have to consider at least the identity of the nucleosides in position 34 to 44 of the tRNA. This conclusion is based both on an artificial model system (tRNA-tRNA interaction) and on the translational efficiency of the tRNA in the living cell (as pointed out by Yarus[147] in his "extended anticodon hypothesis"). Although the extended anticodon is the major determining element in the decoding process, other parts of the tRNA also influence the decoding properties. This aspect is, for example, demonstrated by the altered decoding properties of the Hirsh suppressor, which has a base substitution in the D-stem of tRNA^{Trp} and by some frameshift suppressors.[164,165] Furthermore, although the tRNA-tRNA interaction may be a good model system, it has to be remembered that *in vivo* the ribosome is an active participant in the decoding process.

B. EFFICIENCY OF TRANSLATION

When methyl-deficient tRNA became available,[38] it was used *in vitro* to study the function of methylated nucleosides. Such tRNA is not an ideal substrate, since it is randomly undermethylated to 50% and all the different methylated nucleosides are affected. Nevertheless, the methyl-deficient tRNA^{Phe} and the fully methylated tRNA^{Phe} can be separated and then compared in a polyU mediated synthesis of polyPhe. Only a marginal effect is observed for the efficiency of translation.[42,166,167] tRNA^{Gly} made *in vitro* and totally devoid of modified nucleosides is as efficient as the normal tRNA^{Gly} in decoding a natural mRNA *in vitro*.[59,168] Synthetic yeast tRNA^{Phe}, which lacks all 14 modified nucleosides normally present in native tRNA, is also as efficient *in vitro* as a suppressor prepared by anticodon replacement of the native, modified tRNA^{Phe}.[169] Thus, it was early believed and later confirmed that most methylated nucleosides affect only marginally, if at all, the function of tRNA during *in vitro* translation.

According to the "extended anticodon hypothesis", the major determining element in the decoding process is the anticodon loop and the 3' side of the anticodon stem.[147] This view is also supported by the tRNA-tRNA dimer formation as discussed by Grosjean et al.[149] In fact, Yarus et al.[170] showed by systematic base substitutions in the anticodon arm (positions 29-41) of the amber suppressor Su7 tRNA that changes in the loop at positions 32, 33, 37, and 38 made the greatest impact on the efficiency of suppression but significant differences were also observed upon changes in the anticodon stem (positions 29-31 and 39-41). Consequently, modified nucleosides outside this region should only slightly, if at all, influence the efficiency of translation. Accordingly, a mutant-lacking s^{4}U8 grows as well as the wild-type and cyanogen bromide modification of tRNAs containing s^{4}U8 does not inhibit ribosome binding.[61,171] Although loss of s^{4}U8 modification reduces the ribosomal binding of tRNA^{Tyr}, it does not affect polyUAC directed protein synthesis.[62] However, s^{4}U8 is sensitive to near-UV light and is suggested to be the major chromophore for the growth delay induced by the near UV light irradiation of bacteria.[172,173] Upon irradiation, s^{4}U8 crosslinks with a cytidine in position 13[69] which drastically reduces the acylation rate of the tRNA.[174] This might be the reason for the *relA*$^+$-dependent stringent response observed upon near-UV illumination.[172] Consequently, mutants (*nuv*$^-$) deficient in s^{4}U8 also show a strikingly reduced growth delay and a *nuv*$^-$ mutation reduces the photoprotection efficiency.[175] In fact, mutants resistant to near UV light are of two classes — mutants (*nuvA* or *nuvC*) lacking s^{4}U8 or *relA*$^-$ mutants.[171,172,176] In addition, mnm^{5}s^{2}U34, which is present in tRNAs specific for Glu, Lys, and Gln, may also participate in this phenomenon since illumination with near UV light triggers ppGpp synthesis (a characteristic of a *relA*-dependent stringent response) not only in wild-type *E. coli* but also in *nuv*$^-$ mutants.[177] Thus, thiolated nucleosides, like s^{4}U8 and mnm^{5}s^{2}U34, react with UV light which changes the structure of the tRNA and thus influences the acylation properties and consequently the translation. This aberrant translation induces the *relA*-dependent stringent response, which in turn affects the physiology of the cell in many different ways.[178]

A mutant (*mia*$^-$) of yeast is deficient in D at least in tRNA^{Phe} and tRNA^{Tyr}.[179] The mutant has no clear phenotype and *in vitro*, the tRNA^{Phe} from the mutant is as active as wild-type tRNA in mediating polyPhe synthesis.[179,180]

1. Position 32

Several base substitutions in *Su*7 tRNA affect ribose methylation at position 32 to varying degrees.[170] No correlation exists between the level of ribose (32-0') methylation and the efficiency of suppression. However, position 32 was never completely unmodified in these studies. Thus, an effect of ribose methylation at position 32 on translational efficiency cannot be excluded.

In *E. coli* tRNA^{Arg} (anticodon ICG), s^{2}C is present in position 32. This s^{2}C32 was chemically converted to C32, which changes the structure of the anticodon loop.[181] The functional implication of this demodification of tRNA was tested *in vitro* using MS2 RNA as template. Translation of MS2 RNA *in vitro* using cell free extracts from *E. coli* results in the synthesis of the viral synthetase and the coat protein, but also in several other polypeptides, the latter of which are the results of +1 or −1 frameshifting events.[182] The addition of tRNA$^{Ser}_{3A}$ and tRNA$^{Ala}_{1B}$ promotes and inhibits, respectively, the endogenous −1 frameshifting, which may occur at three sites, of which one is at codons 123-124 (GCA(Ala)-GCA(Ala)).[182,183] It was hypothesized that tRNA$^{Ser}_3$ makes a two-base pair interaction between nucleotides at positions 34 and 35 in the anticodon and the first two bases (underlined) in the codon at this GCA-<u>GC</u>A site and thus promotes a −1 frameshifting event.[184] This results in the appearance of polypeptide 7. The addition of tRNA^{Ala}, which reads GCA, competes efficiently with the frameshifting tRNA$^{Ser}_3$ and thus inhibits frameshifting at this site.[182] Whereas native tRNA^{Arg}(s^{2}C32) inhibits the endogenous −1 frameshifting resulting in polypeptide 7, the unmodified tRNA^{Arg}(C32) does not.[181] No explanation for this experimental result was presented by Bauman et al.[181] However, upstream of the endogenous frameshifting site (codon 123-124), there is the sequence AAC-<u>CCG</u>-<u>A</u>UU (codons 116-118). *E. coli* tRNA^{Arg} (anticodon ICG) reads CGU/C/A. Addition of native tRNA^{Arg} in 15 fold excess might enforce reading of CG-A (underlined) in this sequence, which would result in a +1 frameshift. If this is so, then the native tRNA^{Arg} (s^{2}C32) at this high concentration is able to compete with tRNA^{Pro}(CCG) and tRNA^{Ile}(AUU). This would abolish the −1 frameshift at codons 123-124, i.e., synthesis of polypeptide 7 should be inhibited by the addition of native tRNA^{Arg}(s^{2}C32) as was observed by Bauman et al.[181] If this is so, then the demodified tRNA^{Arg} (C32) would be a less efficient competitor since this inhibition of frameshifting does not occur with the demodified tRNA^{Arg}(C32). If this model is shown to be valid, it suggests that lack of s^{2}C at position 32 lowers the translational efficiency of tRNA^{Arg}.

2. Position 34

Reticulocytes contain two isoaccepting histidine tRNAs, which only differ in the presence (tRNA$^{His}_1$) or absence (tRNA$^{His}_2$) of the hypermodified nucleoside Q in the wobble position. tRNA$^{His}_1$ (Q34) is three times more abundant than tRNA$^{His}_2$ (G34). *In vitro* both incorporate histidine into globin and read the two histidine codons, CAU and CAC, equally well.[185,186] Thus, the modified nucleoside Q does not influence the efficiency of tRNA^{His} from reticulocytes. Another way to study the function of Q is to utilize a mutant, *tgt,* which completely lacks Q in its tRNA.[93] The *his* operon is regulated by an attenuation mechanism, which is sensitive to the charging and translational efficiency of tRNA^{His}.[187] An aberrant cognate decoding of the seven *his*(CAU) codons in the attenuator results in a changed expression of the *his* operon. No difference in the expression of the *his* operon was observed in *tgt*$^+$ and *tgt* cells suggesting that the Q deficient tRNA^{His} is as efficient as the wild-type tRNA in translating the His codons in the leader region of the *his* attenuator. In another study[93] tyrosine tRNAs containing either Q34 or G34 were compared in their ability to bind to the cognate codons. Both tRNAs favor the binding to UAU compared to UAC. However, the binding efficiency of G34-tRNA^{Tyr} to the triplet programmed ribosomes is twofold less than that of Q-tRNA^{Tyr}. Thus, whereas Q34 does not influence the efficiency of tRNA^{His} from both reticulocytes and *E. coli,* the efficiency of *E. coli* tRNA^{Tyr} is reduced twofold. This suggests that the influence of Q34 on individual tRNA species may be different.

Thiolated uridines are present in the wobble position in tRNAs specific for Gln, Lys, and Glu. One such modified nucleoside is mnm^{5}s^{2}U34, which is present in these tRNAs from *E. coli* or *S. typhimurium*. Its synthesis may occur by the following steps:

$$U34 \xrightarrow{asuE} s^2U34 \xrightarrow{trmE} cmnm^5s^2U34 \xrightarrow{trmC1}$$
$$nm^5s^2U34 \xrightarrow{trmC2} mnm^5s^2U34$$

An *asuE* mutant contains a nonthiolated derivative of U, possibly mnm^{5}U34, in position 34. Thus, the modification at position 5 of the uracil ring seems to occur independently of thiolation.[188] If so, then the thiolation reaction does not necessarily have to be the first reaction to occur as depicted. The ochre suppressor *supG,* which is a derivative of tRNA^{Lys}, and *psu2* of phage T4, which is a derivative of tRNA^{Gln}, also contains mnm^{5}s^{2}U34. Mutants defective in *asuE,*[188] *trmE,*[189] and *trmC*[190] all decrease the efficiency of these ochre suppressor tRNAs. Note, tRNA from the *trmC1* mutant has a more extended modification (cmnm^{5}s^{2}U) than the native nucleoside mnm^{5}s^{2}U.[191] Still, the efficiency of the tRNA is decreased. The fact that cmnm^{5}Um is more rigid (see Section V.E.) than mnm^{5}U,[192] suggests that the cmnm^{5}s^{2}U is also more rigid than mnm^{5}s^{2}U. If so, then the reduced suppressor efficiency by *trmC1* tRNA$^{Lys}_{UAA}$ suggests that a nucleoside, which is too rigid, hampers the efficiency of the tRNA to read cognate codons. The *trmC2* mutation results in the accumulation of nm^{5}s^{2}U34 instead of the native mnm^{5}s^{2}U34.[191] Although this mutation decreases the efficiency of the *supG* ochre suppressor (tRNA$^{Lys}_{UAA}$), it does not affect the binding properties of tRNA^{Lys} to AAA and AAG programmed ribosomes.[189] Nevertheless, the results with the *trmC1/C2* mutants demonstrate that an optimally modified nucleoside in the wobble position has evolved and only a slight change in its structure affects the efficiency of the tRNA. A mutation (in *E. coli* strain JF3) at an unknown biosynthetic step and probably defective in the synthesis of mnm^{5}s^{2}U34, also decreases the efficiency of the phage T4 *psu2* ochre suppressor.[193] This bacterial mutant is also temperature sensitive for growth. Since the wobble base seems to be part of the recognition site for these aminoacyl-tRNA ligases, an effect on the charging ability of the tRNA can be envisioned. However, even though the aminoacylation reaction is drastically impaired for a suppressor tRNA, the same tRNA is still active as a suppressor *in vivo*.[194] Therefore, the aminoacylation reaction may not be the rate-limiting step of translation *in vivo*. Furthermore, the effect observed also includes codon context influences[190] as well as efficiency in triplet binding *in vitro trmE*[189] suggesting that the observed defect due to the mnm^{5}s^{2}U34 deficiency is most likely due to an aberrant anticodon:codon interaction.

In many eubacteria a selenium-containing modified nucleoside is present in position 34.[195] In *E. coli* and *S. typhimurium*, tRNA^{Lys} and tRNA^{Glu} contain mnm^{5}Se^{2}U34.[196,197] In these tRNAs, the modified nucleoside present in the wobble position is a mixture of mnm^{5}s^{2}U34 (60%) and mnm^{5}Se^{2}U34 (40%).[197,198] A mutant of *S. typhimurium, selA1*, is unable to incorporate selenium into tRNA.[198] Consequently, tRNA from this mutant only contains the sulfur derivative, mnm^{5}s^{2}U34. The ochre suppressor, *supG*, encodes a tRNA^{Lys}, which has mnm^{5}s^{2}U34 or mnm^{5}Se^{2}U34 as the wobble nucleoside. The efficiency of the sulfur-containing suppressor tRNA is twofold less in some codon contexts as compared to the selenium-containing suppressor.[198] Thus, the substitution of selenium for sulfur in this wobble nucleoside significantly increases the translational efficiency of the tRNA.

A mutant (*sin3*) of *Schizosaccharomyces pombe* is deficient in mcm^{5}s^{2}U34.[199] This mutant has a change in cell size and a reduced growth rate suggesting a reduced efficiency of translation. Furthermore, the *sin3* mutation reduces the efficiency of the UGA and UAA suppressors tRNA^{Ser} and the UGA suppressor tRNA^{Leu}. Since these tRNAs normally do not contain mcm^{5}s^{2}U34 the antisuppressor activity of *sin3* is not easily reconciled with a deficiency of mcm^{5}s^{2}U34. Still these two phenotypes cosegregate which suggests that the pleiotropic phenotype is due to a single genetic event.

The amber suppressor tRNA$_3^{Leu}$ (SUP53) in *S. cerevisiae* contains m^{5}C in the wobble position. The SUP53 tRNA gene encodes a precursor tRNA which contains an intron. Mutations in the intron or removal of it influence the modification pattern of the amber suppressor derivative of tRNA$_3^{Leu}$.[200] Hence, a less efficient suppressor activity is correlated with the absence of m^{5}C34 suggesting that the m^{5}C34 is necessary for an efficient translation.

3. Position 35

Tyrosine tRNAs from yeast and *Drosophila* contain introns, and the mature tRNA^{Tyr} from both organisms have Ψ in position 35. The synthesis of Ψ35 requires the presence of the intron.[201,202] Consequently, a yeast mutant with a precise deletion of the intron in the ochre suppressor SUP6 gene has an ochre suppressor tRNA^{Tyr} completely lacking Ψ35.[201] This undermodified tRNA$_{UAG}^{Tyr}$ has a reduced efficiency of suppression. Although this may be due to a reduced anticodon:codon interaction, the effect observed on suppression may be indirect since it has been shown that Ψ35 interacts directly with the TyrRS.[95] However, this is unlikely since, as previously stated, the aminoacylation reaction may not be the rate-limiting step of the translation process *in vivo*.[194] Furthermore, Ψ is able to pair with A although there is a quantitative difference between a A-Ψ and A-U base pair. In the A-U base pair, the keto group at position 2 is close to the A whereas in the case of the A-Ψ base pair the keto group at position 6 of Ψ is oriented away from A. This may reduce incorrect base pairing without hampering the ability to pair with A.[203] Therefore, presence of Ψ may prevent misreading. However, this aspect was not addressed by Johnson and Abelson.[201] Since Ward and Rich[204] have shown that alternating A-Ψ polynucleotides have a much higher melting temperature (TM) (58°C) as compared to A-U polymers (32°C), the suggestion by Johnson and Abelson[201] that indeed Ψ35 directly influences the anticodon:codon interaction is reasonable.

4. Position 37
a. i^6A derivatives

Mutants (*miaA1*) of *S. typhimurium* and *E. coli*, which lack ms^2io^6A37 and ms^2i^6A37, respectively, in their tRNA, grow significantly slower than the wild-type strain.[28,205] The difference is greater with increased growth rate and this growth rate difference is as much as 50% in rich media. The polypeptide step-time is reduced by 20 to 31% *in vivo* and this reduction is independent of the growth medium.[28,205,205a] Assuming that the reduced polypeptide growth rate is only due to the reduced translational efficiency of those tRNAs normally containing ms^2io^6A37, the average step time for those tRNAs is increased four times in *S. typhimurium*[28] or two and a half times in *E. coli*.[205] ms^2i^6A37 and ms^2io^6A37 are present in tRNAs that read codons starting with U, and among them the amber suppressor tRNAs. The lack of these modified nucleosides dramatically reduce the efficiency of amber suppressors (between 25 to 99% depending on suppressor and codon context, see below).[206,207] Lack of ms^2i^6A37 in Su3 $tRNA^{Tyr}$ results in little or no suppression *in vitro*.[29] The Hirsh suppressor, which has an A24 to G24 base substitution in $tRNA^{Trp}$, is able to read both UGG(Trp) and UGA (nonsense codons). If this UGA suppressor tRNA lacks ms^2i^6A37 it is almost inactive *in vivo* as well as *in vitro*.[207,208] Yeast tRNA normally contains i^6A37 and mutants deficient in this modified nucleoside have been characterized.[209,210] Lack of i^6A37 in yeast also has a profound effect on the efficiency of amber suppressors but only a slight[209] or no[210] effect on growth rate. The first mutant lacking ms^2i^6A37 was isolated as a mutant in which the *trp* operon was derepressed eightfold.[27,211] The derepression is caused by a slower translation of the two trp codon in the *trp* leader sequence by the undermodified $tRNA^{Trp}$, which normally contains this modified nucleoside. A similar effect induced by the lack of ms^2io^6A37 or ms^2i^6A37 has been observed in several other operons known to be regulated through an attenuator mechanism.[28,212-214] Thus, the presence of ms^2i^6A37, ms^2io^6A37, or i^6A37 (yeast) in place of the unmodified A37 influences profoundly the efficiency of normal and suppressor tRNAs as well as cellular growth.

tRNA, which lacks only the ms^2 group and consequently contains only i^6A37, is synthesized either under conditions of limited iron or by overproduction of a specific tRNA.[29,31-33,100] Analyses of kinetic and thermodynamic parameters of tRNA-tRNA complexes differing in their degree of A37 modification suggest that the major stabilizing element of ms^2i^6A37 is the ms^2 group.[161] As previously stated, lack of ms^2i^6 modification has a dramatic effect on the efficiency of suppression *in vivo*. Furthermore, the lack of ms^2i^6 modification almost completely inhibits the activity of $tRNA_{UAG}^{Tyr}$ mediated suppression *in vitro* whereas the lack of ms^2 modification reduces the efficiency of suppression by only 50%.[29] In a polyU mediated polyPhe synthesis, the activity of the ms^2 deficient $tRNA^{Phe}$ is reduced by 70 to 80%.[100] $tRNA^{Trp}$, with i^6A37 instead of ms^2i^6A37, induces a threefold derepression of the *trp* operon, which has two adjacent Trp codons in its leader mRNA.[100] Thus, in a special codon context such as the repeated codons in polyU or in the *trp* leader sequence, the presence of the ms^2 group seems to affect the efficiency of a tRNA. However, no effect due to ms^2 deficiency was observed *in vitro* using a natural mRNA as template.[100] A mutant (*miaB*), which is blocked in the synthesis of the ms^2 group, has recently been isolated.[35] The *miaB1* mutation reduces the efficiency of amber suppressors about two to sixfold but results in no significant difference in growth rate in several different kinds of media between wild-type and mutant. Thus, the efficiency of the tRNA when reading codon contexts present in natural mRNAs seems to be influenced less by the ms^2 group. However, the ms^2 group of ms^2i^6A37 and ms^2io^6A37 does have a significant effect in reading some codon contexts, e.g., iterated codons, and in stabilizing tRNA-tRNA complexes. The latter results suggest that the ribosome must change the conformation of the anticodon such that the stability of the anticodon:codon interaction is different from that in model experiments analyzing the stability of the tRNA-tRNA dimer.

Although it is clear that the presence of ms^2i^6A37 or ms^2io^6A37 is of profound importance for the efficiency of tRNA, two unmodified $tRNA^{Ser}$ species, which read codons starting with U, normally exist in *E. coli*.[215] These tRNAs have a very high stability as a tRNA-tRNA complex suggesting that there is no need for a stabilizing modification at position 37.[161] Furthermore, an efficient nonsense suppressor, which has an unmodified A37, has also been isolated.[216] Thus, the ms^2io^6A37 modification has a differential effect depending on the identity of the tRNA, as also observed with different amber suppressors.[206]

b. t^6A37 derivatives

The modified nucleoside t^6A is present in position 37 in all tRNAs reading codons starting with A. Furthermore, it is one of the eight modified nucleosides, which are present in the same position in tRNAs from all three kingdoms as well as in mitochondria.[146] This universal occurrence suggests that it plays an essential role in the performance of the tRNA. Starvation of a *relA1, thr*⁻

strain of *E. coli* results in the accumulation of tRNA^{Ile} of which 50% is lacking the t^{6}A37 modification.[45] Such undermodified tRNA binds less well to poly AUC and poly AUU programmed ribosomes indicating that the presence of t^{6}A37 is required for proper anticodon:codon interaction. The t^{6}A37-containing tRNA^{Lys} is preferred at most AAG codons tested in globin mRNA *in vitro* compared to a tRNA^{Lys} containing a precursor to t^{6}A. In contrast, the undermodified tRNA^{Lys} is preferred at AAA codons.[217,218] Thus, the t^{6}A37 modification not only improves the efficiency of the tRNA, but also influences codon choice. Brewers yeast contains two forms of tRNA$^{Arg}_{III}$ whose nucleotide sequences are identical except that one form has t^{6}A37 and the other has an unmodified A37. In comparison to the unmodified A37 tRNA species, t^{6}A-containing tRNA^{Arg} binds slightly better to polyAG programmed ribosomes, binds free AGA triplets more efficiently, and forms a more stable complex with a tRNA with a complementary anticodon.[162] It was concluded that the presence of t^{6}A37 stabilizes the adjacent base pair by an improved stacking interaction, consistent with the fact that t^{6}A in a Upt^{6}A dinucleotide stabilizes the stacking interaction compared to UpA.[219] In fact, this stabilizing effect is in the same range as that of a combination of dinucleotides containing ms^{2}i^6 or i^6 modifications of A. Thus, the t^{6}A37 and the i^{6}A derivatives both stabilize the anticodon:codon interaction. This is consistent with the suggestion by Nishimura[10] that the presence of these modified nucleosides in those tRNAs reading codons starting with A or U is due to the stabilization of the intrinsic weak interactions between U36-A/A36-U.

c. Y base

Yeast tRNA^{Phe} contains the modified purine Y (nucleoside yW), which is a derivative of guanine that can be excised by HCl. Yeast tRNA^{Phe} devoid of Y decreases the binding of complementary codons suggesting that the presence of this modified base improves the anticodon:codon interaction.[220,221] The whole base is excised in these experiments and, therefore, the results cannot be taken as support for a modification mediated improvement of the anticodon:codon interaction. tRNA^{Phe} from mouse liver contains yW37 and two methylated nucleosides in the 5' part of its anticodon. tRNA^{Phe} from mouse neuroblastoma cells lacks these three modified nucleosides. These tRNAs were compared in the translation of globin mRNA *in vitro*.[217] The yW37 deficient tRNA^{Phe} donates phenylalanine preferentially in competition with the fully modified tRNA^{Phe} at 13 out of 15 specific sites tested. Thus, the presence of yW37 in some way lowers the efficiency of tRNA^{Phe} at these sites. However, at the two additional sites tested, the fully modified tRNA^{Phe} is preferred indicating an improvement of the function of the tRNA at these sites. The anticodon region (Gm34-A35-A36-yW37) of yeast tRNA^{Phe} was replaced by C34-U35-A36-G37. This converted the tRNA^{Phe} to an amber suppressor, of which the activity is possible to measure *in vitro*.[222] tRNA^{Phe}, which has an unmodified G37, is highly active in reading amber codons *in vitro*. These results show that the Y base is not essential for the activity of a tRNA, but this does not rule out an important role for the yW37 modification in tRNA^{Phe}.

5. Positions 38, 39, and 40 (*hisT* mediated Ψ formation)

Among several derepressed histidine mutants in *S. typhimurium*, one, *hisT*, lacks Ψ at positions 38 and 39 of tRNA^{His}.[223] The *his* operon is preceded by a leader sequence containing seven consecutive histidine codons.[224] Johnston et al.[187] suggested an attenuation mechanism for the regulation of the *his* operon. According to this model, a slowed translation by a tRNA^{His} deficient in Ψ would account for the derepressed phenotype of this *hisT* mutant. In fact, the translation elongation rate *in vivo* is decreased by 20% in an *hisT* mutant.[225] The *supE* amber suppressor, which is a derivative of tRNA^{Gln}, normally contains Ψ in positions 38 and 39. Introduction of an *hisT* mutation in a *supE* strain decreases the efficiency of *supE* mediated suppression 10-fold.[226] Thus, it is clear that the presence of these two Ψ residues in the anticodon stem and loop improves the efficiency of the tRNA dramatically. tRNA^{Tyr} and its amber suppressor derivative *supF30* contain Ψ only in position 39, i.e., in the last base pair of the anticodon stem. The combination of an *hisT* mutation with the *supF30* suppressor decreases the efficiency of the amber suppressor tRNA$^{Tyr}_{UAG}$ only twofold.[227] Thus, the impact of Ψ in these three positions is not the same and quantitatively the modification seems to be more important when present in the loop (position 38) than in the stem (positions 39 and 40). This is reasonable, since Ψ and U forms a similar base pair with A (see Section V.B.3). Thus, the presence of U instead of Ψ does not impair the stability of the stem. However, a minor but significant effect is still observed due to the lack of Ψ, even in the stem (position 39), suggests a perturbation of the anticodon. Mutations in *hisT* has pleiotropic effects of which many, but not all, can be explained by their influence on tRNA mediated attenuation.[105,228-232] One such pleitropic effect not easily reconciled with a tRNA mediated attenuation mechanism is how Ψ deficiency influences the synthesis of ppGpp. Starvation of stringent (*relA$^+$*) *hisT* mutants for histidine does not provoke ppGpp synthesis.[233] Absence of ppGpp synthesis is not observed when the same *hisT* mutant is starved for amino acids whose cognate tRNAs do not contain Ψ in positions 38, 39, or 40. This anomaly may be due to the inability of Ψ deficient uncharged tRNA^{His} to bind to the ribosomal A-site which is a prerequisite for the *relA$^+$* mediated ppGpp synthesis. Nevertheless, such histidine starvation of an *hisT* mutant leads to an abrupt cessation of stable RNA synthesis, i.e., histidine starvation of an *hisT* mutant provokes stringent response without ppGpp accumulation. These interesting properties of the *hisT* mutants have not been further investigated.

6. Position 54 (ribothymidine; m^{5}U54)

All tRNAs from eubacteria and most elongator tRNAs from eukaryotes contain m^{5}U54 (ribothymidine). tRNA from archaebacteria contains m^1 Ψ in position 54. Note, the position of the methyl group in m^{1}Ψ and m^{5}U in position 54 is the same in relation to the ribose moiety and the phosphate-ribose backbone suggesting an evolutionary convergence for the presence of a methyl group in this position of the tRNA.[234] The almost ubiquitous presence of m^{5}U in tRNA suggests an essential role in cellular growth. Indeed, the presence of m^{5}U54 as compared to U54 stabilizes the structure of the tRNA as shown by NMR analysis.[130] This stability is even further increased by the presence of s^{2}m^{5}U54.[130] Since mutants of *E. coli* and *S. cerevisiae* devoid of m^{5}U54 grow well, the presence of this modified nucleoside in tRNA is not essential for cellular growth.[116,235] This conclusion is supported by the existence of bacterial species which completely lack m^{5}U54.[236-240] However, a mutant (*trmA5*) of *E. coli*, which completely lacks m^{5}U54, is outgrown after only 20 generations by wild-type cells in a mixed population experiment.[241] The difference in growth rate between the *trmA5* mutant and wild-type is 4%, which is of evolutionary significance even though it is small.[242] Under conditions where the tRNA concentration *in vivo* is rate limiting, tRNA$_{UAG}^{Tyr}$ lacking m^{5}U54 suppresses as well as tRNA$_{UAG}^{Tyr}$ containing m^{5}U54.[118] Also, *in vitro*, the m^{5}U54-deficient tRNA supports poly U-mediated polyPhe synthesis as well as wild-type tRNA.[117] A deficiency in m^{5}U54 facilitates initiation of protein synthesis in *Streptococcus faecalis* with unformylated tRNA$_f^{Met}$ and a similar mechanism may be operating in *E. coli* protein synthesis.[19,126,127,243] Several eukaryotic tRNAs lack m^{5}U54.[244-246] Some of these tRNAs can be methylated *in vitro*. Thus, methylated and nonmethylated pairs of tRNAs differing only in the content of m^{5}U54 can be tested in protein synthesis *in vitro*. Marcu and Dudock[244] observed that tRNA^{Gly} containing m^{5}U54 inhibits protein synthesis *in vitro* compared to tRNA$_1^{Gly}$ lacking this modified nucleoside. This inhibitory effect of m^{5}U54 is reversed by an optimal concentration of polyamine. Also poly A-mediated polyLys synthesis is more efficient with tRNA lacking m^{5}U54. This was attributed to a decreased A-site binding (P-site binding was not impaired) and an improved translocation efficiency.[125] tRNA^{Phe} from different mammalian tissues contains different amounts of m^{5}U54. Such tRNA^{Phe} can be quantitatively converted *in vitro* to m^{5}U54-containing tRNA^{Phe}.[247] Such pairs of tRNA^{Phe} were compared in poly U-directed polyPhe synthesis. In contrast to eukaryotic tRNA$_1^{Gly}$ and eubacterial tRNA^{Lys}, mammalian tRNA^{Phe} with increased m^{5}U54 content is correlated to an apparent increased V_{max} with no change in K_m.[247] Apparently, presence of m^{5}U54 in tRNA may either increase or decrease the rate of polypeptide synthesis suggesting that this ubiquitous modified nucleoside m^{5}U54 might be involved in regulating protein synthesis. Thus, the presence of m^{5}U54 improves, in a small but significant way, the growth rate of *E. coli*, stabilizes the structure of tRNA, and may be part of a translational regulation mechanism operating in mammalian cells. Although it is clear that the catalytic activity of tRNA(m^{5}U54) methyltransferase is not essential for cellular growth, the polypeptide or transcripts is, as shown by disrupting the *trmA* gene of *E. coli*.[248] These results suggest that the peptide or transcript may have a function other than catalyzing the formation of m^{5}U in tRNA. Interestingly, the tRNA(m^{5}U54)methyltransferase is present in two forms in the cell, one of which is covalently bound to a segment of 16S rRNA.[249] Although the reason for the presence of two populations of the enzyme is unknown, this observation also suggests an alternative function for the TrmA peptide.

C. CODON CONTEXT SENSITIVITY

It has been known for some time that the nucleotides preceding and following a codon influence its efficiency of translation. In fact, there are severe constraints on the nucleotide sequence surrounding a sense codon.[250-254] The strongest context biases in *E. coli* are found in genes with low expression. It was therefore suggested that codon contexts, which reduce expression, have been selected.[251,255] Thus, codon context may be an integral part of the regulatory mechanisms operating to determine the level of expression of a gene. The efficiency of nonsense suppressors is highly dependent on the codon context.[256-259] Of course, the codon context effect of nonsense suppressors is also influenced by the codon context sensitivity of the release factor, with which the nonsense suppressors compete. Since the efficiency of nonsense suppressors varies at least fivefold[258,259] but the codon context sensitivity shown by the RF1 is only twofold,[260] some effect must be attributed to the tRNA itself. Furthermore, missense suppressors also show codon context effects clearly demonstrating an effect due entirely to the sensitivity of the tRNA for the surrounding nucleotides.[261] Moreover, codon context sensitivity has also been observed *in vitro*.[262] Thus, the efficiency of tRNA to translate a specific codon is strongly influenced by the nucleotides surrounding the codon to be read. The codon context effect from nearby nucleotides can be attributed to several mechanisms: (a) a release factor effect in the case of nonsense suppressors, (b) tRNA-tRNA interaction, and (c) a stacking effect by nearby nucleotides. Grosjean and Chantrenne[263] proposed that the anticodon:codon complex is stabilized by stacking interactions involving nucleosides in positions 37 and 34. These two positions in the tRNA are often modified and thus it is reasonable to assume that modification might influence the codon context sensitivity of the tRNA. Since tRNAs in the P- and A-sites are only 18Å apart,[264] tRNA-tRNA interaction is likely to occur at some time during the translation cycle. Indeed, Smith and Yarus[265] have shown that such interactions occur. Changes on the 5'side of the anticodon of the tRNA in the P-site influence strongly the entry of the tRNA into the A-site. Their model proposed close contacts between the wobble base of the tRNA in the P-site and the hypermodified nucleoside at position 37 of tRNA in the A-site. The degree of modification at position 37 varies from no modification at all, to a small methyl group, or to highly bulky substitutions like threonyl or isopentenyl groups. The codon constraints can be partly due to these different size modifications present in the tRNA at positions 34 and 37. Therefore, modifications *per se* can influence the codon context in several ways and, consequently, modulate gene expression.

1. Position 34

The *trmC1* and *trmC2* mutations influence the synthesis of mnm^{5}s^{2}U34, which is present in the ochre suppressor tRNA$^{Lys}_{UAA}$ (*supG* or *supL*). This tRNA contains cmnm^{5}s^{2}U34 and nm^{5}s^{2}U34 in the mutants *trmC1* and *trmC2*, respectively.[191] These altered modifications reduce the efficiency of this ochre suppressor tRNA and a small but significant increase in codon context effect is also observed.[190] Pedersen and Curran[260] have recently shown that the release factor RF1 shows a twofold codon context effect and the preferred 3′nucleotide is U. The affinity of RF1 for UAG-A and UAG-C is the same. Still, a small difference between two such contexts was observed when comparing fully modified and aberrantly modified tRNA$^{Lys}_{UAA}$ [190] suggesting that the effect of mnm^{5}s^{2}U on codon context is at the level of the anticodon:codon interaction.

Mammalian tRNA^{His} normally contains Q34 but undermodified species can be prepared from rabbit reticulocytes. Q34 and G34 containing tRNA^{His} translate *in vitro* globin mRNA at the same overall efficiency.[185,186] Furthermore, they translate CAC or CAU at 11 different sites equally well. Therefore, Q34 has no effect on overall incorporation of histidine or into individual histidine-containing sites in the globin mRNA.

2. Position 37

Fully modified mouse liver tRNA^{Phe} contains a yW37 (nucleoside of the Y base) and two methylated nucleosides (Gm34, Cm32) in the 5′portion of the anticodon. The hypomodified tRNA^{Phe} from mouse neuroblastoma cells lacks these modified nucleosides. These tRNAs were compared in translating phenylalanine codons UUU/C at 15 different sites in globin mRNA.[217] In most sites, the hypomodified tRNA^{Phe} is preferred. There are, however, several sites which deviate from the mean suggesting a codon context effect, although there was no indication that adjacent bases in the mRNA affect the selection of the tRNA. However, a striking pattern of tRNA preference exists at the tandem codon UUC-UUC in which the hypomodified tRNA^{Phe} at the P-site prefers the fully-modified tRNA^{Phe} at the A-site. These results support the notion that the nature of the previously selected tRNA is a determinant for the selection of tRNA at the A-site.[261,265] The size, shape, charge, and hydrophobicity of the modification at position 37 of the tRNA selected at the A-site may influence this preference.

tRNA$^{Lys}_{1,2}$ from rabbit liver contains t^{6}A37 while tRNA$^{Lys}_{4}$ contains an unidentified precursor to t^{6}A37.[217] These tRNAs were compared *in vitro* using globin mRNA as a template. At 13 out of 22 sites tested, the fully-modified species was preferred. However, at some sites, the reverse was observed indicating a functional role of t^{6}A37 in the codon context sensitivity of the tRNA. However, this could not be attributed to a neighboring base or codon. At two sites, the tandem codons AAG-AAG appear in globin mRNA. At both these sites the t^{6}A37 containing tRNA^{Lys} is preferred at the first AAG whereas the hypomodified tRNA^{Lys} is preferred at the second AAG. In contrast to the selection of tRNA^{Phe} at tandem codons, the fully-modified tRNA at the P-site prefers the hypomodifed tRNA^{Lys} at the A-site. These results are also consistent with a tRNA-tRNA interaction on the ribosome. Although the effect of the preceding tRNA was only observed at tandem codons, such interactions are also likely to occur at other codon contexts. tRNAs with a certain modification may have difficulty coexisting on the same ribosome due to the spatial problems discussed previously.

The hypermodified nucleoside ms^{2}io^{6}A37 or ms^{2}i^{6}A37 is found in tRNAs reading codons starting with U and consequently, in nonsense suppressor tRNAs. The efficiency of the amber suppressors is decreased to varying degrees (up to 200-fold) by a deficiency in ms^{2}io^{6}A37 or ms^{2}i^{6}A37.[206] However, the unmodified tRNA is also more sensitive to codon context. These results were obtained using amber codons at different positions in *lacI* mRNA. Therefore, no firm conclusions can be made as how and if some of the neighboring nucleotides influence the activity of the tRNA. Ericson and Björk[266] constructed a system in which the amber codon was at the same site in the *hisD* mRNA of *S. typhimurium* with context difference only in the nucleotide 3′ of the amber codon. Whereas the ratio of the efficiency of modified amber suppressor tRNAs at the two codon contexts (UAG-A versus UAG-C) is about one and a half, the ms^{2}io^{6}A-deficient tRNAs show a ratio of five irrespective of the strength of the suppressor. Thus, the presence of ms^{2}io^{6}A37 specifically counteracts an unfavorable nucleotide at the 3′ side of the amber codon. Similar results have also been obtained by Björnsson and Isaksson[267] utilizing another assay system. These results are not due to a codon context sensitivity of the release factor since RF1 has the same affinity for UAG-A and UAG-C.[260] The results are also not consistent with a tRNA-tRNA interaction since the ms^{2}io^{6}A37 of the A-site tRNA is oriented towards the 5′ side of the tRNA in the P-site. A changed tRNA-tRNA interaction would therefore most likely be mediated by base substitutions 5′ of the test codon as argued by Yarus and Thompson.[268] The ms^{2}i^6 modification stabilizes the anticodon:codon interaction by improved stacking.[208] Ericson and Björk[266] proposed a model in which the ms^{2}i^6 modification stabilizes the anticodon:codon interaction by improved stacking both on A36 in the anticodon and on the first base (U) of the codon. Furthermore, the wobble base (C34) stacks on the base (A or C) 3′ of the codon. The free energy increments for an unpaired dangling base GX/C (where X is A or C) are −1.1 and −0.4, respectively.[269] Thus, the most stable anticodon:codon complex is a fully-modified A in position 37 (A37) and an A on the 3′ side of the codon. The least stable complex would be the unmodified A37 and C3′ of the codon. This was also observed. The ratio of UAG-A to UAG-C is higher (5) for the unmodified tRNAs as compared to the modified tRNA (1.5) and the same irrespective of the strength of the ms^{2}io^{6}A-deficient suppressor. Thus the modification is of greater importance when the 3′ base is C than if it is an A. Thus, the ms^{2}io^{6}A37 senses the identity of the base 3′ of the codon perhaps by inducing a conformational change of the anticodon that also alters the quality of the stacking interaction of the wobble nucleoside. However, other kinetic explanations are not ruled out. Since codon context influences gene expression, the degree of modification as demonstrated by Ericson and Björk[266] might be an important regulatory device as pointed out earlier.[19,105,190,217]

D. FIDELITY

Several kinds of translational errors can be envisioned. A missense error can either be the result of a misacylated tRNA or of an anticodon:codon mismatch (misreading) on the ribosome. The role of modified nucleosides in the aminoacylation reaction was discussed in Section II. In this section, only the roles of the modified nucleosides of tRNA in the misreading of sense codons, reading of stop codons, or ribosomal shifting of the reading frame (frameshifting) will be addressed. Furthermore, premature termination ("drop-off") may also occur resulting in shorter peptides, which are most likely degraded quickly. The missense errors occur at a frequency of 5×10^{-3} to about 1×10^{-5} and depend on both the particular codon and the codon context.[270] The frequency of frameshifting and premature termination may be in the order of 10^{-3} [270] although in some systems frameshifting is very high and approaches 100%.[271] Manley[272] showed that 31% of the initiated β-galactosidase chains terminated prematurely, which corresponds to a "drop-off" frequency of 3.5×10^{-5} per codon.[273] However, *E. coli* cells can be grown for hundreds of generations with a 10 times higher error frequency[274] and Ram mutants of *E. coli,* which also have higher error frequency, do not have abnormal death rates.[275] Thus, *E. coli* cells may tolerate a rather high level of wastage, suggesting that this bacteria has no energy limitation. However, the cell has evolved several mechanisms to control and cope with errors that do occur. These include codon choice and context, proofreading, editing, stringent response and protein turnover, and in several of these mechanisms, modified nucleosides have been shown to play an important role in a direct or indirect way.

1. Missense errors
a. Position 34

The modified nucleoside Q is present in the wobble position of tRNAs encoding Tyr, His, Asn, and Asp. The *E. coli* mutant, *tgt,* lacks the enzyme, which inserts Q base in place of guanine. The *tgt* mutant grows as well as the wild type but is less able to survive in the stationary phase.[93,276] This mutant is also able to weakly suppress some amber mutations suggesting that lack of Q34 may induce some missense errors.[277]

Readthrough at termination codons by naturally occurring suppressors would allow an organism or a virus to produce two proteins from one cistron.[278-280] A classical example for such a readthrough event is found in the TMV virus. Two *Drosophila* tRNA[Tyr] species occur, which differ in the presence or absence of Q34.[281] Only the G34-containing tRNA[Tyr] is able to translate a stop codon in the TMV RNA, resulting in two peptides from the same cistron. Thus, the degree of modification at position 34 controls the ability to translate stop codons. Since stop codons may be a device to regulate formation of two peptides from the same cistron, the level of Q34 in tRNA may play an important role in this regulation. Indeed, the tobacco plant, which is the natural host for the TMV virus, contains in its leaves two tRNA^{Tyr} species. Both isoacceptors have G34 and both promote readthrough of the UAG stop codon.^{282} Thus, the tRNA^{Tyr} present in the leaves lacks Q and consequently translates UAG. In wheat germ, however, tyrosine two tRNAs exist which only differ in the presence or absence of Q34. Only the G34-containing tRNA^{Tyr} is able to read UAG. The mechanism for reading this nonsense codon by the unmodified tRNA^{Tyr} (anticodon GΨA) is not clear. According to Crick's wobble hypothesis, the GΨA-UAG interaction is illegitimate.^{150} These Q-deficient tRNAs do not read UAA codons^{283} implying that G34 of the GΨA anticodon plays a selective role in this recognition process, since the Ψ35-A base pair is the same for reading UAG and UAA. However, since the Ψ-A base pair is more stable than the U-A base pair,^{204} the Ψ-A base pair may be required for UAG suppression. Note, that Q-deficient tRNAs specific for histidine, asparagine, and aspartic acid have so far not been observed to misread XAG codons (X may either be C, A, or G). They all have an unmodified U35. As previously discussed, Ψ35 in a yeast amber suppressor is apparently essential for suppressor activity.^{201} Thus, the presence of Ψ35 may be important for the observed UAG reading of the Q-deficient tRNA^{Tyr}. However, the *tgt* mutant of *E. coli* is able to induce a weak UAG readthrough at some codon contexts.^{277} None of the Q-containing *E. coli* tRNAs, including tRNA^{Tyr}, contain Ψ35. Thus, although Ψ35 might be a necessary requirement for a strong readthrough by Q-deficient tRNA^{Tyr}, it is thus not essential for this misreading. It was suggested that the Q34 in the *syn* conformation may not base pair with the G in UAG but G34 in the same conformation may.^{283} Still, the nature of the recognition of UAG by the GΨA anticodon is obscure. Since the UAG codon is preceded and followed by CAA codons in the replicase gene of TMV, a hopping mechanism was considered. Alterations of either the 5′ or the 3′ CAA codon does not influence the readthrough of the amber codon suggesting that hopping is not a likely mechanism but that the identity of the six nucleotides 3′ of the amber codon is important for readthrough.^{284} Obviously, more studies are needed to unravel the molecular mechanism by which Q-deficient tRNA^{Tyr} is able to read UAG codons.

It is thought that late in the evolution of the code, the AUA codon was reallocated from Met to Ile.^{285} If this is so, then some mechanism must have evolved such that no misreading occurs by tRNA^{Met} and tRNA^{Ile}. This is indeed the case and modification of these tRNAs seems to be of pivotal importance in this aspect. The elongator Met-tRNA$^{Met}_m$ contains ac^{4}C in the wobble position and reads AUG codons. Many tRNAs contain C34 but only the elongator tRNA$^{Met}_m$ in *E. coli* and a few tRNAs in *Halobacterium* contain ac^{4}C34.^{78} This suggests a unique function for the acetyl group present in Met-tRNA$^{Met}_m$. This modification has no influence on the methionylation reaction.^{82} However, upon removal of the ac^4 group by bisulfite, the tRNA gains the ability to read AUA.^{286} Note, that AUA is normally read by a minor tRNA^{Ile}, which has the modified nucleoside k^2C34 (L34) in the wobble position. The ac^4C-lacking $tRNA_m^{Met}$ binds twice as well as wild-type $tRNA_m^{Met}$ to the AUG-programmed ribosome and even stronger to AUG triplets in the absence of ribosomes. Consequently, the undermodified $tRNA_m^{Met}$ incorporates Met more efficiently using Poly(A,U,G)-programmed ribosomes. Thus, ac^4 modification in the CAU anticodon partially prevents the reading of the complementary AUG codon, and increases the efficiency in the reading of the noncomplementary AUA codon. Although A-C base pairing is uncommon and does not adhere to Crick's wobble rules, it is present in some tRNAs, e.g., $tRNA_m^{Met}$.[287] C-A base pairs are formed in double strands[288] although they are much weaker than I-C and G-C base pairs. An A-C base pair might involve the rare imino tautomer of C, which can form a Watson-Crick base pair with A that is stereochemically similar to an A-U base pair.[286] The ability for C to read A most likely requires a specific conformation of the tRNAs, since most amber suppressors, which normally have C34, are unable to read the ochre codon UAA. Thus, the function of the ac^4 modification appears to primarily prevent misreading of AUA(Ile) and this is achieved at the cost of a slight reduction in the ability to read AUG (met) by $tRNA_m^{Met}$.

The AUA(Ile) codon is read by a minor species of $tRNA^{Ile}$ ($tRNA_2^{Ile}$), which constitutes only 5% of the total $tRNA^{Ile}$ in the cell.[289] This tRNA has lysidine (k^2C34,L34) in the wobble position[83] and recognizes only AUA.[289] Thus, the modification C34 to k^2C34 alters the wobble nucleoside so that A is recognized instead of G. This is not wobble, but a complete conversion of a base-pairing specificity.[84] As discussed (Section II), this modification also inhibits the misacylation of methionine. Therefore, a single post-transcriptional modification affects both aminoacylation and codon specificity. During evolution, the appearance of a post-transcriptional modification converted an AUG encoding $tRNA^{Met}$ species to an AUA encoding $tRNA^{Ile}$ species. This might have contributed to the conversion of an AUA Met codon into an Ile codon.

b. Position 37

The influence of ms^2i^6A37 modification on missense errors, which requires mismatch in the wobble position, was measured *in vivo* by Bouadloun et al.[206] Erroneous reading of the Trp codon UGG by $tRNA^{Cys}$ was monitored. $tRNA^{Cys}$ normally reads UGU/C codons and has the anticodon sequence U33-GCA-ms^2i^6A37. When misreading UGG(Trp), $tRNA^{Cys}$ (anticodon GCA) competes with cognate $tRNA^{Trp}$ (anticodon CCA), which also has ms^2i^6A37. This misreading by $tRNA^{Cys}$ requires a G34-G mismatch, i.e., a third codon position mismatch. A 30-fold lower misreading was observed in a *miaA* mutant strain, which is unable to synthesize the ms^2i^6 modification. The misreading of UGA codons, probably by $tRNA^{Trp}$ (anticodon CCA), is also due to a third codon position mismatch. Moreover in this case, the misreading is drastically reduced in the *miaA* strain.[206,207] The A37 base may be in its normal stacking position when a mismatch in the third codon position occurs, and if this is so, then the modification will improve the stability of the anticodon:codon interaction as has been shown for the amber suppressors resulting in an improved efficiency of the cognate interaction. When tRNA^{Cys} misreads the Trp codon UGG, both tRNAs compete for the same codon and normally carry the ms^{2}i^{6}A modification. Since the error rate is lower in the *miaA* strain, the modification deficiency has a larger effect in the weaker, near-cognate interaction between tRNA^{Cys} (anticodon GCA) and UGG than in the stronger, cognate interaction between tRNA^{Trp} (anticodon CCA) and UGG.

In vitro missense errors using poly(U) as mRNA were used to study the influence of ms^{2}i^{6}A modification of tRNA^{Phe}(anticodon GAA) and of the near-cognate tRNA$_4^{Leu}$ (anticodon UAA).[290] Also in this case a third codon position error was monitored, i.e., a U*34-U mismatch [U* is cmnm^{5}Um].[192] Both of these tRNAs carry ms^{2}i^{6}A. In a competition experiment using tRNA$_4^{Leu}$, which lacks ms^{2}i^{6}A, and tRNA^{Phe}, which contains the modification, misreading by undermodified tRNA$_4^{Leu}$ was decreased twofold as compared with the native tRNA$_4^{Leu}$. As part of the ternary complex, undermodified tRNA$_4^{Leu}$ and undermodified tRNA^{Phe} show the same efficiency at the initial selection step. However, the number of GTP molecules hydrolyzed per peptide bond formed is increased twofold for the undermodified species. The authors concluded that the ms^{2}i^6 modification does not exert its effect in the initial selection step but in the proofreading step. Thus, in this case, lack of ms^{2}i^{6}A37 prevents third codon position errors. Ehrenberg et al.[291] suggested that the proofreading step occurs after EF-Tu leaves the ribosome although conflicting results have been obtained.[292] Diaz and Ehrenberg[290] proposed that the EF-Tu masks the effect of the modification during the initial selection step.

Erroneous reading of the Arg codon CGU in the *rplL* gene (encoding ribosomal protein L7/L12) by tRNA^{Cys} (anticodon CGA) was investigated.[206] In this case, a first codon position (A36-C) mismatch was monitored. In addition, the modified or unmodified tRNA^{Cys} was found to compete with tRNA^{Arg}, which normally does not have the ms^{2}i^{6}A37 modification. No effect of the *miaA* mutation was observed.[206] The lack of effect of ms^{2}i^{6}A37 deficiency in a first codon position mismatch suggests that neither modified nor unmodified A37 contribute to the efficiency of reading CGU(Arg) by the tRNA^{Cys}. Perhaps the neighboring mismatch A36-C forces the A37 base, either modified or unmodified, out of the normal 3'-anticodon stack. If this is so, then the ms^{2}i^6 group does not influence the stability of the first codon position base pair. When *E. coli* tRNA^{Phe} lacking the ms^{2}i^{6}A modification is added in a 20-fold excess to an *in vitro* protein-synthesizing system, Phe is incorporated in response to the CUU(Leu) codon as well as to the normal phenylalanine codon, UUU/C.[99] No misincorporation is observed when the fully-modified tRNA^{Phe} is added under similar conditions. This misincorporation also requires a first codon position mismatch and resembles the Cys/Arg misincorporation discussed earlier. Furthermore, the competing tRNA (tRNA^{Leu} anticodon U$_{33}$-GAG-m^{1}G37) does not normally carry the ms^{2}i^{6}A modification. Wilson and Roe[99] suggested that the lack of the ms^{2}i^{6}A modification induces a larger probability for wobble in the third anticodon position. This suggestion is consistent with that of Jukes[293] which proposed that a hypermodified nucleoside located next to the anticodon would prevent a first codon position wobble.

Menninger[294] suggested that the premature release of peptidyl tRNA is a result of an active editing mechanism, in which noncognate tRNAs are released that have escaped the proofreading step in the ribosome A-site. A specific reduction in peptidyl tRNA accumulation is observed for those tRNAs which normally contain ms^2i^6A.[295] The molecular mechanism might be that the primary effect of the ms^2i^6 modification is at the proofreading step in the ribosome A-site, which would result in less substrate to be released. Thus, these results are consistent with the ribosome editor model, which links the release of peptidyl tRNA to mistranslational events.

In conclusion, when mismatch occurs in the wobble position, an ms^2i^637 deficiency prevents such near cognate misreading most likely by enhanced proofreading. This observation is also consistent with a reduced polypeptide elongation rate.[28,205] On the other hand, lack of ms^2i^6A37 modification is neutral or may indeed increase the frequency of errors induced by noncognate tRNAs that requires a mismatch in the first, and perhaps, second codon positions. Hence, near-cognate errors might be more aggressively proofread whereas noncognate errors might be prevented by the presence of the modification due to the stabilization of the base pair between the first codon base and the anticodon position next to the modified nucleoside as suggested by Jukes.[293] These apparently conflicting results are, in fact, correlated with the hypothesis of Ninio,[296] who suggests that modification which increases third position wobble decreases first position error. If this is so, then the effect imposed by ms^2i^6A37 or ms^2io^6A37 on error may be dependent on the error monitored. These results are also consistent with the allosteric three-site model,[296a] which predicts that when the ribosome has an occupied E-site, the A-site is in the low-affinity state. In this ribosomal mode presence of the modification influences the entry of the near-cognate tRNA but probably not the noncognate tRNA, since the latter even fully modified has almost no ability to bind to the low affinity A-site. Furthermore, in the low affinity A-site, the conformation of it may be such that the modification plays no role in the potential binding of the noncognate tRNA. Obviously, more experiments must be done, which address these questions in order to establish the mechanism(s) behind these apparently selective effects imposed by this remarkable modification.

c. Position 38, 39, and 40 (hisT mediated Ψ formation)

The *hisT* gene product catalyzes formation of Ψ in positions 38, 39, and 40 of which only 38 are in the anticodon loop region. The polypeptide elongation rate is reduced by 20%.[225] Accordingly, one would expect that mutations in the *hisT* gene would result in a reduced error level. Indeed, histidine starvation induces mistranslation most likely by misincorporation of glutamine.[297] The mistranslation frequency at the His codons are about 0.1 in the *hisT*$^+$ strain, but this is reduced 10-fold in an *hisT* mutant strain.[297] Both of the histidine tRNAs (anticodon GUG) and the two glutamine tRNAs (anticodon CUG and UUG) have Ψ in position 38 (tRNA^{His} and one of the tRNA^{Gln}s also have Ψ in position 39), i.e., in the anticodon loop. Apparently, the affinities of the near-cognate glutamine tRNAs are affected greater by the lack of Ψ38 than by the affinities of the cognate tRNA^{His}. Asparagine starvation, which results in lysine misincorporation, was unaffected by a *hisT* mutation.[297] In this case, both tRNA^{Asn} and tRNA^{Lys} have Ψ only in position 39, i.e., in the anticodon stem. Therefore, it seems as Ψ, located in the anticodon loop, influences the structure of the anticodon and thus the fidelity more than when it is present in the stem.

d. Position 54 (m^5U54)

Kersten et al.[125] demonstrated that in a polyU system the leucine misincorporation *in vitro* increased tenfold with m^5U deficient tRNA suggesting that such tRNA is error prone.

2. Frameshifting errors

Pieczenik[298] suggested that some modifications at position 37 might prevent base pairing with the nucleotide 5′ of the codon. Such modifications might have evolved to limit overlapping interaction with mRNA and to allow codons to be less context sensitive, thereby, resulting in a more flexible choice of amino acids in the protein. The methylated nucleoside m^1G is present in position 37 in tRNAs that translate codons of the type C$\overset{U}{\underset{G}{C}}$N from all organisms. The methyl group in position 1 of guanine prevents formation of one of the hydrogen bonds formed between G and C in a Watson-Crick base pair. Consequently, the stability of an m^1G-C duplex is lower than the corresponding G-C duplex.[157] A mutant, *trmD3*, of *S. typhimurium* that is deficient in m^1G in its tRNA has been isolated.[299] Lack of m^1G37 suppresses several +1C frameshift mutations in the *his* operon. All three tRNA$^{Pro'}$s present in *S. typhimurium* have m^1G in position 37 and they read CCN codons.[300] Since the *trmD3* mutation induces reading of quadruple bases of the type CCCU but not of the type ACCU, it was suggested that the frameshifting tRNA^{Pro} had the capacity to read four bases via an interaction between the four nucleosides N34GG-G37 in the anticodon region. Although it has been established that the amino acid proline is inserted at the CCCU/A frameshifting site, the actual mechanism has not been established.[301] The classical *sufD41* mutation, which results in an insertion of a C in the anticodon of tRNA^{Gly}, was thought to be a quadruple base reader,[302] but recent experiments have shown that the suppressor specificity is not compatible with a four base anticodon:codon interaction.[303] Whatever the mechanism, the presence of m^1G at position 37 is an important constituent in maintaining the tRNA in the proper reading frame. The conservation of m^1G37 in tRNAs that read CUN(Leu), CCN(Pro), and CGN(Arg) codons in all organisms suggests that the function of m^1G37 is to prevent frameshifting. A lack of m^1G37 reduces the growth rate considerably (up to 24% at 37°C) and the polypeptide elongation rate,[227,304] which demonstrates the importance of this ubiquitous modified nucleoside in cell physiology.

Overlapping genes occur in the *gag-pol-pro* translation unit of several retroviruses. The genes are all expressed from the first AUG codon. The *gag, pol,* and *pro* genes encode a vital structural protein, the reverse transcriptase, and a viral protease, respectively. In some retroviruses, the Gag-Pol polypeptide is generated via readthrough of an in-frame termination codon by a naturally occurring suppressor, while in others, it is generated by a –1 frameshift. In some retroviruses, the synthesis of the Gag-Pol-Pro peptide requires one –1 frameshift whereas others require two –1 frameshifts. The size of the overlapping sequences varies from 13 nucleotides to more than 100 nucleotides. In all of these overlapping sequences, one of three common sequences occur: A-AAC, U-UUA, or U-UUU. In the zero reading frame, these codons are read by tRNAs specific for Asn (AAC), Leu (UUA), and Phe (UUU), respectively. The corresponding tRNAs are thought to signal the frameshifting event.[305] Mutagenic experiments have established that the signal also consists of a heptanucleotide sequence including one codon 5' of these codons as shown in Table 3. In addition, downstream of these heptanucleotides, there is a stem-loop structure, which is also part of the frameshifting signal. The suggested mechanism for this frameshifting is a simultaneous –1 slippage of both tRNAs in the A- and the P-site of the ribosome.[305,306] tRNA^{Asn} and tRNA^{Phe} from mammalian tissues contain Q34 and Wye37, respectively. Most of the tRNA^{Phe} in HIV-infected cells and tRNA^{Asn} in HTLV-1- and BLV-infected cells lack the corresponding modifications.[307] The same tRNA species are not undermodified in noninfected cells. How the degree of tRNA modification is changed upon virus infection is unknown. On the other hand, tRNA^{Leu}, as well as several other tRNA species, are not changed upon infection. Since tRNA^{Phe} translates UUU present in the heptanucleotide of HIV and tRNA^{Asn} translates AAC present in HTLV-1 and BLV, the lack of these hypermodified nucleosides is correlated with the known frameshifting event. Therefore, the "shifty" tRNA proposed to be part of the simultaneous frameshifting may be a hypomodified isoacceptor. Based on the modification pattern of the tRNA in the ribosomal A-site of each frameshifting signal in the zero frame, the signals were grouped into three classes: Class I, tRNA^{Asn}, which normally contains a Q34; Class II, tRNA^{Phe}, which normally contains yW37, and Class III, tRNA^{Leu}, which normally lacks a hypermodified nucleoside (Table 3). Consequently, the three tRNAs in the A-site of the heptanucleotide all lack hypermodified nucleosides in the anticodon loop in the infected cells and this is also true for the tRNAs in the P-site (tRNA^{Lys} reading AAA is also undermodified in HTLV-1-infected cells as well as Q34 and yW37 of the Class III tRNAs in the P-site). Thus, the frameshifting event may be facilitated by a hypomodification making the tRNAs more "shifty". This correlation indeed suggests a functional relationship between the ability to frameshift and hypomodification but a direct test of the functional role of the modified nucleosides is required before a firm conclusion can be reached. The fact that this type of frameshifting exists in the *E. coli* gene (*dnaX*), which encodes both the τ and δ subunits of DNA polymerase III,[309-311] and the fact that the mammalian sequences also frameshift in *E. coli*[306] make it possible to investigate the involvement of hypermodified

TABLE 3
Ribosomal Frameshifting Sites in Viruses and the Occurrence of Hypermodified Nucleosides in the Anticodon Loop of tRNAs at the Ribosomal A- and P-Sites[a]

Class		Sequences at and around frameshift site	Modified nucleoside ribosomal P-site	Modified nucleoside ribosomal A-site	Source and location
Class I	A	A-AAA-AAC-U/G	mcm^{5}s^{2}U34; ms^{2}i^{6}A37	Q34 (Asn)	gag-pro: BLV, HTLV-1 and 2, MMTV, STLV-1; gag-pol, EIAV
	B	U-UUA-AAC-C/U/G	none	Q34 (Asn)	pro-pol: BLV, HTLV-1 and 2, STLV-1, gag-pol: EIAV, IBV
	C	G/A-$^{GG}_{AA}$A-AAC-G/A	?	Q34 (Asn)	gag-pro: MPMV, SRV-1 and 2; gag-pol: VISNA
Class II	A	G-GGU-UUU-C	none	yW37 (Phe)	gag-pol: mouse IAP
	B	A-AAU-UUU-U/C	Q34	yW37 (Phe)	pro-pol: MPMV, SRV-1 and 2; gag-pol: 17.6
Class III	A	A/G-^A_GAU-UUA-U	Q34	none (Leu)	gag-pol: RSV pro-pol: MMTV
	B	U-UUU-UUA-G/U	yW37	none (Leu	gag-pol: HIV-1 and 2; SIV; gypsy

[a] Table modified from Reference 308.

nucleosides in this kind of frameshifting event. If this is so, then the degree of tRNA modification may control the level of frameshifting required for the growth of several retroviruses as well as the synthesis of τ and δ subunits of DNA polymerase III in *E. coli*.

E. CODON CHOICE

The genetic code is highly degenerate. In the three extreme cases, six codons direct the incorporation of serine, leucine, or arginine. The use of the different synonymous codons is highly nonrandom and the bias is most extreme in highly expressed genes.[312-316] The codon choice affects translation efficiency and accuracy both *in vivo* and *in vitro*.[317-319] The individual codons have different affinities for the tRNA and most likely are translated at different rates.[318,319,319a] Therefore, codon choices may modulate the rate of translation and regulate gene expression. Modified nucleosides, especially those in the wobble position of the tRNA, have been shown to be involved in codon choice. Thus, the degree of modification may directly influence gene expression.

The wobble position often contains a modified nucleoside and a correlation exists between the kind of modified nucleoside present in this position and the coding capacity of the tRNA.[10] Uridines in this position are always modified (except in tRNAs from mitochondria and *Mycoplasma*). Two fundamentally different kinds of modified uridines are found: 5-hydroxy uridine derivatives (xo^5U34) and the 5-methyl-2-thiouridine derivatives (xm^5s^2U34). The C2'-endo and the C3'-endo forms of ribose puckering for the unmodified Up are almost equally stable, whereas the C2'-endo form for the xo^5U modification is more stable than the C3'-endo form. On the other hand, the xm^5s^2U modification is more rigid and almost exclusively takes the C3'-endo form. Therefore, the xm^5s^2U or xo^5U modification regulates the rigidity in the first position of the anticodon, which in turn influences correct and efficient translation. These two kinds of uridine modification influence the codon choice in opposing ways. tRNAs having the xo^5U derivatives of modification in the wobble position extend the wobble capacity, whereas the xm^5s^2U derivatives restrict it. The former derivatives (mo^5U, cmo^5U, and $mcmo^5U$) are found in eubacterial tRNAs specific for valine (GUN), serine (UCN), proline (CCN), threonine (ACN), and alanine (GCN), i.e., they are all members of tRNAs that read families of four codons specifying the same amino acid. In triplet-dependent binding and in *in vitro* synthesis of MS2 coat proteins these tRNAs read codons ending with A, G, or U.[320-326] From NMR analysis, it was concluded that an interaction between the xo^5 group and the 5'-phosphate stabilizes the C2'-endo configuration and favors formation of xo^5U34-U and xo^5U34-G base pairs. These uridines are still able to adopt the C3'-endo configuration and they form an xo^5U34-A base pair.[154] Within each codon family the codons ending in U or C are also read by at least another tRNA. Furthermore, tRNA from mitochondria as well as *Mycoplasma* have an unmodified U34 in the corresponding tRNAs. These tRNAs are able to read all four codons in the codon families.[237-239,327-329] Thus, it is not obvious why these modifications are present in this group of tRNAs. An Aro⁻ mutant of *E. coli* completely lacks xo^{5}U modifications and has an unmodified U in the corresponding tRNAs.[330,331] Such an Aro⁻ mutant grows 20% slower than wild-type suggesting that the presence of this group of modified uridines are of importance under some physiological conditions.[242]

The xm^{5}s^{2}U34 nucleosides restrict an intrinsic wobble capacity of the uridine. In triplet-dependent binding and in *in vitro* protein synthesis, these modified nucleosides primarily recognize A as the third letter of the codon.[332,333] One such modified nucleoside is mnm^{5}s^{2}U34, which is found in *E. coli* tRNAs specific for glutamine, lysine, and glutamic acid, i.e., in tRNAs of mixed codon families. These tRNAs only recognize codons ending with A or G but not codons ending in U, which would result in a missense error. The xm^{5}s^2 modification makes the structure rigid and results in a uridine, which can only adopt the C3'-endo form.[154] Thus, contrary to the xo^{5}U derivatives the xm^{5}s^{2}U derivatives cannot adopt the C2'-endo form.[154,334,335] This kind of modified nucleoside has recently been identified in two other tRNAs in mixed codon families. The tRNA$_4^{Leu}$, which reads UUA/G codons, and tRNA$_4^{Arg}$, which reads AGA/G codons, have cmnm^{5}Um and mnm^{5}U, respectively, in the wobble position. The three modified uridines have different conformational rigidities: mnm^{5}s^{2}U > cmnm^{5}Um > mnm^{5}U.[192] tRNA$_4^{Leu}$ recognizes more efficiently UUA than UUG[336] in accordance with the Previously stated rule. *In vitro* in the presence of different amounts of competing cognate tRNAs, modified tRNA$_4^{Leu}$, and tRNA$_4^{Arg}$ were tested for misreading of Phe (UUU/C) and Ser (AGU/C) codons, respectively. The concentration of the competing tRNAs necessary to inhibit misreading is inversely correlated to the rigidity of the modified nucleoside. Thus, a less rigid nucleoside is more prone to misread codons ending with U or C.[192] While the xo^{5}U-modified nucleosides contribute to an *efficient* translation of the codons, the major effect by the xm^{5}s^{2}U modification is to ensure a *correct* translation of the codons perhaps at the expense of an efficient reading of these codons. Consistent with this latter hypothesis is the behavior of the *trmC1* and *trmC2* mutants of *E. coli*. These mutants have cmnm^{5}s^{2}U34 and nm^{5}s^{2}U34, respectively, in the tRNAs instead of the native-modified nucleoside mnm^{5}s^{2}U34[191] (see also Section V.B.2). The *trmC1* mutation decreases the efficiency of reading both UAA and UAG codons about twofold. The *trmC2* mutation reduces this efficiency threefold and reading of UAG is somewhat more affected than UAA.[190] The thio group is the major cause for the stability of the C3'-endo form in xm^{5}s^{2}U derivatives but the 5'-substitution also contributes in stabilizing this conformation.[154,334] Since both *trmC* mutations result in the presence of both a thio group and a 5'-substitution the lack of a preferential effect on codon choice is reasonable, at least, in the case of the *trmC1* mutation. Still a reduced efficiency suggests that the native mnm^{5}s^{2}U is optimal for the efficiency. Mutants (*trmE*) having s^{2}U34 instead of mnm^{5}s^{2}U34 in their tRNA are still able to read UAA more efficiently than UAG codons.[189] Such undermodified tRNA^{Lys} also binds AAA better than AAG *in vitro*,[189] which is consistent with the previously mentioned rules.

As mentioned in Section V.B.2, tRNAs specific for lysine, glutamate, and glutamine contain either mnm^{5}s^{2}U or mnm^{5}Se^{2}U as the wobble nucleoside. In triplet-dependent binding studies, sulfur-containing *E. coli* tRNA^{Glu} binds better to GAA than to GAG whereas the selenium-containing tRNA^{Glu} from *Clostridium sticklandi* binds equally well to both codons.[195,337] Apparently, the presence of selenium increases wobble base pairing, i.e., making the wobble nucleoside less rigid. However, since mnm^{5}Se^{2}U residue but not the sulfur-containing counterpart may be significantly ionized at neutral pH, the near equal recognition of GAA and GAG by the selenium-containing tRNA may be a reflection of the difference in the ribosome-binding properties of the ionized mnm^{5}Se^{2}U versus the unionized sulfur-containing analogue.

Yeast tRNAs that read mixed codon families have mcm^{5}s^{2}U or ncm^{5}U at position 34. Since a methylene group bound to an aromatic ring is oriented in a plane perpendicular to the uracil ring plane,[338] it is likely that both of these nucleosides are more rigid than U in a fashion similar to the xm^{5}U derivatives as previously discussed. This is certainly true for mcm^{5}s^{2}U due to the presence of the thio group. The presence of these nucleosides in yeast ochre suppressors was suggested to explain why these suppressors do not read UAG and why an ncm^{5}U34-containing tRNA^{Ser} only reads UCA and cannot read UCG.[339-341] As previously stated, *E. coli* ochre suppressors in strains also carrying *trmC* or *trmE* mutations, which possess rigid modified nucleosides (cmnm^{5}s^{2}U; nm^{5}s^{2}U; s^{2}U) in position 34, are still able to read UAG codons, albeit at lower efficiency than mnm^{5}s^{2}U-containing suppressors. The different 5'-substitutions (xnm^{5}U versus xcm^{5}U) in eubacteria and yeast may act differently to prevent pairing with G. However, analyses of the crystal structures of these two kinds of modified nucleosides suggest similar conformations[338,342] and a similar stabilization of the C3'-endo form.[154] Although the presence of these rigid modified nucleosides, xcm^{5}U, prevent the pairing with G *in vitro*,[343] other features of the yeast tRNA are important to allow complete prevention of translation of codons ending with G since the eubacterial counterparts are still able to read such codons. The UCA reading serine tRNA from yeast does not read UCG. The *sin3, sin4* mutant has a reduced level of the mcm^{5}s^{2}U nucleoside in bulk tRNA.[136] If the presence of this modified nucleoside is responsible for restrictive reading of UCG, the *sin3, sin4* mutations should prevent this restriction. This was tested *in vivo* but found not to be the case.[136]

The synthesis of Q involves the exchange of the guanine base in position 34 in the tRNA with a queuine base. This reaction is catalyzed by the tRNA guanine transglycosidase, the structural gene which is denoted *tgt* in *E. coli*.[93] The three-dimensional structure of Q shows that the cyclopentenediol group is oriented out of the anticodon, such that it does not interfere with the recognition of the codon of mRNA.[344] This is in contrast to the short s^2 group, which is buried in the anticodon loop.[338,342] This orientation of the cyclopentenediol group may allow it to interact with some macromolecule other than mRNA.[344] The structure of Q suggests coding properties similar to an unmodified G although earlier experiments suggested that the Q-containing tRNA favors binding to NAU codons.[345] Indeed, tRNA[Tyr] with or without Q34 binds to both tyrosine codons but both with some preference to UAU.[93] However, soybean mitochondrial tRNA[Tyr] lacks Q34 and recognizes only UAC and not UAU.[346] Therefore, other structural features of the tRNA than Q in position 34 must also be important in inducing codon preference. Although no difference was observed in the codon choice of tRNA[Tyr] or tRNA[His] from mammals using the rabbit reticulocyte cell-free system,[185,186,347] Meier et al.,[348] using the *Xenopus* oocyte system, could demonstrate a Q-mediated codon choice for tRNA[His] from *Drosophila*. The latter authors injected TYMV RNA into the oocytes together with competing aminoacylated tRNA[His] with or without Q in the wobble position. The TYMV coat protein was isolated and the incorporation of His into CAU and CAC at two different sites in the mRNA was determined. The G34-containing tRNA[His] clearly preferred CAC compared to CAU, whereas the Q34-containing tRNA showed no preference, i.e., a similar codon choice as shown by Q34-lacking soybean tRNA[Tyr]. The presence of this discrimination in the oocyte system and the lack of it in the reticulocyte system may be due to the failure by the latter system to react to small structural differences in the anticodon. However, one cannot rule out a possible codon context effect since the two different His codons were located at two different sites in the TYMV mRNA.

The two tRNAs present on the ribosome in the A/P or in the P/E configuration[296a] are only 18Å apart.[264] Therefore, bulky modifications present at position 37 may influence the choice of the tRNAs at these positions. Some experimental evidence for such a selection was observed by Smith and Hatfield[217] and this aspect is discussed in Section V.B.4.

VII. SUMMARY AND OUTLOOK

It is clear that modified nucleosides are of great importance in the different interactions in which tRNA is involved. In Table 4, the effects of specific modified nucleosides in the different tRNA interactions are summarized. A + sign denotes an experimentally demonstrated effect in the specific interaction and a − sign denotes that the function of the modified nucleoside has been tested but no evidence for its participation was revealed. As shown in Table 4 modified nucleosides in the anticodon region (positions 34, 35, 37, and 38, 39, 40 (*hisT*)) are important for the anticodon:codon interaction. For modified nucleosides in other positions, it has been more difficult to firmly establish a direct function in a specific interaction. This may indicate that the latter modified nucleosides stabilize the structure of the tRNA in a minor but important way or interact more directly with the translation apparatus other than mRNA. If this is so, then the function of these modified nucleosides awaits a more direct assay system to reveal their function. The prospects of obtaining new information about the function of the modified nucleosides in the near future are very good, since more sensitive and more precise assay systems for the different tRNA interactions are rapidly developing. Furthermore, the pres-

TABLE 4
Experimental Evidence for Role of Modified Nucleosides in Different Aspects of tRNA Interactions

Position	Modified nucleoside	I Mod. enz.	II Ligases	III Transl. fact.	IV Not inv. in transl.	V.AC-C A General consideration	B Efficiency	C Codon context	D Fidelity	E Codon choice
8	s^4U		−							
10	m^2G		+							
18	Gm	*								
17,20	D						−			
26	m^2_2G		−							
32	s^2C		−				+?			
	Nm						−			
34	I		−							
	Q		+/−			*	+/−			+/−
	m^5C					*	+		+	++
	mnm^5s^2U		+		+?	*	++			
	mcm^5s^2U					*	++			
	mnm^5Se^2U									
	ac^4C		−			*				
	$k^2C(L)$		+			*				
	cmo^5U					*	+	+		++
	mo^5U					*				
35	Ψ		+			*	+			
37	m^2A					*			++	
	m^6A					*	++	+	+	
	ms^2io^6A	+	−							
	i^6A		−							

Position	Modified nucleoside				
37	t⁶A	+/−	*	+	
	ms²t⁶A		*		+
	m¹G	+?	*	+?	+
	yW	−?	*	+	+
38, 39, 40	Ψ	−	*		+
46	m⁷G	+/−		+	
47	acp³U	−			
54	m⁵U	−		−	
64	A*	+			
55	Ψ	+?			

Note: + denotes a suggested function in the indicated tRNA interaction, whereas − denotes that the function has been experimentally tested but no evidence for an involvement of the modified nucleoside in the indicated tRNA interaction was obtained. * denotes that the function is discussed.

ence of new genetic techniques helps to construct suitable genetic systems to evaluate the function *in vivo* of the modified nucleosides. To do this, it is required to isolate more mutants defective in tRNA modification. Of the about 45 different genes involved in tRNA modification in *E. coli* and *S. typhimurium*, only 13 have so far been identified:[15] *miaB*[35]; *queA*[349]. Only three yeast genes involved in tRNA modification have been identified. Therefore, the majority of genes involved in tRNA modification are still to be identified. This is a future challenge and will be of great importance together with the development of *in vivo* and *in vitro* systems to more precisely determine the role of modified nucleosides in the many different tRNA interactions. Thus, many more surprises about the function and interactions of tRNA are still ahead of us. It will be an exciting time to follow the progress towards a deeper and more detailed knowledge about the different fascinating aspects of tRNA interactions.

ACKNOWLEDGMENTS

This work was supported by the Swedish Cancer Society (Project No. 680), the Swedish Natural Science Research Foundation (BU-2930), and the Swedish Board for Technical Development (Grant No. 4206). The critical reading of the manuscript by A. Byström, C. Gustafsson, T. Hagervall, K. Kjellin-Stråby, and B. Persson are gratefully acknowledged. Thanks are due to D. Milton for linguistic improvements of the manuscript and to K. Thonfors-Olsson for excellent typing.

REFERENCES

1. **Schön, A., Krupp, G., Gough, S., Berry-Lowe, S., Kannangara, C. G., and Söll, D.,** The RNA required in the first step of chlorophyll biosynthesis is a chloroplast glutamate tRNA, *Nature (London),* 322, 281, 1986.
2. **Thorbjarnardottir, S., Björnsson, A., Amundadottir, L., and Eggertsson, G.,** Temperature sensitivity caused by missense suppressor *supH* and amber suppressor *supP* in *Escherichia coli, J. Bacteriol.,* 173, 412, 1991.
3. **Tamura, F., Nishimura, S., and Ohki, M.,** The *E. coli divE* mutation, which differentially inhibits synthesis of certain proteins, is in tRNA$_1^{Ser}$, *EMBO J.,* 3, 1103, 1984.
4. **Garcia, G. M., Mar, P. K., Mullin, D. A., Walker, J. R., and Prather, N. E.,** The *E. coli dnaY* gene encodes an arginine transfer RNA, *Cell,* 45, 453, 1986.
5. **Leclerc, G., Sirard, C., and Drapeau, G. R.,** The *Escherichia coli* cell division mutation *ftsM1* is in *serU, J. Bacteriol.,* 171, 2090, 1989.
6. **Stewart, T. S., Roberts, R. J., and Strominger, J. L.,** Novel species of tRNA, *Nature (London),* 230, 36, 1971.
7. **Feber, S. and Ciechanover, A.,** Transfer RNA is required for conjugation of ubiquitin to selective substrates of the ubiquitin- and ATP-dependent proteolytic system, *J. Biol. Chem.,* 261, 3128, 1986.
8. **Scornic, O. A., Ledbetter, M. L. S., and Malter, J. S.,** Role of aminoacylation of histidyl-tRNA in the regulation of protein degradation in chinese hamster ovary cells, *J. Biol. Chem.,* 255, 6322, 1980.

9. Waters, L. C. and Mullin, B. C., Transfer RNA in tumor viruses, *Prog. Nucl. Acids Res. Mol. Biol.*, 20, 131, 1977.
10. Nishimura, S., Modified nucleosides in tRNA, *Transfer RNA: Structure, Properties, and Recognition*, Schimmel, P. R., Söll, D., and Abelson, J. N., Eds., Cold Spring Harbor Laboratory, Cold Spring Harbor, NY, 1979, 57.
11. Farkas, W. R. and Singh, R., Guanylation of transfer ribonucleic acid by a cell-free lysate of rabbit reticulocytes, *J. Biol. Chem.*, 248, 7780, 1973.
12. Okada, N. F., Harada, F., and Nishimura, S., Specific replacement of Q base in the anticodon of tRNA by guanine catalyzed by a cell-free extract of rabbit reticulocytes, *Nucl. Acids Res.*, 3, 2593, 1976.
13. Elliot, M. S. and Trewyn, R. W., Inosine biosynthesis in transfer RNA by an enzymatic insertion of hypoxanthine, *J. Biol. Chem.*, 259, 2407, 1984.
14. Buck, M., Connick, M., and Ames, B. N., Complete analysis of tRNA-modified nucleosides by high-performance liquid chromatography: The 29 modified nucleosides of *Salmonella typhimurium* and *Escherichia coli* tRNA, *Anal. Biol. Chem.*, 129, 1, 1983.
15. Björk, G. R., Ericson, J. U., Gustafsson, C. E. D., Hagervall, T. G., Jönsson, Y. H., and Wikström, P. M., Transfer RNA modification, *Ann. Rev. Biochem.*, 56, 263, 1987.
16. Komine, Y., Adachi, T., Inokuchi, H., and Ozeki, H., Genomic organization and physical mapping of the transfer RNA genes in *Escherichia coli* K12, *J. Mol. Biol.*, 212, 579, 1990.
17. Björk, G. R., Modification of stable RNA, in *Escherichia coli and Salmonella typhimurium Cellular and Moleculara Biology*, Vol. 1, Neidhardt, F. C., Ingraham, J. L., Brooks Low, K., Magasanik, B., Schaechter, M., and Umbarger, H. E., Eds., American Society for Microbiology, Washington, D. C., 1987, 719.
18. Björk, G. R. and Kohli, J., Synthesis and function of modified nucleosides in tRNA, in *Chromatography and Modification of Nucleosides. Part B. Biological Roles and Function of Modification*, Gehrke, C. and Kuo, K., Eds., Elsevier, Amsterdam, 1990.
19. Kersten, H., On the biological significance of modified nucleosides in tRNA, *Prog. Nucl. Acids Res. Mol. Biol.*, 31, 58, 1984.
20. Nishimura, S., Structure of modified nucleosides found in tRNA, in *Transfer RNA: Structure, properties and recognition*, Schimmel, P. R., Söll, D., and Abelson, J. N., Eds., Cold Spring Harbor Laboratory, Cold Spring Harbor, NY, 1979, 547.
21. Grosjean, H., Haumont, E., Droogmans, L., Carbon, P., Fournier, M., deHenau, S., Doi, T., Keith, G., Gangloff, J., Kretz, K., and Trewyn, R. A., Novel approach to the biosynthesis of modified nucleosides in the anticodon loops of eukaryotic tRNAs, in *Biophosphates and their Analogues: Synthesis, Structure, Metabolism and Activity*, Bruzik, K. S. and Stec, W. J., Eds., Elsevier, Amsterdam, 1987, 355.
22. Sakano, H., Shimura, Y., and Ozeki, H., Selective modification of nucleosides of tRNA precursors accumulated in a temperature sensitive mutant of *Escherichia coli*, *FEBS Lett.*, 48, 117, 1974.
23. Shimura, Y., Sakano, H., Kubokawa, S., Nagawa, F., and Ozeki, H., tRNA precursors in RNase P mutants, in *Transfer RNA: Biological Aspects*, Söll, D., Abelson, J. N., and Schimmel, P. R., Eds., Cold Spring Harbor Laboratory, Cold Spring Harbor, New York, 1980, 43.
24. Connolly, D. M. and Winkler, D. M., Structure of *Escherichia coli* K-12 *miaA* and characterization of the mutator phenotype caused by *miaA* insertion mutations, *J. Bacteriol.*, 173, 1711, 1991.
25. Buck, M., McCloskey, J. A., Basile, B., and Ames, B. N., *Cis*-2-methylthio-ribosylzeatin (ms^2io^6A) is present in transfer RNA of *Salmonella typhimurium*, but not *Escherichia coli*, *Nucl. Acid Res.*, 10, 5649, 1982.
26. Agris, P. F., Armstrong, D. J., Schäfer, K. P., and Söll, D., Maturation of a hypermodified nucleoside in transfer RNA, *Nucl. Acids Res.*, 2, 691, 1975.
27. Eisenberg, S. P., Yarus, M., and Soll, L., The effect of an *Escherichia coli* regulatory mutation on transfer RNA structure, *J. Mol. Biol.*, 135, 111, 1979.

28. Ericson, J. U. and Björk, G. R., Pleiotropic effects induced by modification deficiency next to the anticodon of tRNA from *Salmonella typhimurium* LT2, *J. Bacteriol.*, 166, 1013, 1986.
29. Gefter, M. L. and Russel, R. L., Role of modifications in tyrosine transfer RNA: a modified base affecting ribosome binding, *J. Mol. Biol.*, 39, 145, 1969.
30. Hall, R. H., *The Modified Nucleosides in Nucleic Acids*, Columbia University Press, New York, 1971, 329.
31. Rosenberg, A. H. and Gefter, M. L., An iron-dependent modification of several transfer RNA species in *Escherichia coli*, *J. Mol. Biol.*, 46, 581, 1969.
32. Wettstein, F. O. and Stent, G. S., Physiologically induced changes in the property of phenylalanine tRNA in *Escherichia coli*, *J. Mol. Biol.*, 38, 25, 1968.
33. Griffiths, E. and Humphreys, J., Alterations in tRNAs containing 2-methylthio-N^6-(Δ^2=isopentenyl)-adenosine during growth of enteropathogenic *Escherichia coli* in the presence of iron-binding proteins, *Eur. J. Biochem.*, 82, 503, 1978.
34. Buck, M. and Ames, B. N., A modified nucleotide in tRNA as a possible regulator of aerobiosis: synthesis of cis-2-methyl-thioribosyslzeatin in tRNA of *Salmonella*, *Cell*, 36, 523, 1984.
35. Esberg, B. and Björk, G. R., unpublished results, 1992.
36. Hori, H., Saneyoshi, M., Kumagai, I., Miura, K.-I., and Watanabe, K., Effects of modification of 4-thiouridine in *E. coli* tRNA$_f^{Met}$ on its methyl acceptor activity by thermostable Gm-methylase, *J. Biochem.*, 106, 798, 1989.
37. Hori, H., Watanabe, K., Saneyoshi, M., Kamagai, I., Hiroa, I., and Miura, K.-I., Effects of modification of 4-thiouridine in *E. coli* tRNA$_f^{Met}$ on methylation reaction with a thermostable Gm-methylase, *Nucl. Acids Res.*, 17, 175, 1986.
38. Mandel, L. R. and Borek, E., Variability in the structure of ribonucleic acid, *Biochem. Biophys. Res. Commun.*, 4, 14, 1961.
39. Neidhardt, F. C. and Eidlic, L., Characterization of the RNA formed under conditions of relaxed amino acid control in *Escherichia coli*, *Biochim. Biophys. Acta*, 68, 380, 1962.
40. Starr, J. L., The incorporation of amino acids into "methyl-poor" amino acid transfer ribonucleic acid, *Biochem. Biophys. Res. Commun.*, 10, 181, 1963.
41. Svensson, I., Boman, H. G., Eriksson, K. G., and Kjellin, K., Studies on microbial RNA. I. Transfer of methyl groups from methionine to soluble RNA from *Escherichia coli*, *J. Mol. Biol.*, 7, 254, 1963.
42. Littauer, U. Z., Muench, K., Berg, P., Gilbert, W., and Spahn, F., Studies on methylated bases in transfer RNA, *Cold Spring Harbor Symp. Quant. Biol.*, 28, 157, 1963.
43. Biezunski, N., Giveon, D., and Littauer, U. Z., Purification and properties of *Escherichia coli* methyl-deficient phenylalanine tRNA, *Biochim. Biophys. Acta*, 199, 382, 1970.
44. Shugart, L., Novelli, G. D., and Stulberg, M. P., Isolation and properties of undermethylated phenylalanine transfer ribonucleic acids from a relaxed mutant of *Escherichia coli*, *Biochim. Biophys. Acta*, 157, 83, 1968.
45. Miller, J. P., Hussain, Z., and Schweizer, M. P., The involvement of the anticodon adjacent modified nucleoside N-(9-b-D-ribofuranosyl)-purine-6-ylcarbamoyl)-threonine in the biological function of *E. coli* tRNA^{Ile}, *Nucl. Acids Res.*, 3, 1185, 1976.
46. Agris, P. F., Söll, D., and Seno, T., Biological function of 2-thiouridine in *Escherichia coli* glutamic acid transfer ribonucleic acid, *Biochemistry*, 12, 4331, 1973.
47. Zeevi, M. and Daniel, V., Aminoacylation and nucleoside modification of *in vitro* synthesized transfer RNA, *Nature (London)*, 260, 72, 1976.
48. Sampson, J. R. and Uhlenbeck, O. C., Biochemical and physical characterization of an unmodified yeast phenylalanine transfer RNA transcribed *in vitro*, *Proc. Natl. Acad. Sci. U.S.A.*, 85, 1033, 1988.
49. Hall, K. B., Sampson, J. R., Uhlenbeck, O. C., and Redfield, A. G., Structure of an unmodified tRNA molecule, *Biochemistry*, 28, 5794, 1989.

50. **Perret, V., Garcia, A., Puglisi, J., Grosjean, H., Ebel, J.-P., Florentz, C., and Giege, R.,** Conformation in solution of yeast tRNA^{Asp} transcript deprived of modified nucleosides, *Biochimie,* 72, 735, 1990.
51. **Perret, V., Garcia, A., Grosjean, H., Ebel, J.-P., Florentz, C., and Giegé, R.,** Relaxation of a transfer RNA specificity by removal of modified nucleotides, *Nature (London),* 344, 787, 1990.
52. **Normanly, J. and Abelson, J.,** tRNA identity, *Annu. Rev. Biochem.,* 58, 1029, 1989.
53. **Hasegawa, T., Himeno, H., Ishikura, H., and Shimazu, M.,** Discriminator base of tRNA^{Asp} is involved in amino acid acceptor activity, *Biochem. Biophys. Res. Commun.,* 163, 1534, 1989.
53a. **Tamura, K., Asahara, H., Himeno, H., Hasegawa, T., and Shimizu, M.,** Identity elements of *E. coli* tRNA^{Ala}, *J. Mol. Rec.,* 4, 29, 1991.
54. **Jahn, M., Rogers, M. J., and Söll, D.,** Anticodon and acceptor stem nucleotides in tRNA^{Gln} are major recognition elements for *E. coli* glutaminyl-tRNA synthetase, *Nature (London),* 352, 258, 1991.
55. **Himeno, H., Hasegawa, T., Ueda, T., Watanabe, K., Miura, K.-J., and Shimizu, M.,** Role of the extra G-C pair at the end of the acceptor stem of tRNA^{His} in aminoacylation, *Nucl. Acids Res.,* 17, 7855, 1989.
56. **Yokoyama, S.,** personal communication.
57. **Schulman, L. H. and Pelka, H.,** Anticodon switching changes the identity of methionine and valine transfer RNAs, *Science,* 242, 765, 1988.
57a. **Chu, W.-C. and Horowitz, J.,** 19F NMR of 5-fluorouracil-substituted transfer RNA transcribed in vitro: resonance assignment of fluorouracil-guanine base pairs, *Nucleic Acids Res.,* 17, 7241, 1989.
57b. **Tamura, K., Himeno, H., Asahara, H. Hasegawa, T., and Shimizu, M.,** Identity determinants of *E. coli* tRNA^{Val}, *Biochim. Biophys. Res. Commun.,* 177, 619, 1991.
57c. **Himeno, H., Hasegawa, T. Ueda, T., Watanabe, K., and Shimuzu, M.,** Conversion of aminoacylation specificity from tRNA^{Tyr} to tRNA^{Ser} *in vitro, Nucleic Acids Res.,* 18, 6815, 1990.
58. **Schulman, L. H. and Pelka, H.,** An anticodon change switches the identity of *E. coli* tRNA$^{Met}_m$ from methionine to threonine, *Nucleic Acids Res.,* 18, 285, 1990.
59. **Samuelsson, T., Borén, T., Johansen, T.-I., and Lustig, F.,** Properties of a transfer RNA lacking modified nucleosides, *J. Biol. Chem.,* 263, 13692, 1988.
60. **Saneyoshi, M. and Nishimura, S.,** Selective modification of 4-thiouridylate residue in *Escherichia coli* transfer RNA with cyanogen bromide, *Biochim. Biophys. Acta,* 204, 389, 1970.
61. **Saneyoshi, M. and Nishimura, S.,** Selective inactivation of amino acid acceptor and ribosome-binding activities of *Escherichia coli* tRNA by modification with cyanogen bromide, *Biochim. Biophys. Acta,* 246, 123, 1971.
62. **Walker, R. T. and RajBhandary, U. L.,** Studies on polynucleotides. CI. *Escherichia coli* tyrosine and formylmethionine transfer ribonucleic acids: effect of chemical modification of 4-thiouridine to uridine on their biological properties, *J. Biol. Chem.,* 247, 4879, 1972.
63. **Schwartz, I. and Ofengand, J.,** Photo-affinity labeling of tRNA binding in macromolecules. I. Linking of the Phenyl-p-azid of 4-thiouridine in (*Escherichia coli*) valyl-tRNA to 16S RNA at the ribosomal P site, *Proc. Natl. Acad. Sci. U.S.A.,* 71, 3951, 1974.
64. **Hara, H., Horiuchi, T., Saneyoshi, M., and Nishimura, S.,** 4-thiouridine-specific spin-labeling of *E. coli* transfer RNA, *Biochem. Biophys. Res. Commun.,* 38, 305, 1970.
65. **Yang, C.-H. and Söll, D.,** Studies of transfer RNA tertiary structure by singlet-singlet energy transfer, *Proc. Natl. Acad. Sci. U.S.A.,* 71, 2838, 1974.
66. **Daniel, W. E. and Cohn, M.,** Changes in tertiary structure accompanying a single base change in transfer RNA. Proton magnetic resonance and aminoacylation studies of *Escherichia coli* tRNA$^{Met}_{f1}$ and tRNA$^{Met}_{f3}$ and their spin-labeled (s^{4}U8) derivatives, *Biochemistry,* 15, 3917, 1976.

67. **Wetzel, R. and Söll, D.**, Analogs of methionyl-tRNA synthetase substrates containing photolabile groups, *Nucleic Acids Res.*, 4, 1681, 1977.
68. **Siddiqui, M. A. Q. and Ofengand, J.**, The function of pseudouridylic acid in transfer ribonucleic acid. IV. Cyanoethylation of fragments of *Escherichia coli* formylmethionine transfer ribonucleic acid and reconstitution of acceptor activity, *J. Biol. Chem.*, 245, 4409, 1970.
69. **Favre, A., Yaniv, M., and Michelson, A. M.**, The photochemistry of 4-thiouridine in *Escherichia coli* tRNA$_1^{Val1}$, *Biochem. Biophys. Res. Commun.*, 37, 266, 1969.
70. **Yaniv, M., Favre, A., and Barrel, B. G.**, Structure of transfer RNA, *Nature (London)*, 223, 1331, 1969.
71. **Starzyk, R., Schoemaker, H., and Schimmel, R.**, Covalent enzyme-RNA complex: a tRNA modification that prevents a covalent enzyme interaction also prevents aminoacylation, *Proc. Natl. Acad. Sci. U.S.A.*, 82, 339, 1985.
72. **Horowitz, J., Chu, W.-I., Feiz, V., and Derrick, W. B.**, Recognition of *E. coli* Valine tRNA by its Cognate Synthetase, in 14th International tRNA Workshop, Rydzyna, Poland, 1991, 158.
73. **Roe, B., Michael, M., and Dudock, B.**, Function of N^2 methylguanine in phenylalanine transfer RNA, *Nature (London) New Biol.*, 246, 135, 1973.
74. **Sampson, J. R., DiRenzo, A. B., Behlen, L. S., and Uhlenbeck, O. C.**, Nucleotides in yeast tRNA^{Phe} required for the specific recognition by its cognate synthetase, *Science*, 243, 1363, 1989.
75. **Peterkofsky, A.**, A role for methylated bases in the amino acid acceptor function of soluble ribonucleic acid, *Proc. Natl. Acad. Sci. U.S.A.*, 52, 1233, 1964.
76. **Phillips, J. H. and Kjellin-Stråby, K.**, Studies on microbial RNA. IV. Two mutants of *Saccharomyces cerevisiae* lacking N^2-dimethyl-guanine in sRNA, *J. Mol Biol.*, 26, 509, 1967.
77. **Kjellin-Stråby, K.**, personal communication, 1992.
78. **Sprinzle, M., Hartman T., Weber, J., Blank, J., and Zeidler, R.**, Compilation of tRNA sequences and sequences of tRNA genes, *Nucleic Acids Res.*, 17, 1, 1989.
79. **Kruse, T. A., Clark, B. F. C., and Sprinzl, M.**, The effect of specific structural modification on the biological activity of *E. coli* arginine tRNA, *Nucleic Acids Res.*, 5, 879, 1978.
80. **Schulman, L. D. H. and Pelka, H.**, The anticodon contains a major element of the identity of arginine transfer RNAs, *Science*, 246, 1595, 1989.
81. **Wagner, L. P. and Ofengand, J.**, Chemical evidence for the presence of inosine acid in the anticodon of an arginine tRNA from *Escherichia coli*, *Biochim. Biophys. Acta*, 204, 620, 1970.
82. **Stern, L. and Schulman, L. D. H.**, Role of anticodon bases in aminoacylation of *Escherichia coli* methionine transfer RNAs, *J. Biol. Chem.*, 252, 6403, 1977.
83. **Muramatsu, T., Yokoyama, S., Horie, N., Matsuda, A., Ueda, T., Yamaizumi, Z., Kuchino, Y., Nishimura, S., and Miyazawa, T.**, A novel lysine-substituted nucleoside in the first position of the anticodon of minor isoleucine tRNA from *Escherichia coli*, *J. Biol. Chem.*, 263, 9261, 1988.
84. **Muramatsu, T., Nishikawa, K., Nemoto, F., Kuchino, Y., Nishimura, S., Miyazawa, T., and Yokoyama, S.**, Codon and amino acid specificities of a transfer RNA are both converted by a single post-transcriptional modification, *Nature (London)*, 336, 179, 1988.
85. **Schulman, L. H., Pelka, H., and Susani, M.**, Base substitutions in the wobble position of the anticodon inhibit aminoacylation of *E. coli* tRNA$_f^{Mett}$ by *E. coli* Met-tRNA synthetase, *Nucleic Acids Res.*, 11, 1439, 1983.
86. **Muramatsu, T., Kanno, H., and Yokoyama, S.**, Identity Determinants of Isoleucine tRNA from *Escherichia coli*, in 14th International tRNA Workshop, Rydzyna, Poland, 1991, 191.
87. **Seno, T., Agris, P. F., and Söll, D.**, Involvement of the anticodon region of *Escherichia coli* tRNA^{Gln} and tRNA^{Glu} in the specific interaction with cognate aminoacyl-tRNA synthetase. Alteration of the 2-thiouridine derivatives located in the anticodon of the tRNAs by BrCN or sulfur deprivation, *Biochim. Biophys. Acta*, 349, 328, 1974.

88. **Kern, D. and Lapointe, J.,** Glutamyl transfer ribonucleic acid synthetase of *Escherichia coli*. Effect of alteration of the 5-(methylaminomethyl)-2-thiouridine in the anticodon of glutamic acid transfer ribonucleic acid on the catalytic mechanism, *Biochemistry,* 18, 5819, 1979.
89. **Sen, G. C. and Ghosh, H. P.,** Role of modified nucleosides in tRNA: effect of modification of the 2-thiouridine derivative located at the 5′-end of the anticodon of yeast transfer RNA$_2^{Lys}$, *Nucleic Acids Res.,* 3, 523, 1976.
90. **Willick, G. E. and Kay, C. M.,** Circular dicroism study of the interaction of glutamyl-tRNA synthetase with tRNA$_2^{Glu}$, *Biochemistry,* 15, 4347, 1976.
91. **Rould, N. A., Perona, J. J., Söll, D., and Steitz, T.,** Structure of *E. coli* glutaminyl tRNA synthetase complexed with tRNA^{Gln} and ATP at 2.8Å resolution, *Science,* 246, 1135, 1989.
91a. **Rould, M. A., Perona, J. J., and Steitz, T. A.,** Structural basis of anticodon loop recognition by glutaminyl-tRNA synthetase, *Nature (London),* 352, 213, 1991.
92. **Singhal, R. P. and Vakharia, V. N.,** The role of quenine in the aminoacylation of mammalian aspartate transfer RNAs, *Nucleic Acids Res.,* 11, 4257, 1983.
93. **Noguchi, S., Nishimura, Y., Hirota, Y., and Nishimura, S.,** Isolation and characterization of an *Escherichia coli* mutant lacking tRNA-guanine transglycosylase, *J. Biol. Chem.,* 257, 6544, 1982.
94. **Kane, S. M., Vugrinicic, C., Finbloom, D. S., and Smith, D. W. E.,** Purification and some properties of the histidyl-tRNA synthetase from the cytosol of rabbit reticulocytes, *Biochemisry,* 7, 1509, 1978.
95. **Bare, L. A. and Uhlenbeck, O. C.,** Specific substitution into the anticodon loop of yeast tyrosine transfer RNA, *Biochemistry,* 25, 5825, 1986.
96. **Goddard, J. P. and Lowdon, M.,** The effect upon aminoacylation of bisulfite addition to 2-methylthio-N^6-isopentenyl adenosine of *Escherichia coli* phenylalanine tRNA, *FEBS Lett.,* 130, 221, 1981.
97. **Faulkner, R. D. and Uziel, M.,** Iodine modification of *E. coli* tRNA^{Phe}: reversible modification of 2-methylthio-N^6-isopentenyladenosine and lack of disulfide formation, *Biochim. Biophys. Acta,* 238, 464, 1971.
98. **Hecht, S. M., Kirkegaard, L. H., and Bock, R. M.,** Chemical modification of transfer RNA species. Desulfurization with raney nickel, *Proc. Natl. Acad. Sci. U.S.A.,* 68, 48, 1971.
99. **Wilson, R. K. and Roe, B. A.,** Presence of the hypermodified nucleotide N^6-(Δ^2-isopentenyl)-2-methylthioadenosine prevents codon misreading by *Escherichia coli* phenylalanyl-transfer RNA, *Proc. Natl. Acad. Sci. U.S.A.,* 86, 409, 1989.
100. **Buck, M. and Griffiths, E.,** Iron mediated methylthiolation of tRNA as a regulator of operon expression in *Escherichia coli, Nucleic Acids Res.,* 10, 2609, 1982.
101. **Litwack, M. D. and Peterkofsky, A.,** Transfer ribonucleic acid deficient in N^6-(Δ^2-isopentenyl) adenosine due to mevalonic acid limitation, *Biochemistry,* 10, 994, 1971.
102. **Thiebe, R. and Zachau, H. G.,** A specific modification next to the anticodon of phenylalanine transfer ribonucleic acid, *Eur. J. Biochem.,* 5, 546, 1968.
103. **Giegé, R., Garcia, A., Perret, V., Puglisi, J., Pütz, J., Rudinger, J., Theobald, A., Ebel, J.-P., and Florentz, C.,** Searching Elements in tRNA Responsible for its Aminoacylation and Engineering Specificity Switches, 14th International tRNA Workshop, Rydzyna, Poland, 1991, 59.
104. **Garcia, A.,** Ph.D. Thesis, Université Louis Pasteur, Strasbourg, 1990.
105. **Turnbough, C. L., Jr., Neill, R. J., Landsberg, R., and Ames, B. N.,** Pseudouridylation of tRNAs and its role in regulation in *Salmonella typhimurium, J. Biol. Chem.,* 254, 5111, 1979.
106. **Lewis, J. A. and Ames, B. N.,** Histidine regulation in *Salmonella typhimurium.* XI. The percentage of transfer RNA^{His} charged *in vivo* and its relation to the repression of the histidine operon, *J. Mol. Biol.,* 66, 131, 1972.
107. **Brenner, M. and Ames, B. N.,** Histidine regulation in *Salmonella typhimurium.* IX. Histidine transfer ribonucleic acid of the regulatory mutants, *J. Biol. Chem.,* 247, 1080, 1972.

108. **Brenner, M., Lewis, J. A., Straus, D. S., DeLorenzo, F., and Ames, B. N.,** Histidine regulation in *Salmonella typhimurium*. XIV. Interaction of the histidyl transfer ribonucleic acid synthetase with histidine transfer ribonucleic acid, *J. Biol. Chem.,* 247, 4333, 1972.
109. **Arcari, P. and Hecth, S. M.,** Isoenergetic hydride transfer. A reversible tRNA modification with concomitant alteration of biochemical properties, *J. Biol. Chem.,* 253, 8278, 1978.
110. **Hoburg, A., Aschhoff, H. J., Kersten, H., Manderschied, U., and Gassen, H. G.,** Function of modified nucleosides 7-methylguanosine, ribothymidine, and 2-thiomethyl-N^6-(Isopentenyl)adenosine in procaryotic transfer ribonucleic acid, *J. Bacteriol.,* 140, 408, 1979.
111. **Schwartz, I. and Ofengand, J.,** *E. coli* tRNA^{Phe} modified at the 3-(3-amino-3-carboxypropyl) uridine with a photoaffinity label is fully functional for aminoacylation and for ribosomal interaction, *Biochim. Biophys. Acta,* 697, 330, 982.
112. **Nauheimer, U. and Hedgcoth, C.,** Acylation of several tRNAs of *Escherichia coli* by the phenoxyacetyl derivative of *N*-Hydroxysuccinimide, *Arch. Biochem. Biophys.,* 160, 631, 1974.
113. **Friedman, S.,** The effect of chemical modification of 3-(3-amino-3-carboxypropyl) uridine on tRNA function, *J. Biol. Chem.,* 254, 7111, 1979.
114. **Schiller, P. W. and Schechter, A. N.,** Covalent attachment of fluorescent probes to the X-base of *Escherichia coli* phenylalanine transfer ribonucleic acid, *Nucleic Acids Res.,* 4, 2161, 1977.
115. **Hansske, F., Seela, F., Watanabe, K., and Cramer, F.,** Modification of the rare nucleoside X in *Escherichia coli* tRNAs with antigenic determining, photolabile, and paramagnetic residues, *Methods Enzymol.,* 59, 166, 1979.
116. **Björk, G. R. and Isaksson, L. A.,** Isolation of mutants of *Escherichia coli* lacking 5-methyluracil in transfer ribonucleic acid or 1-methylguanine in ribosomal RNA, *J. Mol. Biol.,* 51, 83, 1970.
117. **Svensson, I., Isaksson, L., and Henningsson, A.,** Aminoacylation and polypeptide synthesis with tRNA lacking ribothymidine, *Biochim. Biophys. Acta,* 238, 331, 1971.
118. **Yang, S., Reinitz, E. R., and Gefter, M. L.,** Role of modifications in tyrosine transfer RNA. II. Ribothymidylate-deficient tRNA, *Arch. Biochem. Biophys.,* 157, 55, 1973.
119. **Jin, Y.-X., Lin, J.-H., and Wang, D.-G. (T. P. Wang),** A73, T54 and/or Ψ55 are Possible Enhancers in Yeast tRNA(Ala), in 14th International tRNA Workshop, Rydzyna, Poland, 1991, 162.
120. **Yokoyama, S., Watanabe, K., and Miyazawa, T.,** Dynamic structures and functions of transfer ribonucleic acids from extreme thermophiles, *Adv. Biophys.,* 23, 115, 1987.
121. **Kiesewetter, S., Ott, G., and Sprinzl, M.,** The role of modified purine 64 in initiator/elongator discrimination of tRNA$_f^{Met}$ from yeast and wheat germ, *Nucleic Acids Res.,* 18, 4677, 1990.
122. **Degrès, J., Keith, G., Kuo, K. C., and Gehrke, C. W.,** Presence of phosphorylated *O*-ribosyl-adenosine in T-Y-stem of yeast methionine initiator tRNA, *Nucleic Acids Res.,* 17, 865, 1989.
123. **Joshi, R., Faulhammer, H. G., Chapeville, F., Sprinzl, M., and Haenni, A.,** Aminoacyl RNA domain of turnip yellow mosaic virus Val-RNA interacting with elongation faactor Tu, *Nucleic Acids Res.,* 12, 7467, 1984.
124. **Wikman, F. P., Siboska, G. E., Petersen, H. U., and Clark, B. F. C.,** The site of interaction of aminoacyl-tRNA with elongation factor Tu, *EMBO J.,* 1, 1095, 1982.
125. **Kersten, H., Albani, M., Männlein, E., Praisler, R., Wurmbach, P., and Nierhaus, K. H.,** On the role of ribosylthymine in prokaryotic tRNA function, *Eur. J. Biochem.,* 114, 451, 1981.
126. **Delk, A. S. and Rabinowitz, J. C.,** Partial nucleotide sequence of a prokaryote initiator tRNA that functions in its non-formylated form, *Nature (London),* 252, 106, 1974.
127. **Samuel, C. E. and Rabinowitz, J. C.,** Initiation of protein synthesis by folate-sufficient and folate-deficient *Streptococcus faecalis* R. Biochemical and biophysical properties of methionine transfer ribonucleic acid, *J. Biol. Chem.,* 249, 1198, 1974.

128. **Delk, A. S. and Rabinowitz, J. C.,** Biosynthesis of ribosylthymine in the transfer RNA of *Streptococcus faecalis:* a folate-dependent methylation not involving S-adenosylmethionine, *Proc. Natl. Acad. Sci. U.S.A.,* 72, 528, 1975.
129. **Kersten, H., Sandig, L., and Arnold, H. H.,** Tetrahydrofolate-dependent 5-methyluracil-tRNA transferase activity in *B. subtilis, FEBS Lett.,* 55, 57, 1975.
130. **Davanloo, P., Sprinzl, M., Watanabe, K., Albani, M., and Kersten, H.,** Role of ribothymidine in the thermal stability of transfer RNA as monitored by proton magnetic resonance, *Nucleic Acids Res.,* 6, 1571, 1979.
131. **Raba, M., Limburg, K., Burghagen, M., Katze, J. R., Simsek, M., Heckman, J. E., RajBhandary, U. L., and Gross, H. J.,** Nucleotide sequence of three isoaccepting lysine tRNAs from rabbit liver and SV40-transformed mouse fibroblasts, *Eur. J. Biochem.,* 97, 305, 1979.
132. **Orthwerth, B. J. and Liu, L. P.,** Correlation between a specific isoaccepting lysyl transfer ribonucleic acid and cell division in mammalian tissues, *Biochemistry,* 12, 3978, 1973.
133. **Orthwerth, B. J., Yanuschot, G. R., and Carlson, J. W.,** Properties of $tRNA_4^{Lys}$ from various tissues, *Biochemistry,* 12, 3985, 1973.
134. **Orthwerth, B. J., Lin, V. K., Lewis, J., and Wang, R. J.,** Lysine tRNA and cell division: A G_1 cell cycle mutant is temperature sensitive for modification of $tRNA_5^{Lys}$ to $tRNA_4^{Lys}$, *Nucleic Acids Res.,* 12, 9009, 1984.
135. **Conlon-Hollingshead, C. and Orthwerth, B. J.,** Lys-tRNA$_4$ levels and cell division in mouse 3T3 cells, *Exp. Cell Res.,* 128, 171, 1980.
136. **Grossenbacher, A.-M., Stadelman, B., Heyer, W.-D., Thuriaux, P., and Kohli, J.,** Antisuppressor mutations and sulfur carrying nucleosides in transfer RNA of *Schizosaccharomyces pombe, J. Biol. Chem.,* 261, 16351, 1986.
137. **Nurse, P. and Thuriaux, P.,** Temperature sensitive allosuppressor mutants of the fission yeast *S. pombe* influence cell cycle control over mitotis, *Mol. Gen. Genet.,* 196, 332, 1984.
138. **Gehrke, C.,** Personal communication, 1990.
139. **Young, P. G. and Fantes, P. A.,** Changed division response mutants function as allosuppressors, in Growth, Cancer and the Cell Cycle, Skehan, P. and Friedman, S. J., Eds., Humana Press, Clifton, NJ, 1984, 221.
140. **Caillet, J. and Droogmans, L.,** Molecular cloning of the *Escherichia coli miaA* gene involved in the formation of Δ^2-isopentenyl adenosine in tRNA, *J. Bacteriol.,* 170, 4147, 1988.
141. **Connolly, D. M. and Winkler, M. E.,** Genetic and physiological relationships among the *miaA* gene, 2-methylthio-N^6-(Δ^2-isopentenyl)-adenosine tRNA modification, and spontaneous mutagenesis in *Escherichia coli* K-12, *J. Bacteriol.,* 171, 3233, 1989.
142. **Scornik, O. A.,** Faster protein degradation in response to decreased steady state levels of aminoacylation of tRNA^{His} in chinese hamster ovary cells, *J. Biol. Chem.,* 258, 882, 1983.
143. **Roberts, R. J.,** Structures of two glycyl-tRNAs from *Staphylococcus epidermidis, Nature (London) New Biol.,* 237, 44, 1972.
144. **Krzyzaniak, A., Gamain, A., Barciszewska, M., Gawronska, I., and Barciszewski, J.,** Specific incorporation of glycine into bacterial lipopolysaccharide. Novel function of specific transfer ribonucleic acids, in 14th International tRNA Workshop, Rydzyna, Poland, 1991, 176.
145. **Kannangara, C. G., Gough, S. P., Bruyant, P., Hoober, J. K., Kahn, A., and von Wettstein, D.,** tRNA^{Glu} as a cofactor in δ-aminolevulinate biosynthesis: steps that regulate chlorophyll synthesis, *Trends Biochem. Sci.,* 13, 139, 1988.
145a. **O'Neill, G. P. and Söll, D.,** Expression of the *Synechocystis* sp. strain PCC 6803 tRNA^{Glu} gene provides tRNA for protein and chlorophyl biosynthesis, *J. Bacteriol.,* 172, 6363, 1990.
146. **Björk, G. R.,** Transfer RNA modification in different organisms, *Chemica Scripta,* 26B, 91, 1986.
147. **Yarus, M.,** Translational efficiency of transfer RNAs: uses of an extended anticodon, *Science,* 218, 646, 1982.

148. **Buckingham, R. H. and Grosjean, H.**, The accuracy of mRNA-tRNA recognition, in *Accuracy in Molecular Processes: Its Control and Relevance to Living Systems*, Galas, D. J., Kirkwood, T. B. L., and Rosenberg, R. F., Eds., Chapman and Hall, London, 1986, 83.
149. **Grosjean, H., Houssier, C., and Cedergren, R.**, Anticodon-anticodon interactions and tRNA sequence comparison: approaches to codon recognition, in *Structure and Dynamics of RNA*, van Knippenberg, P. H. and Hilbers, C. W., Eds., Plenum Press, London, 1986, 161.
150. **Crick, F. C. H.**, Codon-anticodon pairing: the wobble hypothesis, *J. Mol. Biol.*, 19, 548, 1966.
151. **Nishimura, S.**, Minor components in transfer RNA: their characterization, location, and function, *Prog. Nucleic Acid Res. Mol. Biol.*, 12, 49, 1972.
152. **Takamoto, T., Takeshi, K., Nishimura, S., and Ukita, T.**, Transfer of valine into rabbit haemoglobin from various isoaccepting species of valyl-tRNA differing in codon recognition, *Eur. J. Biochem.*, 38, 489, 1973.
153. **Ninio, J.**, *Molecular Approaches to Evolution*, Pitman Books, 1982.
154. **Yokoyama, S., Watanabe, T., Murao, K., Ishikura, H., Yamaizumi, Z., Nishimura, S., and Miyazawa, T.**, Molecular mechanism of codon recognition by tRNA species with modified uridine in the first position of the anticodon, *Proc. Natl. Acad. Sci. U.S.A.*, 82, 4905, 1985.
155. **Singer, B. and Kröger, M.**, Participation of modified nucleosides in translation and transcription, *Prog. Nucleic Acid Res. Mol. Biol.*, 23, 151, 1979.
156. **Ikehara, M., Hattori, M., and Fukui, T.**, Synthesis and properties of poly(2-methyladenylic acid). Formation of a Poly(A)-Poly(U) complex with Hoogsteen-type hydrogen bonding, *Eur. J. Biochem.*, 31, 329, 1972.
157. **Engel, J. D. and von Hippel, P. H.**, Effects of methylation on the stability of nucleic acid conformations: studies at the monomer level, *Biochemistry*, 13, 4143, 1974.
158. **Eisenger, J.**, Visible gel electrophoresis and the determinations of association constants, *Biochem. Biophys. Res. Commun.*, 43, 854, 1971.
159. **Grosjean, H., Söll, D. G., and Crothers, D. M.**, Studies of the complex between transfer RNAs with complementary anticodons. I. Origins of enhanced affinity between complementary triplets, *J. Mol. Biol.*, 103, 499, 1976.
160. **Grosjean, H. J., deHenau, S., and Crothers, D. M.**, On the physical basis for ambiguity in genetic coding interactions, *Proc. Natl. Acad. Sci. U.S.A.*, 75, 610, 1978.
161. **Houssiser, C. and Grosjean, H.**, Temperature jump relaxation studies on the interactions between transfer RNAs with complementary anticodons. The effect of modified bases adjacent to the anticodon triplet, *J. Biomol. Struct. Dyn.*, 3, 387, 1985.
162. **Weissenbach, J. and Grosjean, H.**, Effect of threonylcarbamoyl modification (t^6A) in yeast $tRNA_{III}^{Arg}$ on codon-anticodon and anticodon-anticodon interactions, *Eur. J. Biochem.*, 116, 207, 1981.
163. **Houssier, C., Degrée, P., Nicoghosian, K., and Grosjean, H.**, Effect of uridine dethiolation in the anticodon triplet of tRNA (Glu) on its association with tRNA (Phe), *J. Biolmol. Struct. Dyn.*, 5, 1259, 1988.
164. **Hirsch, D.**, Tryptophan tRNA as the UGA suppressor, *J. Mol. Biol.*, 58, 439, 1971.
165. **Atkins, J. F., Weiss, R. B., Thompson, S., and Gesteland, R. F.**, Towards a genetic dissection of the basis of triplet decoding, and its natural subversion: programmed reading frame shifts and hops, *Annu. Rev. Genet.*, 25, 201, 1991.
166. **Peterkofsky, A., Jesensky, C., Bank, A., and Mehler, A. H.**, Studies on the role of methylated bases in the biological activity of soluble ribonucleic acid, *J. Biol. Chem.*, 239, 2918, 1964.
167. **Fleissner, E.**, Polypeptide synthesis with methyl-deficient soluble ribonucleic acid, *Biochemistry*, 6, 621, 1967.
168. **Claesson, C., Samuelsson, T., Lustig, F., and Borén, T.**, Codon reading properties of an unmodified transfer RNA, *FEBS Lett.*, 273, 173, 1990.

169. **Noren, C. J., Anthony-Cahill, S. J., Suich, D. J., Noren, K. A., Griffith, M. C., and Schultz, P. G.,** In vitro suppression of an amber mutation by a chemically aminoacylated transfer RNA prepared by runoff transcription, *Nucleic Acids Res.,* 18, 83, 1989.
170. **Yarus, M., Cline, S. W., Wier, P., Breeden, L., and Thompson, R. C.,** Actions of the anticodon arm in translation on the phenotypes of RNA mutants, *J. Mol. Biol.,* 192, 235, 1986.
171. **Ramabhadran, T. V., Fossum, T., and Jagger, J.,** *Escherichia coli* mutant lacking 4-thiouridine in its transfer ribonucleic acid, *J. Bacteriol.,* 128, 671, 1976.
172. **Ramabhadran, T. V. and Jagger, J.,** Mechanism of growth delay induced in *Escherichia coli* by near ultraviolet radiation, *Proc. Natl. Acad. Sci. U.S.A.,* 73, 59, 1976.
173. **Thomas, G. and Favre, A.,** 4-Thiouridine as the target for near-ultraviolet light induced growth delay in *Escherichia coli, Biochem. Biophys. Res. Commun.,* 66, 1454, 1975.
174. **Carre, D. S., Thomas, G., and Favre, A.,** Conformation and functioning of tRNA: cross-linked tRNAs as substrate for tRNA nucleotidyl-transferase and animoacyl synthetases, *Biochimie,* 56, 1089, 1974.
175. **Thomas, G. and Favre, A.,** 4-Thiouridine triggers both growth delay induced by near-ultraviolet light and photoprotection, *Eur. J. Biochem.,* 113, 67, 1980.
176. **Ryals, J., Hsu, R.-Y., Lipsett, M. N., and Bremer, H.,** Isolation of single-site *Escherichia coli* mutants deficient in thiamine and 4-thiouridine syntheses: identification of a *nuvC* mutant, *J. Bacteriol.,* 151, 899, 1982.
177. **Thomas, G., Thiam, K., and Favre, A.,** tRNA thiolated pyrimidines as targets for near-ultraviolet-induced synthesis of guanosine tetraphosphate in *Escherichia coli, Eur. J. Biochem.,* 119, 381, 1981.
178. **Cashel, M. and Rudd, K. E.,** The stringent response, in *Escherichia coli and Salmonella typhimurium. Cellular and Molecular Biology,* Vol. 2, Neidhardt, F. C., Ingraham, J. L., Brooks Low, K., Magasanik, B., Schaechter, M., and Umbarger, H. E., Eds., American Society for Microbiology, Washington, D.C., 1987, 1410.
179. **Lo, R. Y. C., Bell, J. B., and Roy, K. L.,** Dihydrouridine-deficient tRNAs in *Saccharomyces cereviciae, Nucleic Acids Res.,* 10, 889, 1982.
180. **Reggie, Y. C. L. and Bell, J. B.,** Characterization of a mutation in *Saccharomyces cerevisiae* that produces mutant isoaccepting tRNAs for several of its tRNA species, *Curr. Genet.,* 3, 73, 1981.
181. **Bauman, U., Fischer, W., and Sprinzl, M.,** Analysis of modification-dependent structural alterations in the anticodon loop of *Escherichia coli* tRNA^{Arg} and their effects on translation of res2 RNA, *Eur. J. Biochem.,* 152, 645, 1985.
182. **Atkins, J. F., Gesteland, R. F., Reid, B. R., and Anderson, C. W.,** Normal tRNAs promote ribosomal frameshifting, *Cell,* 18, 1119, 1979.
183. **Dayhuff, T. J., Atkins, J. F, and Gesteland, R. F.,** Characterization of ribosomal frameshift events by protein sequence analysis, *J. Biol. Chem.,* 261, 7491, 1986.
184. **Bruce, A. G., Atkins, J. F., and Gesteland, R. F.,** tRNA anticodon replacement experiments show that ribosomal frameshifting can be caused by doublet decoding, *Proc. Natl. Acad. Sci. U.S.A.,* 83, 5062, 1986.
185. **Smith, D. W. E. and McNamara, A. L.,** The effect of the Q base modification on the usage of tRNA^{His} in globin synthesis, *Biochem. Biophys. Res. Commun.,* 104, 1459, 1982.
186. **McNamara, A. L. and Smith, D. W. E.,** The function of the histidine tRNA isoaccepting species in hemoglobin synthesis, *J. Biol. Chem.,* 253, 5964, 1978.
187. **Johnston, H. M., Barnes, W. M., Chumley, F. G., Bossi, L., and Roth, J. R.,** Model for regulation of the histidine operon of *Salmonella, Proc. Natl. Acad. Sci. U.S.A.,* 77, 508, 1980.
188. **Sullivan, M. A., Cannon, J. F., Webb, F. H., and Bock, R. M.,** Antisuppressor mutation in *Escherichia coli* defective in biosynthesis of 5-methylaminomethyl-2-thiouridine, *J. Bacteriol.,* 161, 368, 1985.

189. Elseviers, D., Petrullo, L. A., and Gallagher, P. J., Novel *E. coli* mutants deficient in biosynthesis of 5-methylaminomethyl-2-thiouridine, *Nucleic Acids Res.*, 12, 3521, 1984.
190. Hagervall, T. G. and Björk, G. R., Undermodification in the first position of the anticodon of supG-tRNA reduces translational efficiency, *Mol. Gen. Genet.*, 196, 194, 1984.
191. Hagervall, T. G., Edmonds, C. G., McCloskey, J. A., and Björk, G. R., Transfer RNA (5-methylaminomethyl-2-thiouridine) methyltransferase from *Escherichia coli* K-12 has two enzymatic activities, *J. Biol. Chem.*, 262, 8488, 1987.
192. Yokoyama, S. Tukai, K., Sakamoto, K., Kawai, G., Miyazawa, T., Yamaizumi, Z., and Nishimura, S., *Roles of Modified Nucleosides of tRNAs in Codon Recognition,* in 14th International tRNA Workshop, Rydzyna, Poland, 1991, 25.
193. Colby, D. S., Scheele, P., and Guthrie, C., A functional requirement for modification of the wobble nucleoside in the anticodon of a T4 suppressor tRNA, *Cell,* 9, 449, 1976.
194. Jahn, M., Rogers, M. J., and Söll, D., *Transfer RNA Recognition by a E. coli Glutaminyl-tRNA Synthetase,* in 14th International tRNA Worshop, Rydzyna, Poland, 1991, 77.
195. Ching, W.-M., Tsai, L., and Wittwer, A. J., Selenium containing transfer RNAs, *Curr. Top. Cell. Reg.,* 27, 497, 1985.
196. Wittwer, A. J., Specific incorporation of selenium into lysine- and glutamate-accepting tRNAs from *Escherichia coli, J. Biol. Chem.,* 258, 8637, 1983.
197. Wittwer, A. J. and Stadtman, T. C., Biosynthesis of 5-methylaminomethyl-2-selenouridine, a natural occurring nucleoside in *Escherichia coli* tRNA, *Arch. Biochem. Biophys.,* 248, 540, 1986.
198. Kramer, G. F. and Ames, B. N., Isolation and characterization of a selenium metabolism mutant of *Salmonella typhimurium, J. Bacteriol.,* 170, 736, 1988.
199. Heyer, W.-D., Thuriaux, P., and Kohli, J., An antisuppressor mutation of *Schizosaccharomyuces pombe* affects the post-transcriptional modification of the "Wobble" base in the anticodon of tRNAs, *J. Biol. Chem.,* 259, 2856, 1984.
200. Strobel, M.C. and Abelson, J., Effect of intron mutations on processing and function of *Saccharomyces cerevisiae* SUP53 tRNA *in vitro* and *in vivo, Mol. Cell. Biol.,* 6, 2663, 1986.
201. Johnson, P. F. and Abelson, J., The yeast tRNA^{Tyr} gene intron is essential for correct modification of its tRNA product, *Nature (London),* 302, 681, 1983.
202. Choffat, Y., Suter, B., Behra, R., and Kubli, E., Pseudouridine modification in the tRNA^{Tyr} anticodon is dependent on the presence, but independent of the size and sequence, of the intron in eucaryotic tRNA^{Tyr} genes, *Mol. Cell. Biol.,* 8, 3332, 1988.
203. Woese, C. R., *The Genetic Code. The Molecular Basis for Genetic Expression,* Harper & Row, New York, 1967, 134.
204. Ward, D. C. and Rich, E., Conformational properties of polyformycin. A poly ribonucleotide with individual residues in the syn conformation, *Proc. Natl. Acad. Sci. U.S.A.,* 61, 1494, 1968.
205. Diaz, I., Ehrenberg, M., and Kurland, C. G., Effects of *miaA* on translation and growth rates, *Mol. Gen. Genet.,* 208, 373, 1987.
205a. Mikkola, R. and Kurland, C. G., Media dependence of translational mutant phenotype, *FEMS Microbiol. Lett.,* 56, 265, 1988.
206. Bouadloun, F., Srichaiyo, T., Isaksson, L. A., and Björk, G. R., Influence of modification next to the anticodon in tRNA on codon context sensitivity of translational suppression and accuracy, *J. Bacteriol.,* 166, 1022, 1986.
207. Petrullo, L. A., Gallagher, P. J., and Elseviers, D., The role of 2-methylthio-N6-Isopentenyladenosine in readthrough and suppression of nonsense codons in *Escherichia coli, Mol. Gen. Genet.,* 190, 289, 1983.
208. Vacher, J., Grosjean, H., Houssier, C., and Buckingham, R. H., The effect of point mutations affecting *Escherichia coli* tryptophan tRNA on anticodon-anticodon interactions and on UGA suppression, *J. Mol. Biol.,* 177, 329, 1984.

209. **Janner, F., Vögeli, G., and Fluri, R.,** The antisuppressor strain *sinl* of *Schizosaccharomyces pombe* lacks the modification isopentenyladenosine in transfer RNA, *J. Mol. Biol.,* 139, 207, 1980.
210. **Laten, H., Gorman, J., and Buck, R. M.,** Isopentenyladenosine deficient tRNA from an antisuppressor mutant of *Saccharomyces cerevisiae, Nucleic Acids Res.,* 5, 4329, 1978.
211. **Yanofsky, C. and Soll, L.,** Mutations affecting tRNA^{Trp} and its charging and their effect on regulation of transcription termination at the attenuator of the tryptophan operon, *J. Mol. Biol.,* 113, 663, 1977.
212. **Landick, R. and Yanofsky, C.,** Transcription attenuation, in *Escherichia coli and Salmonella typhimurium. Cellular and Molecular Biology,* Vol. 2, Neidhardt, F. C., Ingraham, J. L., Low, K. B., Magasanik, B., Schaechter, M., and Umbarger, H. E., Eds., American Society for Microbiology, Washington, D.C., 1982, 1276.
213. **Golnick, P. and Yanofsky, C.,** tRNA^{Trp} translation of leader peptide codon 12 and other factors that regulate expression of the tryptophanase operon, *J. Bacteriol.,* 172, 3100, 1990.
214. **Mayaux, J. F., Fayat, G., Pauvert, M., Springer, M., Grunberg-Managö, B. I., and Blanquet, S.,** Control of phenylalanyl-tRNA synthetase genetic expression. Site-directed mutagenesis of the pheS,T operon regulating region *in vitro, J. Mol. Biol.,* 184, 31, 1985.
215. **Grosjean, H., Nicoghosian, K., Haumont, E., Söll, D., and Cedergren, R.,** Nucleotide sequences of two serine tRNAs with a GGA anticodon: the structure-function relationships in the serine family of *E. coli* tRNAs, *Nucleic Acid Res.,* 13, 5697, 1985.
216. **Murgola, E. J., Prather, N. E., Pagel, F. T., Mims, B. H., and Hijazi, K. A.,** Missense and nonsense suppressors derived from a glycine tRNA by nucleotide insertion and deletion *in vivo, Mol. Gen. Genet.,* 193, 76, 1984.
217. **Smith, D. W. E. and Hatfield D. L.,** Effects of post-transcriptional base modifications on the site-specific function of transfer RNA in eukaryote translation, *J. Mol. Biol.,* 189, 663, 1986.
218. **Smith, D. W. E., McNamara, A. L., Rice, M., and Hatfield, D.,** The effects of a post-transcriptional modification on the function of tRNA^{Lys} isoaccepting species in translation, *J. Biol. Chem.,* 256, 10033, 1981.
219. **Watts, M. T. and Tinoco, I.,** Role of hypermodified bases in transfer RNA. Solution properties of dinucleoside monophosphates, *Biochemistry,* 17, 2455, 1978.
220. **Pongs, O. and Reinwald, E.,** Function of Y in codon-anticodon interaction of tRNA^{Phe}, *Biochem. Biophys. Res. Commun.,* 50, 357, 1973.
221. **Miller, P. S., Barrett, J. C., and Ts'o, P. O. P.,** Synthesis of oligodeoxyribonucleotide ethyl phosphotriesters and their specific complex formation with transfer ribonucleic acid, *Biochemistry,* 13, 4887, 1974.
222. **Bruce, A. G., Atkins, J. F., Wills, N., Uhlenbeck, O., and Gesteland, R. F.,** Replacement of anticodon loop nucleotides to produce functional tRNAs amber suppressors derived from yeast tRNA^{Phe}, *Proc. Natl. Acad. Sci. U.S.A.,* 79, 7127, 1982.
223. **Singer, C., Smith, G. R., Cortese, C., and Ames, B. N.,** Mutant tRNA ineffective in repression and lacking two pseudouridine modifications, *Nature (London) New Biol.,* 238, 72, 1972.
224. **Barnes, W. M.,** DNA sequence from the histidine operon control region: seven histidine codons in a row, *Proc. Natl. Acad. Sci. U.S.A.,* 75, 4281, 1978.
225. **Palmer, D. T., Blum, P. H., and Artz, S. W.,** Effects of the *hisT* mutation of *Salmonella typhimurium* on translation elongation rate, *J. Bacteriol.,* 153, 357, 1983.
226. **Bossi, L. and Roth, J. R.,** The influence of codon context on genetic code translation, *Nature (London),* 286, 123, 1980.
227. **Hagervall, T. G., Ericson, J. U., Esberg, K. B., Li, J.-N., and Björk, G. R.,** Role of tRNA modification in translational fidelity, *Biochim. Biophys. Acta,*1050, 263, 1990.
228. **Rizzino, A. A., Bresalier, R. S., and Freundlich, M.,** Derepressed levels of the isoleucine-, valine and leucine enzymes in *hisT1504,* a strain of *Salmonella typhimurium* with altered leucine transfer ribonucleic acid, *J. Bacteriol.,* 117, 449, 1974.

229. Cortese, R., Landsberg, R., Vonder Haar, R. A., Umbarger, H. E., and Ames, B. N., Pleitropy of *hisT* mutants blocked in pseudouridine synthesis in tRNA: *leucine* and *isoleucine-valine* operons, *Proc. Natl. Acad. Sci. U.S.A.*, 71, 1857, 1974.
230. Bresalier, R. S., Rizzino, A. A., and Freundlich, M., Reduced maximal levels of derepression of the isoleucine-valine and leucine enzymes in *hisT* mutant of *Salmonella typhimurium*, *Nature (London)*, 253, 279, 1975.
231. Rosenfeld, S. A. and Brenchley, J. E., Regulation of nitrogen utilization in *hisT* mutants of *Salmonella typhimurium*, *J. Bacteriol.*, 143, 801, 1980.
232. Brenchley, J. E. and Williams, L. S., Transfer RNA involvement in the regulation of enzyme synthesis, *Annu. Rev. Microbiol.*, 29, 251, 1975.
233. Spandaro, A. A., Spena, V., Santonastaso, V., and Donini, P., Stringency without ppGpp accumulation, *Nature (London)*, 291, 256, 1981.
234. Pang, H., Ihara, M., Kuchino, Y., Nishimura, S., Gupta, R., Woese, C. R., and McCloskey, J. A., Structure of a modified nucleoside in archaebacterial tRNA which replaces ribosylthymine, 1-methylpseudouridine, *J. Biol. Chem.*, 257, 3589, 1982.
235. Hopper, A. K., Funakawa, A. H., Phau, H. D., and Martin, N. C., Defects in modification of cytoplasmic and mitochondrial transfer RNAs are caused by single nuclear mutations, *Cell*, 28, 543, 1982.
236. Johnson, L., Hayashi, H., and Söll, D., Isolation and properties of a transfer ribonucleic acid deficient in ribothymidine, *Biochemistry*, 9, 2823, 1970.
237. Guindy, Y. S., Samuelsson, T., and Johansen, T.-I., Unconventional codon reading by *M. mycosides* tRNAs as revealed by partial sequence analysis, *Biochem. J.*, 258, 869, 1989.
238. Samuelsson, T., Guindy, Y. S., Lustig, F., Borén, T., and Lagerkvist, U., Apparent lack of discrimination in the reading of certain codons in *Mycoplasma mycoides*, *Proc. Natl. Acad. Sci. U.S.A.*, 84, 3166, 1987.
239. Andachi, Y., Yamao, F., Muto, A., and Osawa, S., Codon recognition patterns as deduced from sequences of the complete set of transfer RNA species in *Mycoplasma capricolum*. Resemblance to mitochondria, *J. Mol. Biol.*, 209, 37, 1989.
240. Vani, B. R. R., Taya, Y., Noguchi, S., Yamaizumi, Z., and Nishimura, S., Occurrence of 1-methyladenosine and absence of ribothymidine in transfer ribonucleic acid from *Mycobacterium smegmalis*, *J. Bacteriol.*, 137, 1084, 1979.
241. Björk, G. R. and Neidhardt, F. C., Physiological and biochemical studies on the function of 5-methyluridine in transfer ribonucleic acid of *Escherichia coli*, *J. Bacteriol.*, 124, 99, 1975.
242. Björk, G. R., unpublished results, 1992.
243. Baumstark, B. R., Spremulli, L. L. RajBhandary, U. L., and Brown, G. M., Initiation of protein synthesis without formylation in a mutant of *Escherichia coli* that grows in the absence of tetrahydrofolate, *J. Bacteriol.*, 129, 457, 1977.
244. Marcu, K. B. and Dudock, B. S., Effect of ribothymidine in specific eukaryotic tRNAs on their efficiency in *in vitro* protein synthesis, *Nature (London)*, 261, 159, 1976.
245. Marcu, K., Marcu, D., and Dudock, B., Wheat germ tRNAs containing uridine in place of ribothymidine: a characterization of an unusual class of eukaryotic tRNAs, *Nucleic Acids Res.*, 5, 1075, 1978.
246. Roe, B. A., Chen, E. Y., and Tsen, H. Y., Studies on the ribothymidine content of specific and human tRNAs: a postulated role for 5-methyl cytosine in the regulation of ribothymidine biosynthesis, *Biochem. Biophys. Res. Commun.*, 68, 1339, 1976.
247. Roe, B. A. and Tsen, H.-Y., Role of ribothymidine in mammalian tRNA^{Phe}, *Proc. Natl. Acad. Sci. U.S.A.*, 74, 3696, 1977.
248. Persson, B., Gustafsson, G., Berg, D., and Björk, G. R., The gene for a tRNA modifying enzyme, m^{5}U54-methyltransferase, is essential for viability in *E. coli*, *Proc. Natl. Acad. Sci. U.S.A.*, in press, 1992.
249. Gustafsson, C. and Björk, G. R., unpublished, 1992.
250. Shpaer, E. G., Constraints on codon context in *Escherichia coli* genes: their possible role in modulating the efficiency of translation, *J. Mol. Biol.*, 188, 555, 1986.

251. **Yarus, M. and Folley, L. S.,** Sense codons are found in specific context, *J. Mol. Biol.,* 182, 529, 1985.
252. **Kolaskar, A. S. and Reddy, B. V. B.,** Contextual constraints on codon pair usage: structural and biological implictions, *J. Biomol. Struct. Dyn.,* 3, 725, 1986.
253. **Gouy, M.,** Codon contexts in enterobacterial and coliphage gene, *Mol. Biol. Evol.,* 4, 426, 1987.
254. **Gutman, G. A. and Hatfield, G. W.,** Nonrandom utilization of codon pairs in *Escherichia coli, Proc. Natl. Acad. Sci. U.S.A.,* 86, 3699, 1989.
255. **Folley, L. S. and Yarus, M.,** Codon contexts from weakly-expressed genes reduce expression *in vivo, J. Mol. Biol.,* 209, 359, 1989.
256. **Salser, W.,** The influence of the reading context upon the suppression of nonsense codons, *Mol. Gen. Genet.,*105, 125, 1969.
257. **Salser, W., Fluck, M., and Epstein, R.,** The influence of the reading context upon the suppression of nonsense codons, *Cold Spring Harbor Symp. Quant. Biol.,* 34, 513, 1969.
258. **Bossi, L.,** Context effects: translation of UAG codons by suppressor tRNA is affected by the sequence following in the message, *J. Mol. Biol.,* 164, 73, 1983.
259. **Miller, J. H. and Albertini, A. M.,** Effects of surrounding sequence on the suppression of nonsense codons, *J. Mol. Biol.,* 164, 59, 1983.
260. **Pedersen, W. T. and Curran, J. F.,** Effects of the nucleotide 3' to an amber codon on ribosomal selection rates of suppressor tRNA and release factor $-1, J. Mol. Biol.,$ 219, 231, 1991.
261. **Murgola, E. J., Pagel, F. T., and Hijazi, K. A.,** Codon context effects in missense suppression, *J. Mol. Biol.,* 175, 19, 1984.
262. **Carrier, M. J. and Buckingham, R. H.,** An effect of codon context on the mistranslation of UGU codons *in vitro, J. Mol Biol.,* 175, 29, 1984.
263. **Grosjean, H. and Chantrenne, H.,** On codon-anticodon interactions, *Mol. Biol. Biochem. Biphys.,* 32, 347, 1980.
264. **Fiarclough, R. H. and Cantor, C. R.,** The distance between the anticodon loops of two tRNAs bound to the 70S *Escherichia coli* ribosome, *J. Mol. Biol.,* 132, 575, 1979.
265. **Smith, D. and Yarus, M.,** tRNA-tRNA interactions within cellular ribosomes, *Proc. Natl. Acad. Sci. U.S.A.,* 86, 4397, 1989.
266. **Ericson, J. U. and Björk, G. R.,** tRNA anticodons with the modified nucleoside 2-methyl-N^6-(4-hydroxyisopentenyl)adenosine distinguish between bases 3' of the codon, *J. Mol. Biol.,* 218, 509, 1991.
267. **Björnsson, A. and Isaksson, L. A.,** Personal communication, 1991.
268. **Yarus, M. and Thompson, R. C.,** Precision of protein biosynthesis, in *Gene Function in Prokaryotes,* Cold Spring Harbor Laboratory Press, Cold Spring Harbor, New York, 1983, 23.
269. **Freier, S. M., Kierzek, R., Jaeger, J. A., Sugimoto, N., Caruthers, M. H., Neilson, T., and Turner, D. H.,** Improved free-energy parameters for predictions of RNA duplex stability, *Proc. Natl. Acad. Sci. U.S.A.,* 83, 9373, 1986.
270. **Parker, J.,** Errors and alternatives in reading the universal genetic code, *Microbiol. Rev.,* 53, 273, 1989.
271. **Atkins, J. F., Weiss, R. B., and Gesteland, R. F.,** Ribosome gymnastics – degree of difficulty 9.5, style 10.0, *Cell,* 62, 413, 1990.
272. **Manley, J. L.,** Synthesis and degradation of termination and premature termination fragments of β-galactosidase *in vitro, J. Mol. Biol.,* 125, 407, 1978.
273. **Kurland, C. G., Jörgensen, F., Richter, A., Ehrenberg, M., Bilgin, N., and Rojas, A.-M.,** Through the accuracy window, in *The Ribosome. Structure, Function and Evolution,* Hill, W. E., Dahlberg, A., Garrett, R. A., Moore, P. B., Schlessinger, D., and Warner, J. R., Eds., American Society for Microbiology, Washington, D.C., 1990, 513.
274. **Gallant, J. and Palmer, L.,** Error propagation in viable cells, *Mech. Ageing Dev.,* 10, 27, 1979.

275. **Jörgensen, F. and Kurland, C. G.**, Death rates of bacterial mutants, *FEMS Microbiol. Lett.*, 40, 43, 1987.
276. **Jänel, G., Michelsen, U., Nishimura, S., and Kersten, H.**, Queuosine modification in tRNA and expression of the nitrate reductase in *Escherichia coli*, *EMBO J.*, 3, 1603, 1984.
277. **Frey, B., Jänel, G., Michelsen, U., and Kersten, H.**, Mutations in the *Escherichia coli fnr* and *tgt* genes: control of molybtate reductase activity and the cytochromal complex by *fnr*, *J. Bacteriol.*, 171, 1524, 1989.
278. **Ryoji, M., Hsia, K., and Kaji, A.**, Read-through translation, *Trends Biochem.*, 8, 88, 1983.
279. **Engelberg-Kulka, H. and Schonlaker-Schwarz, R.**, Stop is not the end: physiological implications of translational readthrough, *J. Theor. Biol.*, 131, 477, 1988.
280. **Valle, R. P. C. and Morch, M.-D.**, Stop making sense or regulation at the level of termination in eukaryotic protein synthesis, *FEBS Lett.*, 235, 1, 1988.
281. **Bienz, M. and Kubli, E.**, Wild-type tRNA_G^{Tyr} reads the TMV RNA stop codon, but Q base-modified tRNA_Q^{TYR} does not, *Nature (London)*, 294, 188, 1981.
282. **Beier, H., Barciszewska, M., Krupp, G., Mitnacht, R., and Gross, H. J.**, UAG readthrough during TMV RNA translation: isolation and sequence of two tRNAs^{Tyr} with suppressor activity from tobacco plants, *EMBO J.*, 3, 351, 1984.
283. **Beier, H., Barciszewska, M., and Sickinger H.-D.**, The molecular basis for the differential translation of TMV RNA in tobacco protoplasts and wheat germ extracts, *EMBO J.*, 3, 1091, 1984.
284. **Skuzeski, J. M., Nichols, L. M., Gesteland, R. F., and Atkins, J. F.**, The signal for a leaky UAG stop codon in several plant viruses includes the two downstream codons, *J. Mol. Biol.*, 218, 365, 1991.
285. **Osawa, S. and Jukes, T. H.**, Evolution of the genetic code as affected by anticodon content, *Trends Genet.*, 4, 191, 1988.
286. **Stern, L. and Schulman, L. D. H.**, The role of the minor base N^4-acetylcytidine in the function of the *Escherichia coli* noninitiator methionine transfer RNA, *J. Biol. Chem.*, 253, 6153, 1978.
287. **Cory, S., Marker, K. A., Dube, S., and Clark, B. F. C.**, Primary structure of a methionine transfer RNA from *Escherichia coli*, *Nature (London)*, 220, 1039, 1968.
288. **Lomant, A. J. and Fresco, J. R.**, Structural and energetic consequences of noncomplementary base oppositions in nucleic acid helices, *Prog. Nucleic Acid Res. Mol. Biol.*, 15, 185, 1975.
289. **Harada, F. and Nishimura, S.**, Purification and characterization of AUA specific isoleucine transfer ribonucleic acid from *Escherichia coli* B, *Biochemistry*, 13, 300, 1974.
290. **Diaz, I. and Ehrenberg, M.**, ms^{2}i^{6}A deficiency enhances proofreading in translation, *J. Mol. Biol.*, 222, 1161, 1991.
291. **Ehrenberg, M., Rojas, A.-M., Diaz, I., Bilgin, N., Weiser, J., Claesens, F., and Kurland, C. G.**, New aspects of elongation factor Tu function in translation, in *The Ribosome. Structure, Function, & Evolution*, Hill, W. E., Dahlberg, A., Garrett, R. A., Moore, P. B., Schlessinger, D., and Warner, J. R., Eds., American Society for Microbiology, Washington, D.C., 1990, 373.
292. **Dix, D. B. and Thompson, R. C.**, Elongation factor Tu guanosine 3'-diphosphate 5'-diphosphate complex increases the fidelity of proofreading in protein biosynthesis: mechanism for reducing translational errors introduced by amino acid starvation, *Proc. Natl. Acad. Sci. U.S.A.*, 83, 2027, 1986.
293. **Jukes, T. H.**, Possibility for the evolution of the genetic code from preceeding form, *Nature (London)*, 246, 22, 1973.
294. **Menninger, J. R.**, Ribosome editing and the error catastrophe hypothesis of cellular ageing, *Mech. Ageing Dev.*, 6, 131, 1977.
295. **Petrullo, L. A. and Elseviers, D.**, Effect of a 2-methylthio-N6-isopentenyl adenosine deficiency on peptidyl-tRNA release in *Escherichia coli*, *J. Bacteriol.*, 165, 608, 1986.
296. **Ninio, J.**, Kinetic devices in protein synthesis, DNA replication, and mismatch repair, *Cold Spring Harbor Symp. Quant. Biol.*, 52, 639, 1987.

296a. **Nierhaus, K. H.**, The allosteric three-site model for the ribosomal elongation cycle: features and future, *Biochemistry*, 29,4997, 1989.
297. **Parker, J.**, Specific mistranslation in *hisT* mutants of *Escherichia coli*, *Mol. Gen. Genet.*, 187, 405, 1982.
298. **Pieczenik, G.**, Predicting coding function from nucleotide sequence or survival of "fitness" of tRNA, *Proc. Natl. Acad. Sci. U.S.A.*, 77, 3539, 1980.
299. **Björk, G. R., Wikström, P. M., and Byström, A. S.**, Prevention of translational frameshifting by the modified nucleoside 1-methylguanosine, *Science*, 244, 986, 1989.
300. **Kuchino, Y., Watanabe, S., Harada, F., and Nishimura, S.**, Primary structure of AUA-specific isoleucine transfer ribonucleic acid from *Escherichia coli*, *Biochemistry*, 19, 2085, 1980.
301. **Hagervall, T. G., Atkins, J., Tuohy, T., and Björk, G. R.**, unpublished results, 1992.
302. **Riddle, D. L. and Carbon, J.**, Frameshift suppression: a nucleotide addition in the anticodon of a glycine transfer RNA, *Nature (London) New Biol.*, 242, 230, 1973.
303. **Weiss, R., Dunn, D., Atkins, J., and Gesteland, R.**, Ribosomal frameshifting from –2 to +50 nucleotides, *Prog. Nucleic Acid Res. Mol. Biol.*, 39, 159, 1990.
304. **Li, J.-N. and Björk, G. R.**, unpublished results, 1992.
305. **Jacks, T., Madhani, H. D., Masiarz, F. R., and Varmus, H. E.**, Signals for ribosomal frameshifting in the Rous sarcoma virus *gag-pol* region, *Cell*, 55, 447, 1988.
306. **Weiss, R. B., Dunn, D. M., Shuh, M., Atkins, J. F., and Gesteland, R. F.**, *E. coli* ribosomes Re-phase on retroviral frameshift signals at rates ranging from 2 to 50 percent, *The New Biologist*, 1, 159, 1989.
307. **Hatfield, D., Feng, Y.-X., Lee, B. J., Rein, A., Levin, J. G., and Oroszlan, S.**, Chromatographic analysis of the aminoacyl-tRNAs which are required for translation of codons at and around the ribosomal frameshift sites of HIV, HTLV-1, and BLV, *Virology*, 173, 736, 1989.
308. **Hatfield, D. L., Smith, D. W. E., Lee, B. J., Worland, P. J., and Oroszlan, S.**, Structure and function of suppressor tRNAs in higher eukaryotes, *Crit. Rev. Biochem. Mol. Biol.*, 25, 71, 1990.
309. **Tsuchihashi, Z. and Kornberg, A.**, Translational frameshifting generates the γ subunit of DNA polymerase III holoenzyme, *Proc. Natl. Acad. Sci. U.S.A.*, 87, 2526, 1990.
310. **Flower, A. M. and McHenry, C. S.**, The γ subunit of DNA polymerase III holoenzyme of *Escherichia coli* is produced by ribosomal frameshifting, *Proc. Natl. Acad. Sci. U.S.A.*, 87, 3713, 1990.
311. **Blinkowa, A. L. and Walker, J. R.**, Programmed ribosomal frameshifting generates the *Escherichia coli* DNA polymerase III γ subunit from within the τ subunit reading frame, *Nucleic Acids Res.*, 18, 1725, 1990.
312. **Grosjean, H. and Fiers, W.**, Preferential codon usage in prokaryotic genes: the optimal codon-anticodon interaction energy and the selective codon usage in efficiently expressed genes, *Gene*, 18, 199, 1982.
313. **Gouy, M. and Gautier, C.**, Codon usage in bacteria: correlation with gene expressivity, *Nucleic Acids Res.*, 10, 7055, 1982.
314. **Ikemura, T.**, Codon usage and tRNA content in unicellular and multicellular organisms, *Mol. Biol. Evol.*, 2, 13, 1985.
315. **Bennetzen, J. L. and Hall, B. D.**, Codon selection in yeast, *J. Biol. Chem.*, 257, 3026, 1982.
316. **deBoer, H. A. and Kastelein, R. A.**, Biased codon usage: an exploration of its role in optimization of translation, in *Maximizing Gene Expression*, Resnikoff, W. S. and Gold, L., Eds., Butterworth Publishing, Stoneham, MA, 1986, 225.
317. **Dix, D. B., Thomas, L. K., and Thompson, R. C.**, Codon choice and gene expression: synonymous codons differ in translation efficiency and translational accuracy, in *The Ribosome. Structure, Function, & Evolution*, Hill, W. E., Dahlberg, A. Garrett, R. A., Moore, P. B., Schlessinger, D., and Warner, J. R., Eds., American Society for Microbiology, Washington, D.C., 1990, 527.

318. **Curran, J. F. and Yarus, M.,** Rates of aminoacyl-tRNA selection at 29 sense codons *in vivo, J. Mol. Biol.,* 5, 65, 1989.
319. **Bonekamp, F., Dalb, H., Christensen, T., and Jensen, K. F.,** Translation rates of individual codons are not correlated with tRNA abundances or with frequencies of utilization in *Escherichia coli, J. Bacteriol.,* 171, 5812, 1989.
319a. **Sörenson, M. A., Kurland, C. G., and Pederson, S.,** Codon usage determines translation rate in *Escherichia coli, J. Mol. Biol.,* 207, 365, 1989.
320. **Ishikura, H., Yamada, Y., and Nishimura, S.,** Structure of serine tRNA from *Escherichia coli*. I. Purification of serine tRNA's with different codon responses, *Biochim. Biophys. Acta,* 228, 471, 1971.
321. **Murao, K., Hasegawa, T., and Ishikura, H.,** Nucleotide sequence of valine tRNA mo^{5}UAC from *Bacillus subtilis, Nucleic Acids Res.,* 10, 715, 1982.
322. **Oda, K., Kimura, F., Harada, F., and Nishimura, S.,** Restoration of valine acceptor activity by combining oligonucleotide fragments derived from a *Bacillus subtilis* ribonuclease digest of valine tarnsfer RNA, *Biochim. Biophys. Acta,* 179, 97, 1969.
323. **Takeishi, K., Takemoto, T., Nishimura, S., and Ukita, T.,** Selective utilization of valyl-tRNA having a particular coding specificity in a rabbit hemoglobin synthesizing system, *Biochem. Biophys. Res. Commun.,* 26, 746, 1972.
324. **Takemoto, T., Takeishi, K., Nishimura, S., and Ukita, T.,** Transfer of valine into rabbit hemoglobin from various isoaccepting species of valyl-tRNA differing in codon recognition, *Eur. J. Biochem.,* 38, 489, 1973.
325. **Mitra, S. K., Lustig, F., Akesson, B., Axberg, T., Elias, P., and Lagerkvist, U.,** Relative efficiency of anticodons in reading the valine codons during protein synthesis *in vitro, J. Biol. Chem.,* 254, 6397, 1979.
326. **Samuelsson, T., Elias, P., Lustig, F., Axberg, T., Fölsch, G., Akesson, B., and Lagerkvist, U.,** *J. Biol. Chem.,* 255, 4583, 1980.
327. **Kilpatrick, M. W. and Walker, R. T.,** The nucleotide sequence of glycine tRNA from *Mycoplasma mycoides* sp. capri, *Nucleic Acids Res.,* 8, 2783, 1980.
328. **Heckman, J. E., Sarnoff, B., Alzner-DeWeerd, S., Yin, S., and RajBhandary, U. L.,** Novel features in the genetic code and codon reading patterns in *Neurospora crassa* mitochondria based on sequences of six mitochondrial tRNAs, *Proc. Natl. Acad. Sci. U.S.A.,* 77, 3159, 1980.
329. **Barrell, B. G., Anderson, S. Bankier, A. T., DeBruijn, M. H. L., Chen, E., Coulson, A. R., Drouin, J., Eperon, I. C., Nierlich, D. P., Roe, B. A., Sanger, F., Schreier, P. H., Smith, A. J. H., Staden, R., and Young, I. G.,** Different patterns of codon recognition by mammalian mitochondrial tRNAs, *Proc. Natl. Acad. Sci. U.S.A.,* 77, 3164, 1980.
330. **Björk, G. R.,** A novel link between the biosynthesis of aromatic amino acids and transfer RNA modification in *Escherichia coli, J. Mol. Biol.,* 140, 391, 1980.
331. **Hagervall, T., Jönsson, Y.H., Edmonds, C. G., McCloskey, J. A., and Björk, G. R.,** Chorismic acid, a key metabolite in modification of tRNA, *J. Bacteriol.,* 172, 252, 1990.
332. **Sekiya, T., Takeishi, K., and Ukita, T.,** Specificity of yeast glutamic acid transfer RNA for codon recognition, *Biochim. Biophys. Acta,* 182, 411, 1969.
333. **Lustig, F., Elias, P., Axberg, T., Samuelsson, T., Tittawella, I., and Lagerkvist, U.,** Codon reading and translational error. Reading of the glutamine and lysine codons during protein synthesis *in vitro, J. Biol. Chem.,* 256, 2635, 1981.
334. **Yokoyama, S., Yamaizumi, Z., Nishimura, S., and Miyazawa, T.,** 1H NMR studies on the conformational characteristics of 2-thiopyrimidine nucleotides found in transfer RNAs, *Nucleic Acids Res.,* 6, 2611, 1979.
335. **Yamamoto, Y., Yokoyama, S., Miyazawa, T., Watanabe, K., and Higuchi, S.,** NMR analyses on the molecular mechanism of the conformational rigidity of 2-thioribothymidine, a modified nucleoside in extreme thermophile tRNAs, *FEBS Lett.,* 157, 95, 1983.
336. **Goldman, E., Holmes, W. M., and Hatfield G. W.,** Specifity of codon recognition by *Escherichia coli* tRNA[Leu] isoaccepting species determined by protein synthesis *in vitro* directed by phage RNA, *J. Mol. Biol.,* 129, 567, 1979.

337. **Wittwer, A. J. and Ching, W.-M.,** Selenium containing tRNA^{Glu} and tRNA^{Lys} from *Escherichia coli*: purification, codon specificity and translational activity, *Biofactors*, 2, 27, 1989.
338. **Hillen, W., Egert, E., Lindner, H. J., and Gassen, H. G.,** Crystal and molecular structure of 2-thio-5-carboxymethyluridine and its methyl ester: helix terminator nucleosides in the first position of some anticodons, *Biochemistry*, 17, 5314, 1978.
339. **Sherman, F., Ono, B., and Stewart, R.,** Use of the iso-1-cytochrome C system for investigating nonsense mutants and suppressors in yeast, in *Nonsense Mutations and tRNA Suppressors*, Celis, J. E. and Smith, J. D., Eds., Academic Press, New York, 1979, 133.
340. **Piper, P. W.,** Characterization of nonsense suppressor tRNAs from *Saccharomyces cerevisiae*: identification of the mutation alterations that give rise to the suppressor function, in *Transfer RNA: Biological Aspects*, Söll, D., Abelson, J. N., and Schimmel, P. R., Eds., Cold Spring Harbor Laboratory, Cold Spring Harbor, New York, 1980, 379.
341. **Waldron, C., Cox, B. S., Wills, N., Gesteland, R. F., Piper, P. W., Colby, D., and Guthrie, C.,** Yeast ochre suppressor SUQ5-ol is an altered tRNA$^{Ser}_{UCA}$, *Nucleic Acids Res.*, 9, 3077, 1981.
342. **Hillen, W., et al.,** Restriction on amplification or wobble recognition. The structure of mam^{5}s^{2}U and its interaction of odd uridines with the anticodon loop backbone, *FEBS Lett.*, 94, 361, 1978b.
343. **Weissenbach, J. and Dirheimer, G.,** Pairing properties of the methylester of 5-carboxymethyl uridine in the wobble position of yeast tRNA$^{Arg}_3$, *Biochim. Biophys. Acta*, 518, 530, 1978.
344. **Yokoyama, S., Miyazawa, T., Iitaka, Y., Yamaizumi, Z., Kasai, H., and Nishimura, S.,** Three-dimensional structure of hypermodified nucleoside Q located in the wobbling position of tRNA, *Nature (London)*, 282, 107, 1979.
345. **Harada, F. and Nishimura, S.,** Possible anticodon sequences of tRNA^{His}, tRNA^{Asn}, and tRNA^{Asp} from *Escherichia coli B*, Universal presence of nucleoside Q in the first position of the anticodons of these transfer ribonucleic acids, *Biochemistry*, 11, 301, 1972.
346. **Swamy, G. S. and Pillay, D. T. N.,** Characterization of *glycine max* cytoplasmic, chloroplastic and mitochondrial tRNAs and synthetases for phenylalanine, tryptophan and tyrosine, *Plant Sci. Lett.*, 25, 73, 1982.
347. **Olsen, C. E. and Penhoet, E. E.,** Chromatographic and functional comparison of human placenta and HeLa cell tyrosine tRNA, *Biochemistry*, 15, 4649, 1976.
348. **Meier, F., Suter, B., Grosjean, H., Keith, G., and Kubli, E.,** Queuosine modification of the wobble base in tRNA^{His} influences '*in vivo*' decoding properties, *EMBO J.*, 4, 823, 1985.
349. **Reuter, K., Slany, R., Ullrich, F., and Kersten, H.,** Structure and organization of *Escherichia coli* genes involved in biosynthesis of the deazaguanine derivative Queuine, a nutrient factor for eukaryotes, *J. Bacteriol.*, 173, 2256, 1991.

Chapter 3

CORRELATION BETWEEN CODON USAGE AND tRNA CONTENT IN MICROORGANISMS

Toshimichi Ikemura

TABLE OF CONTENTS

I. Introduction 88

II. Codon Usage in Genes of *E. coli*, *S. typhimurium*, and
    S. cerevisiae 89

III. Cellular tRNA Contents of *E. coli*, *S. typhimurium*, and
     S. cerevisiae 89

IV. Correlation Between Codon Usage and tRNA Content 95
    A. Frequency of tRNA Usage 95
    B. Frequency of Isoaccepting tRNA Usage 95
    C. Preference Among Multiple Codons Recognized by a Single
       tRNA 97

V. Codons Optimal for an Organism's Translation System 98
    A. Optimal Codons 98
    B. Frequency of Optimal Codon Usage 99

VI. Other Constraints on Codon Choices 101

VII. Mechanisms for the Correlation Between Codon Usage and tRNA
     Content 104
    A. Molecular Mechanisms 104
    B. Evolutionary Viewpoints 106

VIII. Codon Choices in Genes of Other Unicellular Organisms and
      Distinction from Higher Eukaryotes 107
    A. Unicellular Organisms 107
    B. Distinction from Higher Eukaryotes 107

References 108

I. INTRODUCTION

The following few sentences are from the opening speech of Prof. O. Maaløe[1] at the Alfred Benzon Symposium IX, entitled *Control of Ribosome Synthesis:* "As you know, the ribosomes account for nearly $^1/_3$ of the mass of rapidly growing *E. coli* cells, and at least $^3/_4$ of the flow of energy and matter made available by a host of different enzymes pass over the ribosomes on its way into stable macromolecules. This alone shows the general importance of our subject." Although the exact meaning of "the flow of energy and matter ... pass over the ribosomes" is not clear, it is apparent that great amounts of energy and mass are required for translation. This is closely related to the subject matter of this chapter. The choice among synonymous codons in both prokaryotic and eukaryotic genes is clearly nonrandom, although it does not affect the nature of proteins synthesized. On the basis of codon usages in a total of 161 genes compiled, Grantham and his colleagues[2,3] proposed that each type of genome has a particular coding strategy ("genome hypothesis"), and synonymous codons are used differently by different kinds of organisms, i.e., an organism-specific codon-choice pattern. The accumulation of DNA sequence data on diverse organisms has made it clear that codon-choice patterns of genes in a single unicellular organism are actually similar to each other regardless of gene function, even though the patterns are somewhat related to amounts of protein produced from the genes. This is consistent with the genome hypothesis. However, in higher eukaryotes, as in higher vertebrates, codon-choice patterns of different genes in a single organism often differ significantly.[4] Some genes are extremely GC-rich at the third codon position, e.g., 95% G+C, while others are rather AT-rich (e.g., 30% G+C). Thus, it is not generally possible to identify organism-specific codon-choice patterns in higher vertebrates such as mammals. The obvious diversity in the third codon G+C% was found related to long-range G+C% mosaic structures of the genome that are designated "isochores" which are possibly related to chromosomal banding patterns.[4-10]

Codon usage of a wide range of organisms has been extensively studied from various respects and has been reviewed elsewhere.[11-15] Thus, in this chapter, we restrict the subject matter to the correlation of codon usage of three microorganisms to tRNA content and related topics. The cellular tRNA content of *E. coli, Salmonella typhimurium,* and *Saccharomyces cerevisiae* was measured by Hatfield et al.,[16-18] and the organism-specific codon-choice patterns of these microorganisms were attributed to the availability of tRNA isoacceptors within a cell.[16-21] In these studies, the following four deductions listed were made, which, based on codon usage data[22,23] accumulated after the publications, are reexamined in this chapter.

1. There is a close correlation between tRNA content and codon usage in most protein genes sequenced for these organisms. The exceptional cases are largely confined to genes with low expression.

2. There are clear similarities in codon choice among different genes of one microorganism, regardless of gene function. We called this organism-specific codon choice the "codon dialect" of the organism. The codon dialect is related to the specific tRNA isoacceptor population of an organism.
3. The extent of bias in codon choice is related to the protein production level of each gene. Codon usage in genes for abundant protein molecules is always more dependent on tRNA content than that in moderately or poorly expressed genes.
4. Foreign-type genes such as those of transposons, plasmids, and viruses often have quite different codon patterns than those of host organisms, and thus, the above deductions are not necessarily applicable to them.

II. CODON USAGE IN GENES OF *E. COLI*, *S. TYPHIMURIUM*, AND *S. CEREVISIAE*

Codon usage in several highly expressed genes of two Enterobacteriaceae, *E. coli* and *S. typhimurium*, and of the yeast, *S. cerevisiae*, is listed in Table 1. Choices among synonymous codons are evidently biased, and there is a clear similarity of codon choice patterns among the genes of each organism, regardless of gene function, i.e., the codon dialect for unicellular organisms.[4] The characteristics of *E. coli* codon choice (described as the *E. coli* dialect) differ considerably from those of the *S. cerevisiae* dialect, but are similar to those of the *Salmonella* dialect. This is true for a major portion of genes within these organisms, because the codon usage pattern summed for all available genes of each organism has these characteristics (Table 1). The extent of codon bias (which is described as the accent of codon dialect) of the summed pattern is more moderate than that for highly expressed genes. Highly expressed genes almost always have a strong accent, but those with moderate or low expression have only a moderate accent.[16-21] Dialectal codon choice patterns for these three microorganisms were attributed mainly to the availability of tRNA isoacceptors.

III. CELLULAR tRNA CONTENTS OF *E. COLI*, *S. TYPHIMURIUM*, AND *S. CEREVISIAE*

Using two-dimensional polyacrylamide gel electrophoresis, tRNAs can be separated with a high degree of resolution.[24,25] Most tRNA isoacceptor molecules of an organism can be separated by improved methods.[26] The separation pattern for *S. typhimurium*[18] is shown in Figure 1. Since cells were labeled by ^{32}P-orthophosphate for several generations, the radioactivity of each gel spot indicates the cellular content of tRNA molecules, after correction for RNA chain length. Separated tRNA molecules were assigned to known tRNAs by RNA fingerprinting, and relative amounts were measured based on the number

TABLE 1
Codon Usage in *E. coli*, *S. typhimurium*, and *S. cerevisiae* Genes

			ECO					STY				YSC					
		groEL	gap	ompC	tufA	fba	Sum.	ompA	rpoB	prsA	Sum.	G3PD	pyk1	ENO	EF1A	adh1	Sum.
ARG	CGA	0	0	0	0	0	2.7	0	1	0	3.3	0	0	0	0	0	2.2
	CGC *	5	4	1	2	2	23.2	2	31	10	24.3	0	0	0	0	0	2.0
	CGG	0	0	0	0	0	4.3	0	0	0	5.3	0	0	0	0	0	1.1
	CGU *	17	8	12	21	4	25.7	10	57	15	21.9	0	2	1	0	0	7.4
	AGA	0	0	0	0	0	1.2	0	0	0	2.3	11 *	22	13	18	8	23.7
	AGG	0	0	0	0	0	0.8	0	0	0	1.6	0	0	0	0	0	7.5
LEU	CUA	0	0	0	0	0	2.6	0	1	1	4.0	0	0	0	0	3	11.9
	CUC	1	0	1	0	0	9.9	0	14	3	10.0	0	0	0	0	0	4.2
	CUG *	38	19	24	27	26	57.7	21	102	18	54.6	0	0	0	0	0	8.6
	CUU	0	0	0	1	0	9.1	0	6	1	9.8	0	0	0	0	0	9.7
	UUA	1	1	1	0	0	9.3	1	0	0	12.1	0	3	2	3	2	24.4
	UUG	1	0	0	0	1	11.2	1	4	3	12.0	21 *	32	36	21	19	32.0
SER	UCA	0	0	0	0	0	5.2	0	0	2	6.2	0	1	0	0	0	15.6
	UCC	5	9	8	3	8	9.7	6	32	7	11.4	12 *	13	17	6	7	14.8
	UCG	0	0	0	0	1	8.0	0	4	1	8.9	0	0	0	0	0	6.7
	UCU	11	6	6	7	11	9.7	9	23	2	8.4	13 *	13	13	14	14	24.7
	AGC	1	0	2	0	5	15.3	2	18	5	16.7	0	0	0	0	0	7.3
	AGU	0	0	1	0	0	6.4	0	0	0	6.6	0	0	1	0	0	11.5
THR	ACA	0	0	0	1	0	5.0	1	3	0	5.0	0	0	0	0	0	15.5
	ACC *	25	15	12	16	10	25.6	15	36	11	25.0	12 *	28	12	14	9	14.3
	ACG	0	0	0	0	0	12.7	0	11	1	16.6	0	0	0	0	0	6.7
	ACU *	8	12	12	13	7	9.6	7	11	3	7.4	12 *	10	8	14	5	22.2

Correlation Between Codon Usage and tRNA Content

AA	Codon																	
PRO	CCA		3	1	3			1	7.9	2	6	2	5.7		12	24	13	23	10	21.1
	CCC		0	0	0			0	4.0	0	0	1	5.4		0	0	0	0	1	5.8
	CCG	*	10	8	1	19	13	13	25.9	17	40	9	25.4		0	0	0	0	0	4.2
	CCU		1	0	0	1	1	1	6.0	3	10	1	7.6		0	1	0	0	2	12.8
ALA	GCA	*	25	4	0	5	9	9	20.0	7	12	8	11.9		0	0	0	7	0	15.0
	GCC		3	1	8	1	2	2	24.3	3	12	11	27.4		7	12	10	0	16	15.6
	GCG	*	13	2	0	8	8	8	35.7	3	38	5	40.5		0	0	0	0	0	5.3
	GCU	*	34	25	3	13	12	12	17.3	22	15	10	13.6		25	31	48	30	19	28.0
GLY	GGA		0	0	18	0	0	0	5.7	0	0	0	6.3		1	0	0	0	0	8.9
	GGC	*	31	14	0	19	21	8	32.7	22	47	12	36.2		0	0	0	0	3	8.9
	GGG		0	0	19	0	1	0	9.3	0	3	0	10.9		0	0	0	0	0	5.1
	GGU	*	28	19	29	9	19	23	28.3	16	55	8	19.7		25	34	35	42	41	34.1
VAL	GUA	*	14	8	9	10	10	12	11.6	10	17	5	11.8		0	0	0	0	0	9.9
	GUC		2	1	2	0	0	1	14.7	2	22	8	18.4		15	24	20	19	17	14.7
	GUG	*	12	4	2	4	4	3	26.6	3	33	9	25.9		0	0	0	1	0	9.5
	GUU	*	30	21	12	24	24	14	20.1	12	38	13	14.8		22	24	15	26	19	26.4
LYS	AAA	*	37	26	17	18	19	19	36.2	15	52	6	35.3		1	1	3	3	4	38.2
	AAG		3	1	0	5	5	6	11.1	5	27	2	11.8		25	36	32	46	20	35.2
ASN	AAC	*	18	17	32	7	7	18	24.7	16	48	14	23.2		12	26	21	16	11	25.8
	AAU		1	1	0	0	0	0	13.7	2	3	3	17.9		0	0	0	0	0	31.6
GLN	CAA		0	0	1	0	0	2	12.9	2	6	3	12.2		5	10	9	12	9	29.3
	CAG	*	16	5	20	8	8	13	30.8	16	52	5	32.7		0	0	0	0	0	10.6
HIS	CAC		1	5	1	10	10	11	11.3	3	16	2	9.2		8	7	10	6	10	8.3
	CAU		0	1	0	1	1	2	11.4	2	4	2	11.2		0	0	0	5	1	12.4
GLU	GAA	*	43	13	11	30	30	19	44.8	9	85	13	40.7		12	28	28	30	20	49.0
	GAG		3	2	0	7	7	3	19.2	3	34	1	21.5		2	0	0	1	0	17.5
ASP	GAC		25	18	23	20	20	13	23.0	16	62	11	22.9		16	19	23	16	14	22.5
	GAU		10	7	9	4	4	8	31.9	9	33	15	33.8		9	13	7	8	2	37.1

TABLE 1 (continued)
Codon Usage in *E. coli*, *S. typhimurium*, and *S. cerevisiae* Genes

		ECO						STY				YSC					
		groEL	gap	ompC	tufA	fba	Sum.	ompA	rpoB	prsA	Sum.	G3PD	pyk1	ENO	EF1A	adh1	Sum.
TYR	UAC *	7	6	24	8	10	13.7	10	32	3	12.5	11	15	9	8	13	16.4
	UAU	0	2	5	2	3	14.0	5	10	1	15.6	0	0	1	0	0	16.5
CYS	UGC	3	3	0	2	4	6.4	1	3	3	5.7	0	0	0	1	0	3.7
	UGU	0	0	0	1	0	4.5	1	4	1	4.1	2	6	1	6	8	7.6
PHE	UUC *	7	10	17	13	11	18.6	8	30	6	15.8	10	15	14	16	8	20.0
	UUU	0	1	2	1	2	17.5	3	15	5	20.5	0	0	1	1	0	23.2
ILE	AUA	0	0	0	0	0	2.4	0	0	0	4.6	0	0	0	0	0	13.0
	AUC *	30	18	10	26	20	28.3	14	64	16	27.0	11	26	9	14	12	18.7
	AUU	3	1	0	3	1	26.6	3	20	9	27.5	9	11	11	16	9	30.9
MET	AUG	22	9	4	10	8	26.9	5	36	9	25.8	7	11	10	8	7	21.4
TRP	UGG	0	3	4	1	4	12.7	5	4	0	11.0	3	1	5	6	5	10.2
TER	UAA	1	1	1	1	1	1.8	1	1	0	1.9	1	1	1	1	1	1.0
	UAG	0	0	0	0	0	0.1	0	0	0	0.10	0	0	0	0	0	0.4
	UGA	0	0	0	0	0	0.7	0	0	1	0.7	0	0	0	0	0	0.6

Note: The name of each species is listed at the top of the column in an abbreviated form: ECO (*E. coli*), STY (*S. typhimurium*), and YSC (*S. cerevisiae*). Usages for individual genes are listed by the actual number of codons. Codon usage summed for each organism is listed as frequency/1000 (Sum.): in *E. coli*, 937 intrinsic genes were summed; in *S. typhimurium*, 130 genes; and in *S. cerevisiae*, 581 genes. Optimal codons of each organism were deduced by combining the predictions from Rules 1–4. An example of this deduction for *E. coli* arginine is as follows. The most abundant *E. coli* arginine isoacceptor responds to CGU, CGC, and CGA (Table 2). Therefore, Rule 1 predicts the preference "CGU,CGC,CGA>CGG,AGA,AGG". Because this tRNA has inosine in the anticodon, Rule 3 predicts "CGU,CGC>CGA". Then the combination of these preferences is "CGU,CGC>CGA,CGG,AGA,AGG," and thus, both CGU and CGC are the optimal codons in *E. coli* for arginine. The optimal codons of these organisms deduced previously[17,18,20,21] are specified by asterisks. No asterisks were used for amino acids whose isoacceptors have not yet been quantified.

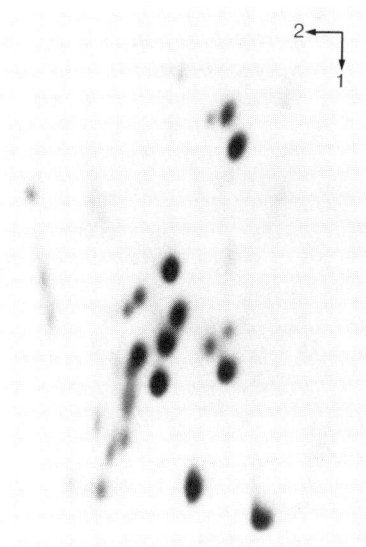

FIGURE 1. Two-dimensional gel electrophoresis of ^{32}P-labeled *S. typhimurium* tRNAs.[18] Electrophoresis in the first dimension was from top to bottom on a 10% (w/v) polyacrylamide gel, and that in the second dimension was from right to left on a 22% (w/v) polyacrylamide gel containing 7 *M* urea. The conditions for electrophoresis were presented previously.[26]

of molecules in cells. The amount of each tRNA measured previously by these procedures from *E. coli*[16,21] and *S. typhimurium*,[18] as well as the codons recognized by each tRNA, is listed in Table 2. Eight tRNA species were newly quantified (see the legend in Table 2). The amount of tRNA$_1^{Leu}$ was normalized to 1.0. Clearly, the tRNA contents of the two Enterobacteriaceae are quite similar, indicating that the populations of tRNA molecules have been conserved during evolution. It should be pointed out that these tRNA populations differ greatly from that of *S. cerevisiae*.[17,21]

Recently, Komine et al.[27] mapped all the tRNA genes in the *E. coli* genome, and determined the exact gene copy numbers (Table 2). Figure 2 shows that the contents of individual *E. coli* tRNAs are closely related to the gene copy number (correlation coefficient r = 0.73). This observation confirms our previous proposal that the gene number is a major factor in determining *E. coli* tRNA content.[16] The amounts of tRNAs encoded within ribosomal RNA operons designated as spacer or distal tRNAs (e.g., Ala1B, Trp, Thr1+3) appear somewhat higher than that expected from the copy number. This may be related to the strong expression of rRNA operons. Dispensable tRNAs appear to be present in lower amounts than what would be expected from the corresponding gene copy number, e.g., RNA$_2^{Ser}$, tRNA$_1^{Gly}$, tRNA$_2^{Thr}$.

TABLE 2
Content and Codon Recognition of tRNAs in *E. coli* and *S. typhimurium* tRNAs

aa	tRNA	Codon	No. of tRNA genes *E. coli*	tRNA content *E. coli*	tRNA content *S. typhimurium*
Leu	1	CUG	4	1.00	1.0
	2	CUU,CUC	1	0.30	0.2
	4	UUA,UUG	1	0.25	0.2
	5(6)	UUG	1	0.20	0.2
	3	CUA,CUG	1	Minor	Minor
Val	1	GUA,GUG,GUU	4	1.05	0.9
	2A+2B	GUC,GUU	2	0.40	0.2
Gly	3	GGU,GGC	4	1.10	0.9
	2	GGA,GGG	1	0.15	0.2
	1	GGG	1	0.10	(0.1)
Ala	1	GCU,GCA,GCG	3	1.00	1.0
	2	GCC	2	(0.3)	0.3
Arg	2(1)	CGU,CGC,CGA	4	0.90	0.7
	CGG	CGG	1	Minor	(Minor)
	AGR	AGA,AGG	1	Minor	(Minor)
	AGG	AGG	1	Minor	(Minor)
Ile	1	AUU,AUC	3	1.00	1.0
	2	AUA	1	0.05	(0.05)
Lys		AAA,AAG	2	1.00	0.9
Glu	2	GAA,GAG	4	0.90	0.9
Asp	1	GAU,GAC	3	0.80	1.0
Thr	1+3	ACU,ACC	2	0.80	0.6
	2	ACG	1	(0.1)	0.1
	4	ACA,ACG	1	(0.1)	0.1
Asn		AAU,AAC	3	0.60	0.5
Gln	2	CAG	2	0.40	0.4
	1	CAA	2	0.30	0.3
Tyr	1+2	UAU,UAC	3	0.50	0.3
Ser	3	AGU,AGC	1	0.25	0.2
	1	UCU,UCA,UCG	1	0.25	0.3
	5	UCU,UCC	2	0.25	0.2
	2	UCG	1	(0.05)	0.05
His		CAU,CAC	1	0.40	0.3
Trp		UGG	1	0.30	0.2
Pro	3	CCG,CCA,CCU	1	(0.3)	0.3
	1	CCG	1	(0.3)	0.3
	2	CCC,CCU	1	Minor	Minor
Phe		UUU,UUC	2	0.35	0.2
Cys		UGU,UGC	1	Minor	(Minor)
Met	m	AUG	2	0.30	0.4
	f	AUG,GUG,UUG	3	0.50	0.3

TABLE 2 (continued)
Content and Codon Recognition of tRNAs in *E. coli* and *S. typhimurium* tRNAs

Note: The relative abundances of individual tRNAs were measured on the basis of molecular numbers in cells using two-dimensional polyacrylamide gel electrophoresis. tRNA in column two denotes the specific tRNA designation for the isoacceptor(s). The amount of tRNA$_1^{Leu}$ is normalized to 1.0. The tRNA contents of the two organisms were measured by different gel systems.[16,18] Because of the general similarity of the two tRNA populations, contents of several tRNAs not quantified by the respective separation system are tentatively estimated from data of the other organism and listed in brackets. Gene copy numbers of *E. coli* tRNAs are included according to the data of Komine et al.[27] Based on their RNA fingerprints, eight gel spots in our previous work[16,18] can be newly assigned to known tRNAs.

IV. CORRELATION BETWEEN CODON USAGE AND tRNA CONTENT

A. FREQUENCY OF tRNA USAGE

To examine the correlation between frequency of codon usage and tRNA content, the frequency of tRNA usage, i.e., anticodon usage, was determined as follows. The usage frequency of a tRNA responding to a single codon was defined as the occurrence of the codon itself, and that of a tRNA responding to multiple codons as the total occurrence of the corresponding codons.[16] The content of tRNA and frequency of its usage calculated for the summed codon usage are roughly proportional (see Figure 3a), indicating a close correlation between the two variables.[10,16,20,28,29] Biological significance of this correlation is discussed later in this chapter. The degree of the correlation was previously found to depend on the expressivity of genes [a regression line (y = a + bx) was drawn in the graph in Figure 3]. High slope values as well as negative y intercepts of the regression line were found for genes that encode abundant protein species. This is clear from the example of *groEL* in Figure 3b. High slope values indicate strong dependence of tRNA use on its content, and negative y intercept has been attributed to "upward concavity" of functions deduced from curvilinear regression.[16,17,20,21] To confirm the "upward concavity", the data points in Figure 3b were analyzed by quadratic curve fitting as shown by the dotted line.

B. FREQUENCY OF ISOACCEPTING tRNA USAGE

By comparing synonymous codon choices with tRNA isoacceptor content, the correlation between codon usage and tRNA content can be examined without the influence of amino acid composition of individual protein species. In Figure 4, both content and usage frequency of the most abundant isoacceptor of each amino acid are normalized to 1.0. Clearly, the most abundant

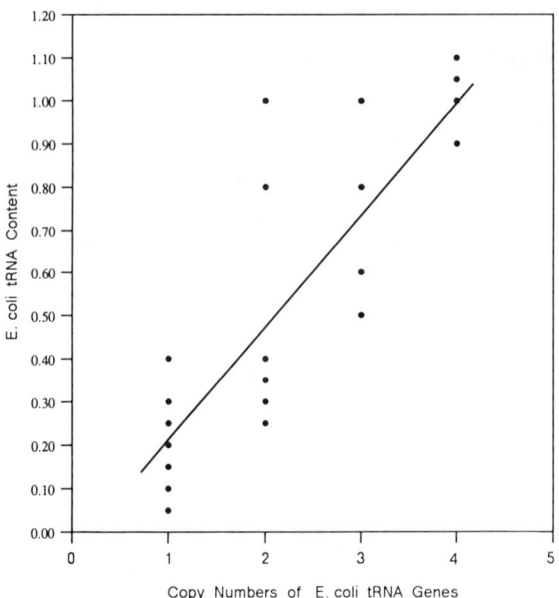

FIGURE 2. Correlation between gene copy number and tRNA content of *E. coli*. The line corresponds to the regression line (correlation coefficient r = 0.73). The data from Table 2 are shown graphically. Redundant tRNAs (Val 2A+2B, Metf 1+2, or Tyr 1+2) were treated as a single tRNA species.

isoacceptor is always used at the highest frequency. The line in each graph indicates the function when the frequency of isoacceptor use is directly proportional to content. In modestly expressed genes, the data points were found to be represented roughly by this linear function.[16-21] This notion was confirmed here by analysis of the summed codon usage pattern (Figure 4a). Data points of three highly expressed genes are situated far away from the line, and are drawn as probable concave curves (dashed lines; Figure 4 b–d). The dependence of isoacceptor usage on its content is thus much greater than that expected from the proportionality of the two values. That is, the highly expressed genes selectively use codons recognized by the most abundant isoacceptor but almost completely avoid codons for other isoacceptors.[11] Essentially the same results were obtained for most of highly expressed *E. coli*, *S. typhimurium*, and *S. cerevisiae* genes.[16-21] Based on these findings, we proposed that codon choice in these organisms is constrained by tRNA availability, and this is particularly evident for highly expressed genes. It should be noted again that the population of *S. cerevisiae* tRNA isoacceptors is very different from that of *E. coli*.[4,21] The differential isoacceptor population is a cause of the organism-specific codon-choice pattern.

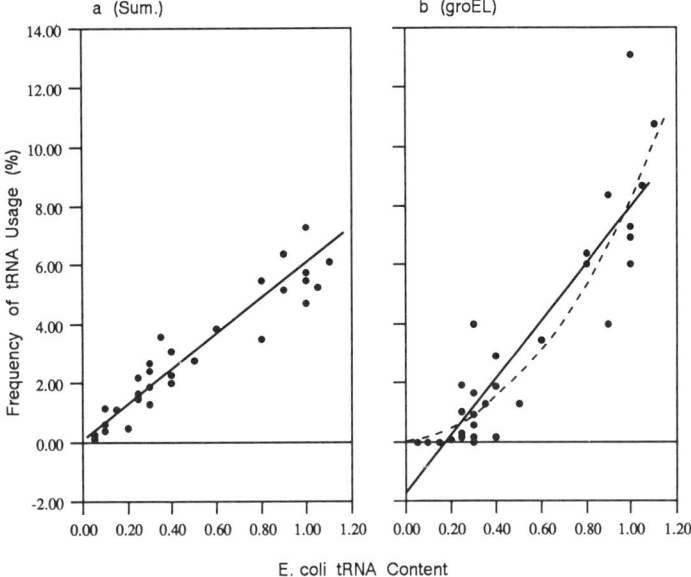

FIGURE 3. Correlation between tRNA content and frequency of codon usage summed for 937 *E. coli* intrinsic genes (a) and that of *groEL* (b). The data from Table 1 are shown graphically.

C. PREFERENCE AMONG MULTIPLE CODONS RECOGNIZED BY A SINGLE tRNA

Figure 4 shows that the codon choice in *E. coli* genes is strongly constrained by tRNA availability. This observation provided the basis for **Rule 1** in a previous report.[21] The following definite constraints on the choices among codons recognized by a single tRNA (Rules 2–4) were defined:[20,21]

- **Rule 2** — Introduction of a thiolated uridine or of 5-carboxymethyl uridine at the anticodon wobble position leads to a preference for an A-terminated codon over a G-terminated codon.[30] This rule is mainly applicable for highly expressed genes, and is affected by the "context effect",[31] most likely due to the modified nucleotide. For example, AAA is preferred if G is 3′ to the Lys codon, i.e., AAA-G, while AAG is used quite often if C is 3′ to the Lys codon, i.e., AAG-C. General features of the context effect have also been reported.[32]
- **Rule 3** — Introduction of inosine at the anticodon wobble position produces preference for U- and C-terminated codons over an A-terminated codon, which leads to purine-purine wobble pairing. The tRNA isoacceptor populations as well as the modification patterns in the anticodon and anticodon loop often differ for taxonomically distant organ-

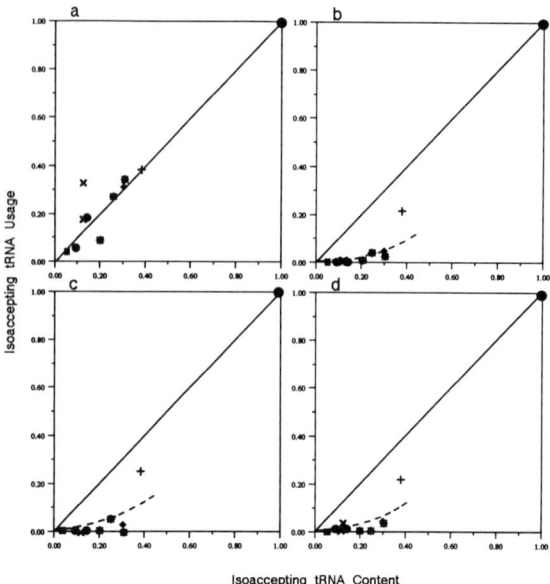

FIGURE 4. Correlation between tRNA isoacceptor content and frequency of tRNA isoacceptor usage in decoding *E. coli* mRNAs. Both the amount and the frequency of use of the most abundant tRNA for each amino acid are normalized to 1.0. The straight line in each graph is that predicted if the use of isoacceptors is proportional to its content. The most abundant isoacceptor of each amino acid is indicated by (●). Other isoacceptors are Leu (*), Gly (·), Thr (X), Ala (♦), Val (+) and Ile (■). (a) Sum. in Table 1; (b) *groEL*; (c) *gap*; and (d) *tufA*. Details as well as results for other *E. coli, S. typhimurium*, and *S. cerevisiae* genes have been presented previously.[16-21]

isms. This is also a factor responsible for organism-specific codon-choice patterns.[20,21]

- **Rule 4** — Codons of the (A/U)-(A/U)-pyrimidine type, i.e., AAX, AUX, UAX, and UUX where X is a pyrimidine, would support an optimal interaction strength between a codon and an anticodon when the third letter of the codon is C.[33,34] This rule is mainly applicable for highly expressed genes, but not for genes with moderate or low expression, showing its weaker constraint. For example, compare the summed codon patterns of Phe, Tyr, and Asn with those of highly expressed genes (Table 1). This rule may possibly be related to fidelity in translation.[35]

V. CODONS OPTIMAL FOR AN ORGANISM'S TRANSLATION SYSTEM

A. OPTIMAL CODONS

Codon choices in *E. coli, S. typhimurium*, and *S. cerevisiae* genes were found to conform well to expectations based on Rules 1–4, and exceptional cases were mostly confined to genes of low expression.[4,21] However, it is not an easy task to understand this conclusion directly from the data in Table 1 or those in the Codon Compilation Database.[22,23] The codon that is translated by the most abundant isoacceptor of each amino acid (Rule 1) and also conforms to Rules 2–4 was previously designated as the "optimal codon", i.e., the codon optimal for translational efficiency of the respective organism as discussed in detail in our previous papers.[20,21] The optimal codons of the Enterobacteriaceae and *S. cerevisiae* are indicated by an asterisk in the first column below the respective organism in Table 1. Each optimal codon corresponds strikingly well to the preferred codon of each organism. We proposed the four rules and the optimal codons derived from them when only small numbers of *E. coli* and *S. cerevisiae* gene sequences had been determined.[16,17,19,20] It has become clear that these are applicable for a major portion of *E. coli, S. typhimurium*, and *S. cerevisiae* genes (refer to the summed codon patterns; Table 1). This again supports the idea that the synonymous codon choices in genes of these organisms are under constraints related to translation efficiency. The difference in the spectrum of the optimal codons used by *E. coli* and *S. cerevisiae* is due to that in the actual population of isoacceptors and in the modified nucleotides at the anticodon wobble position.[4,21]

B. FREQUENCY OF OPTIMAL CODON USAGE

Figure 5 shows the distribution of optimal (o) and nonoptimal (X) codons in *E. coli* and *S. typhimurium* genes.[18] For Met and Trp, a single codon corresponds to the amino acid and is indicated by the symbol "–". Highly expressed genes, i.e., the *E. coli* ribosomal protein gene, *rplA*, and the *S. typhimurium* membrane protein gene, *ompA*, use mostly optimal codons, showing their codon choices to be highly constrained by tRNA availability. In moderately expressed genes, such as amino acid synthesis genes, e.g., *E. coli thrA* and *S. typhimurium trpA*, optimal codon usage clearly decreases. To examine this feature quantitatively, the frequency of use of optimal codons was defined as

$$F_{op} = \frac{\text{the numbers of o}}{\text{sum of the numbers of o and X}}$$

We previously listed the F_{op} calculated for all available *E. coli* and *S. cerevisiae* genes, and found that this parameter is high for highly expressed genes and low for weakly expressed genes.[4,18,21] The cellular content of a wide variety of *E. coli* protein molecules has been measured by Neidhardt and his colleagues[36] by two-dimensional gel electrophoresis. Their cellular content is thought to correspond to the degree of gene expression, although the stability of protein molecules has to be taken into consideration. It is thus possible to examine the correlation between the frequency of optimal codon use (F_{op}) and gene expressivity. For all *E. coli* genes whose nucleotide sequences and relative protein contents are known (Table 3), the F_{op} and protein content

a) E. coli

rplA (ribosome protein L1) ; Fop=0.93
-ooooXo-ooooXoo ooooo ooooooooooXoooooooo o ooooXoo ooo oooooooo
oo oooo oooXoXoXoooooooooooooooo-o oo ooXooo-oX ooXo o o-oooooooo
oooooooo-XoooooooXoooooooooooooooo ooooo oooooo X o ooooooooo
ooooooooooooooooXoo o oo-oooooo oooo o oo

thrA ; Fop=0.61
-XoXXoooX XoXoooXoooX Xoo XXXoXooXoXX XXXoooo ooo-XoooX oo oXXXo
   XooXXXoXXXXXXXXooXooXoXXoXoX ooXXXXo Xo oX XXXo o oooooX ooXX
- oXX-Xooooooo ooooo oXoooooooX oXo oX XXX oooXoo oXoo -oo-ooooX
oXXoooooXXooo o oooooX XoX   XX-X ooXXXo oooX oXXXX - oooo-X
X oooooX XooXoXoXoooX oXoXoXXXXoXoXXXoo o o oXoXXoX XoXo-o-o o
  ooX-oX-Xo-oooXXoo- oXoX oooXXX   oo o o oXX oXooXo-ooXooXoooo
oXoXooooooXoXXo ooo oXooXoXo oooXXooXoXXooXXXXooX oo o XooXo o
oooooooo o-ooXo oooooXoXoXoooooooooXXoXooX -oXXo o XoX ooXo XooXoX
o oXXooo-oXoooXXoXoXXXXoXXoXoooX oooXXoo o oooo XXX oooooo oXX
oooXXoo - oo oXoXoooo XoooXX oooXoXXooXXooXXoXXoo oX-oo oXX o X
XoooXX oo- o XooXooXo-oXooo oX X o- oooXXXXoooXXoooXoo XoXoXo
oXoXXoXXo oXoX-oXo XX XXXoooXXo oXooXoXooXX o oX ooXXXoo oX oo
ooooXoooXoXoX XXooooXoooXXoooX oXoXoXXo oXooX -XXXX

b) S. typhimurium

ompA ; Fop=0.88
-oXXoooXooooooooooooXooo oo-oooooooo- oo oooX X ooo ooXoooooXooo
ooooXooXoo-oo -Xoo-oooo ooXooXoooooXoooooXXoo o XXoooo-o-oo oX
oXXooo oo ooo ooooooooXXoooXooooooooo-ooooo Xoooooo oooo ooo
ooooXoooooooooXooooooooX oooX oooooooo ooXoooooo oooooo oo oo
o oXooooo ooo oooooo Xoooo oo ooo ooXo oo ooo-oo oooooo ooo
XooXoo ooo oooXoooooo oooooooo

trpA ; Fop=0.64
-oooXXXXXXo oXooXXXooooo XoXoo ooXX XoX XoX oooXoXo ooX oXooo
XooXoXoXooXXooo Xo-oooXooo ooXoXooX-ooXoooXooX ooXXo ooooo ooXo
Xoooo XXoooooXX Xoooooo oXXo XoooXo ooooooooX o oXoooooooooXoX
XXXXXoXo XoXoXoooooo oooo oXooXXoooo o XXXXXoXooXo Xoo-XoXXX XX
X-oXX oo

FIGURE 5. Occurrence of optimal (o) and nonoptimal (X) codons in *E. coli* and *S. typhimurium* genes.[18] Codons of Met and Trp are indicated by a dash (–). Blank spaces are used for several amino acids either because the contents of their tRNA isoacceptors have not been clarified or because no criteria were deduced. (a) *E. coli* genes; and (b) *S. typhimurium* genes. Results for other *E. coli* genes have been presented previously.[20,21]

are plotted in Figure 6. The definite correlation between the two values shows optimal codon frequency in each gene to be actually related to its production level. A similar type of correlation between codon bias and protein production has also been noted by other groups.[37-39] Experimental evidence which shows that codon choice affects translational efficiency has been reported by many groups.[40-45]

Figure 7a and 7b show histograms of F_{op} for *E. coli* intrinsic genes and foreign-type genes (genes for plasmids, transposons, enterotoxins, and restriction-modification systems), respectively. F_{op} of foreign-type genes is significantly lower than those of intrinsic genes, showing that the former genes do not necessarily use the *E. coli* dialect. *E. coli* genes with low F_{op} were found to be mostly confined to regulatory genes with low expression.[4,21] *S. cerevisiae* genes with high F_{op} correspond to genes with high expression,[17,21] and those with very low F_{op} values often correspond to nuclear genes encoding mitochondrial proteins, e.g., PET122, MRS3, PET54, or mating-related factors, e.g., STE4, HML.

VI. OTHER CONSTRAINTS ON CODON CHOICES

To explain the nonrandom patterns of synonymous codon choices in the three microorganisms, four constraints related to translation efficiency (**Rules**

TABLE 3
Frequency of Optimal Codon Usage (F_{op}) and Numbers of Protein Molecules per *E. coli* Genome

Gene	Locus	F_{op}	No. of molecules Average	Rich
tufA	ECOSTR3	0.93	60567	86606
rpsA	ECORPSA	0.92	11504	19003
dnaK	ECODNAK	0.89	6078	8968
atpA	ECOUNC#6	0.87	6143	4786
aceE	ECOACE#2	0.86	5403	10844
lpd	ECOACE#4	0.85	6152	5645
fabB	ECOFABB	0.85	21043	28285
atpD	ECOUNC#8	0.85	5408	4650
ssb	ECOSSB	0.84	1452	1725
rpoB	ECORPLRPO#5	0.83	1551	2173
carB	ECOCARAB#2	0.82	1009	321
sucB	ECOGLTA#7	0.82	3747	243
valS	ECOVALS	0.82	550	859
glyS	ECOGLYS#2	0.80	1130	1576
trxA	ECOTRXA	0.80	2269	2254
metK	ECOMETK	0.79	3578	4211
sdhA	ECOGLTA#4	0.79	2013	158
ilvE	ECOILVE#2	0.78	4227	2666
glnS	ECOGLNS	0.78	794	1041
pheT	ECOTHRINF#6	0.77	890	1138
pheS	ECOTHRINF#5	0.77	1050	1350
nusB	ECONUSB	0.77	684	839
purC	ECOPURCA#2	0.77	6167	2755
sucA	ECOGLTA#6	0.76	1616	250
tyrS	ECOTYRS	0.76	1563	1649
livJ	ECOLIVHMGF#1	0.75	14662	1254
grpE	ECOGRPE	0.74	2236	2905
glpK	ECOGLYK	0.73	14652	2045
hisS	ECOHISS	0.73	1027	1414
ppc	ECOPPCG	0.72	903	351
trxB	ECOTRXB	0.72	1496	1726
rpoA	ECORPOA#1	0.71	4664	6377
gor	ECOGOR	0.71	536	517
thrS	ECOTHRINF#1	0.71	553	855
lon	ECOLON	0.71	1319	816
folA	ECOFOLA	0.68	7459	9388
polA	ECOPOLA#1	0.67	283	432
hisP	ECOHISMP	0.64	282	211
tyrA	ECOPHEAB#4	0.61	222	120

Note: The relative numbers of *E. coli* protein molecules analyzed in three different growth conditions (acetate-, glycerol-, and rich-medium) were obtained from VanBogelen et al.,[36] and the number of protein molecules per genome was calculated with their equation. Average (column 4) corresponds to the average values observed with the three growth conditions, and Rich (column 5) corresponds to those observed with the rich-medium. Proteins that gave multiple spots in their gel system were omitted from the analysis. Locus (column 2) corresponds to the GenBank Locus name.

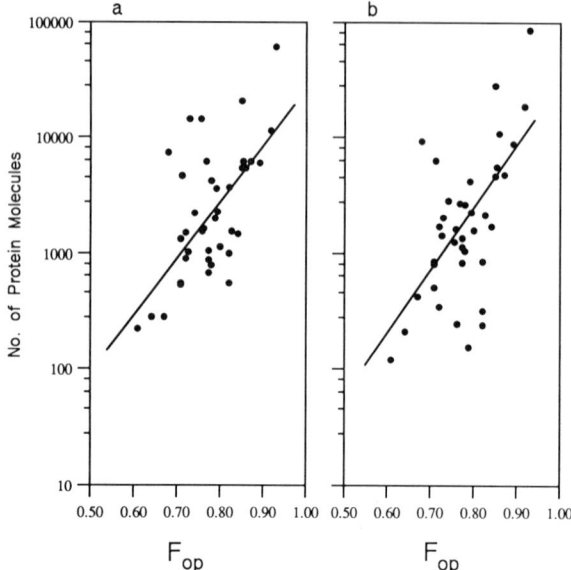

FIGURE 6. Correlation between frequency of optimal codon use (F_{op}) and number of *E. coli* protein molecules per genome. The data of Table 3 are shown graphically. (a) Average and (b) Rich.

1–4) are proposed. When a wide range of microorganisms is concerned, other factors must be considered and these have been summarized previously.[21] For organisms with a base-biased genome, the bias of the genome G+C% is an important factor; the G+C% becomes **Rule 5**.[46-54] Figure 8 shows histograms of G+C% at the third codon position of genes of a GC-rich microorganism, *Streptomyces* (70–75% G+C; see graph a), and of an AT-rich microorganism, *Clostridium* (30–35% G+C; see graph b), as well as of *S. cerevisiae* (ca. 40% G+C; see graph c) and *E. coli* (ca. 50% G+C; see graph d). The effect from the genome G+C% is clearly less evident for *S. cerevisiae* and *E. coli* than for the G+C%-biased organisms, *Streptomyces* and *Clostridium* (see graphs a and b). In the foreign-type *E. coli* genes (graph e), there is a significant number of AT-rich ones, and the distribution is much wider than that of the intrinsic genes (graph d). This may be related to their evolutionary origin. The third codon G+C% of higher vertebrates is known to be distributed in a wide range; a histogram of 1882 human genes is listed (see graph f). This wide distribution was attributed to the long range G+C% mosaic structures of the genome (as discussed below).

Codon choices of individual microorganisms are determined by combined effects of translational efficiency (Rules 1–4) and genome G+C% (Rule 5). It should be noted that T- and C-terminated codons are often recognized by a G-starting anticodon, and A- and G-terminated codons by a U-starting anticodon.

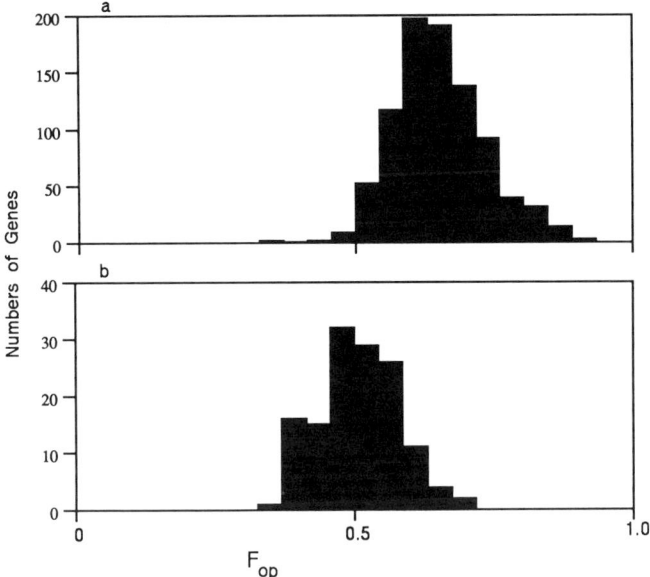

FIGURE 7. Histograms for F_{op} of *E. coli* genes: (a) 937 intrinsic genes are represented and (b) 133 foreign-type genes.

Because of this wobble pairing, constraints due to tRNA content and genome G+C% are fairly compatible with each other. In organisms with an evidently biased genome, e.g., ≥70% or ≤30% G+C, these two factors appear usually harmonious, since tRNA content has been well adapted to the spectrum of biased codons.[15,53] For organisms with moderate genome G+C%, e.g., 40–60% G+C, however, there are cases where the two factors direct the opposite preference. For example, AAG is the optimal Lys codon of *S. cerevisiae* (the CUU anticodon tRNA is the most abundant $tRNA^{Lys}$ isoacceptor),[17] while the genome G+C% is about 40%. By examining such cases for enterobacterial genes, it was found that the bias of codon choices due to the genome G+C% is evident for low or modestly expressed genes, but not for highly expressed genes.[46,47,55] That is, the contribution of base composition increases as gene expressivity decreases. This can also be observed by analyzing codon choices of *S. cerevisiae* genes for such amino acids encoded by two codons as Lys, Phe, Asn, and Tyr, in which constraints due to translational efficiency cause a tendency toward using G- or C-terminated codons, i.e., the tendency of using such codons is opposite to the genome G+C%. In Figure 9, the two values, G+C% at the third codon position of the four amino acids and F_{op} (used for indices of gene expressivity), are plotted for individual *S. cerevisiae* genes. A clear correlation (r = 0.81) can be seen: genes with low F_{op} (thus with low expression) use AT-rich codons conforming to its genome G+C%, while those with high F_{op} use GC-rich codons conforming to translation efficiency.

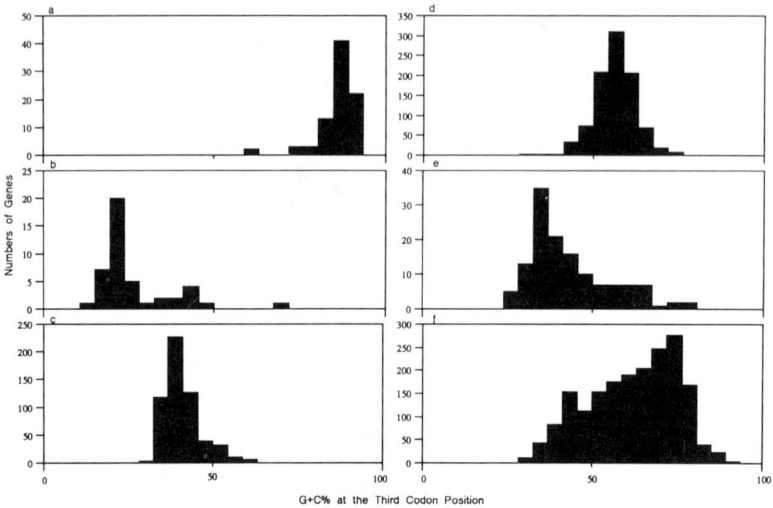

FIGURE 8. Histograms for G+C% at the third codon position in genes of different organisms. The G+C% at the third position was calculated using codon usage data compiled by our group.[23] (a) *Streptomyces*, 84 genes; (b) *Clostridium*, 44 genes; (c) *S. cerevisiae*, 560 genes; (d) and (e) *E. coli* intrinsic (937 genes) and foreign-type (133 genes), respectively; and (f) Human, 1882 genes. Since the number of genes differed significantly among organisms, the scale in the graph was changed accordingly.

VII. MECHANISMS FOR THE CORRELATION BETWEEN CODON USAGE AND tRNA CONTENT

A. MOLECULAR MECHANISMS

A close correlation between codon usage and tRNA population should have resulted from the accumulation of a great number of mutations and successive base substitutions in both protein and tRNA genes during evolution. Figure 3a shows that the *E. coli* tRNA population and its summed codon spectrum are closely correlated. One clear factor that may be deduced from this correlation is the constraints placed on the tRNA content by the amino acid composition of cellular proteins.[16,20] However, the correlation between synonymous codon usage and tRNA isoacceptor content (Figure 4) should not be due to the amino acid composition. For organisms with strong G+C%-biased genomes, the adaptation of the isoacceptor content to the biased codon spectrum should be a major mechanism used in producing the correlation.[15,53] In the case of organisms with moderate G+C%, e.g., *E. coli*, 50% G+C, we proposed that the codon usage was adapted to the tRNA isoacceptor content.[16-21] The similarity in codon-choice patterns among different genes of one organism with a moderate genome G+C% is considered as evidence for the constraint imposed by the tRNA content on codon choice, although the adjustment of the tRNA content to the codon spectrum may also be an important factor. If only the adjustment of isoacceptor content to the codon spectrum is considered, one

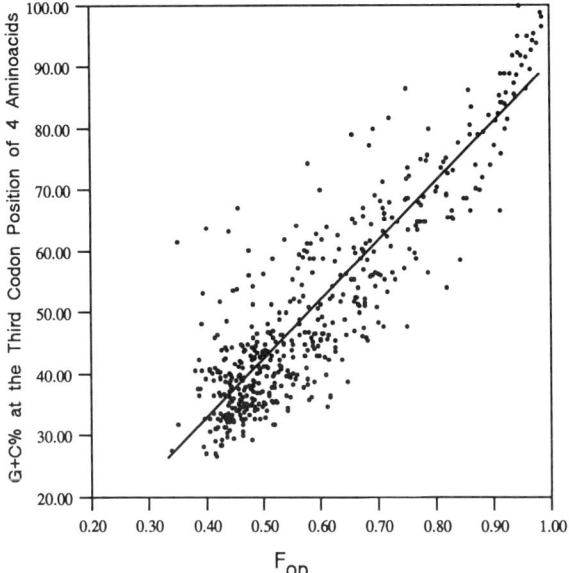

FIGURE 9. Correlation between *S. cerevisiae* F_{op} and G+C% at the third codon position for four amino acids (Lys, Phe, Asn, and Tyr).

must postulate an unknown mechanism by which codon-choice patterns are roughly equalized among different genes (codon dialect) of the organism with the moderate G+C% genome.

On the basis of the fact that cellular processes of protein synthesis require large amounts of energy and mass, a molecular mechanism was proposed by which the codon choice has been constrained during evolution by the tRNA content. This can be further explained regarding the correlation between the extent of codon bias (accent of dialect) and protein production level. Highly expressed genes use mostly the codons optimal for translation efficiency. As protein production decreases, so does optimization (Table 3, Figure 6). This correlation can be explained on the basis of the following two viewpoints.

First, as pointed out at the beginning of this chapter, cellular processes of protein synthesis require large amounts of energy and mass. Based on analysis of codon usages of *E. coli* and *S. cerevisiae*, we proposed that a codon dialect should reflect an organism's strategy for producing large amounts of proteins with a minimal load.[4,21] The amount of protein production is known to be closely related with the amount of mRNA production. If codons translated by minor isoacceptors are frequently used in a highly expressed gene, ribosomes perform the uneconomical task of finding the proper tRNA present in small quantities for many mRNA molecules of this gene. Energy (GTP) seems to be consumed during this waiting period for "proofreading": a certain level of GTP hydrolysis is involved in rejecting tRNAs incorrectly bound to ribosomes, especially near-cognate tRNAs whose anticodon sequences are similar to and confused with that of the proper tRNA.[56-58] If a synonymous mutation from an optimal codon (recognized by a major tRNA) to a nonoptimal codon (recognized by a minor tRNA) occurs, the collision frequency of ribosomes increases with incorrect, but near-cognate tRNAs, e.g., major isoacceptors, at the respective codon; and therefore, the level of GTP hydrolysis would also increase. The resultant loss of GTP energy and of productive ribosome working time would result in phenotypic effects such as decrease in growth rate and/or viability. This should become important as the protein production levels (and mRNA levels) increase, which would result in the obvious codon bias.[4,21]

Second, a codon choice pattern may be involved in determining the amount of protein produced from a gene. In addition to the regulatory portions of a gene such as the promoter or operator, the coding region may also be involved in determining gene expressivity. For example, there are significant numbers of *E. coli* genes in which nonoptimal codons characteristically cluster at the beginning portion of the gene.[20,59-61] This may be related to a regulatory mechanism of gene expression in which the translation efficiency in the beginning portion of a gene is involved, e.g., autogeneous repression. Ribosomal frameshifting[62] may also be related to the clustered occurrence of nonoptimal codons.

Both of these views are related in part to each other. However, we think that the first possibility is more important in establishing the general correlation between codon bias and gene expressivity, since gene expressivity can be regulated more efficiently at the transcriptional level, except in cases of translation-transcription coupled regulation. In the following section, the first possibility will be discussed from an evolutionary viewpoint. The correlation between codon bias and gene expressivity was discussed from various view points by many groups.[37-39,55,63] Dynamic aspects of protein synthesis, e.g., tRNA cycling, has also been discussed on a quantitative basis.[64-66]

B. EVOLUTIONARY VIEWPOINTS

In the previous section, a synonymous change from the optimal to nonoptimal codon is postulated to be slightly deleterious and that in the opposite direction to be slightly advantageous. The level of the fitness change should decrease as the protein production decreases (because of the decrease in amounts of mRNA). If the absolute value of fitness change is below a certain level, the mutation can be regarded as neutral,[67] even when it causes change in translation efficiency. The proportion of mutations that can be regarded as neutral or nearly neutral should be larger for genes with low gene expressivity than for genes with high expressivity. The present DNA sequence may represent an equilibrium or near equilibrium between the selective force and random drift of a neutral mutation, and F_{op} (extent of codon bias) may be an index of this balancing point.[20,21] The mathematical evaluation of codon usage and tRNA content from an evolutionary view has been published by other groups.[68,69]

We previously proposed that the evolutionary view can be examined by analyzing the rate of synonymous substitution (nucleotide substitution that does not cause amino acid replacement).[4,18] Table 2 shows that tRNA populations of *E. coli* and *S. typhimurium* are essentially the same. We can, therefore, examine the effect of this common constraint imposed by the tRNA population on the rate of synonymous substitution between these organisms. That is, we can examine whether this constraint accelerates or decelerates the substitution rate. By analyzing all pairs of protein genes thus far available for both organisms, this constraint was found to decelerate, rather than accelerate, the synonymous substitution rate; i.e., synonymous substitution in moderately or poorly expressed genes is clearly higher (e.g., by severalfold) than that of highly expressed genes.[4,18] This is consistent with the prediction of the view based on the neutral theory of evolution.[67] A similar conclusion has been reported by analyzing more than 20 pairs of enterobacterial genes.[70] In the case of synonymous substitution between taxonomically distant organisms, changes in the tRNA population during evolution should be taken into account. The effects of such change on codon usage have been discussed previously.[16,20,69]

VIII. CODON CHOICES IN GENES OF OTHER UNICELLULAR ORGANISMS AND DISTINCTION FROM HIGHER EUKARYOTES

A. UNICELLULAR ORGANISMS

Since the same conclusions concerning codon choice were drawn for the prokaryotes, *E. coli* and *S. typhimurium,* and the eukaryote, *S. cerevisiae,* these observations should be true for a wide range of organisms.[21] When codon usage in various prokaryotes was examined, the synonymous codon-choice patterns of *E. coli* were found to be fairly similar to those of other Enterobacteriaceae, but significantly different from those of *Anabaena, Bacillus, Clostridium, Halobacterium,* and *Streptococcus.*[22,23] The latter organisms are all taxonomically distant from Enterobacteriaceae. These variations should be mainly due to differences in genome G+C% and tRNA isoacceptor populations.

B. DISTINCTION FROM HIGHER EUKARYOTES

The general characteristics of codon usage of unicellular organisms are significantly different from those of higher eukaryotes, especially those of higher vertebrates.[4] When codon choice patterns of higher vertebrate genes (even those of one organism) are compared, they are often shown to be quite different, and thus, the organism-specific codon-choice pattern (codon dialect) is usually difficult to detect. For example, by extensively searching the GenBank database, we calculated codon usage in about 1900 human sequences (see Figure 8f). The highest G+C% in the third codon position was 97%, and that of about 250 sequences was 80% or more. The lowest G+C% was 27%, and that in about 150 sequences was 40% or less.[10] This evident diversity in the third codon G+C% was attributed to long range G+C% mosaic structures of the genome that are possibly related to chromosomal banding patterns.[5-10] Regarding codon choice patterns, the distinction between unicellular microorganisms and multicellular higher eukaryotes, rather than the distinction between prokaryotes and eukaryotes, is emphasized.[4] The reason for this is as follows. In unicellular organisms, most, if not all, genes are expressed in individual cells. To maintain the efficiency of translation processes, e.g., to save GTP energy used in the proofreading process, these genes have an analogous codon-choice pattern (codon dialect) that fits the tRNA population. In the case of higher eukaryotes, each organism is composed of an enormous number of cells. Many of the cells are highly differentiated, and a restricted spectrum of genes is expressed in individual cells. The target of Darwinian selection is the organism instead of individual cells. Thus, a fitness change caused by a synonymous change in an ordinary gene should be extremely smaller than that for unicellular organisms. A synonymous change to a codon with higher tRNA availability may save a certain amount of GTP energy and effective working time of ribosomes *in a restricted portion of cells*. However, the contribution of this change *in overall fitness of an organism* should be extremely small, so it would presumably be counted as a neutral mutation. This contrasts with the case of unicellular organisms where the cell itself is the direct target of selection. Codon usage in ordinary genes of multicellular organisms is therefore thought to be less stringently constrained by tRNA availability than that in genes of unicellular organisms. When some constraints from factors other than translation efficiency exist, e.g., long range G+C% mosaic structures of the genome, their codon choices presumably follow the constraints. The evident diversity of the third codon G+C% and the correlation of this diversity with the G+C% mosaic structures may reflect the weaker constraint imposed by tRNA content.[4]

REFERENCES

1. **Maaløe, O.,** Past, present and future trends, in *Control of Ribosome Synthesis (Alfred Benzon Symposium IX)*, Kjeldgaard, N. C. and Maaloe, O., Eds., Munksgaard, Copenhagen, 1976, 15.
2. **Grantham, R., Gautier, C., Gouy, M., Mercier, R., and Pave, A.,** Codon catalog usage and the genome hypothesis, *Nucl. Acids Res.*, 8, r49, 1980.
3. **Grantham, R., Gautier, C., Gouy, M., Jacobzone, M., and Mercier, R.,** Codon catalog usage is a genome strategy modulated for gene expressivity, *Nucleic Acids Res.*, 9, r43, 1981.
4. **Ikemura, T.,** Codon usage and tRNA content in unicellular and multicellular organisms, *Mol. Biol. Evol.*, 2, 13, 1985.
5. **Bernardi, G., Olofsson, B., Filipski, J., Zerial, M., Salinas, J., Cuny, G., Meunier-Rotival, M., and Rodier, F.,** The mosaic genome of warm-blooded vertebrates, *Science*, 228, 953, 1985.

6. **Bernardi, G.,** The isochore organization of the human genome, *Annu. Rev. Genet.,* 23, 637, 1989.
7. **Aota, S. and Ikemura, T.,** Diversity in G+C content at the third position of codons in vertebrate genes and its cause, *Nucleic Acids Res.,* 14, 6345, 8702 (Erratum), 1986.
8. **Ikemura, T. and Aota, S.,** Global variation in G+C content along vertebrate genome DNA: possible correlation with chromosome band structures, *J. Mol. Biol.,* 203, 1, 1988.
9. **Ikemura, T., Wada, K., and Aota, S.,** Giant G+C% mosaic structures of the human genome found by arrangement of GenBank human DNA sequences according to genetic positions, *Genomics,* 8, 207, 1990.
10. **Ikemura, T. and Wada, K.,** Evident diversity of codon usage patterns of human genes with respect to chromosome banding patterns and chromosome numbers: relation between nucleotide sequence data and cytogenetic data, *Nucleic Acids Res.,* 19, 4333, 1991.
11. **Post, L. E. and Nomura, M.,** DNA sequences from the *str* operon of *Escherichia coli, J. Biol. Chem.,* 255, 4660, 1980.
12. **Li, W.-H., Luo, C.-C., and Wu, C.-I.,** Evolution of DNA sequences, in *Molecular Evolutionary Genetics,* MacIntyre, R. J., Ed., Plenum Press, New York, 1, 1985.
13. **Sharp, P. M., Cowe, E., Higgins, D. G., Shields, D. C., Wolfe, K. H., and Wright, F.,** Codon usage patterns in *Escherichia coli, Bacillus subtilis, Saccharomyces cerevisiae, Schizosaccharomyces pombe, Drosophila melanogaster* and *Homo sapiens;* A Review of the Considerable Within-Species Diversity, *Nucleic Acid Res.,* 16, 8207, 1988.
14. **Andersson, S. G. E. and Kurland, C. G.,** Codon preferences in free-living microorganisms, *Microbiol. Rev.,* 54, 198, 1990.
15. **Osawa, S. and Jukes, T. H.,** Evolution of the genetic code as affected by anticodon content, *Trends Genet.,* 4, 191, 1988.
16. **Ikemura, T.,** Correlation between the abundance of *Escherichia coli* transfer RNAs and the occurrence of the respective codons in its protein genes, *J. Mol. Biol.,* 146, 1, 1981.
17. **Ikemura, T.,** Correlation between the abundance of yeast transfer RNAs and the occurrence of the respective codons in protein genes, *J. Mol. Biol.,* 158, 573, 1982.
18. **Ikemura, T.,** Codon usage, tRNA content, and rate of synonymous substitution, in *Population Genetics and Molecular Evolution,* Ohta, T. and Aoki, K., Eds., Japan Sci. Soc. Press, Tokyo, 1985, 385.
19. **Ikemura, T.,** The frequency of codon usage in *E. coli* genes: correlation with abundance of cognate tRNA, in *Genetics and Evolution of RNA polymerase, tRNA and Ribosomes,* Osawa, S., Ozeki, H., Uchida, H., and Yura,T., Eds., University of Tokyo Press, Tokyo, Elsevier/North Holland, Amsterdam, 1980, 519.
20. **Ikemura, T.,** Correlation between the abundance of *Escherichia coli* transfer RNAs and the occurrence of the respective codons in its protein genes: a proposal for a synonymous codon choice that is optimal for the *E. coli* translational system, *J. Mol. Biol.,* 151, 389, 1981.
21. **Ikemura, T. and Ozeki, H.,** Codon usage and transfer RNA contents: organism-specific codon-choice patterns in reference to the isoacceptor contents, *Cold Spring Harbor Symp. Quant. Biol.,* 47, 1087, 1983.
22. **Wada, K., Aota, S., Tsuchiya, R., Ishibashi, F., Gojobori, T., and Ikemura, T.,** Codon usage tabulated from the GenBank genetic sequence data, *Nucleic Acids Res.,* 18, 2367, 1990.
23. **Wada, K., Wada, Y., Doi, H., Ishibashi, F., Gojobori, T., and Ikemura, T.,** Codon usage tabulated from the GenBank genetic sequence data, *Nucleic Acids Res.,* 19, 1981, 1991.
24. **Ikemura, T. and Dahlberg, J. E.,** Small ribonucleic acids of *Escherichia coli.* I. Characterization by polyacrylamide gel electrophoresis and fingerprint analysis, *J. Biol. Chem.,* 248, 5024, 1973.
25. **Ikemura, T. and Nomura, M.,** Expression of spacer tRNA genes in ribosomal RNA transcription units carried by hybrid ColE1 plasmid in *E. coli, Cell,* 11, 779, 1977.

26. **Ikemura, T.,** Purification of RNA molecules by gel techniques, *Methods Enzymol.,* 180, 14, 1989.
27. **Komine, Y., Adachi, T., Inokuchi, H., and Ozeki, H.,** Genomic organization and physical mapping of the transfer RNA genes in *Escherichia coli* K12, *J. Mol. Biol.,* 212, 579, 1990.
28. **Varenne, S., Buc, J., Lloubes, R., and Lazdunski, C.,** Translation is a non-uniform process; effect of tRNA availability on the rate of elongation of nascent polypeptide chains., *J. Mol. Biol.,* 180, 549, 1984.
29. **Holm, L.,** Codon usage and gene expression, *Nucleic Acids Res.,* 14, 3075, 1986.
30. **Nishimura, S.,** Modified nucleosides and isoaccepting tRNA, in *Transfer RNA,* Altman, S., Ed., MIT Press, Cambridge, MA, 1978, 168.
31. **Shpaer, E. G.,** Constraints on codon context in *Escherichia coli* genes: their possible role in modulating the efficiency of translation, *J. Mol. Biol.,* 188, 555, 1986.
32. **Yarus, M. and Folley, L. S.,** Sense codons are found in specific contexts, *J. Mol. Biol.,* 182, 529, 1985.
33. **Grosjean, H., Sankoff, D., Min Jou, W., and Cedergren, R. J.,** Bacteriophage MS2 RNA: a correlation between stability of the codon: anticodon interaction and the choice of code words, *J. Mol. Evol.,* 12, 113, 1978.
34. **Grosjean, H. and Fiers, W.,** Preferential codon usage in prokaryotic genes: the optimal codon-anticodon interaction energy and the selective codon usage in efficiently expressed genes, *Gene,* 18, 199, 1982.
35. **Parker, J.,** Errors and alternatives in reading the universal genetic code, *Microbiol. Rev.,* 53, 273, 1989.
36. **VanBogelen, R. A., Hutton, M. E., and Neidhardt, F. C.,** Gene-protein database of *Escherichia coli* K-12: edition 3, *Electrophoresis,* 11, 1131, 1990.
37. **Bennetzen, J. L. and Hall, B. D.,** Codon selection in yeast, *J. Biol. Chem.,* 257, 3026, 1982.
38. **Gouy, M. and Gautier, C.,** Codon usage in bacteria: correlation with gene expressivity, *Nucleic Acids Res.,* 10, 7055, 1982.
39. **Sharp, P. M. and Li, W.-H.,** The codon adaptation index—a measure of directional synonymous codon usage bias, and its potential applications, *Nucleic Acids Res.,* 15, 1281, 1987.
40. **Robinson, M., Lilley, R., Little, S. Emtage, J. S., Yarranton, G., Stephens, P., Millican, A., Eaton, M., and Humphreys, G.,** Codon usage can affect efficiency of translation of genes in *Escherichia coli, Nucleic Acids Res.,* 12, 6663, 1984.
41. **Pedersen, S.,** *Escherichia coli* ribosomes translate in vivo with variable rate, *EMBO J.,* 3, 2895, 1984.
42. **Bonekamp, F., Andersen, H. D., Christensen, T., and Jensen, K. F.,** Codon-defined ribosomal pausing in *Escherichia coli* detected by using the *pyrE* attenuator to probe the coupling between transcription and translation, *Nucleic Acids Res.,* 13, 4113, 1985.
43. **Bonekamp, F. and Jensen, K. F.,** The AGG codon is translated slowly in *E. coli* even at very low expression levels, *Nucleic Acids Res.,* 16, 3013, 1988.
44. **Williams, D. P., Regier, D., Akiyoshi, D., Genbauffe, F., and Murphy, J. R.,** Design, synthesis and expression of a human interleukin-2 gene incorporating the codon usage bias found in highly expressed *Escherichia coli* genes, *Nucleic Acids Res.,* 16, 10453, 1988.
45. **Sørensen, M. A., Kurland, C. G., and Pedersen, S.,** Codon usage determines translation rate in *Escherichia coli, J. Mol. Biol.,* 207, 365, 1989.
46. **Nichols, B. P., Blumenberg, M., and Yanofsky, C.,** Comparison of the nucleotide sequence of *trpA* and sequences immediately beyond the Trp operon of *Klebsiella aerogenes, Salmonella typhimurium* and *E. coli, Nucleic Acids Res.,* 9, 1743, 1981.
47. **Nichols, B. P., Miozzari, G. F., vanCleemput, M., Bennet, G. N., and Yanofsky, C.,** Nucleotide sequences of the *trpG* regions of *Escherichia coli, Shigella dysenteriae, Salmonella typhimurium* and *Serratia marcescens, J. Mol. Biol.,* 142, 503, 1980.

48. **Yanofsky, C. and vanCleemput, M.**, Nucleotide sequence of *trpE* of *Salmonella typhimurium* and its homology with the corresponding sequence of *Escherichia coli*, *J. Mol. Biol.*, 155, 235, 1982.
49. **Wada, A. and Suyama, A.**, Third letters in codons counterbalanced the (G+C)–content of their first and second letters, *FEBS Lett.*, 188, 291, 1985.
50. **Bernardi, G. and Bernardi, G.**, Compositional constraints and genome evolution, *J. Mol. Evol.*, 24, 1, 1986.
51. **Muto, A. and Osawa, S.**, The guanine and cytosine content of genomic DNA and bacterial evolution, *Proc. Natl. Acad. Sci. U.S.A.*, 84, 166, 1987.
52. **Sueoka, N.**, Directional mutation pressure and neutral molecular evolution, *Proc. Natl. Acad. Sci. U.S.A.*, 85, 2653, 1988.
53. **Osawa, S., Ohama, T., Yamao, F., Muto, A., Jukes, T. H., Ozeki, H., and Umesono, K.**, Directional mutation pressure and transfer RNA in choice of the third nucleotide of synonymous two-codons sets, *Proc. Natl. Acad. Sci. U.S.A.*, 85, 1124, 1988.
54. **Filipski, J.**, Evolution of DNA sequence contribution of mutational bias and selection to the origin of chromosome compartments, in *Advances in Mutagenesis Research*, Vol. 2, Obe, G., Ed., Springer-Verlag, New York, 1990, chap. 1.
55. **Shields, D. C. and Sharp, P. M.**, Synonymous codon usage in Bacillus subtilis reflects both translational selection and mutational biases, *Nucleic Acids Res.*, 15, 8023, 1987.
56. **Hopfield, J. J.**, Kinetic proofreading: a new mechanism for reducing errors in biosynthetic processes requiring high specificity, *Proc. Natl. Acad. Sci. U.S.A.*, 71, 4135, 1974.
57. **Thompson, R. C. and Stone, P. J.**, Proofreading of the codon-anticodon interaction on ribosomes, *Proc. Natl. Acad. Sci. U.S.A.*, 74, 198, 1977.
58. **Thompson, R. C., Dix, D. B., Gerson, R. B., and Karim, A. M.**, A GTPase reaction accompanying the rejection of Leu-tRNA$_2$ by UUU-programmed ribosomes, *J. Biol. Chem.*, 256, 81, 1981.
59. **Nomura, M., Yates, J. L., Dean, D., and Post, L. E.**, Feedback regulation of ribosomal protein gene expression in *Escherichia coli*: structural homology of ribosomal RNA and ribosomal protein mRNA, *Proc. Natl. Acad. Sci. U.S.A.*, 77, 7084, 1980.
60. **Burns, D. M. and Beacham, I. R.**, Rare codons in *E. coli* and *S. typhimurium* signal sequences, *FEBS Lett.*, 189, 318, 1985.
61. **Chen, G. T. and Inouye, M.**, Suppression of the negative effect of minor arginine codons on gene expression; preferential usage of minor codons within the first 25 codons of the *Escherichia coli* genes, *Nucleic Acids Res.*, 18, 1465, 1990.
62. **Weiss, R. B., Dunn, D. M., Atkins, J. F., and Gesteland, R. F.**, Ribosomal frameshifting from -2 to +50 nucleotides, *Prog. Nucl. Acid Res. Mol. Biol.*, 39, 159, 1990.
63. **Dix, D. B. and Thompson, R. C.**, Codon choice and gene expression: synonymous codons differ in translational accuracy, *Proc. Natl. Acad. Sci. U.S.A.*, 86, 6888, 1989.
64. **von Heijne, G. and Blomberg, C.**, The concentration dependence of the error frequencies and some related quantities in protein synthesis, *J. Theor. Biol.*, 78, 113, 1979.
65. **Gouy, M. and Grantham, R.**, Polypeptide elongation and tRNA cycling in *Escherichia coli*: a dynamic approach, *FEBS Lett.*, 115, 151, 1980.
66. **Varenne, S. and Lazdunski, C.**, Effect of distribution of unfavorable codons on the maximum rate of gene expression by an heterologous organism, *J. Theor. Biol.*, 120, 99, 1986.
67. **Kimura, M.**, *The Neutral Theory of Molecular Evolution*, Cambridge University Press, Cambridge, 1983.
68. **Bulmer, M.**, Coevolution of codon usage and transfer RNA abundance, *Nature (London)*, 325, 728, 1987.
69. **Shields, D. C.**, Switches in species-specific codon preferences: the influence of mutation biases, *J. Mol. Evol.*, 31, 71, 1990.
70. **Sharp, P. M. and Li, W.-H.**, The rate of synonymous substitution in enterobacterial genes is inversely related to codon usage bias, *Mol. Biol. Evol.*, 4, 222, 1987.

Chapter 4

AMINOACYL-tRNA (ANTICODON):CODON ADAPTATION IN HIGHER EUKARYOTES

Dolph L. Hatfield, Jae-Eon Jung, Byeong J. Lee, and In Soon Choi

TABLE OF CONTENTS

I. Introduction 114

II. Adaptation of tRNA Content to the Requirements of
    Protein Synthesis 114

III. Aminoacyl-tRNA: Protein Amino Acid Compositions 115

IV. Aminoacyl-tRNA (Anticodon):Codon Compositions 117

V. Possible Mechanisms for Determining Cellular Amounts of
   Isoacceptors 120

References 122

I. INTRODUCTION

It has been recognized for some time that the amounts of aminoacyl-tRNAs in microorganisms (see Chapter 3 for review), in cellular organelles (see Chapter 5 for review), and in cells and tissues that are specialized in making large quantities of a single protein or a few proteins of higher eukaryotes are adapted to the requirements of protein synthesis. This phenomenon has been described in a variety of ways including

1. The correlation between aminoacyl-tRNA pool and amino acid composition of proteins[1]
2. The (functional) adaptation of tRNA population to protein synthesis[2] or to codon usage (see Chapter 5 and Reference 3)
3. The specialization of tRNA for protein synthesis[4]
4. The correlation between codon usage and tRNA population or tRNA content (see Chapter 3)
5. The adaptation of the isoacceptor tRNA concentration to mRNA codon frequency[5]
6. Aminoacyl-tRNA(anticodon):codon adaptation[6]

The purpose of the present chapter is to examine this phenomenon in the cells and tissues of higher eukaryotes. Since this phenomenon has been most extensively studied in the red blood cells of higher vertebrates and in the silk gland of the silkworm moth, *Bombyx mori*, we will discuss these two systems in greatest detail. Adaptation of the amounts of individual species of tRNA (or the tRNA content) to protein synthesis in microorganisms and in cellular organelles is reviewed elsewhere in this book (see Chapters 3 and 5, respectively).

II. ADAPTATION OF tRNA CONTENT TO THE REQUIREMENTS OF PROTEIN SYNTHESIS

The adaptation of the cellular tRNA population to its requirements for protein synthesis has been examined at two different cellular levels. One level involves determination of the relative amounts of individual aminoacyl-tRNAs within a tRNA population and comparing these values to the amounts of the corresponding amino acids within a major protein, e.g., within hemoglobin in reticulocytes,[4,7] or major proteins, e.g., within the lactoproteins of the mammary gland,[8] synthesized in the same cell. Such a study may be described as an examination of the aminoacyl-tRNA:protein amino acid compositions within a particular cell. The second level of examining the adaptation of the tRNA population to its requirements for protein synthesis involves the determination of the relative amounts of individual isoacceptors and/or their gene copy number for decoding synonymous codons and comparing these values to the frequency of the cognate codon usage within the corresponding mRNA(s) or gene(s) encoding the protein(s) from the same cell (see Chapters 3 and 5 for review in microorganisms and cellular organelles, respectively, and References 5 and 6 for studies in higher eukaryotes). The latter study, which provides the more appropriate molecular comparison, may be described as an examination of the aminoacyl-tRNA(anticodon): codon compositions within a particular cell or cellular organelle. Studies involving the adaptation of tRNA for protein biosynthesis at both these levels are described below.

III. AMINOACYL-tRNA:PROTEIN AMINO ACID COMPOSITIONS

The earliest studies which examined the aminoacyl-tRNA content of cells in relation to the amounts of the corresponding amino acids in protein(s) were carried out with *E. coli* cells,[1] with the internal cortex of calf lens,[9] with the silk gland of *Bombyx mori*[10] and with mammalian reticulocytes.[4,7,11,12] Each of these cell types in higher eukaryotes is specialized in making large quantities of a single protein or only a few proteins. Interestingly, a correlation was observed between the amounts of certain aminoacyl-tRNAs and the amounts of the corresponding amino acids in the major protein or proteins synthesized within the same cell type. For example, a highly significant correlation was observed between the relative amounts of the individual aminoacyl-tRNAs within the rabbit reticulocyte tRNA population and the relative amounts of the corresponding amino acids within the α and β chains of rabbit globin[4,7] (see Figure 1). The percent of aminoacyl-tRNAs are plotted along the ordinate and the percent of amino acid residues in hemoglobin are plotted along the abscissa. The correlation coefficient (*r* value) between these two parameters was found to be 0.88. However, it should be noted that methionine and leucine tRNAs were not considered in the determination of this correlation coefficient. The level of methionine tRNA in reticulocytes was in excess of the amount of methionine present in rabbit hemoglobin, while that of leucine tRNA was less than the amount of leucine present in hemoglobin. Smith has proposed that these two aminoacyl-tRNAs may have an important role in controlling the rate of globin synthesis[7] (also see following). In any case, the highly significant correlation coefficient of 0.88 strongly suggests that the tRNA population in rabbit reticulocytes, at least for most aminoacyl-tRNAs, is adapted to its requirements for hemoglobin synthesis.[7]

The tRNA population in sheep reticulocytes was found to change developmentally as the type of hemoglobin synthesized by the cell changed. Furthermore, a correlation was observed between the changes in specific isoacceptors and the number of the corresponding amino acid residues in the different globin types.[12] That is, hemoglobin B contains eight methionine and no isoleucine residues, while hemoglobin C contains two methionine and two isoleucine residues. The isoleucine tRNA levels were enriched two- to threefold in

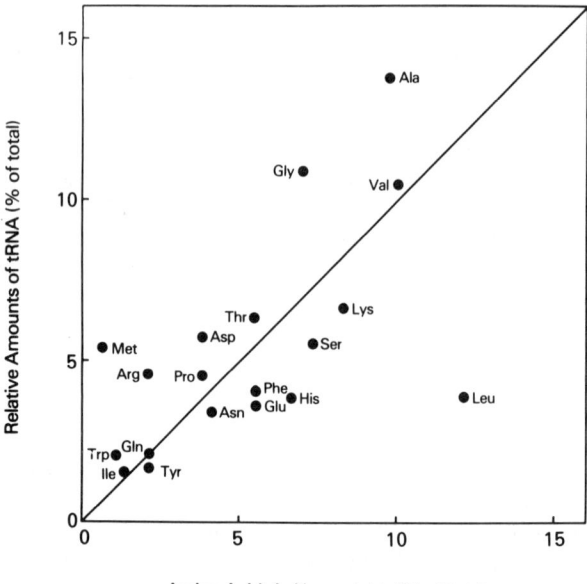

FIGURE 1. Relation between the relative amounts of aminoacyl-tRNAs in rabbit reticulocytes and the amino acid composition of rabbit hemoglobin. The percent of each aminoacyl-tRNA within the total tRNA population (with the exception of cysteine tRNA) of rabbit reticulocytes is plotted along the ordinate and the percent of each amino acid within hemoglobin along the abscissa.[4,7]

reticulocytes synthesizing hemoglobin C as compared to the corresponding cells synthesizing hemoglobin B. Likewise, one of the methionine isoacceptors was substantially reduced in amounts in reticulocytes synthesizing hemoglobin C.

Specific tRNAs involved in the synthesis of silk proteins in *Bombyx mori* have been shown to be induced as the silk gland expresses the corresponding proteins.[2] Silk consists of two fibrous proteins. One is fibroin which is synthesized in the posterior portion of the silk gland and the other is sericin which is synthesized in the middle portion of the gland. More than 90% of the amino acid composition of fibroin is glycine (46%), alanine (29%), serine (12%) and tyrosine (6%), while more than 75% of the amino acid composition of sericin is serine (37%), aspartic acid (17%), glycine (16%) and glutamic acid (7%).[2] A strong correlation has been shown to exist between the increase in amounts of the various specific aminoacyl-tRNAs required for expression of fibroin and sericin (in the respective parts of the gland) and in relative amounts of these silk proteins as they reach maximal expression and secretion.[2]

As previously noted, the studies described in this section have shown that the cellular tRNA population is adapted to the immediate requirements of protein synthesis. This was demonstrated by determining the relative amounts of aminoacyl-tRNAs and comparing these levels to those of the cognate amino acids within protein. Most of these studies have focused on cells which are specialized in making large quantities of one or only a few proteins. It should also be noted that a correlation exists between these two parameters in higher eukaryotic cells making a large number of many different kinds of proteins, e.g., in mammalian liver.[2] The more appropriate comparison, however, is between the levels of individual isoacceptors in cells and the frequency of usage of the corresponding cognate codons (see following).

IV. AMINOACYL-tRNA(ANTICODON):CODON COMPOSITIONS

The most extensively studied cell types in higher eukaryotes with respect to comparing the isoacceptor levels and the cognate codon frequencies within mRNA(s) are the silk gland of *Bombyx mori*[3,5,13] and the reticulocytes of rabbits[6,14,15] and humans.[6,16] Before considering the studies involving *Bombyx mori* and reticulocytes, it should also be noted that the relative amounts of the isoacceptors for glycine and alanine tRNAs and their cognate codon frequencies in collagen mRNA have been examined in collagenous and noncollagenous chicken tissues.[17] Interestingly, a correlation was observed between the amounts of these isoacceptors in tissues which synthesize collagen, but not in tissues which do not synthesize collagen, and the frequencies of their cognate codons in collagen mRNA. In another study, the frequencies of synonymous codon usage in a number of vertebrate genes from muscle and liver have been compared.[18] The fact that no significant differences in the patterns of synonymous codon usage were observed in this study suggested that the isoacceptor tRNA (anticodon) levels in vertebrate tissues may be adapted to codon usage.

As noted, fibroin is synthesized in the posterior portion of the silk gland of *Bombyx mori*. Glycine comprises 46% of the amino acid residues in fibroin and is coded principally by GGU and GGA, while alanine comprises 29% of the amino acid residues and is coded for the most part by GCU.[19] The relative amounts of the corresponding glycine and alanine isoacceptors in the posterior silk gland of *Bombyx mori* during the period of fibroin secretion were found to be correlated closely to the frequency of their cognate codons in fibroin mRNA.[3,5,13] Two alanine tRNAs have been isolated and sequenced from the posterior silk gland.[20-22] They have identical sequences with the exception of one nucleotide in the anticodon stem which is either Ψ or C. The species with Ψ in the anticodon stem appears to be specific to the silk gland and its level is dramatically increased during the period of fibroin secretion. Similarly, a tissue specific tRNA^{Ala} has been observed to accumulate in large amounts in the silk gland of the spider, *Nephila clavipes*, when the gland is actively making fibroin.[23] The interesting feature of the latter system is that the gland can be cultured *in vitro* and its biosynthetic activity stimulated. These findings may well provide an experimental system for understanding the underlying mechanisms responsible for adapting the tRNA isoacceptor levels to their requirements for protein synthesis (see following).

The tRNA populations in rabbit[4,6,7,14,15] and in human[6,16] reticulocytes have also been studied in detail. Aminoacyl-tRNAs were fractionated on reverse phase columns from the two tissues and the amounts and distributions of the various isoacceptors examined.[14-16] The relative amounts of isoacceptors were found to vary widely in reticulocytes from those which were highly enriched to others which were not detected or were present in very low levels. Comparison of the elution profiles of aminoacyl-tRNAs from rabbit reticulocytes to the corresponding aminoacyl-tRNA profiles from rabbit liver also manifested numerous differences in the amounts and distributions of isoacceptors in the two tissues.[14,15] These studies in conjunction with earlier studies showing a strong correlation between the amounts of aminoacyl-tRNAs and the corresponding amino acids in hemoglobin in rabbit reticulocytes[4,7] suggested that the amounts of the individual isoacceptors in reticulocytes may also be adapted to their requirements for the synthesis of hemoglobin. Determination of the codon recognition properties of asparagyl- and histidyl-tRNAs[14] and of the other 18 aminoacyl-tRNAs fractionated from rabbit reticulocytes, as well as those from rabbit liver,[15] confirmed this speculation. Similarly, comparison of all 20 aminoacyl-tRNAs from human reticulocytes to those of rabbit reticulocytes by reverse phase chromatography permitted codon assignments to most of the fractionated human reticulocyte isoacceptors[6,16] by analogy to the codon assignments determined for the corresponding isoacceptors in rabbit reticulocytes.[15] In addition, the coding properties of several other isoacceptors were also determined.[6,16] The relative amounts of many isoacceptors in human and rabbit reticulocytes are compared to the frequency of their cognate codons in the corresponding globin mRNA in Figure 2. The percent of isoacceptors to which codon assignments were made are plotted along the ordinate and the percent of their cognate codons in the respective globin mRNAs are plotted along the abscissa for human (upper graph) and for rabbit (lower graph) reticulocytes. The correlation coefficient (r value) was determined between these two parameters and found to be 0.78 for human and 0.88 for rabbit reticulocytes. The p value was less than 0.001 for each comparison. These comparisons show a highly significant correlation between two parameters which are nonrandom with respect to the codons used and with regard to the relative amounts of isoacceptors that translate these codons. Thus, the tRNA isoacceptor population in human and rabbit reticulocytes is adapted to the cognate codon frequency in the corresponding globin mRNAs.

Clearly, some isoacceptors are more abundant in human and rabbit reticulocytes than is required for translating their cognate codons in globin, while others are less abundant for this function. For example, the glycine isoacceptor which decodes GGU/C and the arginine isoacceptor which decodes CGU/C/A are present in greater amounts in both human and rabbit reticulocytes than is expected for translating their cognate codons in the respective globin mRNAs (see Figure 2).[6] Even though these isoacceptors may be used extensively in synthesis of proteins other than globin, the amounts of these isoacceptors in reticulocytes seem excessive even for these requirements. They may possibly, therefore, have a function or functions in reticulocytes other than that of

Aminoacyl-tRNA (Anticodon):Codon Adaptation

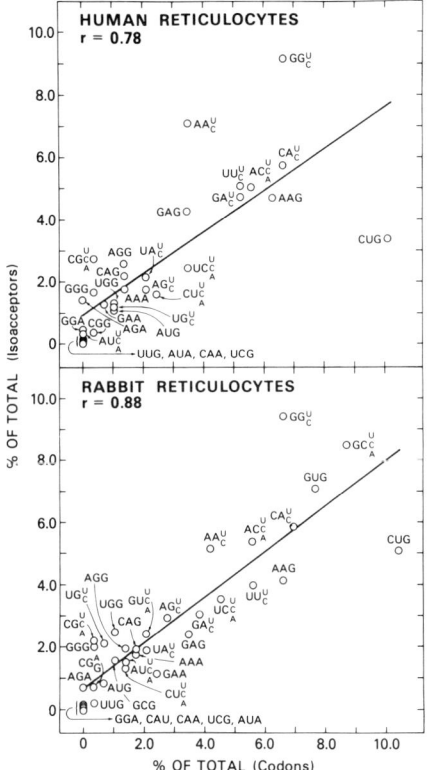

FIGURE 2. Linear correlation between the relative amounts of isoacceptor tRNAs from human and rabbit reticulocytes and codon usage in the corresponding α and β globin mRNAs. The frequencies of isoacceptor aminoacyl-tRNAs in human (upper graph) and rabbit (lower graph) reticulocytes, expressed as a percent of the total amino acid acceptance, are plotted along the ordinate, while the frequencies of cognate codon usage in the corresponding α and β globin mRNAs, expressed as a percent of the total codons, are plotted along the abscissa. (From Hatfield, D. and Rice, M., *Biochem. Int.*, 13, 835, 1986. With permission.)

translating their cognate codons. Another possibility is that such isoacceptors may be required in higher amounts than many other isoacceptors to satisfy their demands for translating cognate codons in protein synthesis.

Isoacceptors which occur in reticulocytes in lower quantities than expected for optimal synthesis of globin present another problem than those isoacceptors which occur in higher levels. The most obvious tRNA species in human and rabbit reticulocytes which occurs in lower amounts than expected is the leucine isoacceptor which decodes CUG (see Figure 2). It might be concluded that since this isoacceptor occurs in limiting amounts, it may function to regulate the overall rate of protein synthesis. However, a question may be raised as why are the amounts of virtually all isoacceptors so carefully adjusted to the demands of protein synthesis when only one species needs to occur in limiting amounts to become the regulator of all the others. It may be that since the leucyl-tRNA which decodes CUG is the most abundant leucine isoacceptor in reticulocytes and CUG is the most frequently used codon in globin, then this isoacceptor may occur in sufficient levels for decoding CUG codons at maximal rates. Thus, some isoacceptors may be required in lesser amounts than expected, e.g., tRNA^{Leu} that decodes CUG, while others may be required in greater amounts than expected, e.g., tRNA^{Gly} that decodes GGU/C and tRNA^{Arg} that decodes CGU/C/A (see previous discussion), to optimally translate their cognate codons.

V. POSSIBLE MECHANISMS FOR DETERMINING CELLULAR AMOUNTS OF ISOACCEPTORS

Clearly, the aminoacyl-tRNA population in cells is adapted to its requirements for protein synthesis. This is true, not only in microorganisms and cellular organelles, but in specific cells and tissues of higher eukaryotes specialized in making large quantities of one or a few proteins. Furthermore, as noted previously, this observation might well be true in cells and tissues of higher eukaryotes involved in making many different kinds of proteins[2,18] as is known to occur in microorganisms. A question arises then as to how a cell can adjust its tRNA population to fit the demands for protein synthesis when different cells and tissues have different requirements for protein synthesis. This question has been addressed in detail in Chapter 3 of this book in relation to the adaptation of the tRNA population to synonymous codon usage in microorganisms. Whether similar mechanisms as those proposed in microorganisms occur in cells and tissues of higher eukaroytes remain to be established. It should be noted, however, that the possible means by which the amounts of aminoacyl-tRNAs are determined in higher eukaryotes has been addressed experimentally. For example, Litt and Howell-Litt[24] examined the levels of certain tRNAs in Friend leukemia cells and found that differential degradation rates of aminoacyl-tRNAs and the corresponding deacylated tRNAs play a role in determining their levels in these cells. On the other hand, Chevallier and Garel[25] reported that some tRNAs in the silk glands of *Bombyx mori* have the same rates of degradation, but different rates of accumulation. Thus, these silk gland tRNAs may be under transcriptional control.

The tRNA population in Friend leukemia cells has been examined before and after induction of erythroid differentiation.[26] The tRNA population increased about twofold after induction, but the increase did not result in an enrichment of specific tRNAs required for hemoglobin synthesis as is known to occur in normal development of erythroid cells.[4,6,7,12] It would appear, therefore, that the tRNA population in Friend leukemia cells is established and is not altered to meet the demands of synthesis of a specific protein. Prolactin, which is known to induce lactation in the mammary gland, was found to induce the tRNA population two- to threefold in mouse mammary explants.[27] The tRNA enrichment, however, did not appear to be selective for specific tRNAs, but rather to be uniform for the entire tRNA population. It would seem, therefore, that the amounts of specific isoacceptors in the tRNA population of the mammary gland, which are adapted to the requirements for casein synthesis,[8] are determined as the gland develops and prepares for lactation.

Zasloff[28] has proposed that the transport of mRNA and tRNA from the nucleus to the cytoplasm is coupled. This is indeed a most intriguing possibility to account for the aminoacyl-tRNA(anticodon):codon adaptation in eukaryotic cells, since the levels of tRNAs transported to the cytoplasm would be expected to correlate with the frequency of their cognate codons. Even though the studies discussed above, which show a relative increase in the entire level of the tRNA population in Friend leukemia cells during erythroid differentiation[26] and in mammary cells during prolactin treatment,[27] do not appear to support this proposal, it is most certainly worthy of further consideration and experimentation.

The possible mechanisms discussed above which may account for the various levels of specific isoacceptors within the tRNA population of higher eukaryotic cells and tissues have dealt largely with the adjustment of the tRNA population to cellular regulatory mechanisms invoked by a particular cell or tissue. We have not considered the adaptation of the tRNA population to its requirements for protein synthesis from an evolutionary standpoint as is discussed in detail in Chapter 3 of this book. It would seem that the maximal amount of any given isoacceptor synthesized in a cell would be determined by gene copy number, promoter efficiency, availability of RNA polymerase III, cofactors, and/or other regulatory elements or proteins as is discussed in Chapter 6 of this book. The availability or amounts of each of these gene and regulatory elements may largely be established by evolutionary mechanisms. Thus, the actual amounts of tRNA isoacceptors present in any given cell would then be determined by the demands of protein synthesis which would then invoke another set of regulatory mechanisms within that particular cell. Whatever the mechanism or mechanisms that determine the different amounts of tRNA isoacceptors in cells and tissues of higher eukaryotes, it is extremely important to understand these mechanisms as they, of course, must account for the means by which the relative amounts of so many isoacceptors are adapted to their requirements for translating cognate codons.

REFERENCES

1. **Yamane, T.,** Correlation between aminoacyl-sRNA pool and amino acid composition of proteins, *J. Mol. Biol.,* 14, 616, 1965.
2. **Garel, J.-P.,** Functional adaptation of tRNA population, *J. Theor. Biol.,* 43, 211, 1974.
3. **Garel, J.-P.,** Quantitative adaptation of isoacceptor tRNAs to mRNA codons of alanine, glycine and serine, *Nature (London),* 260, 805, 1976.
4. **Smith, D. W. E. and McNamara, A. L.,** Specialization of rabbit reticulocyte transfer RNA content for hemoglobin synthesis, *Science,* 171, 577, 1971.

5. **Chavancy, G., Chevallier, A., Fournier, A., and Garel, J.-P.**, Adaptation of iso-tRNA concentration to mRNA codon frequency in the eukaryote cell, *Biochimie*, 61, 71, 1979.
6. **Hatfield, D. and Rice, M.**, Aminoacyl-tRNA(anticodon):codon adaptation in human and rabbit reticulocytes, *Biochem. Int.*, 13, 835, 1986.
7. **Smith, D. W. E.**, Reticulocyte transfer RNA and hemoglobin synthesis, *Science*, 190, 529, 1975.
8. **Hentzen, D.**, Analysis of the transfer RNA population of mouse mammary glands infected with a latent mammary tumor virus, *Cancer Res.*, 36, 3082, 1976.
9. **Garel, J.-P., Virmaux, N., and Mandel, P.**, Adaptation fonctionelle des tRNA a la biosynthese protéique dans un systeme cellulaire hautement différencié, *Bull. Soc. Chimie Biol.*, 52, 987, 1970.
10. **Garel, J.-P. and Mandel, P.**, Functional adaptation of tRNAs to fibroin biosynthesis in the silkgland of *Bombyx môri* L, *FEBS Lett.*, 7, 327, 1970.
11. **Yang, W.-K.**, Isoaccepting transfer RNA's in mammalian differentiated cells and tumor tissues, *Cancer Res.*, 31, 639, 1971.
12. **Litt, M. and Kabat, D.**, Studies of transfer ribonucleic acids and hemoglobin synthesis in sheep reticulocytes, *J. Biol. Chem.*, 237, 6659, 1972.
13. **Garel, J.-P., Hentzen, D., and Daillie, J.**, Codon responses of tRNA^{Ala}, tRNA^{Gly} and tRNA^{Ser} from the posterior part of the silkgland of *Bombyx mori* L, *FEBS Lett.*, 39, 359, 1974.
14. **Smith, D. W. E., Meltzer, V. N., and McNamara, A. L.**, A comparison of rabbit liver and reticulocyte transfer RNA: evidence of unique species in reticulocytes, *Biochim. Biophys. Acta*, 349, 366, 1974.
15. **Hatfield, D., Matthews, C. R., and Rice, M.**, Aminoacyl-transfer RNA populations in mammalian cells: chromatographic profiles and patterns of codon recognition, *Biochim. Biophys. Acta*, 564, 414, 1979.
16. **Hatfield, D., Varricchio, F., Rice, M., and Forget, B. G.**, The aminoacyl-tRNA population of human reticulocytes, *J. Biol. Chem.*, 257, 3183, 1982.
17. **Quenzar, B., Agoutin, B., Reinisch, F., Weill, D., Perin, F., Keith, G., and Heyman, T.**, Distribution of isoaccepting tRNAs and codons for proline and glycine in collagenous and noncollagenous chicken tissues, *Biochem. Biophys. Res. Commun.*, 150, 148, 1988.
18. **Hastings, K. E. M. and Emerson, C. P.**, Codon usage in muscle genes and liver genes, *J. Mol. Evol.*, 19, 214, 1983.
19. **Suzuki, Y. and Brown, D. D.**, Isolation and identification of the messenger RNA for silk fibroin from *Bombyx mori*, *J. Mol. Biol.*, 63, 409, 1972.
20. **Meza, L., Araya, A., Leon, G., Krauskopf, M., Siddiqui, M. A. Q., and Garel, J.-P.**, Specific alanine-tRNA species associated with fibroin biosynthesis in the posterior silkgland of *Bombyx mori* L, *FEBS Lett.*, 77, 255, 1977.
21. **Garel, J.-P., Garber, R. L., and Siddiqui, M. A. Q.**, Transfer RNA in posterior silk gland of *Bombyx mori*: polyacrylamide gel mapping of mature transfer RNA, identification and partial structural characterization of major isoacceptor species, *Biochemistry*, 16, 3618, 1977.
22. **Sprague, K. U., Hagenbüchle, O., and Zuniga, M. C.**, The nucleotide sequence of two silk gland alanine tRNAs: implications for fibroin synthesis and for initiator tRNA structure, *Cell*, 11, 561, 1977.
23. **Candelas, G. C., Arroyo, G., Carrasco, C., and Dompenciel, R.**, Spider silkglands contain a tissue-specific alanine tRNA that accumulates *in vitro* in response to the stimulus for silk protein synthesis, *Dev. Biol.*, 140, 215, 1990.
24. **Litt, M. and Howell-Litt, R.**, Control of specific transfer RNA concentrations in amino acid-deprived Friend leukemia cells operates at the level of RNA degradation, *J. Biol. Chem.*, 255, 375, 1980.
25. **Chevallier, A. and Garel, J.-P.**, Differential synthesis rates of tRNA species in the silk gland of *Bombyx mori* are required to promote tRNA adaptation to silk messages, *Eur. J. Biochem.*, 124, 477, 1982.

26. **Weil, S., Hirata, R., McNamara, A., and Smith, D. W. E.,** Changes in tRNA levels during the induction of hemoglobin synthesis in Friend leukemia cells, *Biochem. Biophys. Res. Commun.,* 120, 707, 1984.
27. **Green, M. R., Hatfield, D., Miller, M. J., and Peacock, A. C.,** Prolactin homogeneously induces the tRNA population of mouse mammary explants, *Biochem. Biophys. Res. Commun.,* 129, 233, 1985.
28. **Zasloff, M.,** tRNA transport from the nucleus in a eukaryotic cell: carrier mediated translocation process, *Proc. Natl. Acad. Sci. U.S.A.,* 80, 6436, 1983.

Chapter 5

ADAPTATION OF tRNA POPULATION TO CODON USAGE IN CELLULAR ORGANELLES

Laurence Maréchal-Drouard, André Dietrich, and Jacques H. Weil

TABLE OF CONTENTS

I.	Number of Mitochondrial or Chloroplast tRNAs in Different Organisms	126
II.	Genetic Code	130
III.	Evolution of the Genetic Code: the Codon Capture Hypothesis	131
IV.	Anticodon Content and Codon Recognition Mechanism	132
	A. Codon Recognition in Yeast	133
	B. Codon Recognition in Chloroplasts	134
	C. Codon Recognition in Mammalian Mitochondria	135
V.	Levels of Isoaccepting tRNAs and Codon Usage in Chloroplasts and Mitochondria	135
VI.	Mitochondrial tRNAs of Nematodes	137
VII.	tRNA Species in *Mycoplasma capricolum*	138
	References	139

I. NUMBER OF MITOCHONDRIAL OR CHLOROPLAST tRNAs IN DIFFERENT ORGANISMS

It is striking that the number of tRNA species present in mitochondria from fungi,[1] mammals,[2] amphibians,[3] echinoderms,[4] or nematodes[5] is within an average of 22 to 24 (see Table 1). This is less than the minimum of 32 tRNAs needed to read all the 61 sense codons of the universal genetic code using the codon-anticodon pairing rules proposed by Crick.[6]

In agreement with the endosymbiotic hypothesis,[7] it can be speculated that the mitochondrial (mt) genetic codes of nonplant organisms have evolved from the code of the eubacterial endosymbiont which was the progenitor of mitochondria and represent an evolutionary simplification in which a minimum number of tRNAs (corresponding to 22 anticodons in the vertebrate mt code for instance) have been conserved to allow functional efficiency. These minimal codes are probably the result of AT or GC pressure and genomic economization.[8]

In the case of chloroplasts, 30 tRNAs are used for translation. This number is higher than that in nonplant mitochondria, but smaller than that in bacteria, as most of the CNN anticodons present in the ancestral eubacterial code have disappeared during evolution of the chloroplast (cp) code, presumably under AT pressure.

The evolutionary pathway of plant mitochondria has been different from that of animal and fungal mitochondria: on the one hand plant mt genomes have inserted "promiscuous" cp DNA sequences containing tRNA genes and, on the other hand, plant mitochondria import tRNAs from the cytosol. Therefore, the plant mt tRNA population is a mosaic of species transcribed from nuclear (nu), cp, and mt genes.[9,10] Because of these different genetic origins, the number of plant mt tRNAs has not yet been accurately determined but, according to the results we have obtained with potato mitochondria,[9] this number should be around 35. Plant mitochondria appear therefore to contain more tRNAs than mitochondria of other organisms (see previous discussion). We have shown that the potato mt genome contains at least 15 "native" and 5 "chloroplast-like" tRNA genes and that 11 potato mt tRNAs are coded for by the nu genome and imported from the cytosol. The presence of "chloroplast-like" tRNAs and the import of tRNAs from the cytosol seem to be two general features of plant mitochondria, but differences in the chloroplast-like and imported tRNA species appear among plants, especially between monocots and dicots (Figure 1),[9-11] suggesting that the evolutionary flux is still in progress.

The tRNA population of animal and fungal mitochondria is usually different from that of the nucleus. However, import of nu-encoded tRNAs must occur in protozoan mitochondria, as *Tetrahymena* and *Paramecium* mt genomes contain only eight and three tRNA genes, respectively.[12,13] The import of a tRNA^{Lys} of nu origin into yeast mitochondria has also been reported,[14] al-

TABLE 1
Anticodon Content in Genetic Codes of Vertebrate, Yeast, Nematode, and Plant Mitochondria and of Chloroplasts

Universal Code		Anticodon					
Codon	Amino acid	Minimal code	Mito vertebrate	Mito yeast	Mito nematode	Mito plant	Chloro
UUU	Phe	GAA	GAA	GAA	GAA	GAA	GAA
UUC	Phe	GAA	GAA	GAA	GAA	GAA	GAA
UUA	Leu	UAA	U*AA	U*AA	TAA	C*AA	C*/U*AA(a)
UUG	Leu	UAA	U*AA	U*AA	TAA	C*AA	C*/U*AA(a)
CUU	Leu	UAG	UAG	*UAG* Thr	TAG	NAG	UAG*
CUC	Leu	UAG	UAG	*UAG* Thr	TAG	NAG	UAG*
CUA	Leu	UAG	UAG	*UAG* Thr	TAG	NAG	UAG*
CUG	Leu	UAG	UAG	*UAG* Thr	TAG	NAG	UAG*
AUU	Ile	GAU	GAU	GAU	GAT	GAU	GAU
AUC	Ile	GAU	GAU	GAU	GAT	GAU	GAU
AUA	Ile	*UAU* Met	*CAU* Met	*CAU* Met	*CAT* Met	L*AU	C*AU
AUG	Met	UAU	CAU	CAU	CAU	CAU	CAU
GUU	Val	UAC	UAC	TAC	TAC	?	GAC
GUC	Val	UAC	UAC	TAC	TAC	?	GAC
GUA	Val	UAC	UAC	TAC	TAC	?	U*AC
GUG	Val	UAC	UAC	TAC	TAC	?	U*AC
UCU	Ser	UGA	TGA	UGA	TGA	GGA	GGA
UCC	Ser	UGA	TGA	UGA	TGA	GGA	GGA
UCA	Ser	UGA	TGA	UGA	TGA	TGA	TGA
UCG	Ser	UGA	TGA	UGA	TGA	TGA	TGA
CCU	Pro	UGG	TGG	UGG	TGG	UGG	U*GG
CCC	Pro	UGG	TGG	UGG	TGG	UGG	U*GG
CCA	Pro	UGG	TGG	UGG	TGG	UGG	U*GG
CCG	Pro	UGG	TGG	UGG	TGG	UGG	U*GG
ACU	Thr	UGU	UGU	TGT	TGT	?	GGU
ACC	Thr	UGU	UGU	TGT	TGT	?	GGU
ACA	Thr	UGU	UGU	TGT	TGT	?	TGT
ACG	Thr	UGU	UGU	TGT	TGT	?	TGT
GCU	Ala	UGC	TGC	TGC	TGC	IGC	U*GC
GCC	Ala	UGC	TGC	TGC	TGC	IGC	U*GC
GCA	Ala	UGC	TGC	TGC	TGC	IGC	U*GC
GCG	Ala	UGC	TGC	TGC	TGC	IGC	U*GC
UAU	Tyr	GUA	GTA	GUA	GTA	NUA	GUA
UAC	Tyr	GUA	GTA	GUA	GTA	NUA	GUA
UAA	–	–	–	–	–	–	–
UAG	–	–	–	–	–	–	–

TABLE 1 (continued)
Anticodon Content in Genetic Codes of Vertebrate, Yeast, Nematode, and Plant Mitochondria and of Chloroplasts

Universal Code		Anticodon					
Codon	Amino acid	Minimal code	Mito vertebrate	Mito yeast	Mito nematode	Mito plant	Chloro
CAU	His	GUG	GTG	GUG	GTG	GTG	GTG
CAC	His	GUG	GTG	GUG	GTG	GTG	GTG
CAA	Gln	UUG	TTG	TTG	TTG	TTG	U*UG
CAG	Gln	UUG	TTG	TTG	TTG	TTG	U*UG
AAU	Asn	GUU	GTT	GTT	GTT	GTT	GTT
AAC	Asn	GUU	GTT	GTT	GTT	GTT	GTT
AAA	Lys	UUU	U*UU	U*UU	TTT	TTT	TTT
AAG	Lys	UUU	U*UU	U*UU	TTT	TTT	TTT
GAU	Asp	GUC	GUC	GUC	GTC	GTC	GTC
GAC	Asp	GUC	GUC	GUC	GTC	GTC	GTC
GAA	Glu	UUC	U*UC	U*UC	TTC	TTC	UU*C
GAG	Glu	UUC	U*UC	U*UC	TTC	TTC	UU*C
UGU	Cys	GCA	GCA	GCA	GCA	GCA	GCA
UGC	Cys	GCA	GCA	GCA	GCA	GCA	GCA
UGA	–	*UCA Trp*	*U*CA Trp*	*U*CA Trp*	*TCA Trp*	–	–
UGG	Trp	UCA	U*CA	U*CA	TCA	CCA	CCA
CGU	Arg	UCG	UCG	ACG	ACG	ICG	ICG
CGC	Arg	UCG	UCG	ACG	ACG	ICG	ICG
CGA	Arg	UCG	UCG	ACG	ACG	ICG	ICG
CGG	Arg	UCG	UCG	ACG	ACG	ICG	ICG
AGU	Ser	GCU	GCU	GCU	TCT	GCT	GCT
AGC	Ser	GCU	GCU	GCU	TCT	GCT	GCT
AGA	Arg	UCU	–	U*CU	*TCT Ser*	NCU	TCT
AGG	Arg	UCU	–	U*CU	*TCT Ser*	NCU	TCT
GGU	Gly	UCC	UCC	U*CC	TCC	GCC	GCC
GGC	Gly	UCC	UCC	U*CC	TCC	GCC	GCC
GGA	Gly	UCC	UCC	U*CC	TCC	GCC	TCC
GGG	Gly	UCC	UCC	U*CC	TCC	GCC	TCC

Note: Deviations from the universal genetic code are in italics and asterisked. T is written when the sequence of the gene was determined. The two tRNAs (a), with anticodon C*AA and U*AA, found in chloroplasts are both able to recognize the UUA and UUG codons.

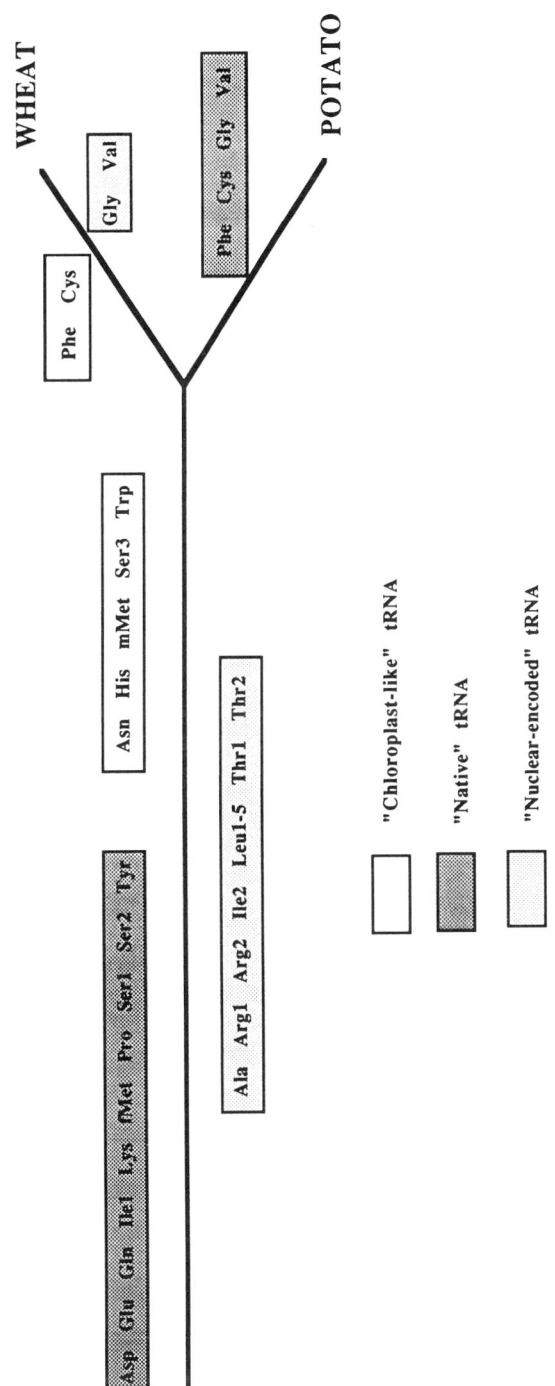

FIGURE 1. Different genetic origins of plant mitochondrial tRNAs in potato and wheat. tRNA isoacceptors were numbered according to Maréchal-Drouard et al.[9]

TABLE 2
Deviations from the Universal Genetic Code Observed in Various Mitochondria

Codon in universal code		Changes to organisms	Change possibly caused by
UGA, Stop	Trp	All except plants	Anticodon CCA to UCA
AUA, Ile	Met	Yeast, metazoa except Echinoderms	tRNA(CAU) changed from pairing with AUG to AUR
AGR, Arg	Ser	Metazoa, except vertebrates and insects	
	Stop	Vertebrates	tRNA(GCU) changed from pairing with AGN to AGY
AGA, Arg	Ser	Insects	tRNA(GCU) changed from pairing with AGN to AGY/A
AAA, Lys	Asn	Flatworms, echinoderms	Anticodon GUU to IUU Anticodon UUU disappeared
CUN, Leu	Thr	Yeast	tRNA(UAG) changed from amino-acylating Leu to Thr

From Jukes, T.H., *Experientia,* 46, 1149, 1990. With permission.

though the involvement of this tRNA in mt protein synthesis has not been clearly established.

As there seems to be no tRNA import from the cytosol (except in protozoans) and no restricted codon usage in nonplant mitochondria and in chloroplasts, other possibilities, such as unusual codon-reading abilities, have to be considered to explain how the apparently incomplete tRNA populations of these organelles can read all the codons of the universal genetic code.

II. GENETIC CODE

In 1979, it was first discovered that the genetic code in human mitochondria differs from the universal code by using AUA as a codon for methionine rather than for isoleucine, and UGA for tryptophan rather than as a stop codon.[15] It is now evident that the genetic code is not "frozen", as was first stated by Crick.[6] Nine exceptions to the universal code have been reported within the last few years: among them three concern the nu code and six the mt code.[16,17] Variations of the genetic code found in mitochondria of various nonplant organisms are summarized in Table 2. It should be noted that the universal genetic code is used without any change in chloroplasts and mitochondria of green plants. For instance, unlike animal and fungal mitochondria, plant mitochondria and chloroplasts use UGA as a stop codon and AUA as an isoleucine codon.

III. EVOLUTION OF THE GENETIC CODE: THE CODON CAPTURE HYPOTHESIS

As described earlier, variation of the genetic code is a common feature in mitochondria, at least in nonplant organisms. Osawa and Jukes[18] have postulated that during evolution a codon may disappear from the coding sequences. The corresponding tRNA then also disappears, as it is not useful anymore. Later in evolution, the same codon reappears, together with a tRNA which is able to read it. This tRNA may be specific for the original amino acid or for a different one. In the latter case, the codon has been reassigned or "captured". Codon or anticodon changes may result from directional mutation pressure towards either AT or GC during evolution.

According to such an hypothesis, evolution of the mt genetic codes can be explained by a number of different codon "captures":

1. Reassignment of AGR codons from arginine to either serine or stop codons in metazoan mitochondria.[19]
2. Reassignment of the AUA codon from isoleucine to methionine.[20] In mitochondria of protozoans, echinoderms, filamentous fungi, and green plants, as well as in chloroplasts, AUA is an isoleucine codon, like in eubacteria. By contrast, in mitochondria of yeast, nematodes, flies, and vertebrates, AUA encodes methionine.
3. Reassignment of CUN codons from leucine to threonine. Acquisition of CUN (normally leucine) by threonine in yeast mitochondria[21] could have resulted from AT pressure, converting CUN leucine codons to UUR. This reassignment has implied the switch of the aminoacylation specificity of yeast mt $tRNA^{Leu}_{UAG}$ from leucine to threonine, so that CUN codons became codons for threonine.
4. Reassignment of the AAA codon from lysine to asparagine. GC pressure could have led to the switch of AAA from lysine to asparagine in some animal mitochondria[22] by conversion of all AAA codons to AAG, followed by disappearance of the UUU anticodon.
5. Reassignment of stop codons: when chain termination is confined to one or two of the stop codons, the other(s) can be captured by an amino acid if a tRNA appears which has an anticodon able to pair with the currently unassigned codon. Such a case was first found by Yamao et al.[23] in *M. capricolum*.

In the first two cases 1 and 2, the codon "capture" has involved the progressive disappearance of a specific tRNA, while the corresponding codons, AGR or AUA, were converted to AGN or AUY, respectively. At a later stage, AGR and AUA codons could have reappeared, due to changes in the codon-anticodon pairing possibilities of other tRNAs. In the universal code, AUA is an isoleucine codon, which pairs with LAU (L = Lysidine, a derivative of C) and IAU anticodons in bacteria and eukaryotes, respectively. If the modifying enzyme which converts C to L is deleted, the anticodon remains simply CAU, which pairs only with the AUG methionine codon. The AUA isoleucine codon then disappears, but can reappear as a methionine codon, if a tRNA^{Met} with a CAU anticodon becomes able to pair with both AUG and AUA. This is the case in yeast and metazoan mitochondria, with the exception of echinoderms. It was indeed shown by Sibler et al.[24] that the AUA codon is recognized by the CAU anticodon of methionine tRNA during protein synthesis in yeast mitochondria; the authors proposed that the unusual presence of an extra, unpaired nucleotide in the TΨC stem of tRNA^{Met} enhances C:A wobble, thus allowing the reading of both AUA and AUG codons by the same tRNA.

In plant mitochondria and in chloroplasts, AUA appears to be still recognized as an isoleucine codon, thanks to the presence of a tRNA^{Ile} with an anticodon similar to the LAU anticodon used in bacteria. We have found in potato mitochondria a tRNA^{Ile} with an L*AU anticodon (L* = derivative of lysidine) which is encoded by a mt gene possessing a CAT (methionine) anticodon.[25] This potato mt tRNA$^{Ile}_{L*AU}$ can be aminoacylated with isoleucine, but not with methionine. In the same way, tobacco chloroplasts contain a tRNA$^{Ile}_{N*AU}$ which can be charged with isoleucine and not with methionine.[26] The structure of the nucleotide N* was not determined, but it is probably a derivative of lysidine, as it has a chromatographic behavior similar to that of L*.

In metazoan mitochondria, with the exception of echinoderms, the AUA codon has been "captured" by methionine. A reassignment of AUA from methionine back to isoleucine in echinoderm mitochondria has been considered as the most plausible hypothesis to explain this divergence with the other metazoans.[20] Sea urchin mitochondria contain a tRNA$^{Met}_{CAU}$ and a tRNA$^{Ile}_{GAU}$ to read the AUN codons; in nonechinoderm metazoan mitochondria, the tRNA$^{Met}_{CAU}$ is able to recognize both AUA and AUG codons. In echinoderms, the mt tRNA$^{Met}_{CAU}$ has probably changed so as to read only codon AUG and, unless a nuclear-encoded tRNA is imported from the cytosol for that purpose, the mt tRNA$^{Ile}_{GAU}$ would translate AUA, in addition to AUY, as isoleucine. Pairing of tRNA$^{Ile}_{GAU}$ with AUA and AUY codons could result either from a modification of the G at the first position of the anticodon (perhaps to inosine, as in eukaryotes), or from some structural change(s) of the tRNA allowing a G:A wobble base-pairing, although such a pairing has not been described yet.[20]

IV. ANTICODON CONTENT AND CODON RECOGNITION MECHANISM

In nonplant mitochondria, as well as in chloroplasts, 61 to 62 (depending whether UGA is used or not as a stop codon) sense codons are found in DNA sequences coding for proteins, except in vertebrate mitochondria, where AGA

TABLE 3
Anticodon:Codon Pairing[a]

Anticodon first base	Codon third base	Examples
U	U, C, A, G	Mitochondrial genetic code in family boxes
U modified	A, G	Mitochondrial genetic code in two-codon sets
U,2-thiolated or its equivalent	A	Eukaryotes
Uridine 5-oxyacetic acid	U, A, G	Eubacteria in family boxes
C	G	All genetic codes
C*[b]	A	Bacteria, plant mitochondria chloroplasts
G	U, C	All genetic codes
A	U	Rare
I	U, C, A	Eukaryotes

[a] Table adapted from References 6 and 8.
[b] Lysidine or its derivative.

and AGG codons are not used in coding sequences. Considering the number of tRNAs present in these organelles, the "wobble" rules for codon-anticodon pairing[6] have to be expanded (Table 3). It is now generally accepted that in the sets of four synonymous codons ("family boxes"), an unmodified U in the first position of the anticodon "wobble-pairs" with all four bases U, C, A, and G in the third position of the codon. According to Grosjean et al.,[27] an unmodified uridine is actually the only base that can possibly form a more or less stable pair with all four nucleotides in the "wobble" position of the codon. On the other hand, in two-codon sets (pairs of synonymous codons not included in "family boxes"), post-transcriptional modifications of U restrict its pairing to A and G. Different examples of such cases are shown in Table 1.

A. CODON RECOGNITION IN YEAST

Among the yeast mt tRNAs sequenced, $tRNA^{Gly}_{UCC}$, $tRNA^{Pro}_{UGG}$, $tRNA^{Ser}_{UGA}$, and $tRNA^{Thr}_{UGU}$, which recognize sets of four synonymous codons (GGN, CCN, UCN, and ACN, respectively) all contain an unmodified U in the first position of the anticodon.[24] On the other hand, yeast mt $tRNA^{Arg}_{U*CU}$, $tRNA^{Leu}_{U*AA}$, $tRNA^{Lys}_{U*UU}$, and $tRNA^{Trp}_{U*CA}$, which recognize sets of two codons ending with a purine (AGR, UUR, AAR, and UGR, respectively) all contain a modified U in the "wobble" position of the anticodon.[24] Moreover, this modified nucleotide appears to be the same in these four tRNAs and is most likely to be $cmnm^5U$ (5-carboxymethylaminomethyluridine).[24] It has been suggested that this modified nucleotide in the "wobble" position is in a conformation that enhances the rigidity of the anticodon, so as to prohibit the misrecognition of codons terminating with a pyrimidine.[28]

There is, however, one exception to these rules: yeast mt tRNA$^{Arg}_{ACG}$, which corresponds to the four-codon set CGN, has an unmodified A at the first position of the anticodon. All four CGN arginine codons are present in yeast mt protein genes, although at a lower level than the AGR arginine codons. An unmodified A can pair only with U (Table 2), which means that only the CGU codon can establish three base pairs with the anticodon of tRNA$^{Arg}_{ACG}$. As there is no other tRNA^{Arg} able to compete with tRNA$^{Arg}_{ACG}$ for the decoding of CGN codons, a "two out of three" recognition mechanism[29] (only two bases of the anticodon get paired with two bases of the codon) between tRNA$^{Arg}_{ACG}$ and the CGA, CGG, and CGC codons remains the most plausible.

B. CODON RECOGNITION IN CHLOROPLASTS

As mentioned earlier, there are only 30 different tRNA species encoded by the higher plant cp genomes,[30,31] whereas all 61 sense codons are found in cp protein genes and whereas, according to the "wobble" hypothesis, a minimum of 32 tRNA species would be required to decode these 61 code words. Furthermore, since tRNA$^{Leu}_{UAA}$ and tRNA$^{Leu}_{CAA}$, which are present in chloroplasts, should be able to decode the same leucine codon UUG, it appears that there is an apparent deficit of three tRNA species in higher plant chloroplasts. In fact, the alignment of the anticodons of the higher plant cp tRNAs with all sense codons shows an apparent deficit of tRNAs for the following codons: GCU/C (Ala), CGC/A/G (Arg), CUU/C (Leu), and CCU/C (Pro). Since it is generally assumed that there is no import of tRNAs into chloroplasts, these observations suggest that a special mechanism of codon recognition must operate in chloroplasts, at least in some cases.

A "two out of three" mechanism has been shown to operate in an *in vitro* protein-synthesizing system from *E. coli*.[29] *In vitro* translation of rabbit globin mRNA and maize cp glyceraldehyde 3-phosphate dehydrogenase mRNA in a tRNA-dependent wheat germ protein-synthesizing system that has been developed in our laboratory,[32] has shown that cp tRNA$^{Ala}_{U*GC}$, tRNA$^{Arg}_{ICG}$, and tRNA$^{Pro}_{U*GG}$ can read all four alanine (GCN), arginine (CGN), and proline (CCN) codons, respectively.[33] In the presence of these three cp tRNAs, a "two out of three" mechanism involving two strong G:C base pairs in the codon-anticodon interaction is operating. We also demonstrated by *in vitro* translation of synthetic polynucleotides that a "U:U" or a "U:C" pair enables cp tRNA$^{Leu}_{UAm7G}$ to read the CUU/C leucine codons. In this case, the m^{7}G present at the third position of the anticodon seems to facilitate the U:U and U:C "wobble" interactions. Thus, using a "two out of three" mechanism, as well as a "U:N wobble" mechanism, the 30 chloroplast-encoded tRNAs should be able to read all 61 sense codons. It is also interesting to note that a comparison of total and polysome-bound tRNAs has shown that all higher plant cp tRNA species are apparently used in protein biosynthesis.[34]

C. CODON RECOGNITION IN MAMMALIAN MITOCHONDRIA

Mammalian mt DNA encodes 23 tRNAs (Table 1), which are sufficient to translate all the codons of the mt genetic code. This number is much lower than the minimum of 32 required according to the "wobble" hypothesis.[6] For each of the eight genetic code boxes containing four synonymous codons, a single specific tRNA gene with a T in the first position of the anticodon is found in the mammalian mt genome.[35] This suggests that, as described earlier for yeast mitochondria and for chloroplasts, these mammalian mt tRNAs with a U in the "wobble" position can recognize all four synonymous codons of a box, either by a "two out of three" mechanism or by a "U:N wobble". In the case of tRNA genes corresponding to the sets of two synonymous codons ending with a purine, a T is also found at the first position of the anticodon and one can speculate that the corresponding U in the tRNAs is modified to restrict the codon recognition to two codons. In agreement with this hypothesis, it can be noted that in bovine liver mitochondria, $tRNA^{Lys}_{U*UU}$, $tRNA^{Glu}_{U*UC}$, $tRNA^{Leu}_{U*AA}$, and $tRNA^{Trp}_{U*CA}$, which correspond to two-codon sets (UGA is used as a Trp codon in mammalian mitochondria, see Table 1), all contain a modified U at the first position of the anticodon.[36]

V. LEVELS OF ISOACCEPTING tRNAs AND CODON USAGE IN CHLOROPLASTS AND MITOCHONDRIA

It has been shown in *E. coli* and in *S. cerevisiae* that differences in the levels of isoaccepting tRNAs reflect differences in usage of synonymous codons, leading to a strong correlation between codon usage and tRNA content.[37,38] By comparing on the one hand the codon usage in chloroplast protein genes from maize, spinach, and tobacco and on the other hand the tRNA content in bean chloroplasts,[34] we clearly showed that also in chloroplasts there is a strong correlation between the relative abundance of each amino acid in cp proteins and the relative abundance of the corresponding tRNAs. For example, leucine codons and tRNAs^{Leu} are the most abundant in chloroplasts, while the tryptophan codon and the tRNA^{Trp} are the least abundant. Furthermore, we could show for arginine, glycine, isoleucine, and leucine that there is also a good correlation between the relative frequency of synonymous codons in cp protein genes and the relative proportion of the corresponding isoacceptor tRNAs in chloroplasts (Table 4).

It is likely that this adjustment reflects the search for an optimal translation rate in chloroplasts. It should also be noted that the tRNA species encoded by genes present in two copies in the cp genomes (in the inverted repeats) are not found in higher amounts than the tRNAs encoded by single copy genes,[34] indicating that in chloroplasts the relative concentrations of the various tRNAs are not controlled by gene dosage.

In the case of mitochondria, it remains to be determined whether there is

TABLE 4
Synonymous Codon Usage and Relative Content of Isoaccepting Species in Chloroplasts

	\multicolumn{8}{c}{tRNA corresponding to}							
	Arg	Arg	Gly	Gly	Ile[c]	Ile	Leu	Leu	Leu[a]
Anticodons[b]	UCU	ACG	GCC	UCC	CAU[c]	GAU	UAA	CAA	UAG
Codons	AGA AGG	CGU CGC CGA CGG	GGU GGC	GGA GGG	AUA	AUU AUC	UUA	UUG	CUU CUC CUA CUG
Codon usage %	31	69	49	51	26	74	36	21	43
Percentage of each isoacceptor in its tRNA family (in bean)[d]	26	74	50	50	32	68	36	22	42

[a] Codon reading is according to Pfitzinger et al.[34]
[b] Possible nucleotide modifications have not been considered.
[c] Modification of C allows recognition of the isoleucine codon AVA and not of the methionine codon AUG.[26]
[d] Codon usage in the chloroplasts of tobacco, maize, and spinach was taken into account. The values represent the percentage of each codon in the family of synonymous codons.

From Pfitzinger, H. et al., *Nucleic Acids Res.*, 15, 1377, 1987. With permission.

also a correlation between codon usage and tRNA content. The fact that tRNA$_{ACG}^{Arg}$ is a minor species in yeast mitochondria, as compared to tRNA$_{U*CU}^{Arg}$, is in good agreement with the low frequency of CGN codons, as compared to AGR, in mt genes, and suggests that, at least in this case, there is a mechanism which allows adaptation of the levels of expression of the isoacceptor tRNAs to their respective utilization in mt protein synthesis.[24]

VI. MITOCHONDRIAL tRNAs OF NEMATODES

The folded structures of most metazoan mt tRNAs resemble those of prokaryotic tRNAs and eukaryotic nu-encoded tRNAs. However, metazoan mt tRNAs show considerable variations in the size and sequence of the dihydrouridine (DHU) and TΨC loops and in the occurrence of nucleotides that are invariant or semi-invariant in prokaryotic species or eukaryotic nuclear-encoded species. The most extreme case reported concerns the dramatic structural changes found in nematode mt tRNAs.

In the mt DNA of the nematode parasite worm, *Ascaris suum*, the number of putative tRNA genes, 22, is the same as that found in vertebrate or insect mt DNA.[5] In all but one [tRNA$_{AGN}^{Ser}$] of these *A. suum* mt tRNA genes, the TΨC arm and variable loop are replaced by a loop of 4 to 12 nucleotides called TV-replacement loop. In the tRNA$_{AGN}^{Ser}$ gene, the DHU arm is replaced by a loop of 5 nucleotides,[39] as in mammalian mt tRNA$_{AGY}^{Ser}$ and insect mt tRNA$_{AGY/A}^{Ser}$ genes. The same situation has been described for the free-living aerobic nematode, *Caenorhabditis elegans*.[39] In this case, 20 out of the 22 putative mt tRNA genes also have a TV-replacement loop, but both tRNA$_{AGN}^{Ser}$ and tRNA$_{UCN}^{Ser}$ genes show a DHU arm-replacement loop. By hybridizing oligonucleotide probes specific for some of these putative mt tRNA genes to nematode RNAs, Okimoto and Wolstenholme[39] obtained evidence for the transcription of at least nine *C. elegans* and three *A. suum* mt tRNA genes. These results strongly suggest that the anomalous tRNA genes found in nematode mt DNA really encode the species involved in mt protein synthesis. This means that in nematode mitochondria, the complete set of tRNAs lacks either both the TΨC arm and variable loop or the DHU arm. Despite these dramatic structural changes of the tRNAs, the only modification in the nematode mt genetic code, as compared to that in vertebrate or insect mitochondria, seems to be the use of all four AGN codons to specify serine[5] (Table 1).

The structural characteristics of nematode mt tRNAs imply that neither the TΨC arm, nor the DHU arm, is essential for protein synthesis. Furthermore, the nematode mt tRNAs approach the level of simplicity proposed by Crick[6] for the original RNA adaptor: a short RNA complementary to an mRNA codon and capable of attaching to a specific amino acid. It would be interesting to know whether some functional tRNAs exist that lack both the TΨC and DHU arms.[39]

VII. tRNA SPECIES IN *MYCOPLASMA CAPRICOLUM*

The complete set of tRNA species of *Mycoplasma capricolum*, a derivative of Gram-positive eubacteria, has been sequenced.[40] This prokaryotic organism contains only 29 tRNA species and is one of the smallest genetic systems with the exception of mitochondria. Furthermore, a small number of modified nucleotides (only 13) is found in all *M. capricolum* tRNAs. As in mitochondria, the *M. capricolum* genome seems to have discarded the genes for many tRNAs and for many tRNA nucleotide modifying enzymes, as compared to eubacteria.

In the case of alanine, glycine, leucine, proline, serine, and valine, there is only one tRNA species for each of the four-codon sets and all these tRNAs have an unmodified U at the first position of the anticodon. By contrast, $tRNA^{Arg}_{U*CU}$, $tRNA^{Gln}_{U*UG}$, and $tRNA^{Glu}_{U*UC}$, which correspond to two-codon sets, contain a modified U, namely $cmnm^5U$, at the first position of the anticodon. Another interesting feature of *M. capricolum* is the presence of $tRNA^{Arg}_{ICG}$, as found in eukaryotes. According to the "wobble" hypothesis, this tRNA can translate CGU, CGC, and CGA codons. As there is no other $tRNA^{Arg}$ able to recognize the codon CGG, either the $tRNA^{Arg}_{ICG}$ is able to read this codon using a "two out of three" mechanism, as in chloroplasts (see previous discussion), or CGG is not used in this bacterium. Finally, *M. capricolum* contains a $tRNA^{Trp}_{UCA}$ in addition to the normal $tRNA^{Trp}_{CmCA}$. The $tRNA^{Trp}_{UCA}$ is able to read the UGA codon (normally a stop codon in the universal genetic code) as tryptophan. All these characteristics of *M. capricolum* tRNAs resemble those of nonplant mt or of cp tRNAs.

The similarities in tRNA content and tRNA structural features between nonplant mitochondria and *M. capricolum* suggest that both nonplant mt and *Mycoplasma* genomes have evolved under the influence of similar evolutionary constraints, genomic economization and AT-pressure. Since *Mycoplasma* are parasitic in eukaryotes, Andachi et al.[40] have proposed that "*Mycoplasma*-like" organisms have been the eubacterial ancestors of nonplant mitochondria.

REFERENCES

1. Bonitz, S. G., Berlani, R., Coruzzi, G., Li, M., Macino, G., Nobrega, F. G., Nobrega, M. P., Thalenfeld, E., and Tzagoloff, A., Codon recognition rules in yeast mitochondria, *Proc. Natl. Acad. Sci. U.S.A.*, 77, 3167, 1980.
2. Anderson, S., Bankier, A. T., Barell, B. G., deBruijn, M. H. L., Coulson, A. R., Drouin, J., Eperon, I. C., Nierlich, D. P., Roe, B. A., Sanger, F., Schreier, P. H., Smith, A. J. H., Staden, R., and Young, I. G., Sequence and organization of the human mitochondrial genome, *Nature (London)*, 290, 457, 1981.

3. Dawid, I. B., Klukas, C. K., Ohi, S., Ramirez, J. L., and Upholt, W. B., Structure and evolution of animal mitochondrial DNA, in *The Genetic Function of Mitochondrial DNA*, Saccone, C. and Kroon, A. M., Eds, North-Holland, Amsterdam, 1976, 3
4. Jacobs, H. T., Elliot, D., Math, V. V., and Farquharson, A., Nucleotide sequence and gene organization of sea urchin mitochondrial DNA, *J. Mol. Biol.*, 202, 185, 1988.
5. Wolstenholme, D. R., Macfarlane, J. L., Okimoto, R., Clary, D. O., and Wahleitner, J. A., Bizarre tRNAs inferred from DNA sequences of mitochondrial genomes of nematode worms, *Proc. Natl. Acad. Sci. U.S.A.*, 84, 1324, 1987.
6. Crick, F. H. C., Codon-anticodon pairing: the wobble hypothesis, *J. Mol. Biol.*, 19, 548, 1966.
7. Küntzel, H. and Köchel, H. G., Evolution of rRNA and origin of mitochondria, *Nature (London)*, 293, 751, 1981.
8. Osawa, S. and Jukes, T. H., Evolution of the genetic code as affected by anticodon content, *Trends Genet.*, 4, 191, 1988.
9. Maréchal-Drouard, L., Guillemaut, P., Cosset, A., Arbogast, M., Weber, F., Weil, J. H., and Dietrich, A., Transfer RNAs of potato (*Solanum tuberosum*) mitochondria have different genetic origins, *Nucleic Acids Res.*, 18, 3689, 1990.
10. Joyce, P. B. M. and Gray, M. W., Chloroplast-like transfer RNA genes expressed in wheat mitochondria, *Nucleic Acids Res.*, 17, 5461, 1989.
11. Izuchi, S., Terachi, T., Sakamoto, M., Mikami, T., and Sugita, M., Structure and expression of tomato mitochondrial genes coding for tRNA^{Cys}(GCA), tRNA^{Asn}(GUU) and tRNA^{Tyr}(GUA): a native tRNA^{Cys} gene is present in dicot plants but absent in monocot plants, *Curr. Genet.*, 18, 239, 1990.
12. Suyama, Y., Two-dimensional polyacrylamide gel electrophoresis of *Tetrahymena* mitochondrial tRNA, *Curr. Genet.*, 10, 411, 1986.
13. Seihamer, J. J. and Cummings D. J., Altered genetic code in *Paramecium* mitochondria: possible evolutionary trends, *Med. Gen. Genet.*, 187, 236, 1982.
14. Martin, R. P., Schneller, J. M., Stahl, A. J. C., and Dirheimer, G., Import of nuclear deoxyribonucleic acid coded Lysine-accepting transfer ribonucleic acid (anticodon CUU) into yeast mitochondria, *Biochemistry*, 18, 4600, 1979.
15. Barrell, B. G., Bankier, A. T., and Drouin J., A different genetic code in human mitochondria, *Nature (London)*, 282, 189, 1979.
16. Jukes, T. H., Genetic code 1990. Outlook, *Experientia*, 46, 1149, 1990.
17. Jukes, T. H. and Osawa, S., The genetic code in mitochondria and chloroplasts, *Experientia*, 46, 1117, 1990.
18. Osawa, S. and Jukes, T. H., Codon reassignment (codon capture) in evolution, *J. Mol. Evol.*, 28, 271, 1989.
19. Osawa, S., Ohama, T., Jukes, T. H., and Watanabe, K., Evolution of the mitochondrial genetic code. I. Origin of AGR serine and stop codons in metazoan mitochondria, *J. Mol. Evol.*, 29, 202, 1989.
20. Osawa, S., Ohama, T., Jukes, T. H., Watanabe, K., and Yokoyama, S., Evolution of the mitochondrial genetic code II. Reassignment of codon AUA from isoleucine to methionine, *J. Mol. Evol.*, 29, 373, 1989.
21. Osawa, S., Collins, D., Ohama, T., Jukes, T. H., and Watanabe, K., Evolution of the mitochondrial genetic code. III. Reassignment of CUN codons from leucine to threonine during evolution of yeast mitochondria, *J. Mol. Evol.*, 30, 322, 1990.
22. Ohama, T., Osawa, S., Watanabe, K., and Jukes, T. H., Evolution of the mitochondrial genetic code. IV. AAA as an asparagine codon in some animal mitochondria, *J. Mol. Evol.*, 30, 329, 1990.
23. Yamao, F., Muto, A., Kawauchi, Y., Iwami, M., Iwagami, S., Azumi, Y., and Osawa, S., UGA is read as tryptophan in *Mycoplasma capricolum*, *Proc. Natl. Acad. Sci. U.S.A.*, 82, 2306, 1985.

24. **Sibler, A. -P., Dirheimer, G., and Martin, R. P.,** Codon reading patterns in *Saccharomyces cerevisiae* mitochondria based on sequences of mitochondrial tRNAs, *FEBS Lett.*, 194, 131, 1986.
25. **Weber, F., Dietrich, A., Weil, J. H., and Maréchal-Drouard, L.,** A potato mitochondrial isoleucine tRNA is coded for by a mitochondrial gene possessing a methionine anticodon, *Nucleic Acids Res.*, 18, 5027, 1990.
26. **Francis, M. and Dudock, B.,** Nucleotide sequence of a spinach chloroplast isoleucine tRNA, *J. Biol. Chem.*, 257, 11195, 1982.
27. **Grosjean, H. J., De Henau, S., and Crothers, D. M.,** On the physical basis for ambiguity in genetic coding interactions, *Proc. Natl. Acad. Sci. U.S.A.*, 75, 610, 1978.
28. **Hillen, W., Egert, E., Lindner, H. J., and Gassen, H. G.,** Restriction or amplification of wobble recognition, *FEBS Lett.*, 94, 361, 1978.
29. **Samuelsson, T., Elias, P., Lustig, F., Axberg, T., Fölsch, G., Akesson, B., and Lagerkvist, U.,** Aberration of the classic codon reading scheme during protein synthesis *in vitro*, *J. Biol. Chem.*, 255, 4583, 1980.
30. **Sugiura, M. and Wakasugi, T.,** Compilation and comparison of transfer RNA genes from tobacco chloroplasts, *Crit. Rev. Plant Sci.*, 8, 89, 1989.
31. **Hiratsuka, J., Shimada, H., Whittier, R., Ishibashi, T., Sakamoto, M., Mori, M., Kondo, C., Honji, Y., Sun, C. R., Meng, B. Y., Li, Y. Q., Kanno, A., Nishizawa, Y., Hirai, A., Shinozaki, K., and Sugiura, M.,** The complete sequence of the rice (*Oryza sativa*) chloroplast genome: intermolecular recombination between distinct tRNA genes accounts for a major plastid DNA inversion during the evolution of the cereals, *Mol. Gen. Genet.*, 217, 185, 1989.
32. **Pfitzinger, H., Weil, J. H., Pillay, D. T. N., and Guillemaut, P.,** Preparation of a tRNA-dependent wheat germ protein synthesizing system, *Plant Mol. Biol.*, 12, 301, 1989.
33. **Pfitzinger, H., Weil, J. H., Pillay, D. T. N., and Guillemaut, P.,** Codon recognition mechanisms in plant chloroplasts, *Plant Mol. Biol.*, 14, 805, 1990.
34. **Pfitzinger, H., Guillemaut, P., Weil, J. H., and Pillay, D. T. N.,** Adjustment of the tRNA population to the codon usage in chloroplasts, *Nucleic Acids Res.*, 15, 1377, 1987.
35. **Barrell, B. G., Anderson, S., Bankier, A. T., De Bruijn, M. H. L., Chen, E., Coulson, A. R., Drouin, J., Eperon, I. C., Nierlich, D. P., Roe, B. A., Sanger, F., Schreier, P. H., Smith, A. J. H., Staden, R., and Young, I. G.,** Different pattern of codon recognition by mammalian mitochondrial tRNAs, *Proc. Natl. Acad. Sci. U.S.A.*, 77, 3164, 1980.
36. **Sprinzl, M., Hartmann, T., Weber, J., Blank, J., and Zeidler, R.,** Compilation of tRNA sequences and sequences of tRNA genes, *Nucleic Acids Res.*, 17(Suppl.), r1, 1989.
37. **Ikemura, T.,** Codon usage, tRNA content, and rate of synonymous substitution, in *Population Genetics and Molecular Evolution*, Ohta, T. and Aoki, K., Eds., Japan Sci. Soc. Press, Tokyo/Springer-Verlag, Berlin, 385, 1985.
38. **Holm, L.,** Codon usage and gene expression, *Nucleic Acids Res.*, 14, 3075, 1986.
39. **Okimoto, R. and Wolstenholme, D. R.,** A set of tRNAs that lack either the TΨC arm or the dihydrouridine arm: towards a minimal tRNA adaptor, *EMBO J.*, 9, 3405, 1991.
40. **Andachi, Y., Yamao, F., Muto, A., and Osawa, S.,** Codon recognition patterns as deduced from sequences of the complete set of transfer RNA species in *Mycoplasma capricolum*. Resemblance to Mitochondria, *J. Mol. Biol.*, 209, 37, 1989.

Chapter 6

DIFFERENTIAL tRNA GENE EXPRESSION IN EUKARYOTES

Robert M. Pirtle and Irma L. Pirtle

TABLE OF CONTENTS

I.	Introduction	142
II.	Tissue-specific tRNA Usage and Gene Expression	142
III.	Developmental Regulation of tRNA Gene Transcription	145
IV.	Species-specific tRNA Gene Transcription	148
V.	Differential Expression of tRNA Gene Families	150
VI.	Conclusions	151
	Acknowledgments	152
	References	153

I. INTRODUCTION

Eukaryotic tRNA genes and 5S rRNA genes, transcribed by eukaryotic RNA polymerase III (polIII), are well known to have split intragenic promoter sequences, called either the A and B boxes (or A and C boxes for 5S rRNA genes) or the 5'- and 3'-internal control regions (5'- and 3'-ICRs). Specific transcription of the genes transcribed by polIII requires the participation of a number of transcription factors.[1-3] The transcription assembly factor IIIC (TFIIIC or τ factor in yeast) is the primary multisubunit protein binding directly to the internal control regions (ICRs) of eukaryotic tRNA genes during formation of the preinitiation complex. TFIIIC1 appears to interact with the A box,[4] and TFIIIC2 has been shown to interact with the B box (3'-ICR).[5,6] Transcription factor IIIB (TFIIIB) binds to the TFIIIC-DNA complex and occupies a position just upstream of the transcription initiation site, and is apparently the actual transcription initiation factor.[7] Both TFIIIB and TFIIIC may induce the bending of DNA of tRNA genes, thereby facilitating the initiation of transcription by allowing possibly more polIII-DNA contacts to be made.[8] Recently, another transcription factor, TFIIIR, a RNA molecule, has been shown to be involved in tRNA gene transcription.[9] The transcription factor IIIA (TFIIIA) only binds to 5S rRNA genes, and is not implicated in tRNA gene transcription.[1-3]

In the last several years, many surprising discoveries have occurred in this area of research, as for example, the extragenic promoter elements for the U6 snRNA and 7SK RNA genes, the mixed RNA polymerase II (polII) and III (polIII) promoter motifs for several other genes, such as the c-*myc* proto-oncogene,[10] and very recently, the elegant work by Sprague and co-workers[9] on the polIII transcription factor composed solely of RNA (TFIIIR). A number of well-documented and lucid review articles have been written describing the nature and characteristics of the promoter elements involved in the transcription of genes by polIII and its ancillary transcription factors.[1-3,11-15] Thus, the primary emphasis of this brief review will be on aspects of differential tRNA gene expression involved in tissue-specific, developmental-specific, and species-specific tRNA gene expression, and the regulation of expression of tRNA gene families.

II. TISSUE-SPECIFIC tRNA USAGE AND GENE EXPRESSION

Specific tRNA isoacceptors for glycine, serine, and alanine, whose codon usage is commensurate with the predominant amino acids in the secretory proteins fibroin and sericin, undergo dramatic adaptive changes in tRNA gene expression in the posterior silkgland of *Bombyx mori* (reviewed by Goldsmith and Kafatos[16]). The hallmark example of tissue- and developmental-specific tRNA gene expression is the elegant work of Sprague and coworkers[17-19] on the accumulation of a specific alanine tRNA isoacceptor ($tRNA_1^{Ala}$) during the terminal fifth instar stages of development in the silkgland of *Bombyx mori*.

Underwood et al.[19] found that the approximately 20 silkgland-specific alanine tRNA genes are tightly clustered and tandemly repeated at a single locus in the *Bombyx* genome, whereas the 20 to 30 constitutive alanine tRNA genes are widely dispersed. It would appear that differential transcription is the mechanism by which the silkgland-specific alanine tRNA is produced, since the silkgland-specific genes undergo no amplification or other DNA rearrangements. The transcriptional properties of constitutive and silkgland-specific alanine tRNA genes in the silkworm are strikingly different, since the tissue-specific alanine tRNA occurs only in the portion of the silkgland where fibroin is synthesized, and the constitutive tRNA is distributed throughout all cell types in the organism.[17-19] In an analogous system, using the constitutive silkworm alanine tRNA ($tRNA_2^{Ala}$) as probe, the silkglands of the orb-web spider *Nephila clavipes* were also found to express a tissue-specific alanine tRNA by Northern blotting.[20] It would appear that the spider glands apparently use the same mechanism of tRNA adaptation by differential expression for the synthesis of fibroin as the silkworm glands.

The primary differences between the two silkworm alanine tRNA transcription units are in the 5'-flanking sequences, such that the sequences of the upstream regions differ in three corresponding places. The most obvious of these differences is the presence of a polII-like TATA box at about −30 upstream from the constitutive tRNA gene, absent in the upstream region of the silkgland-specific tRNA gene.[18] Rather than simply the binding of the transcription factors TFIIIB and IIIC to the A and B box intragenic promoter elements for the activation of transcription of tRNA genes, apparently a much larger region surrounding tRNA genes binds the polIII transcription factors, perhaps about 160 contiguous bp, from about −13 to +146.[21,22] In the earlier paper[21] it was shown that the promoter control region is comprised of two functionally distinct domains, the basically 5'-flanking domain from −13 to someplace between the transcription start point and the A box, and the essentially 3'-flanking domain, from about +5 to +146, including both the A and B elements and sequences further downstream from the transcription termination site. In fact, the recent gel shift and nuclease protection experiments of tRNA gene:transcription factor complexes,[22] indicate that the silkworm transcription factors TFIIIC and TFIIID (the analogy of these factors to human and other eukaryotic TFIIIC1 and C2 being unknown at present), when combined stoichometrically, interact with the 3'-flanking domain. As determined by mutational alteration of both halves of the large promoter domain, both the 5'- and 3'-flanking domains are essential for binding TFIIIC and TFIIID in the silkworm. The authors concluded that flanking sequences could very well be responsible for differential transcription of tRNA genes, since they have different sequences of sufficient intrinsic heterogeneity to have a wide affinity range for binding transcription factors, leading to variations in promoter strength among tRNA genes.

As a consequence of studying the tissue-specific expression of the silkworm alanine tRNA, Sprague and co-workers[9] remarkably have identified an unusual polIII transcription factor derived from TFIIIB preparations, and have actually shown it to be a RNA molecule (TFIIIR). It is unknown if TFIIIR is a general eukaryotic transcription factor restricted to the silkworm species, whether the factor is an assembly factor or is an actual catalyst in the initiation of transcription, or if TFIIIR participates in a specific phase, such as elongation, of the transcriptional process.

Recently, the regulation of glycine tRNA gene expression was investigated in nuclear extracts from the posterior silkgland of fifth instar larvae.[23] In contrast to the silkgland-specific alanine tRNA, there are no silkgland-specific glycine tRNA species, and so it was necessary to compare the relative transcriptional efficiencies of several glycine tRNA genes. Three sequence motifs, TATAT, AATTTT, and TTC occur within 30-40 bp upstream from the constitutive alanine tRNA gene,[18] the silkworm 5S rRNA gene,[24] and one of four analyzed silkworm glycine tRNA genes.[23,25] Two of the other four glycine tRNA genes have the AATTTT motif, and are transcribed quite efficiently, but not as well as the glycine $tRNA_1$ gene with all three sequence elements, and a gene construct lacking the silkworm flanking region up to –2 is not transcribed. The presence of a protein factor binding to this motif was confirmed by gel retardation experiments and transcription competition assays. The AATTTT motif is apparently the most crucial positive modulatory element of the three sequences, since it occurs in all four of the glycine tRNA genes analyzed in this study, and the authors believe that the *trans*-acting factor could be a transcription factor such as TFIIIB.[23] Furthermore, as shown by copper phenanthroline footprinting, a protein factor binds to a weak inhibitory sequence from –276 to –270 with an alternating purine-pyrimidine tract (TATATAA) at about this position in the 5'-flanking regions of three of the four glycine tRNA genes examined. Alternating purine-pyrimidine tracts also occur in the upstream regions flanking a *Xenopus* Met-B tRNA gene, and appears to be a negative transcription regulatory element.[26] Thus, there appears to be a variety of mechanisms for the tissue-specific regulation of tRNA gene expression in the silkgland.

Lin and Agris[27] have shown that alterations in the levels of tRNA isoaccepting species occur during erythroid differentiation of murine Friend erythroleukemia cells, during induction of hemoglobin synthesis by dimethylsulfoxide, reflecting the general idea that the content of the tRNA species in a given cell type appears to be adapted for specific tRNA usage for synthesis of the major protein products of that cell (see Chapter 4 of this book). In particular, the quantities of the four threonine tRNA isoacceptors exhibited extensive alterations during the differentiation event. The levels of two isoacceptors increased in relative amounts constantly, while the amounts of the other two isoacceptors decreased throughout the time course. In addition, the relative amounts of two proline tRNA isoacceptors decreased to barely detectable levels, whereas two other proline isoacceptors became the predominant species. This adaptation of the tRNA population would most likely occur by differentiation-specific tRNA gene expression, in line for synthesis of the globins and spectrin for this particular differentiated cell line. However, it is possible that the changes in these tRNA isoacceptor populations could occur by posttranscriptional modifications of the specific tRNA isoacceptors. It has been speculated[28] that the observed high transcriptional activity of a threonine tRNA$_{UGU}$ gene (in a tRNA gene heterocluster which also has proline and valine tRNA genes) could be related to the variations in threonine tRNA isoacceptor levels observed by Lin and Agris.[27] Along these same lines, it has also been shown that the relative synthesis of the initiator methionine tRNA in both the steady-state and the newly synthesized tRNA populations remains fairly constant during erythroid differentiation, but that the relative concentration of asparagine tRNA undergoes fluctuations in the steady-state and nascent tRNA populations between days 2 and 3 of induction, as measured by a dot-blot hybridization assay procedure developed for the measurement of the methionine and asparagine tRNA isoacceptors by binding to the cognate tRNA genes immobilized on a membrane support.[29,30] However, no specific conclusions were reached regarding the regulation of asparagine tRNA gene expression.

Bovine lens fiber cells contain two major isoacceptors of phenylalanine tRNA, but the lens epithelial cells primarily contain only one of the isoacceptors.[31] The two phenylalanine tRNAs only differ by a G to A transversion at position 57 of the nucleotide sequence, meaning that the differentiation-specific phenylalanine tRNA isoacceptor has TΨCA in the ribothymidine loop, instead of TΨCG, as does the somatic cell-specific phenylalanine tRNA. Thus, this may be a demonstration of the possible activation of a second phenylalanine tRNA gene during development of mammalian lens cells to facilitate the synthesis of the lens crystallins with their relatively high content of phenylalanine. Lin et al.[32] mention that it is also possible that a tRNA insertase could replace A for G at position 57 for the generation of the developmental change posttranscriptionally, but this is considered less likely.

III. DEVELOPMENTAL REGULATION OF tRNA GENE TRANSCRIPTION

Clarkson et al.[32] found that the clawed frog *Xenopus laevis* has about 8000 tRNA genes per haploid genome, and many are clustered in complex multigene arrangements.[33,34] The most thoroughly analyzed such tRNA gene heterocluster has 8 tRNA genes encoding 7 tRNA isoacceptors encompassed in a 3.18-kb DNA segment tandemly repeated at least 150 times at a single chromosomal site.[33,35,36] It is very probable that this highly clustered tRNA gene arrangement in a tandemly repeated DNA segment is associated with the developmental regulation of differential tRNA gene expression in oocytes. This kind of expression would contrast with tRNA gene expression in somatic cells of the frog, in which different tRNA isoacceptor genes with a different manner of gene dispersion and organization may occur.[37,38]

The developmental regulation of the 5S rRNA genes in *Xenopus* is the best known example of this type of differential gene expression for polIII-transcribed genes, reviewed by Wolffe and Brown[39], and more recently by Wolffe.[40] In brief, around 20,000 oocyte 5S rRNA genes occur mainly at the telomeres of most *Xenopus* chromosomes in clusters of 1000 repeats, whereas about 400 somatic 5S rRNA genes are localized in a single cluster on one *Xenopus* chromosome. The transcription factor TFIIIA interacts with the somatic 5S rRNA gene relatively weakly, but the interaction of TFIIIA with the somatic gene is stabilized by the binding of TFIIIC by both protein-protein contacts with TFIIIA and protein-DNA contacts with both the A and C promoter elements. Taken together, TFIIIA and TFIIIC bind very tightly to the somatic 5S rRNA genes, and do not usually dissociate. Since the TFIIIA and TFIIIC concentrations decrease during embryogenesis, the diminution in relative amounts of the factors apparently restricts transcription from the relatively unstable oocyte 5S rRNA gene transcription complex. Since TFIIIA binds with equal affinity to both 5S rRNA gene types, the sequence differences between the oocyte and somatic 5S rRNA genes apparently have a greater effect on the binding (or activity) of TFIIIC than on the binding of TFIIIA, leading to the continued formation of the relatively more stable and active somatic 5S rRNA gene transcription complexes during embryogenesis.

Clarkson and co-workers[33,37,38] identified four types of tyrosine tRNA genes in *Xenopus,* two oocyte-specific and two somatic-specific. The precursor tRNA transcripts are readily identifiable from each other, since they have different 5'-leader and intron sequences, easily identifiable by Northern blotting and primer extension analyses.[38] The tyrosine tRNA gene in the tandemly repeated 3.18-kb tRNA gene herocluster occurring at a single chromosomal locus is one of the oocyte-specific tRNA genes, and the authors speculate that the other oocyte-specific tyrosine tRNA gene is in a second herocluster containing several tRNA genes in a 3.1-kb DNA fragment, tandemly repeated 800-fold, originally characterized by Rosenthal and Doering.[34] These tRNA genes are designated as oocyte-specific since their transcripts occur in oocytes, appear transiently during early embryogenesis, and are absent from postembryonic somatic cells. One of the types of somatic-specific tRNA genes, the TyrD gene,[37] was found to occur in only several copies per genome, and gives rise to only a small portion of the somatic cell transcription products. The major fraction of somatic cell-specific tyrosine precursor tRNA transcripts are derived from a second type of tyrosine tRNA gene with unique 5'-terminal and intron sequences, and likely to be transcribed from only 20 or 30 tRNA genes. The precursor transcripts of the somatic-specific tRNA genes are not observed in oocytes, but are found in gastrula and later developmental stages of embryogenesis and in somatic cells.[38] The switch from oocyte-specific tRNAs to somatic-specific tRNAs occurs in early embryogenesis, between the midblastula transition and the onset of neurulation, but with some oocyte-specific precursor tRNAs being detectable in tadpoles. The 5'-flanking sequences of the two oocyte-specific genes have significant identity, as do the 5'-flanks of the somatic-specific tRNA genes. The introns have an even stronger degree of identity, with 11 of 13 bp for the two somatic-type genes and 10 of 13 bp for the oocyte-type genes. Thus, the overall pattern of developmental regulation of tyrosine tRNA gene expression in *Xenopus* parallels that of the 5S rRNA genes, in which the RNA population switches from that of oocyte-specific to somatic-specific by the end of gastrulation, but with one or two differences.[38] The authors conclude by specifically speculating that the oocyte-to-somatic specific switch of the tyrosine tRNA genes may depend on the interactions of either transcription factors or polIII with upstream DNA sequences. Thus, results from two major tRNA gene expression groups (Sprague's and Clarkson's) suggest that most developmental- or tissue-specific regulation of tRNA gene transcription could be attributed to differential tRNA gene expression by the interactions of transcription factors or polIII with flanking DNA sequences.

Oei and Pieler[41] have partially purified a putative stimulatory transcription factor that presumably binds to flanking sequences upstream from *Xenopus* tRNA and 5S rRNA genes, and which is chromatographically distinct from TFIIIA, TFIIIB, or TFIIIC. In fact, the purification of proteins potentially involved in the modulation of transcription of *Xenopus* tRNA and 5S rRNA genes revealed the presence of at least two chromatographic components necessary for the stimulation of transcription. One component is a DNA-binding protein that interacts with the upstream regions flanking *Xenopus* tRNA and 5S rRNA genes, and can be separated by gel filtration from a second component required for the stimulation of transcription by the first component. Recently, Cohen and Reynolds[42] screened a *Xenopus* ovary cDNA expression library to identify proteins binding the B box promoter element with oligonucleotide probes containing the B box of the OAX gene (satellite I) and adenovirus VAII genes. A cDNA clone encoding a protein designated YB3 that indeed specifically bound the B box. This protein was found to occur in oocyte S150 extracts fractionated by phosphocellulose ion-exchange chromatography (fraction IIIC), as well as by B box DNA affinity chromatography.[42] The protein occurs in a wide variety of adult tissues. Since the phosphocellulose fraction IIIC and the fraction isolated by B box DNA affinity chromatography are routinely used as the source of TFIIIC activity for *in vitro* transcription studies, Cohen and Reynolds[42] suggest that an abundant B box-binding protein other than TFIIIC could effect transcriptional activity by competing for binding by TFIIIC *in vitro*, and perhaps *in vivo*, as for example, a B box-binding protein could compete with TFIIIC templates of low affinity for the factor. The authors also suggest other roles for YB3, either as a subunit of TFIIIC, similar to a 55K B box-binding protein from a fraction enriched in HeLa TFIIIC, isolated by Seifart and co-workers,[43] or as a polIII regulatory protein completely unrelated to TFIIIC.

In addition to the tyrosine tRNA gene, there are 2 initiator methionine tRNA genes in the *Xenopus* 3.18-kb DNA repeat, the Met-A gene encoding the initiator tRNA found *in vivo*, and a second, the Met-B gene, differing by a C → T transition at position 65 of the coding region, whose transcripts have not been detected *in vivo* and which is poorly expressed *in vitro*.[26,36] The 5'-

flanking region of the Met-B gene contains alternating purine-pyrimidine tracts, from −20 to −12 and from −43 to −32, potentially capable of adopting the Z-DNA conformation, which could preclude the gene from interacting with any potential DNA-binding proteins for activation of transcription of the tRNA Met-B allogene.

IV. SPECIES-SPECIFIC tRNA GENE TRANSCRIPTION

A number of laboratories[44-49] have shown that, in addition to the need for the internal split promoter elements, the 5′-flanking DNA sequences of insect tRNA genes are apparently necessary for transcriptional activity *in vitro*. For example, deletion of the pentanucleotide sequence TCGCT and replacement with vector sequences from −38 to −34 upstream from a *Drosophila melanogaster* valine tRNA gene resulted in a 90% diminution in its template activity.[48] In a subsequent paper,[49] site-directed changes in the upstream TCGCT sequence of the same tRNA gene resulted in significant decreases in template activity when the nucleotides −38 (T), −35 (C), or −34(T) were changed, but no change in transcriptional activity resulted upon mutation of −36 (G). Mutation of −37 (C) actually gave a slight 13% increase in the observed template activity. Upon comparison of the 5′-flanking sequences of many characterized *Drosophila* tRNA genes and their respective transcriptional efficiencies, Sajjadi and Spiegelman[49] proposed that the sequence TNNCT between −30 and −40 upstream from the insect tRNA genes are efficiently transcribed, and furthermore, that this consensus sequence is a positive modulatory element for a large number of *Drosophila* tRNA genes. The upstream regions flanking an arginine tRNA gene[46] and an asparagine tRNA gene[47] from *Drosophila* both have the TNNCT motif,[49] and the 5′-flanking sequences are required for transcription of these genes in homologous *Drosophila* extracts. However, conversely, when the 5′-flanking region of the *Drosophila* arginine tRNA gene was deleted, there was no effect on the observed template activity in heterologous HeLa cell extracts.[46]

Recently, Johnson et al.[50] studied the reasons underlying the differences in transcriptional activity of *Drosophila* tRNA genes in both *Drosophila* and human cell extracts. By interchanging the *Drosophilia* and human transcription factors TFIIIB and TFIIIC in assays for the formation of stable transcription complexes, it was shown that the *Drosophila* TFIIIC and human HeLa cell TFIIIB proteins are unable to form stable heterologous transcription complexes with a *Drosophila* arginine tRNA gene template, whereas the human HeLa cell TFIIIC and *Drosophila* TFIIIB do form stable, but nonproductive, transcription complexes with the *Drosophila* gene substrate. It would appear from these results that the correct positioning of the TFIIIB:TFIIIC protein-protein interactions have a major function in the formation of active polIII transcription complexes, and may be largely responsible for the species-specificity of tRNA gene transcription observed between insects and humans. Thus, Johnson et al.[50] speculate that, since the internal promoter elements are highly conserved due to their importance for tRNA function corresponding to the D- and T-stem-loop regions in tRNA, the specific TFIIIC:DNA interactions involved in transcription of tRNA genes have probably been maintained throughout evolution. However, the TFIIIB:TFIIIC protein:protein contact regions apparently must have undergone significant changes, since they fail to interact between the transcription factors of lower and higher eukaryotic species.[50]

Saccharomyces cerevisiae has about nine leucine tRNA$_3$ genes, and the tRNA coding regions of four of these genes are identical, and the introns and the proximal 5'- and 3'-flanking regions have a great degree of sequence identity (see reference 51), with an almost perfectly conserved pentadeca-nucleotide sequence (TTTCAACAAATAAGT) contiguous with the 5'-end of the genes.[51] This 15-bp sequence apparently serves as a positive modulatory element for transcription of the yeast leucine tRNA$_3$ gene, since transcription is essentially eliminated when the 15-bp sequence is replaced with vector DNA sequences to position −2. When the yeast tRNA gene is transcribed in heterologous *Xenopus* oocyte or HeLa cell extracts, the modulating effect of the 5'-flanking region is lost, and constructs with deleted 5'-flanks are transcribed almost as efficiently as the unmodified gene.[52] DNA sequences from other yeast tRNA gene families also have 15-bp segments highly similar to that of the leucine tRNA$_3$ gene in sequence and relative position,[51] and thus, this motif may be involved in modulating the expression of other yeast tRNA gene families. The 5'-flanking pentadecanucleotide sequence was also shown to be modulating *in vivo*.[51] Using template exclusion assays with mutants of the 15-bp sequence as compared with the normal flanking sequence of the leucine tRNA gene, it was shown that the conserved sequence effects formation of stable preinitiation complexes, with no effect of the upstream sequence on the apparent K$_m$ of the transcription apparatus, but probably influences the rate of precursor tRNA synthesis by changing the rate of initiation of transcription.[53] Raymond et al.[51] speculate that the sequence may be one mechanism by which yeast cells adapt tRNA biosynthesis to match the demand created by codon use preferences.

A human initiator methionine tRNA with its own flanking sequences was shown not to be expressed in yeast cells.[54] By switching the 5'-flanking sequence of this human gene with that of a yeast arginine tRNA gene, the human tRNA product could be expressed *in vivo* in yeast cells.[55] This functionality was demonstrated by using a slow-growing yeast strain in which three of four endogenous yeast initiator methionine tRNA genes had been disrupted. When the yeast strain was transformed with a multicopy plasmid harboring the human initiator tRNA gene, the degree of growth rate enhancement was correlated with the steady-state levels of human tRNA in the transformants. Thus, this provides further evidence for the importance of 5'-flanking sequences on tRNA gene transcription in yeast cells, by promoting the stable binding of the yeast factor τ (corresponding to TFIIIC) that recognizes the A and B boxes, and TFIIIB that binds to the 5'-flanking sequence.[7]

A total of 68 different tRNA genes representing 20 tRNA gene families have been analyzed from the slime mold *Dictyostelium discoideum,* and several unique characteristics were found that distinguish this organism's tRNA genes from the nuclear tRNA genes of other organisms.[56] The proximal 5'-flanking regions have a consensus sequence of about 13 nucleotides occurring in many of the isolated *Dictyostelium* tRNA genes. This 5'-consensus box fits nearly perfectly into the core of the 15-bp motif that is a positive modulatory element in the transcription of yeast tRNA genes previously mentioned.[51] Several subclasses of *Dictyostelium* tRNA genes can be distinguished by consideration of other conserved 5'-flanking elements. More than 80% of the tRNA genes have a sequence feature designated as the ex-B motif, characterized by a strong resemblance to a T-loop core region of a tRNA molecule, and this core is generally extended in most genes to include one to three base pairs which could form a corresponding stem region. This basically delineates the B box of the split internal promoter that is the major binding site for TFIIIC. About 50% of the tRNA genes also occur associated with one of two types of repetitive elements unique to *Dictyostelium,* either a retrotransposon about 50 bp upstream or a transposon about 100 bp downstream from the tRNA gene. Thus, although the tRNA genes in *Dictyostelium* apparently have unique features, it has as yet to be determined if the observed similarities would lead to species-specific gene expression. Since members of the *Dictyostelium* valine tRNA$_{GUU}$ gene family are actively transcribed *in vivo* in yeast,[45] this would probably not be the case in these two lower eukaryotes, since they both have the consensus positive modulatory element in the 5'-flanking regions of many of their tRNA genes.

V. DIFFERENTIAL EXPRESSION OF tRNA GENE FAMILIES

There are a number of important questions concerning the relative differential expression of the different members of a redundant tRNA gene family encoding isoacceptors for a specific amino acid. Schmutzler and Gross[57] summarized some of the general issues as concerned with: (a) the regulation of gene expression of an entire gene family in general, and of the individual family members in particular, and the manner in which the family members contribute to the pool of specific tRNA isoacceptors in different cell or tissue types or stages of differentiation; (b) the role of any variant genes or (allogenes),[58] and tissue- or differentiation-specific tRNA gene expression, the best known example being the *Bombyx mori* alanine tRNA gene;[18] and (c) the role and function of tRNA pseudogenes.

Schmutzler and Gross[57] analyzed the differential expression of the human valine tRNA gene family, of which 11 to 12 (of a possible total of 15) different family members have been characterized by Gross' laboratory and by our laboratory,[28,59-62] including *bona fide* tRNA genes, tRNA allogenes, and tRNA pseudogenes. This was elegantly done by analyzing the precursor transcription products from individual tRNA genes by primer extension analysis, using unique primer oligonucleotides complementary to the D-stem-loop region of the mammalian valine tRNAs (or to the anticodon stem-loop region). The precursor tRNA transcripts were shown to occur about 500-fold less than the mature valine tRNA background, since the 5′-flanking regions, and hence the 5′-leader sequences, differ considerably, and, the precursor cDNA products of even the tRNA allogenes were capable of being analyzed by the reverse transcription technique. A comparison of HeLa cell and human placenta primer extension products indicates that some tRNA genes are expressed in HeLa cells or placenta, and *vice versa*. At least four *bona fide* valine tRNA genes encoding the mature mammalian major valine tRNA$_{IAC}$ and minor valine tRNA$_{CAC}$ were shown to be active *in vivo* by this procedure. Two other valine tRNA allogenes with variant nucleotides are expressed in placenta, raising the possibility of tissue- and/or developmental-specific valine tRNA expression. The precursor products of several uncharacterized tRNA genes were also detected in HeLa cell and placenta tRNA preparations. Amazingly, the analysis revealed the presence of the precursor products of three valine tRNA pseudogenes.[60,61] Thus, the results of Schmutzer and Gross[57] clearly demonstrate different valine tRNA populations between placenta and HeLa cells, and that variant tRNA genes are differentially expressed may suggest the presence of tRNA gene regulation for tissue-specific gene expression in higher mammalian systems.

There are about 14 nonallelic tRNA genes encoding the valine tRNA$_{GUU}$ isoacceptor in *Dictyostelium discoideum,* and 13 of these have been characterized.[45] The genes are identical, except for the 5′- and 3′-flanking regions, and the genes appear to be randomly dispersed among the 7 *Dictyostelium* chromosomes. The expression of each family member was analyzed in yeast, and all but one of the genes was actively transcribed. The primer extension technique was used to study the expression of the individual tRNA gene family members in this organism, since the 5′-flanking regions of the genes are different. A cDNA derived from a unique precursor tRNA was detected, being reverse-transcribed from a tRNA only present in late preaggregation phase, but not detected from growing cells, stationary phase cells, or early developmental cycle cells. The cognate precursor tRNA seems to be transcribed from an unusual tRNA gene that is composed of nucleotides 1 to 54 of a 3′-truncated tRNA gene linked to a *bona fide* tRNA gene.

VI. CONCLUSIONS

As discussed, it is well documented that the 5′- and 3′-extragenic regions of eukaryotic tRNA genes are implicated in positively (or sometimes negatively) modulating transcriptional activities of tRNA genes. From the results obtained from the laboratory groups of Sprague[17-19,21-22,24,44] and Clarkson,[32-33,36-38] and from a number of other research groups, it would appear that, in most cases, the regulation of tissue-specific or developmental-specific tRNA gene expression may be largely attributed to differences in the flanking sequences, especially in upstream motifs, of tRNA genes for interacting with RNA polymerase III and its transcription factors. As suggested by the results of Johnson et al.,[50] transcription factor:tRNA gene flanking region interactions probably underlie species-specific tRNA gene expression observed between the insect and higher eukaryotic RNA polymerase III transcription systems. The results of Gross and co-workers[57] strongly suggest that tRNA genes, tRNA allogenes, and even tRNA pseudogenes are differentially expressed *in vivo,* and again, that the flanking regions are probably implicated in the observed transcriptional differences between the members of a eukaryotic tRNA gene family. To date, no unique *trans*-acting transcription factors have been isolated for the transcription of a specific tRNA gene or battery of tRNA genes, for binding to particular *cis*-acting DNA sequence motifs. From the conclusions drawn by Sprague and co-workers,[21,22] the region of the tRNA transcription unit must be considerably larger than that previously considered in interactions of the RNA polymerase III transcription apparatus with the intragenic control regions. Thus, the mechanisms by which the flanking regions of tRNA genes affect their transcriptional efficiencies are still largely speculative. In fact, the results of some of these studies may be misleading, especially *in vitro* transcription studies, since the cell extracts used may vary with respect to content and/or presence of inhibitory/stimulatory components.

Intriguingly, recent findings indicate that promoter/enhancer elements usually associated with RNA polymerase II-directed transcription, can also occur upstream from a number of genes transcribed by RNA polymerase III (reviewed in References 13 and 14). For example, TATA or TATA-like motifs, usually associated with binding of the RNA polymerase II transcription factor IID (TFIID) for formation of the basal level transcription complex[63] also occur upstream from a number of genes transcribed by RNA polymerase III, including a 7SK RNA gene, U6 snRNA gene, and a number of tRNA genes.

TFIIIB, TFIIIC, and RNA polymerase III are heterogeneous multisubunit protein complexes, and only a few of their subunits have been characterized thus far.[41-43] Thus, it is possible that the regulation of differential tRNA gene expression in eukaryotes could be attributed to having a number of different TFIIIB, TFIIIC, or RNA polymerase III complexes, due to changes in either the subunit composition or perhaps covalent/noncovalent modifications of constitutive transcription factor or polIII subunits, or even due to interactions between the RNA polymerase II and polymerase III transcription systems. As aptly said by Sollner-Webb,[13] there remain many "surprises in polymerase III transcription."

ACKNOWLEDGMENTS

The authors wish to acknowledge support by the Texas Advanced Research Program Grant 003594-049 and University of North Texas Organized Research Funds.

REFERENCES

1. **Geiduschek, E. P. and Tocchini-Valentini, G. P.,** Transcription by RNA polymerase III, *Ann. Rev. Biochem.,* 57, 873, 1988.
2. **Palmer, J. M. and Folk, W. R.,** Unraveling the complexities of transcription by RNA polymerase III, *Trends Biol. Sci.,* 15, 300, 1990.
3. **Gabrielsen, O. S. and Sentenac, A.,** RNA polymerase III (C) and its transcription factors, *Trends Biol. Sci.,* 16, 412, 1991.
4. **Yoshinaga, S. K., Boulanger, P. A., and Berk, A. J.,** Resolution of human transcription factor TFIIIC into two functional components, *Proc. Natl. Acad. Sci. U.S.A.,* 84, 3585, 1987.
5. **Boulanger, P. A., Yoshinaga, S. K., and Berk, A. J.,** DNA-binding properties and characterization of human transcription factor TFIIIC2, *J. Biol. Chem.,* 262, 15098, 1987.
6. **Dean, N. and Berk, A.,** Separation of TFIIIC into two functional components by sequence-specific DNA affinity chromatography, *Nucl. Acids Res.,* 15, 9895, 1987.
7. **Kassavetis, G. A., Braun, B. R., Nguyen, L. H., and Geiduschek, E. P.,** S. cerevisiae TFIIIB is the transcription initiation factor proper of RNA polymerase III, while TFIIIA and TFIIIC are assembly factors, *Cell,* 60, 235, 1990.
8. **Leveillard, T., Kassavetis, G. A., and Geiduschek, E. P.,** *Saccharomyces cerevisiae* transcription factors IIIB and IIIC bend the DNA of a tRNA^{Gln} gene, *J. Biol. Chem.,* 266, 5162, 1991.
9. **Young, L. S., Dunstan, H. M., Witte, P. R., Smith, T. P., Ottonello, S., and Sprague, K. U.,** A class III transcription factor composed of RNA, *Science,* 252, 542, 1991.
10. **Chung, J., Sussman, D. J., Zeller, R., and Leder, P.,** The c-*myc* gene encodes superimposed RNA polymerase II and III promoters, *Cell,* 51, 1001, 1987.
11. **Ciliberto, G., Castagnoli, L., and Cortese, R.,** Transcription by RNA polymerase III, in *Current Topics in Developmental Biology,* Vol. 18., Moscona, A. A. and Monroy, A., Eds., Academic Press, New York, 1983, 59.
12. **Sharp, S. J., Schaack, J., Cooley, L., Burke, D. J., and Söll, D.,** Structure and transcription of eukaryotic tRNA genes, *CRC Crit. Rev. Biochem.,* 19, 107, 1985.
13. **Sollner-Webb, B.,** Surprises in polymerase III transcription, *Cell,* 52, 153, 1988.
14. **Murphy, S., Moorefield, B., and Pieler, T.,** Common mechanisms of promoter recognition by RNA polymerase II and III, *Trends Genet.,* 5, 122, 1989.
15. **Kunkel, G. R.,** RNA polymerase III transcription of genes that lack internal control regions, *Biochim. Biophys. Acta,* 1088, 1, 1991.
16. **Goldsmith, M. R. and Kafatos, F. C.,** Developmentally regulated genes in silkmoths, *Ann. Rev. Genet.,* 18, 443, 1984.
17. **Larson, D., Bradford-Wilcox, J., Young, L., and Sprague, K.,** A short 5'-flanking region containing conserved sequences is required for silkworm alanine tRNA gene activity, *Proc. Natl. Acad. Sci. U.S.A.,* 80, 3416, 1983.
18. **Young, L. S., Takahashi, N., and Sprague, K. U.,** Upstream sequences confer distinctive transcriptional properties on genes encoding silkgland-specific tRNA^{Ala}, *Proc. Natl. Acad. Sci. U.S.A.,* 83, 374, 1986.
19. **Underwood, D. C., Knickerbocker, H., Gardner, G., Condliffe, D. P., and Sprague, K. U.,** Silk gland-specific tRNA^{Ala} genes are tightly clustered in the silkworm genome, *Mol. Cell. Biol.,* 8, 5504, 1988.
20. **Candelas, G. C., Arroyo, G., Carrasco, C., and Dompenciel, R.,** Spider silkglands contain a tissue-specific alanine tRNA that accumulates *in vitro* in response to the stimulus for silk protein synthesis, *Dev. Biol.,* 140, 215, 1990.
21. **Wilson, E. J., Larson, D., Young, L. S., and Sprague, K. U.,** A large region controls tRNA gene transcription, *J. Mol. Biol.,* 183, 153, 1985.
22. **Young, L. S., Rivier, D. H., and Sprague, K. U.,** Sequences far downstream from the classical tRNA promoter elements bind RNA polymerase III transcription factors, *Mol. Cell. Biol.,* 11, 1382, 1991.

23. **Taneja, R., Gopalkrishnan, R., and Gopinathan, K. P.**, Regulation of glycine tRNA gene expression in the posterior silk glands of the silkworm *Bombyx mori, Proc. Natl. Acad. Sci. U.S.A.*, 89, 1070, 1992.
24. **Morton, D. G. and Sprague, K. U.**, *In vitro* transcription of a silkworm 5S RNA gene requires an upstream signal, *Proc. Natl. Acad. Sci. U.S.A.*, 81, 5519, 1984.
25. **Fournier, A., Guerin, M.-A., Corlet, J., and Clarkson, S. G.**, Structure and *in vitro* transcription of a glycine tRNA gene from *Bombyx mori, EMBO J.*, 3, 1547, 1984.
26. **Hipskind, R. A. and Clarkson, S. G.**, 5′-flanking sequences that inhibit *in vitro* transcription of a *Xenopus laevis* tRNA gene, *Cell*, 34, 881, 1983.
27. **Lin, V. K. and Agris, P. F.**, Alterations in tRNA isoaccepting species during erythroid differentiation of the Friend leukemia cell, *Nucl. Acids Res.*, 8, 3467, 1980.
28. **Shortridge, R. D., Johnson, G. D., Craig, L. C., Pirtle, I. L. and Pirtle, R. M.**, A human tRNA gene heterocluster encoding threonine, proline and valine tRNAs, *Gene*, 79, 309, 1989.
29. **Schmedt, E. and Kleiman, L.**, The measurement of the production of $tRNA_i^{Met}$ during erythroid differentiation of the Friend erythroleukemia cell, *Biochem. Cell Biol.*, 65, 188, 1987.
30. **Kleiman, L., Schmedt, E., and Miller, H.**, The independent regulation of $tRNA_i^{Met}$ and $tRNA^{Asn}$ synthesis during Friend cell erythroid differentiation, *Biochem. Cell Biol.*, 66, 772, 1988.
31. **Lin, F.-K., Furr, T. D., Chang, S. H., Horwitz, J., Agris, P. F., and Ortwerth, B. J.**, The nucleotide sequence of two bovine lens phenylalanine tRNAs, *J. Biol. Chem.*, 255, 6020, 1980.
32. **Clarkson, S. G., Birnstiel, M. L., and Serra, V.**, Reiterated transfer RNA genes of *Xenopus laevis, J. Mol. Biol.*, 79, 391, 1973.
33. **Müller, F. and Clarkson, S. G.**, Nucleotide sequence of genes coding for $tRNA^{Phe}$ and $tRNA^{Tyr}$ from a repeating unit of *X. laevis* DNA, *Cell*, 19, 345, 1980.
34. **Rosenthal, D. S. and Doering, J. L.**, The genomic organization of dispersed tRNA and 5S RNA genes in *Xenopus laevis, J. Biol. Chem.*, 258, 7402, 1983.
35. **Fostel, J., Narayanswami, S., Hamkalo, B., Clarkson, S. G., and Pardue, M. L.**, Chromosomal location of a major tRNA gene cluster of *Xenopus laevis, Chromosoma*, 90, 254, 1984.
36. **Müller, F., Clarkson, S. G., and Galas, D. J.**, Sequence of a 3.18 kb tandem repeat of *Xenopus laevis* DNA containing 8 tRNA genes, *Nucl. Acids Res.*, 15, 7191, 1987.
37. **Gouilloud, E. and Clarkson, S. G.**, A dispersed tyrosine tRNA gene from *Xenopus laevis* with high transcriptional activity *in vitro, J. Biol. Chem.*, 261, 486, 1986.
38. **Stutz, F., Gouilloud, E., and Clarkson, S. G.**, Oocyte and somatic tyrosine tRNA genes in *Xenopus laevis, Gen. Dev.*, 3, 1190, 1989.
39. **Wolffe, A. P. and Brown, D. D.**, Developmental regulation of two 5S ribosomal RNA genes, *Science*, 241, 1626, 1988.
40. **Wolffe, A. P.**, *Xenopus* transcription factors: key molecules in the developmental regulation of differential gene expression, *Biochem. J.*, 278, 313, 1991.
41. **Oei, S.-L. and Pieler, T.**, A transcription stimulatory factor binds to the upstream region of *Xenopus* 5S RNA and tRNA genes, *J. Biol. Chem.*, 265, 7485, 1990.
42. **Cohen, I. and Reynolds, W. F.**, The Xenopus YB3 protein binds the B box element of the class III promoter, *Nucl. Acids Res.*, 19, 4753, 1991.
43. **Schneider, H. R., Waldschmidt, R., and Seifart, K. H.**, Human transcription factor IIIC contains a polypeptide of 55 kDa specifically binding to Pol III genes, *Nucl. Acids Res.*, 18, 4743, 1990.
44. **Sprague, K. U., Larson, D., and Morton, D.**, 5′ Flanking sequence signals are required for activity of silkworm alanine tRNA genes in homologous in vitro transcription systems, *Cell*, 22, 171, 1980.

45. Dingermann, T., Amon-Böhm, E., Bertling, W., Marschalek, R., and Nerke, K., A family of non-allelic tRNA$_{GUU}^{Val}$ genes from the cellular slime mold *Dictyostelium discoideum*, *Gene*, 73, 373, 1988.
46. Schaack, J, Sharp, S, Dingermann, T., Burke, D. J., Cooley, L., and Söll D., The extent of a eukaryotic tRNA gene, *J. Biol. Chem.*, 259, 1461, 1984.
47. Lofquist, A. and Sharp, S., The 5′-flanking sequences of *Drosophila melanogaster* tRNA$_5^{Asn}$ genes differentially arrest RNA polymerase III, *J. Biol. Chem.*, 261, 14600, 1986.
48. Sajjadi, F. G., Miller, R. C., Jr., and Spiegelman, G. B., Identification of sequences in the 5′ flanking region of a *Drosophila melanogaster* tRNA$_4^{Val}$ gene that modulate its transcription in vitro, *Mol. Gen. Genet.*, 206, 279, 1987.
49. Sajjadi, F. G. and Spiegelman, G. B., Modulation of a *Drosophila melanogaster* tRNA gene transcription in vitro by a sequence TNNCT in its 5′ flank, *Gene*, 60, 13, 1987.
50. Johnson, D. L., Fan, R. S., and Treinies, M. L., Analysis of the molecular mechanisms for the species-specific transcription of *Drosophila* and human tRNA gene transcription components, *J. Biol. Chem.*, 266, 16037, 1991.
51. Raymond, K. C., Raymond, G. J., and Johnson, J. D., In vivo modulation of yeast tRNA gene expression by 5′-flanking sequences, *EMBO J.*, 4, 2649, 1985.
52. Johnson, J. D. and Raymond, G. J., Three regions of a yeast tRNA$_3^{Leu}$ gene promote RNA polymerase III transcription, *J. Biol. Chem.*, 259, 5990, 1984.
53. Raymond, G. J. and Johnson, J. D., The 5′-flanking sequence of yeast tRNA^{Leu3} genes enhances the rate of transcription from stable preinitiation complexes, *Nucl. Acids Res.*, 15, 9881, 1987.
54. Drabkin, H. J. and RajBhandary, U. L., Attempted expression of a human initiator tRNA gene in *Saccharomyces cerevisiae*, *J. Biol. Chem.*, 260, 5596, 1985.
55. Francis, M. A. and RajBhandary, U. L., Expression and function of a human initiator tRNA gene in the yeast *Saccharomyces cerevisiae*, *Mol. Cell. Biol.*, 9, 4486, 1990.
56. Hofmann, J., Schumann, G., Borschet, G., Gösseringer, R., Bach, M., Bertling, W. M., Marschalek, R., and Dingermann, T., Transfer RNA genes from *Dictyostelium discoideum* are frequently associated with repetitive elements and contain consensus boxes in their 5′ and 3′-flanking regions, *J. Mol. Biol.*, 222, 537, 1991.
57. Schmutzler, C. and Gross, H. J., Genes, variant genes, and pseudogenes of the human tRNA^{Val} gene family are differentially expressed in HeLa cells and in human placenta, *Nucl. Acids Res.*, 18, 5001, 1990.
58. Leung, J., Sinclair, D. A. R., Hayashi, S., Tener, G. M., and Grigliatti, T. A., Informational redundancy of tRNA$_4^{Ser}$ and tRNA$_7^{Ser}$ genes in *Drosophila melanogaster* and evidence for intergenic recombination, *J. Mol. Biol.*, 219, 175, 1991.
59. Arnold, G. J., Schmutzler, C., Thomann, U., van Tol, H., and Gross, H. J., The human tRNA^{Val} gene family: organization, nucleotide sequences and homologous transcription of three single-copy genes, *Gene*, 44, 287, 1986.
60. Kahnt, B., Frank, R., Blöcker, H., and Gross, H. J., An efficiently transcribed human tRNA^{Val} gene variant produces a stable pre-tRNA: repair of the processing defect by *in vitro* mutations, *DNA*, 8, 51, 1989.
61. Thomann, H.-U., Schmutzler, C., Hüdepohl, U., Blow, M., and Gross, H. J., Genes, variant genes and pseudogenes of the human tRNA^{Val} gene family: expression and pre-tRNA maturation *in vitro*, *J. Mol. Biol.*, 209, 505, 1989.
62. Craig, L. C., Wang, L. P., Lee, M. M., Pirtle, I. L., and Pirtle, R. M., A human tRNA gene cluster encoding the major and minor valine tRNAs and a lysine tRNA, *DNA*, 8, 457, 1989.
63. Pugh, B. F. and Tjian, R., Diverse transcriptional functions of the multisubunit eukaryotic TFIID complex, *J. Biol. Chem.*, 267, 679, 1992.

Chapter 7

CODON PAIR UTILIZATION BIAS IN BACTERIA, YEAST, AND MAMMALS

G. Wesley Hatfield and George A. Gutman

TABLE OF CONTENTS

I. Introduction 158

II. Codon Context and Translational Efficiency 158
    A. Codon Context – Nonsense versus Missense Suppression 158
    B. Bases Adjacent to the Codon Triplet Influence Nonsense Suppression 159
    C. tRNA-tRNA Interactions on the Surface of a Translating Ribosome 160
    D. Translational Efficiency of Frequently and Infrequently Used Codons 162

III. Biases in Codon Pair Utilization 164
    A. Codon Pair Bias in *E. coli* 164
    B. Codon Pair Bias in Yeast, Human, Rat, and Mouse 165
    C. Nonadjacent Codon Pair Bias 171

IV. Functional Significance of Codon Pair Utilization Patterns 173
    A. Codon Pairs and Translational Step-time 173
    B. Protein Folding 175
    C. Interactions between Nascent Polypeptide Chains and Cellular Factors 177
    D. Protein Stability 178
    E. Regulation of Protein Expression 178
    F. Messenger RNA Stability 179
    G. Metabolic Growth Rate Control in *E. coli* 180
    H. Practical Applications of Codon Pair Rules 182
        1. Optimizing Genetically Engineered Systems 182
        2. Identification of Protein-Coding Regions 183

Acknowledgments 184

References 184

I. INTRODUCTION

While it is generally agreed that the translational efficiency of a codon is affected by neighboring mRNA sequences, the physicochemical basis of these context effects is more controversial. Evidence supporting the view that neighboring bases influence codon:anticodon interactions has been supplied by studies on nonsense suppressor tRNAs. It has also been argued that these bases are important because they are part of a neighboring codon, and that codon context effects are the consequence of the compatibilities of adjacent tRNAs on the surface of a translating ribosome. In either case, nonsense suppression efficiencies may not be necessarily related to translational efficiencies since nonsense suppression is dependent upon the outcome of a competition between polypeptide chain release factors and suppressor tRNAs at the nonsense codons. Missense suppression studies, however, are not affected by this criticism and may provide a better measure of translational efficiency, but fewer studies involving context effects on missense suppression have been performed. Attempts to directly measure the translational efficiencies of selected codons have provided contradictory results. The results of some experiments support the contention that infrequently used codons are translated slowly while frequently used codons are translated rapidly. Other experiments support the conclusion that there is no relationship between the translatability of a codon and the frequency of its usage or the abundance of its cognate tRNA.

In Section II of this chapter we discuss these differing views of codon context and we conclude our discussion in support of the idea that interactions between adjacent tRNA molecules at the A(minoacyl)- and P(eptidyl)-sites of a translating ribosome affect translational step-times and are a major contributor to codon context bias. In Section III we present and discuss the evidence that codon pair utilization patterns are highly biased, above and beyond previously noted biases in protein coding sequences, and that they differ between organisms. This observation fulfills a prediction of the notion that tRNA-tRNA interactions at adjacent codons influence translational step-times. In Section IV we suggest that translational step-times in *Escherichia coli* are related, perhaps in a predictable way, to the species-specific patterns of codon pair utilization and we discuss the possible functional and practical significance of these observations.

II. CODON CONTEXT AND TRANSLATIONAL EFFICIENCY

A. CODON CONTEXT – NONSENSE VERSUS MISSENSE SUPPRESSION

Many studies support the idea that the ability of a suppressor tRNA to facilitate the translation of a nonsense codon is influenced by nucleotide sequences on either side of the nonsense codon.[1-13] Indeed, such studies defined the phenomenon commonly referred to as "codon context". However, because the efficiency of nonsense suppression reflects the competition, on the surface of a translating ribosome, between aminoacyl-tRNA-mediated polypeptide chain propagation and release factor-mediated peptidyl-tRNA release, sequences surrounding a nonsense triplet could also influence the recognition of release factors. Therefore, nonsense suppression efficiencies might not be a true measure of translation efficiencies.[3,14-16] The experiments of Martin et al.[17] confirmed that this is, in fact, the case. They demonstrated a difference in the relative activities of release factors RF1 and RF2 of *E. coli* at UAA nonsense codons located at 13 separate sites in the *lacI* gene and showed a correlation between the activities of the release factors at these sites and the suppression efficiency of each site. Thus, a significant part of nonsense suppression can be attributed to context-specific effects on polypeptide chain termination.

However, this cannot be the whole story. If nonsense suppression efficiency were determined solely by polypeptide chain termination efficiencies, then all tRNA suppressors would behave in a similar manner with respect to context, i.e., they would be efficient at sites where the chain termination mechanism is weak and inefficient where the chain termination mechanism is optimized. This is not always the case, as evidenced by Bossi's demonstration[2] that four different tRNA suppressors behave differently with respect to the context of many nonsense codons in *lacI*. The complications associated with polypeptide chain termination are not encountered when missense suppression is used to examine codon context effects, and Murgola et al.[18,19] have demonstrated that codon context can affect the efficiency of suppression of missense mutations. They showed that the efficiency of suppression of missense codons at two positions in the *trpA* gene can vary over a 50-fold range when adjacent sequences are changed. These measurements may more closely reflect the translational efficiency of natural tRNAs during polypeptide chain elongation than those of nonsense suppressors.

B. BASES ADJACENT TO THE CODON TRIPLET INFLUENCE NONSENSE SUPPRESSION

The base on the 5' side of the anticodon triplet of all known elongator tRNAs is uracil, and exhaustive studies by Miller and Albertini[13] and Bossi[2] have demonstrated that suppressor tRNA efficiency is often, but not always, higher when the codon 3' of the suppressed nonsense codon begins with A or G. This led to speculation that efficient suppressor tRNAs, and perhaps other tRNAs as well, form base pairs not only with the three bases of the anticodon and the codon but also with a fourth base, the invariant U on the 5' side of the anticodon, as well as the first base of the next codon if it is A or G. This observation raised the possibility that interactions between mRNA and nucleotides outside the anticodon might improve the codon:anticodon interaction and, by extension, translational efficiency. Indeed there are examples of tRNAs that cause +1 frameshifts by reading four bases in the mRNA, and some of these have an additional base on the 3' side of the anticodon which participates in the extended base pairing.[20] Also, Taniguchi and Weissmann[21] have shown that the affinity of fMet-tRNA for the initiator codon of the Qβ coat cistron is increased threefold when the nucleotide on the 3' side of the AUG codon is changed from a G to an A, and Uhlenbeck et al.[22] have shown a tenfold better binding of fMet-tRNA to an AUGA tetra-nucleotide (in the absence of ribosomes) than to the AUGU, AUGC, or AUGG tetra-nucleotides or to the AUG tri-nucleotide. Furthermore, Grosjean et al.[23] have demonstrated that the nucleotide adjacent to the 3' side of a codon influences the stacking energy of anticodon/anticodon complexes of tRNA pairs. While these studies suggest that codon:anticodon stacking energies are an important determinant for codon:anticodon interactions, others have found that codon composition does not seem to influence translational efficiency. Andersson et al.[24] have shown that there is no preferential use of G+C-rich over A+U-rich mRNAs as templates for *in vitro* protein synthesis reactions. This result, taken together with the fact that different organisms use different subsets of preferred codons, argues that the energetics of codon:anticodon base-pair interactions is not the decisive factor in determining either codon usage or translational efficiency. Bossi[2] has also proposed that it is unlikely that fourth base interactions are important for codon context effects. He points out that tRNAs which read codons starting with A have the same modified adenine on the 3' side of the anticodon while a different adenine modification is present at the same position of most tRNAs that read codons starting with U.[25] Therefore, because the modified base of the aminoacyl-tRNA in the A-site is juxtaposed to the invariant U on the 5' side of the anticodon of the peptidyl-tRNA in the P-site, an alternative explanation for the strong, but not absolute, correlation between efficient suppression and an A in the first position of the codon 3' to the nonsense codon might involve tRNA-tRNA interactions.

C. tRNA-tRNA INTERACTIONS ON THE SURFACE OF A TRANSLATING RIBOSOME

Convincing arguments in favor of the idea that the compatibility of adjacent tRNA molecules in the A- and P-sites on the surface of a translating ribosome is a primary contributor to codon context effects have been advanced by Bossi and Roth.[6] Their experiments suggest that tRNA interactions at the A- and P-sites of a translating ribosome are more important than codon usage, or perhaps even codon:anticodon base pair interactions, in determining the translational efficiency of a given codon in a given context. They showed that certain amber mutations in the *hisD* gene of *Salmonella typhimurium* which were efficiently suppressed by the glutamine-inserting *supE* tRNA were not suppressed in strains containing a *hisT* mutation. The *hisT* gene encodes an enzyme, pseudouridylate synthetase, that modifies U residues in the anticodon region of several tRNAs (the *supE* suppressor tRNA contains two pseudouridine residues in its anticodon region). Consequently, these *hisT* and *hisD* strains are histidine auxotrophs. Reversion to histidine prototrophy yielded a mutant strain that retained the amber mutation in the *hisD* gene and the *hisT* genotype, but changed the sequence of the codon on the 3′ side of the amber codon. This result demonstrated that the interaction of the undermodified suppressor tRNA with the nonsense codon was influenced by mRNA sequences outside of the triplet codon, a result consistent with an increased compatibility between the structurally-altered suppressor tRNA and the tRNA isoacceptor that decodes the new adjacent codon.

Genetic evidence for interactions between adjacent tRNAs in the A- and P-sites of a translating ribosome has been provided by Smith and Yarus.[26] They devised a clever *in vivo* assay to demonstrate that mutation-directed structural alterations on the 5′ side of the anticodon loop of tRNA in the P-site of a translating ribosome can affect the translational efficiency of the neighboring 3′ codon. The ability of a glutamine-inserting $tRNA^{Trp}_{UAG}$ suppressor to facilitate the expression of a β-galactosidase gene containing the sequence GCT(Val)-TAG(Trp)-TGG(Trp)-CGA(Arg) was compared with the ability of this same $tRNA^{Trp}$ whose anticodon was altered to suppress a TGA nonsense codon in the same coding context. In each case, $tRNA^{Trp}$ occupies adjacent codons. The authors reasoned that if the $tRNA^{Trp}$ molecules do not interact with each other or with adjacent tRNA molecules, then the suppression efficiency of a third test gene containing both of the nonsense codons, GCT(Val)-TAG(Trp)-TGA(Trp)-CGA(Arg), in a strain containing both $tRNA^{Trp}$ suppressors should be equal to the product of the suppression efficiencies of the two test genes containing one or the other of the nonsense codons. This "expected" suppression efficiency, measured for the unaltered $tRNA^{Trp}_{UAG}$ and $tRNA^{Trp}_{UGA}$ suppressors, was compared with the suppression efficiencies of these tRNA molecules with site-directed mutations on the 3′ or 5′ side of the anticodon loop. Large effects (greater than 30-fold) were observed when purines were substituted for pyrimidines at base positions 32 and 33 on the 5′ side of the tRNA reading the 5′ codon of the tandem nonsense codon pair, i.e., the tRNA in the P-site. Smaller but measurable effects were observed when the tRNAs were altered on the 3′ side of the anticodon. The authors suggest that this functional interference between tRNAs at adjacent codons is due to physical interactions between the bases on the 5′ side of the anticodon loop of the tRNA in the P-site and bases on the 3′ side of the anticodon loop of the tRNA in the A-site. The substitution of the larger purine bases may partially occlude the A-site and reduce the rate at which the A-site tRNA is accepted at the A-site codon. The authors further suggest that this observation might explain the nearly universal occurrence of pyrimidines on the 3′ side of the anticodons of naturally occurring tRNAs. These results support the idea that tRNA-tRNA interactions on the surface of a ribosome can affect translational step-times and contribute in a major way to codon context effects.

D. TRANSLATIONAL EFFICIENCY OF FREQUENTLY AND INFREQUENTLY USED CODONS

Two general hypotheses have been advanced to explain how codon context effects might affect translation rates. One is that the translation rates of individual codons reflect the abundance of their cognate tRNAs, i.e., the more frequently used codons are translated faster than the more infrequently used codons. Another more recent hypothesis favors the idea that translation rates are influenced by the compatibilities of adjacent tRNAs in the A- and P-sites on the surface of translating ribosomes.

It has been known for some time that codon usage is highly biased and varies considerably from organism to organism. It is also known that the number, abundance, primary sequence, and chemical modifications of tRNAs vary among organisms. Within each organism, however, there exists a close correspondence between codon usage and tRNA abundance, and infrequently used codons are disproportionately represented in the coding regions of genes expressed at high levels.[27,28] However, these observations do not directly address the question of whether translational efficiency is related to tRNA abundance and codon usage or to tRNA-tRNA interactions.

The possibility that biases in codon usage can influence polypeptide elongation rates has been widely discussed and, although direct effects of codon choice on translation rates are difficult to demonstrate, Sørensen et al.[29] have presented data in support of the idea that infrequently used codons are translated more slowly than frequently used codons in *E. coli*. In one construct they inserted a 24-codon coding region of the highly expressed *rpsA* gene in-frame into the 3' end of the *lacZ* gene; in another construct they inserted 19 codons from the same region of the *rpsA* gene, this time in a +1 *rpsA* reading frame, into the same site in the 3' end of the *lacZ* gene. The 24 codons of the *rpsA* gene inserted in-frame were all frequently used codons, whereas 11 of the 19 newly created codons of the +1 *rpsA* reading frame were frequently used codons and the remaining 8 codons were infrequently used codons. The time required to translate the wild type *lacZ* mRNA, measured by pulse labeling with ^{35}S-methionine, was determined to be 82 seconds, an average translation time of 12.5 amino acids per second. This translation time for the *lacZ* message was increased by 2 seconds when the 24 common codons were inserted and by 4.7 seconds when the 11 common and 8 infrequent codons were inserted. On the basis of these results, Sørensen et al.[29] calculated that the average translation rate of the 24 common codons of the highly expressed *rpsA* gene was 12 amino acids per second and the average translation rate for the 19-codon sequence, containing 8 rare codons, was 5 amino acids per second. They reasoned that since the common codons were translated at a rate of 12 amino acids per second then the infrequently used codons in this sequence must have been translated at a rate of 2.1 amino acids per second. They concluded from these experiments that rare codons are translated slowly, abundant codons are translated rapidly and the difference in translation rate between infrequently and frequently used codons is approximately sixfold.

Attenuation is a genetic regulatory mechanism employed by bacteria that is sensitive to the relative rates of transcription and translation through the attenuator region of genes regulated by this mechanism. Jensen[30] showed that transcription through the translation rate-sensitive attenuators of the *pyrE* and *pyrBI* operons was decreased in streptomycin-dependent *E. coli* cells containing a mutation in the *rpsL* gene. This mutation reduces the average rate of translation measured in wild-type cells, 16 amino acids per second in these experiments, to about 5 amino acids per second in the mutant cells. Bonekamp et al.[31] used the translation rate-sensitive *pyrE* attenuator system to obtain evidence in support of the idea that translation rates of individual codons are *not* correlated with tRNA abundance or codon usage. They replaced three codons in the leader polypeptide coding sequence of the leader-attenuator region with sets of three, tandem, frequently, or infrequently used codons. While these experiments showed that codon choice can influence translation elongation rates and, thereby, exert large effects on *pyrE* and *pyrBI* attenuator function, they failed to provide any evidence for a correlation between this measurement and codon usage. Specifically, Bonekamp et al.[31] demonstrated that codons served by minor tRNA species, such as the UGU codon for cysteine and the AUA codon for isoleucine, can be translated as rapidly as codons served by abundant tRNA species, such as the CGU codon for arginine and the AUU and AUC codons for isoleucine. In fact, these infrequently used codons were translated even faster than the most frequently used codon in *E. coli*, the CUG codon for leucine. These results show that tRNA abundance is not the sole determinant of translational step-times and support the idea that tRNA-tRNA interactions are important components of codon context.

Robinson et al.[32] have demonstrated that, under special circumstances, rare codons can influence translational efficiency. However, their results suggest that the rate limiting step at a rare codon during polypeptide chain elongation is at the level of aminoacyl-tRNA availability rather than the nature of the codon context, an argument supported in recent reviews by Jensen and Pedersen[33] and by Andersson and Kurland.[34] Robinson et al.[32] point out that approximately 25% of the codons of the chloramphenicol acetyltransferase (CAT) gene are codons infrequently used by *E. coli*, yet this gene can be expressed at high levels in this organism. However, when they inserted four *consecutive* rare AGG arginine codons into the CAT coding region, translation of the altered CAT mRNA was decreased, an effect which was not observed when the major arginine codon CGT was used in the place of the rare AGG codons. Thus, the availability of a minor arginyl-tRNA isoacceptor species can become rate limiting for protein synthesis, but only under extreme circumstances, i.e., when the rare codon is tandemly repeated and the gene is expressed at very high levels. It is significant that the normal CAT gene can be expressed at very high levels, suggesting that the cellular concentrations of aminoacyl-tRNA isoacceptors for most rare codons do not normally fall below levels that compromise the translation of even highly expressed messages. It is also of interest to note that the codon pair AGG-AGG has not appeared in the protein coding region of any *E. coli* gene in our database of approximately 75,000 codon pairs.[35]

A study by Sharp and Li[36] shows that regulatory genes of *E. coli* which are expressed at low levels do not possess a significantly higher level of rare or infrequently used codons than a large number of other genes expressed at moderate to low levels. They suggest that the pattern of codon usage in genes expressed at low levels could simply arise from a comparatively low level of selection against infrequently used codons, and that, while codon usage is correlated with the level of gene expression, it need not be correlated with translation rates. This suggestion is consistent with the conclusions of Robinson et al.[32] and Bonekamp et al.,[31] and has been supported by Jensen and Pedersen[33] and Andersson and Kurland.[34]

On the basis of this brief examination of the literature pertaining to the nature of context effects, we are convinced that a major cause of varying translational step-times in a growing polypeptide chain relates to the structural compatibility of adjacent tRNA molecules on the surface of a translating ribosome. We do not mean, however, to imply that other factors, including mRNA secondary structures, tRNA:ribosome interactions, tRNA:protein interactions, and codon:anticodon interactions, are not also involved in tRNA isoacceptor selection at a given codon in a particular codon context.

III. BIASES IN CODON PAIR UTILIZATION

If tRNA compatibilities play a major role in determining translational step-times, and if the regulation of translational step-times is an important aspect of protein synthesis, as we believe it is, then a clear prediction emerges. The use of adjacent codons (codon pairs) that determine the positioning of tRNAs next to one another on the ribosome should be subject to a high degree of selection, and there should be a substantially nonrandom pattern of utilization of the 3721 (61^2) possible pairs of nonterminating codons in protein coding sequences. In this section we review published and unpublished evidence that codon pair utilization patterns are indeed highly biased, and that this bias is independent of biases in codon usage and in dinucleotide and amino acid pair frequencies. We also report that codon pair utilization patterns differ markedly between organisms, and discuss an intriguing nonadjacent bias in codon pair utilization most strongly apparent in mammalian genes.

A. CODON PAIR BIAS IN *E. COLI*

Our analyses[35] of codon pair bias in *E. coli* were based on a nonredundant collection of protein-coding regions taken from the GENBANK sequence database (Release 40.0) consisting of 75,403 codon pairs in 237 sequences. Codon usage frequencies for the 61 codons were determined for each sequence independently, and used to calculate the expected values for each of the 3721 codon pairs (EXP1). The use of codon frequencies in each sequence, as opposed to using global values, minimizes the contribution to the overall bias of differing codon usage between genes, which is known to be substantial.[37] Comparison of these expected values with the observed values yielded a set of chi-square values which we refer to as CHISQ1. We then calculated a new set of expected values for each codon pair (EXP2) in such a manner as to remove that component of the codon pair bias associated with bias in amino acid nearest neighbors; this yielded a new set of chi-square values (CHISQ2). Thus, the bias represented by CHISQ2 cannot be the consequence either of bias in codon usage *per se* (since the actual codon frequencies were used to calculate the expected values), or of bias in amino acid nearest neighbors. Prompted by our more recent studies on eukaryotic databases, we have applied an additional correction for bias in dinucleotide frequencies, yielding a new set of chi-square values (CHISQ3); as to be discussed in some detail on the following pages, this is of minimal importance for *E. coli*, although it is a very significant factor for the mammalian databases.

Our initial studies[35] were carried out independently of the earlier work on this same subject by Kolaskar and Reddy.[38] They analyzed a database similar in size to ours, consisting of protein-coding sequences from *E. coli* together with nine coliphages, and they corrected for bias in amino acid nearest neighbors in a manner identical to ours. In the discussion to follow, unreferenced assertions are drawn from our own work.

Several conclusions were drawn from our analyses of the *E. coli* database. First, as shown in Figure 1, there is clearly a high degree of bias in codon pair utilization, confirming the results of Kolaskar and Reddy.[38] The sum of CHISQ3 is more than 120 standard deviations removed from its expected value, and the manner in which these calculations were carried out ensures that this was not the simple consequence of nonrandomness in codon usage, amino acid nearest neighbors, or dinucleotides. Second, also shown in Figure 1, the vast majority of this bias represents a *short-range* effect; analyzing codon pairs separated by two or three intervening codons removes more than 95% of the excess chi square. Third, the bias is relatively independent of the *abundance* of codon pairs, i.e., there is little correlation between codon pair usage and codon pair bias (we define the *abundance* of a codon pair simply as the number of times it is observed in the database; this, in turn, is related to the relative frequency of its constituent codons, since frequently used codons will tend to be members of abundant pairs). Abundant pairs can show high degrees of either over or underrepresentation, as can rarely used pairs, and the correlation between the chi-square and the abundance of codon pairs is poor. Fourth, there is a high degree of *directionality* evident in this bias, there being very little correlation between the values of chi-square for any given codon pair and its reverse counterpart, i.e., pair A-B vs. B-A. Kolaskar and Reddy[38] additionally showed that codon pair bias in *E. coli* is independent of differences in stacking energies and in the secondary structures attributed to the codon pairs.

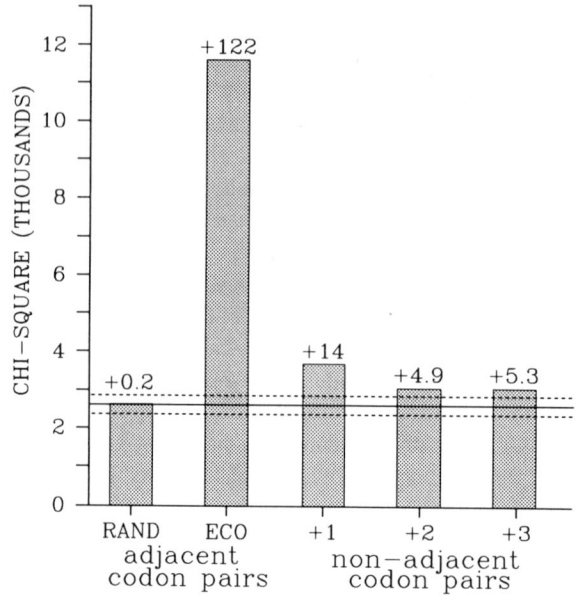

FIGURE 1. Total CHISQ3 (corrected for both amino acid pair and dinucleotide biases) for various modes of analysis of the *E. coli* database. The numbers above each bar represent the number of standard deviations (SD) by which the value exceeds its expectation. The solid horizontal line represents the expected value, and the broken lines represent +/−3 S.D. Adjacent codon pairs were evaluated for the *E. coli* database (ECO) as well as for a randomized version of the same sequences (RAND). Nonadjacent codon pairs were evaluated at separations of 1, 2, and 3 intervening codons, as indicated.

All of these features are consistent with the hypothesis that codon pair bias is a consequence of tRNA-tRNA interactions. In addition, we found that genes expressed at high versus low levels use very different proportions of highly overrepresented versus underrepresented codon pairs; specifically, those *genes expressed at high levels tend to avoid highly overrepresented pairs* (in addition to the well-known avoidance of infrequently used codons). This suggests that (at least a portion of) codon pair bias is related to effects on the translation process.

While the codon pair bias we presented clearly reflects some of the features of "codon context" involving adjacent nucleotides, which have been described by others[39-41] (and discussed in Section II), these two effects are *not* equivalent, despite suggestions to the contrary.[42] As examples, we described the fact that only 5 or 6 codon pairs contribute to the general excess of the 16 pairs of the form AAA-GXX,[39] and that the excess of pairs of the form GCC-XGX[40] is contradicted by a marked *deficit* of the particular pair GCC-GGC (CHISQ2 = −39). One of the striking features described by Gouy[41] in his extensive analysis of codon contexts in *E. coli* is the "III-I" bias, which involves the 3′ nucleotide of one codon and the 5′ nucleotide of its 3′ neighbor. We have more recently evaluated the contribution of such dinucleotide bias to our analyses, and have shown that it represents a negligible component of the *E. coli* codon pair data (full discussion to follow).

"Codon context", in fact, may simply be a description of the average behavior of a set of codon pairs of a given context class, each pair subject to natural selection (as we suggest) on the basis of tRNA-tRNA interactions. However, it certainly remains possible that codon pair bias may be reflecting, at least in part, the consequences of independent selection for particular "contexts". Although the two effects are not equivalent, they might both contribute to a complex balance of selection for a combination of common and distinct features.

B. CODON PAIR BIAS IN YEAST, HUMAN, RAT, AND MOUSE

All of the data we have discussed so far are derived from *E. coli* protein-coding sequences. It is of interest to know if codon pair usage is similarly biased in other organisms. If such bias exists, does it show the same properties as in *E. coli*, and is it species-specific? Interspecific comparisons should contribute to an understanding of the physiological basis of codon pair bias, and will clearly be of considerable practical importance if these effects are related to translational efficiency. To address these questions, we have recently undertaken an analysis of codon pair utilization in a nonredundant collection of coding sequences of the yeast *Saccharomyces cerevisiae* (YST), human (HUM), the laboratory rat *Rattus norvegicus* (RAT), and the laboratory mouse *Mus musculus* (MUS). Our YST database consists of 75,096 codon pairs in 177 sequences, and HUM contains 175,927 codon pairs in 506 sequences, both from GenBank Release 54.0; MUS (200,585 codon pairs in 537 sequences) and RAT (238,672 codon pairs in 660 sequences) were both drawn from GenBank Release 63.0.

Analyses of these eukaryotic sequences, particularly those of the mammals, immediately revealed the importance of a major new element, namely dinucleotide bias. Some (although far from all) of this phenomenon is the consequence of well-known dinucleotide biases, e.g., the striking deficit of the dinucleotide CpG in the genome of higher eukaryotes. We have corrected for dinucleotide biases in a fashion analogous to that described for amino acid pair bias,[35] which yielded a new set of expected (EXP3) and chi-square (CHISQ3) values. As was the case with amino acid pair bias, we make no assumptions about the direction of causality, but simply remove that portion of the codon pair bias associated with bias in dinucleotides. It is important to note that we are concerned with (and correct for) only those dinucleotides formed *between* adjacent codon pairs; any bias of dinucleotides *within* codons (codon triplet positions I-II and II-III) will directly affect codon usage and is, therefore, automatically taken into account in our calculations.

Table 1 illustrates the effects of our corrections for both amino acid pair bias and dinucleotide bias, using representative data drawn from the HUM data-

TABLE 1
Selected Records from the Human (HUM) Database Illustrating Codon Pair Bias

AA1	AA2	COD1	COD2	OBS	EXP1	CHIQS1	EXP2	CHISQ2	EXP3	CHISQ3
Met	Lys	ATG	AAG	198	139.0	25.1	183.7	1.1	182.5	1.3
Met	Asn	ATG	AAT	83	64.2	5.5	69.4	2.7	68.9	2.9
Met	Thr	ATG	ACA	49	53.8	-0.4	54.2	-0.5	53.8	-0.4
Met	Thr	ATG	ACC	81	91.6	-1.2	92.2	-1.4	91.6	-1.2
Met	Thr	ATG	ACG	20	24.7	-0.9	24.9	-1.0	24.7	-0.9
Met	Thr	ATG	ACT	70	48.5	9.6	48.8	9.3	48.5	9.6
Asp	Ala	GAC	GCC	68	161.2	-53.8	146.4	-42.0	62.2	0.5
Asp	Ala	GAC	GCG	17	36.1	-10.1	32.8	-7.6	13.9	0.7
Asp	Ala	GAC	GCT	21	89.0	-51.9	80.8	-44.3	34.4	-5.2
Asp	Gly	GAC	GGA	13	77.6	-53.8	81.0	-57.1	34.4	-13.4
Asp	Gly	GAC	GGC	98	137.9	-11.5	144.0	-14.7	61.2	22.1
Asp	Gly	GAC	GGG	42	84.9	-21.7	88.6	-24.5	37.7	0.5
Cys	Gly	TGT	GGA	58	34.5	16.0	32.5	20.0	52.4	0.6
Cys	Gly	TGT	GGC	64	44.2	8.9	41.6	12.1	67.1	-0.1
Cys	Gly	TGT	GGG	73	29.3	65.4	27.6	74.9	44.4	18.4
Cys	Gly	TGT	GGT	27	20.6	2.0	19.4	3.0	31.2	-0.6

Note: AA1, AA2 – Left- and right-hand encoded amino acids of each codon pair; COD1, COD2 – Left- and right-hand codons of each pair; OBS – Total observed number of occurrences of each pair; EXP1,2,3 – Expected values (EXP) and chi-square (CHISQ) values, uncorrected ("1"), corrected for bias in amino acid nearest neighbors ("2"), and additionally corrected for bias of dinucleotides ("3"), as described in the text. All CHISQ values are represented as *positive* values for overrepresented pairs (OBS > EXP), and as *negative* values for underrepresented pairs (OBS < EXP).

TABLE 2
Chi-Square Values for *E. coli* (ECO), *S. cerevisiae* (YST), and Human (HUM) Codon Pair Databases

	ECO	YST	HUM
CHISQ1[a]	13365 (+131 SD)	6812 (+45 SD)	30822 (+314 SD)
CHISQ2[b]	12066 (+128 SD)	5773 (+40 SD)	28826 (+313 SD)
CHISQ3[c]	11617 (+122 SD)	5048 (+31 SD)	11843 (+105 SD)

Note: It should be noted that the values of CHISQ for the different species do not represent differences in the *magnitude* of the bias, but only the probability of observing such a bias; this, in turn, is directly related to the relative sizes of the databases (the ECO and YST databases are approximately of equal size, while HUM is about 2.3 times larger than the other two).

[a] Without correction.
[b] Corrected for amino acid pair bias.
[c] Additionally corrected for dinucleotide bias.

base. Our initial analysis shows the first codon pair in this table (ATG-AAG) to be highly overrepresented (CHISQ1=25). However, this bias is associated with a high degree of overrepresentation of the amino acid pair Met-Lys, and it virtually disappears after correcting for this factor (CHISQ2 = 1.1). The dinucleotide formed between codons of this first group of six pairs (G-A) shows little overall bias in the HUM database, and this is reflected in the fact that there is little difference between the values of CHISQ2 and CHISQ3 in this group.

The second group of six codon pairs all form the highly underrepresented C-G dinucleotide between them, and they are all substantially underrepresented even after correcting for amino acid pair bias (CHISQ2 ranges from –7.6 to –57). Following correction for the dinucleotide bias, only one pair remains highly underrepresented (GAC-GGA, CHISQ3 = –13), but another is now substantially overrepresented (GAC-GGC, CHISQ3 = 22), while three others are found at levels close to their new expected values.

The third group of codon pairs all form the highly overrepresented dinucleotide T-G between them. Three of the four pairs are highly overrepresented before correcting for this dinucleotide bias, while only one remains substantially overrepresented after correction (TGT-GGG, CHISQ3 = 18).

Table 2, which summarizes the results of our analyses of the ECO, YST and HUM databases, illustrates two important features. First, there is a highly significant codon pair bias in all three species, even after dinucleotide bias is discounted (CHISQ3). Second, the effect associated with dinucleotide bias, i.e., the difference between CHISQ2 and CHISQ3, is much more pronounced in eukaryotes than in *E. coli*. It is by far the *predominant* effect in mammals, representing two thirds of the amount of CHISQ2 in excess of its expectation in HUM; MUS and RAT (not shown) exhibit a very similar

TABLE 3
Correlation Coefficients (r) of Pairwise Comparisons of CHISQ3 Values for Five Databases Using Signed Values CHISQ3

	YST[a]	HUM[b]	MUS[c]	RAT[d]
ECO[e]	0.08	0.04	−0.01	0.01
YST		0.06	0.07	0.08
HUM			0.55	0.58
MUS				0.64

[a] *S. cerevisiae*
[b] *Human*
[c] *M. musculus.*
[d] *R. norvegicus.*
[e] *E. coli.*

pattern. Dinucleotide bias represents a much smaller effect in YST, and only a very minor one in ECO.

Although the predominant dinucleotide bias in HUM is the well-known CpG deficit ($\chi^2 = 5980$), other dinucleotides are also very highly biased. There is a striking deficit of TA [$\chi^2 = 2149$], as well as an excess of TG [$\chi^2 = 4180$], CA [$\chi^2 = 1911$] and CT [$\chi^2 = 1090$]. Overall, the deficit of CpG contributes only 35% of the total χ^2 associated with dinucleotide bias in the HUM database, and 17% in YST.

It is interesting to note that the degree of bias in amino acid pairs (represented by the difference between CHISQ1 and CHISQ2) is almost the same in ECO and YST (~1050), although the magnitudes of the remaining chi-square (CHISQ2) are very different. In HUM, this difference is about twice that value (~2000) in a database slightly more than twice the size of the other two (see footnote to Table 2). This suggests that the bias of amino acid pairs may be quite independent of the codon pair bias. We also find that the bias associated with each amino acid appears to be highly specific to each organism, suggesting it is *not* a simple consequence of global constraints on protein structure.

Table 3 shows the correlation coefficients of CHISQ3 values for all codon pairs, compared between the different species. While all species show a very high degree of bias, the values of CHISQ3 for each of the 3721 codon pairs show virtually no correlation between ECO, YST, and mammals. In view of the fact that the complex relationship between codon usage and the translational machinery differs between organisms, one might expect codon pair bias to also have evolved in a species-specific manner. Comparison of HUM with the two rodents yields correlation coefficients of 0.55–0.58; thus the codon pair bias of humans and rodents have diverged substantially, although they still show a degree of similarity. Perhaps more surprising is that even MUS and

RAT show a substantial degree of dissimilarity (r = 0.64). Thus, whatever the selective basis of codon pair bias in these species, it clearly can affect substantial changes over only moderate periods of evolutionary time (no more than 24 million years, and as few as 10, are thought to separate the genera Mus and Rattus[43,44]).

C. NONADJACENT CODON PAIR BIAS

We have already pointed out that less than 5% of the value of CHISQ3 for ECO remains when codon pairs separated by two or three intervening codons are evaluated (Figure 1). The results we obtained with the eukaryotic databases have prompted us to analyze further the dependence of codon pair bias on adjacency in all our databases. Figure 2 shows the values of CHISQ3 for ECO, YST, and HUM (represented as the number of standard deviations in excess of its expected value), for separations ranging from 0 codons (adjacent pairs) up to 20. The data for ECO (solid bars) reflects our earlier findings, namely that the value of CHISQ3 for adjacent pairs decreases to +14 SD for pairs separated by 1 codon, and to about +5 SD for pairs separated by 2 or 3 codons. In addition, it can be seen that this "residual" value remains at levels between +5 to +7 SD for separations of up to 20 intervening codons. The data for the eukaryotes, on the other hand, show somewhat different behaviors. For YST, the value of CHISQ3 for adjacent pairs is +31 SD; for pairs separated by 1 or more codons the value remains between +5 and +9 SD, generally (but not uniformly) higher than ECO. In the case of HUM, the adjacent CHISQ3 value of +109 SD is reduced to +43SD for a separation of 1, but fluctuates between +30 and +40 SD for separations of 3 to 20, without exhibiting any strong downward trend. Thus, in the case of ECO, only 5% of the excess CHISQ3 value associated with adjacent pairs still remains when pairs separated by 2 or more codons are evaluated, even out to separations of 20 (or more). In the case of YST, this value is about 20%, while in HUM, the nonadjacent bias remains at a level close to 35% of the adjacent value.

Our hypothesis, namely that codon pair bias is the result of the biological consequences of adjacent tRNA interactions on the surface of a translating ribosome, requires that such bias be dependent on *adjacency*. The discovery of a third ribosomal tRNA binding site[45] may broaden this expectation to include interactions between tRNA molecules separated by a single intervening codon, but cannot readily account for significant interactions over a range of 10–20 codons. Although the ECO data also exhibit a statistically significant nonadjacent effect (+5 SD), this nevertheless represents less than 5% of the adjacent CHISQ value, a pattern largely consistent with the requirement for adjacency. However, at least 20% of the value of CHISQ3 of YST and 35% of that of HUM evidently is *not* associated with adjacency, and these substantial components are, therefore, unlikely to be the consequence of tRNA-tRNA interactions.

We have considered the possibility of a relationship between this distant bias and mRNA higher-order structure, but it is difficult to find support for

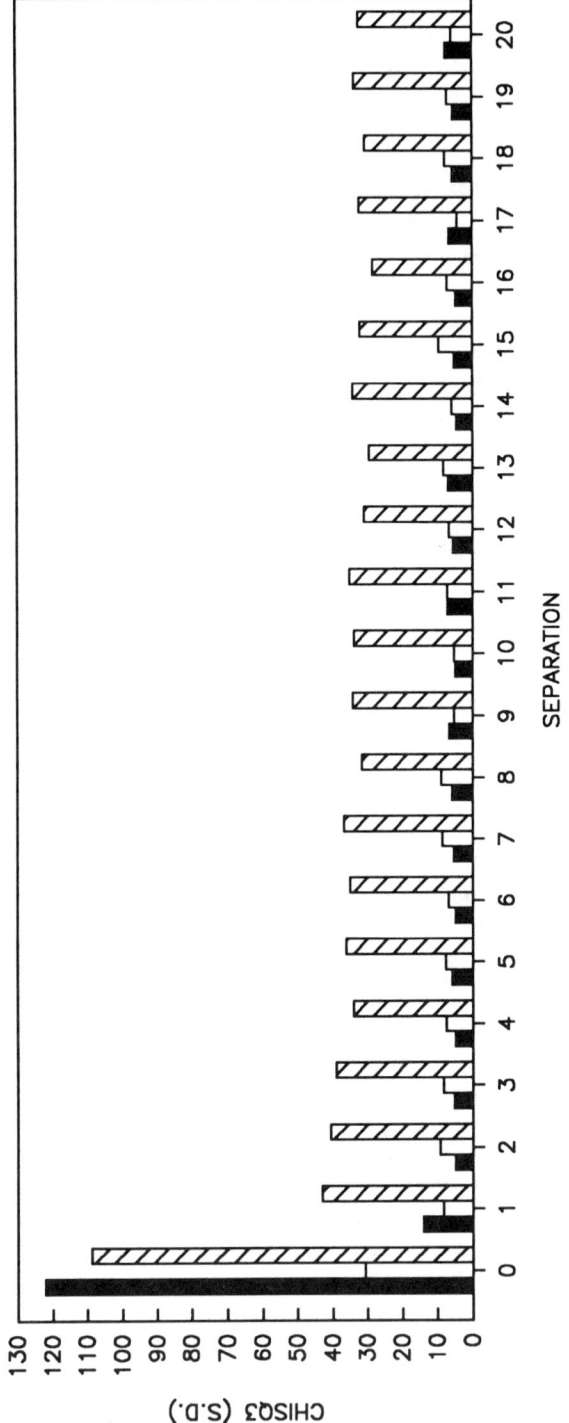

FIGURE 2. Total CHISQ3 for adjacent and nonadjacent codon pairs in *E. coli* (ECO, solid bars), *S. cerevisiae* (YST, open bars), and human (HUM, cross-hatched bars) databases. Values on the abscissa indicate the number of codons separating the pairs under consideration, zero representing adjacent pairs. CHISQ3 is plotted

such a concept. For one thing, we are dealing with global biases in large databases; if mRNA structures were responsible for this bias they would have to be structures common to many different messages, since any message-specific features would not be expected to contribute significantly to the overall bias. In an attempt to develop testable predictions in this regard, we have imagined two properties of mRNA that could contribute to global codon pair bias. First, mRNAs might generally interact in an ordered fashion with structural proteins to form nucleosome-like structures, in which case some periodicity in the nonadjacent bias might be expected. However, our initial analyses indicate a complete lack of correlation, within each species, between values of CHISQ3 at one degree of separation and those at any other, i.e., no evidence for periodicity. Second, mRNAs might require common structures associated with either translation initation or termination, structures more extensive than those known to be associated, for example, with ribosome-binding sites. However, we have not been able to discern any disproportionate contribution to this "distant" CHISQ3 by either the 5′ or 3′ portions of coding regions, which might be expected under such circumstances.

At this juncture, we have no explanation for the distant codon pair bias. This bias appears to exhibit as high a degree of species specificity as does the adjacent bias; it is not dependent on the amino acid sequence of the encoded proteins (thus arguing against a relationship with constraints on protein structure), and we argue above against a relationship with mRNA structure. While the basis for this effect remains a mystery, the differing behaviors of codon pair utilization patterns in prokaryotes and eukaryotes, and some of their practical consequences, are discussed further in the following section.

IV. FUNCTIONAL SIGNIFICANCE OF CODON PAIR UTILIZATION PATTERNS

A. CODON PAIRS AND TRANSLATIONAL STEP-TIME

Many reports have provided evidence that ribosomes translate mRNA with a variable rate.[29-33,46,47,60] Translational pauses which result in the *in vivo* accumulation of nascent polypeptide chains have been observed during the translation of rabbit reticulocyte globins,[61] fibroins in the silk gland of *Bombyx mori* larvae[62] and the ampullate gland of a spider,[63] the colicins A, E1, E2, and E3 in *E. coli*,[64] the bacteriophage MS2 coat protein,[65] the *E.coli* maltose-binding protein,[66] and *in vitro*, for the translation of rabbit globin and tobacco mosaic virus mRNAs.[67] In addition to the tRNA-codon and tRNA-tRNA interactions discussed in Section II, it has been suggested that mRNA secondary structures also might impede the progress of ribosomes.[68] However, Sørensen et al.[29] have demonstrated that even very stable mRNA secondary structures in *lacZ* mRNA do not affect the *in vivo* rate of β-galactosidase production. Furthermore, the ability of translating ribosomes to efficiently disrupt mRNA secondary structures lies at the heart of the attenuation mechanism.[50] Thus, it appears unlikely that mRNA secondary structure alone can explain the discontinuity of *in vivo* translation rates.

The idea that the compatability of adjacent tRNAs on the surface of a translating ribosome is a primary factor responsible for variations in translation rates in *E. coli* is supported by several lines of evidence. Early experimental evidence was provided by the isolation of context mutants by Bossi and Roth[6] which relieved the inability of an undermodified *supE* suppressor tRNA to facilitate the suppression of some amber codons in the *hisD* gene of *Salmonella typhimurium*. More recently, Smith and Yarus[26] have shown that changes in the structure of a tRNA at one codon affect the translational efficiencies of adjacent codons (see Section II). Other studies have shown that the chemical modifications of bases in tRNA molecules can affect their translational efficiencies. As early as 1972 Singer et al.[69] described a tRNA structural mutation which caused the derepression of the genes of the *his* operon. It was subsequently shown that this same mutation, which lacks the enzymatic activity to convert uridine residues to pseudouridine in the anticodon region of several tRNA species,[70-72] also causes a derepression of the genes of the *ilvGMEDA* operon,[73] which encodes enzymes required for the biosynthesis of isoleucine, valine, and leucine. When it was demonstrated that both of these operons are regulated by attenuation and that the regulatory histidyl- and leucyl-tRNA molecules for both operons normally contain pseudouridines in their anticodon regions, it became clear that these operons were derepressed because the undermodified tRNAs could not efficiently translate tandem cognate codons in the leader polypeptide coding regions of these operons. More recently, it has been shown that hypomodifications of tryptophanyl- and phenylyalanyl-tRNAs can affect the attenuator-mediated regulation of the *trp*[74] and *phe*[75] operons as well. Others have also demonstrated that undermodified suppressor tRNAs exhibit altered, context specific, suppression efficiencies, as, for example, in the work of Björk and his colleagues.[76,77] These results further support the idea that tRNA-tRNA interactions on the surface of a translating ribosome control translational step-times.

As we have reviewed in Section III, computer analyses of protein coding sequences of prokaryotes and eukaryotes show a high degree of bias in codon pair usage; some pairs are used many more times than expected and others many fewer times than expected. This observation, coupled with the assumption that tRNA-tRNA interactions on the surface of a translating ribosome influence translational elongation rates, suggests that there might exist a relationship between codon pair bias and the translational step-times of individual codon pairs. The use of one codon next to another may have co-evolved with the abundance and structure of tRNA isoacceptors in order to allow control of the kinetics of translation of a growing polypeptide chain without simultaneously imposing constraints on amino acid sequence or protein structure. It is well-documented that translational elongation rates through a messenger RNA are variable.[46,47] Because some segments of a message are translated rapidly and other segments are translated slowly, it has been suggested that translational pause sites might be functionally important for a number of reasons, as, for example, directing the folding of nascent polypeptide chains.[48,49]

In order to determine if the translational step-times of over and underrepresented codon pairs are demonstrably different, an *in vivo* assay capable of measuring the relative translational step-times of selected codon pairs in a growing polypeptide chain is required. We are currently evaluating two such assays. One assay takes advantage of the fact that the transit time of a ribosome through the leader polypeptide coding sequence, and the ribosome release time, from the leader RNA of the leaded-attenuator regions of the *trp* and *ilvGMEDA* amino acid operons of *E.coli* sets the basal level of transcriptional read-through at these attenuators.[78-81] The other assay takes advantage of the fact that ribosome pausing at a site near the beginning of an mRNA coding sequence can result in ribosome queuing and, thereby, inhibit translation initiation by physically interfering with the attachment of a new ribosome to the message.[59] Preliminary data suggests that the placement of several different overrepresented codon pairs in the stem one region of the *trp* and *ilvGMEDA* leader RNAs, at sites not expected to alter attenuator function, results in an increased basal level of attenuation. Likewise, the placement of several different overrepresented codon pairs early in the coding region of the *lacZ* gene results in decreased levels of translation as well as transcriptional polarity. With both assays it is observed that the more overrepresented the codon pair the more severely it slows the ribosome. While only a small number of codon pairs have been examined, this correlation is intriguing and suggests that overrepresented codon pairs are generally translated more *slowly* than underrepresented codon pairs.

If tRNA-tRNA interactions are important for determining translational step-times and if translational step-times are related to codon pair utilization then we should be able to use the values we have generated for codon pair bias to exploit this relationship. However, before we can effectively use these codon pair rules, we need to understand the significance of variable translation rates, i.e., why have codon pairs with very different translational step-times evolved, and what are the physiological consequences of their presence in protein coding regions. A variety of possibilities exist, such as control of *in vivo* protein folding, regulation of nascent polypeptide interactions with cellular components or of protein stability, controlling the level of protein expression and of mRNA stability, and coordination of levels of protein synthesis with the growth rate of the cell.

B. PROTEIN FOLDING

The idea that all of the information required for protein structure resides in the amino acid sequence has been around for a long time. In a recent commentary on protein folding, Szulmajster[82] points out that it was in 1925 when Anson and Mirsky[83] demonstrated that alkaline denatured hemoglobin reassumes its native conformation when the pH is adjusted to neutrality. This observation was supported by the pioneering studies of Anfinsen[84] in the 1960s on the refolding properties of ribonuclease. But even in these early studies, Anfinsen[85] noted substantial temporal differences between the rate of ribonuclease folding *in vitro* (which takes several hours) and its rate of synthesis *in vivo* (which takes one to two minutes). He later accounted for this discrepancy by discovering an enzyme in the ribonuclease-producing cells which catalyzes the formation of the correct disulfide pairings in ribonuclease in less than two minutes.[85]

To date, scores of proteins have been refolded correctly *in vitro*, even without the assistance of catalysts. Nevertheless, two lines of evidence are consistent with the idea that there are important temporal contributions to protein folding. First, it is still not possible to predict the complete three-dimensional structure of any protein from its amino acid sequence, and second, many denatured proteins cannot spontaneously refold *in vitro* or *in vivo*.[86-90] Efforts to correctly predict complete protein structures have dealt largely with the energetic and spatial interactions of amino acids along an intact polypeptide chain. While a variety of factors may contribute to the failure of efforts to predict complete protein structures, one important possibility is that protein folding, at least for some proteins, is determined by temporal as well as spatial considerations. While it has been acknowledged that the structure of proteins is surely influenced by the interactions of amino acids during the synthesis of the nascent polypeptide chain, little attention has been paid to this other important facet of protein folding. This apparent oversight may not be surprising, however, since there has been little rational basis available for incorporating the kinetic parameters of polypeptide chain elongation (even if they could be measured) into an analysis of protein structures.

During the synthesis of a nascent polypeptide chain, secondary structures which depend on interactions between amino acids located near one another, such as α-helices, can form rapidly, on the order of microseconds (μs).[86-90] Therefore, since the average translational step-time in *E. coli* is about 65 milliseconds (ms), the formation of such structures in the nascent polypeptide chain is not likely to be affected by varying translation rates. However, the establishment of higher-order interactions involving distant secondary structures, such as the positioning of one newly synthesized α-helical region next to another can take much longer, up to several minutes *in vitro*.[86,88] This is because there are many ways that the nascent structures can associate with one another, and each of these alternative structures competes with the formation of the most thermodynamically stable folding intermediate. At this juncture in polypeptide synthesis, translational pausing for a time sufficient to allow the proper higher-order intermediate structure to form might contribute significantly to the final structure of the protein. Because of the temporally controlled interactions of structural domains, the structure of native *in vivo* synthesized protein can be very different from the structure of the same polypeptide chain following denaturation and refolding of the completed polypeptide.

We have suggested that overrepresented codon pairs are translated slowly in *E. coli*. Therefore, the occurrence of overrepresented pairs in the coding sequences of *E. coli* genes should, in effect, represent translational pause sites. Is the distribution of such translational pause sites correlated with particular features of protein structure? Since it seems unlikely that *all* pause sites are a consequence of their effects on protein folding, we are currently developing computer programs to distinguish codon usage patterns which are associated with various features thought to be relevant for folding, such as domain boundaries, from the "background" occurrence of pause sites which may not be related to protein folding.

C. INTERACTIONS BETWEEN NASCENT POLYPEPTIDE CHAINS AND CELLULAR FACTORS

It has long been known that low molecular weight ligands can influence the refolding of denatured proteins *in vitro*. It is possible, therefore, that the protein coding regions of some genes contain translational pause signals in order to facilitate the binding of small molecular weight ligands such as cofactors to the nascent polypeptide chain. Such a ligand may influence the folding of the remainder of the protein chain as it is synthesized, as is known to occur *in vitro*.

Translational pause sites might also be distributed in protein coding regions to facilitate the interaction of nascent polypeptide chains with other proteins. Obvious cases might involve the directed association of proteins into multienzyme complexes or highly ordered protein structures. Along this line, King and his colleagues[91,92] have isolated temperature sensitive folding mutants in the gene for the tail spike endorhamnosidase of phage 22 which interfere with the subunit association pathway, but not with the structure or the activity of the completed polypeptide chain.

In other cases, translational pauses might facilitate the interaction of the nascent polypeptide chain with cellular proteins required for directing its folding or for its proper trafficking. The list of proteins that are known to carry out such functions, termed chaperones or chaperonins,[93,94] is growing at a rapid pace.[49] These proteins interact with polypeptide chains to influence either their folding, their subunit assembly, their patterns of glycosylation, or their cellular localization. The first protein to be recognized as a molecular chaperone was the immunoglobulin heavy chain binding protein, BiP.[95-97] BiP comprises about 1% of the total protein in the lumen of the endoplasmic reticulum of pre-B cells, and binds to nascent immunoglobulin heavy chains to prevent heavy chain aggregation and to facilitate the association of heavy chains with light chains during immunoglobulin assembly. BiP has also been implicated in fostering the transport of completed immunoglobulin molecules out of the endoplasmic reticulum. In addition to its role in immunoglobulin assembly and transport, BiP has been implicated in facilitating the oligomeric assembly and glycosylation of hemagglutinin subunits[98-100] and the assembly of vesicular stomatitis virus Protein G trimers.[101] It has been postulated by Rothman[49] that

BiP, as well as other molecular chaperones, must recognize some consensus amino acid sequence that is either transiently expressed on the surface of growing polypeptide chains or permanently expressed on the surface of denatured proteins. He supposes that there must be signals, perhaps encoded in the protein synthetic machinery, that expose these sequences in a temporally regulated fashion during translation. Programmed translational pausing would, of course, nicely serve such a purpose. Properly spaced pause sites could transiently expose intermediate structures to attract chaperones to bind to the surface of nascent polypeptide chains to direct their subunit assembly, transport across a membrane, glycosylation, or any of a number of other purposes.

Other known proteins that might bind to structures transiently exposed on the surface of nascent polypeptide chains by translational pausing are the enzymes that put the "finishing touches" on protein maturation, protein disulfide-isomerase,[102,103] and peptidyl-prolyl-*cis-trans* isomerase.[103]

D. PROTEIN STABILITY

Another potential role for translational pause signals derives from the observation that many proteins are degraded by cellular proteases when they are expressed in a heterologous host; the host's own proteins, on the other hand, are quite stable. We would suggest that, just as certain consensus amino acid sequences might be temporally exposed to molecular chaperones during the synthesis of a polypeptide chain, some translational pause sites may direct the folding of a protein in such a manner as to hide amino acid sequences that would serve as substrates for the host's cellular proteases. Thus, since the array of cellular proteases varies from one organism to another, foreign proteins would be more susceptible to proteolysis.

E. REGULATION OF PROTEIN EXPRESSION

The level of expression of a protein is proportional to the number of ribosomes that complete the translation of its mRNA per unit time, which, in turn, depends on the number of ribosomes that *initiate* translation of the mRNA per unit time. A variable elongation rate through the message will not affect the steady-state level of protein expression (the number of ribosomes that eventually complete translation of the message) unless the rate of translational initiation is affected. This could happen only if translational pause sites were located sufficiently close to the beginning of a message, or were so numerous, that pausing ribosomes caused queuing to the beginning of the message and physically inhibited ribosome attachment and initiation. Failing this, internal pause sites will merely delay the appearance of the first protein molecules following induction of translational initiation; and thereafter, the rate of synthesis of the protein will be proportional to the rate of translational initiation.[59] Because of this fact, and because of the lack of compelling evidence that rare codons are translated slowly (see Section II), the idea that rare codons are generally used to regulate the expression of proteins encoded by genes expressed at low levels does not seem attractive. Indeed, the work of Sharp and Li[36] suggests that the relationship between codon usage and level of gene expression results from the *avoidance* of rare codons in highly expressed genes rather than a *preference* for rare codons in poorly expressed genes. We suggest that these well-known biases in codon usage may be simply a bookkeeping phenomenon related to matching codon usage with substrate availability, and have no direct connection with translational speeds or control of levels of protein expression. This conclusion agrees well with our finding that there is no correlation between codon usage and codon pair bias.

Our experimental results suggest that the placement of an overrepresented codon pair early in the coding region of the *lacZ* gene can dramatically reduce the translatability of this message, presumably by blocking new translation initiation. It therefore seems likely that *E. coli* and other organisms may have exploited this consequence of translational pausing to regulate the expression of proteins. However, given the wide latitudes available to a cell for regulating mRNA synthetic rates, it is likely that the regulation of translation initiation is reserved for the fine tuning of protein levels under special circumstances. (In eukaryotes, as an intriguing example, alternative mRNA splicing might yield mature messages with different 5' coding regions that are translated at different rates resulting from codon pair utilization differences limiting translation initiation.) However, because different organisms differ markedly in their patterns of codon pair utilization, as discussed in Section III, the fortuitous appearance of one or more highly overrepresented codon pairs near the beginning of the mRNA of a cloned gene might severely compromise the level of protein produced by that gene when it is expressed in a heterologous system.

F. MESSENGER RNA STABILITY

It is well known that protein levels are correlated with message turnover rates in growing cells. It is also known that the uncoupling of translation and transcription in bacteria can lead to increased rates of message degradation, due to the activities of cellular nucleases and to rho-dependent, premature transcriptional termination, referred to as transcriptional polarity.[104] It is reasonable to postulate that one use for translational pause sites might be to distribute ribosomes along the mRNA molecule in such a way as to mask nuclease sensitive and rho-dependent termination sites and, therefore, to protect the message from degradation.

If translational pause sites are important for distributing ribosomes along the message, what prevents the transcribing RNA polymerase from racing ahead of the leading ribosome while it is pausing? It seems likely that, just as with the attenuators for amino acid operons, *transcriptional* pause sites may be important for maintaining the coupling of transcription and translation during protein synthesis. Therefore, it is tempting to speculate that transcriptional pause sites are coordinated with translational pause sites in the coding regions of prokaryotic genes. While we have argued that translational pause sites are determined by appropriate selection of codon pairs, an additional level of complexity which involves RNA secondary structures important for transcriptional pausing might also be part of the basic design of prokaryotic protein coding sequences.[50-52,88-90] We might, therefore, envision the transcription complex and the leading ribosome proceeding through a protein coding region in an "inch worm" fashion, each progressing at a varying rate, but the two never becoming widely separated.

G. METABOLIC GROWTH RATE CONTROL IN *E. COLI*

Jensen and Pedersen[33] have recently published an insightful analysis of the importance of the relative rates of DNA, RNA, and protein synthesis in determining the growth rate of *E. coli*. Their basic premise is that the growth of *E. coli* is normally limited by the nutritional content of its growth medium; if the substrates for macromolecular synthesis are limiting, then the richer the medium, the more rapidly cells can grow. In a rich medium more energy can be devoted to macromolecular synthesis, since many of the metabolic intermediates (amino acids, purines, pyrimidines, vitamins, etc.) are provided. When cells grow rapidly they contain more DNA replication forks, but they are also larger, so that the DNA/mass and protein/mass ratios remain constant. On the other hand, the RNA/mass ratio increases in a manner roughly proportional to the growth rate, due to an increased production of ribosomes.

The mechanism by which ribosomal RNA synthesis is coupled to the growth rate of *E. coli* cells has been a subject of investigation for nearly 40 years. In 1961 Stent and Brenner[105] demonstrated that starving cells for an amino acid resulted in an immediate inhibition of stable RNA synthesis. Cashel and Gallant[106] subsequently discovered that this inhibition, termed the "stringent" response, was correlated with the intracellular accumulation of guanosine tetraphosphate, termed "magic spot". It was subsequently shown that magic spot synthesis is stimulated by the presence of a non-aminoacylated tRNA in the acceptor site of an idling ribosome.[107] Efforts to determine how magic spot inhibits transcription at stringently controlled promoters have met with mixed results; it has been demonstrated, however, that magic spot can enhance transcriptional pausing.[52,108-110]

Jensen and Pedersen[33] suggest the following model for metabolic growth rate control. At all growth rates, the "housekeeping" genes which are expressed at low to moderate levels are potentially able to sequester more RNA polymerase in transcription elongation complexes than is present in the cell; therefore, the transcriptional efficiency of any given promoter will depend on its ability to compete for a limited supply of RNA polymerase molecules. Jensen and Pedersen make the assumption that stringent promoters have a lower intrinsic affinity for RNA polymerase; they are therefore difficult to saturate at low RNA polymerase concentrations, although, once a closed complex is formed, they initiate transcription very efficiently. Thus, if the concentration of free cellular RNA polymerase were to increase with increasing growth rate, then the activities of these low affinity promoters would increase, whereas the activity of the high affinity housekeeping promoters (which are already saturated at lower RNA polymerase concentrations) would not change. Previous reports had indicated that translation rates do not vary with growth rate. However, Jensen and Pedersen reinterpreted these data to suggest that there is, in fact, a 30 to 50% increase in translation elongation rates when cells increase their growth rate fourfold. Thus, their model states that at slow growth rates the substrates for protein synthesis are present at low concentration, which accounts for the lower rate of protein synthesis, and suggests that the basal level of magic spot in these more slowly growing cells would be high. This high level of magic spot would increase transcriptional pausing and result in a larger number of RNA polymerase molecules being sequestered in housekeeping genes. The decrease in the availability of free RNA polymerase would have the greatest effect on promoters with the lowest affinity, namely the stringently controlled promoters. As the nutrient level of the growth medium increases and the substrates for protein synthesis become less limiting, the growth and protein synthesis rates increase and the basal level of magic spot decreases. This drop in the level of magic spot allows transcription elongation to proceed at a higher rate, more RNA polymerase molecules are freed per unit time, and the greater availability of free RNA polymerase allows more efficient transcription of the less competitive, stringently controlled promoters at the higher growth rate; hence, metabolic growth rate control.

This is an ingenious model which possesses a great deal of merit, although some of the assumptions of the model, particularly those relating to changes in rates of protein elongation at differing growth rates as a consequence of nutrient limitations and differences in promoter affinities for RNA polymerase, need to be experimentally verified. Interestingly, Cashel has recently demonstrated that the growth rate is linearly related and the transcriptional activities of the *rrna P1* promoters are exponentially related to the *in vivo* manipulated levels of magic spot, even when the cells are growing in a nutrient sufficient medium.[111] This observation suggests that growth rate control is not affected by limiting nutrients as suggested by Jensen and Pedersen.[33] However, our analyses of codon pair bias suggest an alternative explanation for the mechanism by which RNA polymerase molecules are sequestered in transcription elongation complexes at low growth rates, to be subsequently freed to transcribe stringently controlled promoters at higher growth rates.

We have already reported[35] that genes expressed at low levels in *E. coli* tend to favor overrepresented (slowly translated) codon pairs compared with genes expressed at high levels. The most highly expressed genes in *E. coli* are, in large part, genes of the protein synthesis system, the stringently controlled genes, whereas genes involved in producing metabolic intermediates for macromolecular synthesis (housekeeping genes) are among those expressed at moderate to low levels. Since these housekeeping genes tend to contain a higher proportion of overrepresented codon pairs (translational pause sites), we would expect them also to contain more *transcriptional* pause sites in order to maintain the normal tight coupling between translation and transcription (as discussed earlier). Thus, at low growth rates the high affinity promoters of these genes would compete effectively for RNA polymerase (as suggested by Jensen and Pedersen), and the presence of many transcriptional pause sites would result in effective sequestration of RNA polymerase in transcription elongation complexes, thus limiting the pool of free polymerase available for the low affinity, stringently controlled promoters. The response to a shift to higher growth rates would proceed as already described.

One central difference distinguishes Jensen and Pedersen's views from our alternative. While Jensen and Pedersen propose that the slower transcription rate operating at low growth rates (which encourages polymerase sequestration) is a consequence of limiting substrate availability, we suggest that slow transcription rates are directly "hard-wired" as transcriptional pause sites into the general category of housekeeping genes. The higher translation rates which characterize higher growth rates would then simply be the consequence of the relative dominance of highly expressed genes under these conditions, i.e., genes which have fewer translational pause sites. While these two ideas are certainly not mutually exclusive, specific and testable predictions can distinguish them. In particular, under our view, housekeeping genes should be found to have intrinsically slower transit times, from the beginning to the end of the gene, for both translation and transcription, compared with stringent genes, resulting from an increased density of both translational and transcriptional pause sites. On the other hand, it should be noted that the results of Nomura and his colleagues suggest that the intracellular pool of RNA polymerase molecules is not limiting at slow growth rates.[113]

H. PRACTICAL APPLICATIONS OF CODON PAIR RULES
1. Optimizing Genetically Engineered Systems

If codon pair bias is a consequence of variations in translational step-times, such variation must be of considerable physiological importance, and there would be clear applications of our data to manipulating and optimizing protein synthesis in genetically engineered systems. A "faulty" distribution of pause sites in a protein synthesized in a heterologous system might result in a lowered yield of biologically active product due to defective folding, mRNA degradation, suppression of translation initiation, or increased proteolysis. Knowing the rules which relate codon pair bias and translational step-times could allow a rational approach to redistributing such pause sites in order to prevent such problems.

The effectiveness of such an approach depends on our being able to determine the "optimal" distribution of step-times in any given protein coding region. One intuitively attractive solution would be to determine the distribution of chi-square values in the native gene using the *donor* organism's "rules", then modify the sequence of the gene to match this same distribution of chi-squares using the *host* organism's "rules". A computer program has, in fact, been developed which carries out such a process.

However, a critical assumption implicit in this straightforward approach is that the functional basis of codon pair bias, e.g., variations in translational step-time, is the *same in different species*. While we have preliminary evidence that there exists a strong relationship between codon pair bias and step-times in *E. coli*, we have not yet developed a system to examine this question in any other organism. It is certainly tempting to assume that a similar relationship exists in eukaryotes, but even if such an overall relationship does hold, the existence of the substantial component of nonadjacent bias which we have described complicates any such assumption, since this component of codon pair bias is unlikely to be related to translation rates. In addition, the markedly different translation elongation rates which characterize eukaryotes and prokaryotes suggest that the physiological significance of translational step-times may be manifested in different ways in different organisms

There are, however, a variety of possible approaches to applying codon pair bias rules in order to improve expression of heterologous genes in *E. coli*, even without knowing the rules which may relate such bias to protein structure, and without making any assumptions about codon pair bias in other species. For example, translational pausing at or near the ribosome binding site (as modeled in our β-galactosidase fusion constructs) is likely to severely affect translation initiation, thereby lowering the yield of the gene product. Such pauses could be identified by their codon pair chi-square values (using only the *E. coli* values), and the responsible codon pairs replaced with non-pausing pairs encoding identical amino acids. Gene expression can also be severely limited by mRNA degradation, which in turn can be encouraged by "naked" regions of message devoid of ribosomes (as exemplified by the phenomenon of transcriptional polarity[104]). Redistributing translational pause sites could be used, in general, to create a more even distribution of ribosomes on the message, or more specifically, to increase the occupancy of particular regions known to be sensitive to degradation.

Inappropriate translational pausing might also result in premature translation termination, and the production of peptides truncated at predictable sites. Such pausing and/or termination could also result in frame-shifting or internal reinitiation of translation if a cryptic initation site happened to be located nearby, further increasing the yield of inappropriate products. The products of both of these processes, it should be noted, can be identified and characterized experimentally, and the sites of inappropriate termination or initiation closely defined. In either case, such undesired pause sites could be readily located and removed by suitable modification of the gene sequence.

2. Identification of Protein-Coding Regions

A variety of computer methods have been described to identify protein coding regions in DNA sequences of unknown function.[59] Most of these methods rely on analyzing various combinations of known constraints on amino acid composition, codon preference, and "positional" biases in base composition (which may simply be a reflection of biases in "codon context"). However, none of these methods has proven entirely satisfactory, and more powerful methods are continually being sought. A general weakness of many of these methods is that they make the assumption (explicitly pointed out by Staden[112]) that *codons are used in random order* within a sequence. The strong bias in codon pair utilization which characterizes both prokaryotes and eukaryotes violates this assumption. Thus, there is good reason to expect that by adding a consideration of the species-specific codon pair bias, we might increase the power of algorithms used to identify protein-coding regions. The rapidly increasing size of the nucleotide sequence database, and in particular the development of the current project to sequence the entire human genome, should provide many valuable opportunities for applying such algorithms.

ACKNOWLEDGMENTS

We are grateful to Stuart Arfin, Denis Heck, Elaine Ito, and Suzanne Sandmeyer for reviewing this manuscript. This work was supported in part by gifts from Amgen Inc., Thousand Oaks, California, and Monsanto, St. Louis, Missouri.

REFERENCES

1. **Akaboshi, E., Inouye, M., and Tsugita, A.**, Effect of neighboring nucleotide sequences on suppression efficiency in amber mutants of T4 phage lysozyme, *Mol. Gen. Genet.*, 149, 1, 1976.
2. **Bossi, L.**, Context effects: translation of UAG codon by suppressor tRNA is affected by the sequence following UAG in the message, *J. Mol. Biol.*, 164, 73, 1983.
3. **Fluck, M. M., Salser, W., and Epstein, R. H.**, The influence of the reading context upon the suppression of nonsense codons, *Mol. Gen. Genet.*, 151, 137, 1977.
4. **Comer, M. M., Guthrie, C., and McClain, W. H.**, An ochre suppressor of bacteriophage T4 that is associated with a transfer RNA, *J. Mol. Biol.*, 90, 665, 1974.
5. **Comer, M. M., Foss, K., and McClain, W. H.**, A mutation of the wobble nucleotide of a bacteriophage T4 transfer RNA, *J. Mol. Biol.*, 99, 283, 1975.
6. **Bossi, L. and Roth, J. R.**, The influence of codon context on genetic code translation, *Nature (London)*, 286, 123, 1980.
7. **Colby, D. S., Schedl, P., and Guthrie, C.**, A functional requirement for modification of the wobble nucleotide in the anticodon of a T4 suppressor tRNA, *Cell*, 9, 449, 1976.
8. **Fluck, M. M. and Epstein, R. H.**, Isolation and characterization of context mutations affecting the suppressibility of nonsense mutations, *Mol. Gen. Genet.*, 177, 615, 1980.
9. **Feinstein, S. I. and Altman, S.**, Coding properties of an ochre-suppressing derivative of *Escherichia coli* tRNA^{Tyr}, *J. Mol. Biol.*, 112, 453, 1977.
10. **Engelberg-Kulka, H.**, UGA suppression by normal tRNA^{Trp} in *Escherichia coli*: codon context effects, *Nucleic Acids Res.*, 9, 983, 1981.

11. **Murgola, E. J.**, Restricted wobble in UGA codon recognition by glycine tRNA suppressors of UGG, *J. Mol. Biol.,* 149, 1, 1981.
12. **Traboni, C., Ciliberto, G., and Cortese, R.,** A novel method for site-directed mutagenesis: its application to an eukaryotic tRNA^Pro gene promoter, *EMBO J.,* 1, 415, 1982.
13. **Miller, J. H. and Albertini, A. M.,** Effects of surrounding sequence on the suppression of nonsense codons, *J. Mol. Biol.,* 164, 59, 1983.
14. **Beaudet, A. L. and Caskey, C. T.,** Release factor translation of RNA phage terminator codons, *Nature (London),* 227, 38, 1970.
15. **Ganoza, M. X. and Tomkins, J. K. M.,** Polypeptide chain termination *in vitro*: competition for nonsense codons between a purified release factor and suppressor tRNA, *Biochem. Biophys. Res. Commun.,* 40, 1455, 1970.
16. **Murgola, E. J., Pagel, F. T., and Hijazi, K. A.,** Codon context effects in missense suppression, *J. Mol. Biol.,* 175, 19, 1984.
17. **Martin, R., Weiner, M., and Gallant, J.,** Effects of release factor context at UAA codons in *Escherichia coli, J. Bact.,* 170, 4714, 1988.
18. **Murgola, E. J., Hijazi, H. A., Gröinger, H. U., and Dahlberg, A. E.,** Mutant 16S ribosomal RNA: a codon specific translational suppressor, *Proc. Natl. Acad. Sci. U.S.A.,* 85, 4162, 1988.
19. **Murgola, E. J.,** Suppression and the code: beyond codons and anticodons, *Experientia,* 46, 1990, 1134.
20. **Culbertson, M. R., Leeds, P., Sandbaken, M. G., and Wilson, P. G.,** Frameshift suppression, in *The Ribosome: Structure, Function, and Evolution,* Hill, W. E., Dahlberg, A., Garrett, R. A., Moore, P. B., Schlessinger, D., and Warner, J. R., Eds., American Society for Microbiology, Washington, D. C., 1990, 559.
21. **Taniguichi, T. and Weissman, C.,** Inhibition of Qβ RNA 70S ribosome initiation complex formation by an oligonucleotide complementary to the 3′ terminal region of *E. coli* 16S ribosomal RNA, *Nature (London),* 275, 770, 1978.
22. **Uhlenbeck, O. C., Baller, J., and Doty, P.,** Complementary oligonucleotide binding to the anticodon loop of fMet-transfer RNA, *Nature (London),* 225, 508, 1970.
23. **Grosjean, H., Söll, D. G., and Crothers, D. M.,** Studies of the complex between transfer RNAs with complementary anticodons. I. Origins of enhanced affinity between complementary triplets, *J. Mol. Biol.,* 103, 499, 1976.
24. **Andersson, S. G. E., Buckingham, R. H., and Kurland, C. G.,** Does codon composition influence ribosome function?, *EMBO J.,* 3, 91, 1984.
25. **Nishimura, S.,** Minor components in transfer RNA: their characterization, location, and function, *Progr. Nucl. Acid Res. Mol. Biol.,* 12, 49, 1972.
26. **Smith, D. and Yarus, M.,** tRNA-tRNA interactions within cellular ribosomes, *Proc. Natl. Acad. Sci. U.S.A.,* 86, 4397, 1989.
27. **Post, L. E. and Nomura, M.,** DNA sequences from the *str* operon of *Escherichia coli, J. Biol. Chem.,* 255, 4660, 1980.
28. **Ikemura, T.,** Correlation between the abundance of *Escherichia coli* transfer RNAs and the occurrence of the respective codons in its protein genes, *J. Mol. Biol.,* 146, 1, 1981.
29. **Sørensen, M., Kurland, C. G., and Pedersen, S.,** Codon usage determines translation rate in *Escherichia coli, J. Mol. Biol.,* 207, 365, 1989.
30. **Jensen, K. F.,** Hyperregulation of *pyr* gene expression in *Escherichia coli* cells with slow ribosomes. Evidence for RNA polymerase pausing *in vivo?*, *Eur. J. Biochem.,* 175, 587, 1988.
31. **Bonekamp, F., Dalbøge, H., Christensen, T., and Jensen, K. F.,** Translation rates of individual codons are not correlated with tRNA abundances or with frequencies of utilization in *Escherichia coli, J. Bact.,* 171, 5812, 1989.
32. **Robinson, M., Lilley, R., Little, S., Emtage, J. S., Yarranton, G., Stephens, P., Millican, A., Easton, M., and Humphreys, G.,** Codon usage can affect efficiency of translation of genes in *Escherichia coli, Nucleic Acids Res.,* 12, 6663, 1984.

33. **Jensen, K. F. and Pedersen, S.**, Metabolic growth rate control in *Escherichia coli* may be a consequence of subsaturation of the macromolecular biosynthetic apparatus with substrates and catalytic components, *Micro. Rev.,* 54, 89, 1990.
34. **Andersson, S. G. E. and Kurland, C. G.**, Codon preferences in free-living microorganisms, *Micro. Rev.,* 54, 198, 1990.
35. **Gutman, G. A. and Hatfield, G. W.**, Nonrandom utilization of codon pairs in *Escherichia coli, Proc. Natl. Acad. Sci. U.S.A.,* 86, 3699, 1989.
36. **Sharp, P. M. and Li, W.-H.**, Codon usage in regulatory genes in *Escherichia coli* does not reflect selection for 'rare' codons, *Nucleic Acids Res.,* 14, 7737, 1986.
37. **Gouy, M. and Gautier, C.**, Codon usage in bacteria: correlation with gene expressivity, *Nucleic Acids Res.,* 10, 7055, 1982.
38. **Kolaskar, A. S. and Reddy, B. V. B.**, Contextual constraints on codon pair usage: structural and biological implications, *J. Biomol. Struct. Dynam.,* 3, 725, 1986.
39. **Shpaer, E. G.**, The secondary structure of mRNAs from *Escherichia coli*: its possible role in increasing the accuracy of translation, *Nucleic Acids Res.,* 13, 275, 1985.
40. **Yarus, M. and Folley, L. S.**, Sense codons are found in specific contexts, *J. Mol. Biol.,* 182, 529, 1985.
41. **Gouy, M.**, Codon contexts in enterobacterial and coliphage genes, *Mol. Biol. Evol.,* 4, 426, 1987.
42. **Buckingham, R. H.**, Codon context, *Experientia,* 46, 1126, 1990.
43. **Sarich, V.**, Rodent macromolecular systematics, in *Evolutionary Relationships Among Rodents: a Multidisciplinary Analysis,* Luckett, W. P. and Hartenberger, J.-L., Eds., Plenum Press, New York, 1984.
44. **Catzeflis, F. M., Sheldon, F. H., Ahlquist, J. E., and Sibley, C. G.**, DNA-DNA hybridization evidence of the rapid rate of Muroid rodent DNA evolution, *Mol. Biol. Evol.,* 4, 242, 1987.
45. **Nierhaus, K. H.**, The allosteric three-site model for the ribosomal elongation cycle: features and future, *Biochemistry,* 29, 4997, 1990.
46. **Varenne, S., Buc, J., Lloubes, R., and Lazdunski, C.**, Translation is a non-uniform process: effect of tRNA availability on the rate of elongation of nascent polypeptide chains, *J. Mol. Biol.,* 180, 549, 1984.
47. **Varenne, S. and Lazdunski, C.**, Effect of distribution of unfavorable codons on the maximum rate of gene expression by a heterologous organism, *J. Theor. Biol.,* 120, 99, 1986.
48. **Purvis, I. J., Bettany, A. J. E., Santiago, T. C., Coggins, J. R., Duncan, K., Eason, R., and Brown, A. J. P.**, The efficiency of folding of some proteins is increased by controlled rates of translation *in vivo*: a hypothesis, *J. Mol. Biol.,* 193, 413, 1987.
49. **Rothman, J. E.**, Polypeptide chain binding proteins: catalysts of protein folding and related processes in cells, *Cell,* 59, 591, 1989.
50. **Landick, R. and Yanofsky, C.**, Transcription attenuation, in *Escherichia coli and Salmonella typhimurium: Cellular and Molecular Biology,* Vol. 2, Neidhardt, F. C., Ingraham, J. L., Low, K. B., Magasanik, B., Schaechter, M., and Umbarger, H. E., Eds., American Society for Microbiology, Washington D. C., 1987, 1276.
51. **Gardner, J. F.**, Initiation, pausing and termination of transcription in the threonine operon regulatory region of *Escherichia coli, J. Biol. Chem.,* 257, 3896, 1982.
52. **Hauser, C. A., Sharp, J. A., Hatfield, L.K., and Hatfield, G. W.**, Pausing of RNA polymerase during *in vitro* transcription through the *ilvB* and *ilvGEDA* attenuator region of *Escherichia coli* K12, *J. Biol. Chem.,* 260, 1765, 1985.
53. **Landick, R. and Yanofsky, C.**, Stability of an RNA secondary structure affects *in vitro* transcription pausing in the *trp* operon leader region, *J. Biol. Chem.,* 259, 11550, 1984.
54. **Landick, R., Carey, J., and Yanofsky, C.**, Translation activates the paused transcription complex and restores transcription of the *trp* operon leader region, *Proc. Natl. Acad. Sci. U.S.A.,* 82, 4663, 1985.

55. **Lawther, R. P. and Hatfield, G. W.,** Multivalent translational control of transcription termination at the attenuator of *ilvGEDA* of *Escherichia coli*, *Proc. Natl. Acad. Sci. U.S.A.*, 77, 1862, 1980.
56. **Nargang, F. E., Subrahmanyam, C. S., and Umbarger, H. E.,** Nucleotide sequence of *ilvGEDA* operon attenuator region of *Escherichia coli*, *Proc. Natl. Acad. Sci. U.S.A.*, 77, 1823, 1980.
57. **Oxender, D. L., Zurawski, G., and Yanofsky, C.,** Attenuation in the *Escherichia coli* tryptophan operon: role of RNA secondary structure involving the tryptophan codon region, *Proc. Natl. Acad. Sci. U.S.A.*, 76, 5524, 1979.
58. **Kuroda, M. I. and Yanofsky, C.,** Evidence for the transcript secondary structures predicted to regulate transcription attenuation in the *trp* operon, *J. Biol. Chem.*, 259, 12838, 1984.
59. **Liljenström, H. and von Heijne, G.,** Translation rate modification by preferential codon usage: intragenic position effects, *J. Theor. Biol.*, 124, 43, 1987.
60. **Pedersen, S.,** *E. Coli* ribosomes translate *in vivo* with variable rate, *EMBO J.*, 3, 2895, 1984.
61. **Protzel, A. and Morris, A. J.,** Gel chromatographic analysis of nascent globin chains. Evidence of nonuniform size distribution, *J. Biol. Chem.*, 249, 4594, 1974.
62. **Lizardi, P. M., Mahdari, V., Shields, D., and Candelas, G.,** Discontinuous translation of silk fibroin in a reticulocyte cell-free system and in intact silk gland cells, *Proc. Natl. Acad. Sci., U.S.A.*, 76, 6211, 1979.
63. **Candelas, G., Candelas, T., Ortiz. A., and Rodriguez, O.,** Translational pauses during a spider fibroin synthesis, *Biochem. Biophys. Res. Commun.*, 116, 1033, 1983.
64. **Varenne, S., Knibiehler, M., Cavard, D., Morlon, J., and Lazdunski, C.,** Variable rate of polypeptide elongation for colicins A, E2, and E3, *J. Mol. Biol.*, 159, 57, 1982.
65. **Chaney, W. and Morris, A.,** Nonuniform size distribution of nascent peptides. The effect of messenger RNA structure upon the rate of translation, *Arch. Biochem. Biophys.*, 194, 283, 1979.
66. **Randall, L. L., Josefsson, L.-G., and Hardy, S. J. S.,** Novel intermediates in the synthesis of maltose-binding protein in *Escherichia coli*, *Eur. J. Biochem.*, 107, 375, 1980.
67. **Abraham, A. K. and Pihl, A.,** Variable rate of polypeptide chain elongation *in vitro*. Effect of spermidine, *Eur. J. Biochem.*, 106, 257, 1980.
68. **Baim, S. B., Pietras, D. F., Eustice, D. C., and Sherman, F.,** A mutation allowing an mRNA secondary structure diminishes translation of *Saccharomyces cerevisiae* iso-1-cytochrome c, *Mol. Cell. Biol.*, 5, 1839, 1985.
69. **Singer, C. E., Smith, G. R., Cortese, R., and Ames, B. N.,** Mutant tRNA [His] ineffective in repression and lacking two pseudouridine modifications, *Nature (London)*, 238, 72, 1972.
70. **Cortese, R., Kammen, H. O., Spengler, S. J., and Ames, B. N.,** Biosynthesis of pseudouridine in transfer ribonucleic acid, *J. Biol. Chem.*, 249, 1103, 1974.
71. **Ciampi, M. S., Arena, F., Cortese, R., and Daniel, V.,** Biosynthesis of pseudouridine in the *in vitro* transcribed tRNA[Tyr] precursor, *FEBS Lett.*, 77, 75, 1977.
72. **Turnbough, C. L., Neill, R. J., Landsberg, R., and Ames, B. N.,** Pseudouridylation of tRNAs and its role in regulation in *Salmonella typhimurium*, *J. Biol. Chem.*, 254, 5111, 1979.
73. **Cortese, R., Landsberg, R., Van der Haar, R. A., Umbarger, H. E., and Ames, B. N.,** Pleiotropy of *hisT* mutants blocked in pseudouridine synthesis in tRNA: leucine and isoleucine-valine operons, *Proc. Natl. Acad. Sci. U.S.A.*, 71, 1857, 1974.
74. **Yanofsky, C. and Soll, L.,** Mutations affecting tRNA[Trp] and its charging and their effect on regulation of transcription termination at the attenuator of the tryptophan operon, *J. Mol. Biol.*, 113, 663, 1977.
75. **Gowishankar, J. and Pittard, J.,** Regulation of pheylalanine biosynthesis in *Escherichia coli* K-12: control of transcription of the *pheA* operon, *J. Bacteriol.*, 150, 1130, 1982.

76. **Hagervall, T. G. and Björk, G. R.**, Undermodification in the first position of the anticodon of *supG*-tRNA reduces translational efficiency, *Mol. Gen. Genet.*, 196, 194, 1984.
77. **Bouadloun, F., Srichaiyo, T., Isaksono, L. A., and Björk, G. R.**, Influence of modification next to the anticodon in tRNA on codon context sensitivity of translational suppression and accuracy, *J. Bact.*, 166, 1022, 1986.
78. **Roesser, J. R. and Yanofsky, C.**, Ribosome release modulates basal level expression of the *trp* operon of *Escherichia coli.*, *J. Biol. Chem.*, 263, 14251, 1988.
79. **Roesser, J. R., Nakamura, Y., and Yanofsky, C.**, Regulation of basal level expression of the *trp* operon of *Escherichia coli, J. Biol. Chem.*, 264, 12284, 1989.
80. **Roesser, J. R. and Yanofsky, C.**, The RNA antiterminator causes transcription pausing in the leader region of the tryptophan operon, *J. Biol. Chem.*, 265, 14251, 1990.
81. **Pagel, J. M., Parekh, B., and Hatfield, G. W.**, Ribosome release influences the basal level of transcription termination at the attenuator of the *ilvGMEDA* operon of Escherichia coli, (unpublished results).
82. **Szulmajster, J.**, Protein folding, *Bioscience Rep.*, 8, 645, 1988.
83. **Anson, M. L. and Mirsky, A. E.**, On some general properties of proteins, *J. Gen. Physiol.*, 9, 169, 1925.
84. **Anfinsen, C. B., Haber, M., Sela, M., and White, F. H., Jr.**, The kinetics of formation of native ribonuclease during oxidation of the reduced polypeptide chain, *Proc. Natl. Acad. Sci. U.S.A.*, 47, 1309, 1961.
85. **Anfinsen, C. B.**, Principles that govern the folding of protein chains, *Science*, 181, 223, 1973.
86. **Touchette, N., Perry, K., and Matthews, C. R.**, Folding of dihydrofolate reductase from *Escherichia coli, Biochemistry*, 25, 5445, 1986.
87. **Ostermann, J., Horwich, A. L., Neupert, W., and Hartl, F.-U.**, Protein folding in mitochondria requires complex formation with hsp60 and ATP-hydrolysis, *Nature (London)*, 341, 125, 1989.
88. **Kim, P. S. and Baldwin, R. L.**, Specific intermediates in the folding reactions of small proteins and the mechanism of protein folding, *Annu. Rev. Biochem.*, 51, 459, 1982.
89. **Goldberg, M. E.**, The second translation of the genetic message: protein folding and assembly, *Trends Biochem. Sci.*, 10, 388, 1985.
90. **Baldwin, R. L.**, How does protein folding get started?, *Trends Biochem. Sci.*, 14, 291, 1989.
91. **King, J.**, Genetic analysis of protein folding pathways, *Biotechnology*, 4, 297, 1986.
92. **King, J., Haase, C., and Yu, M.**, Temperature-sensitive mutations affecting kinetic steps in protein-folding pathways, in *Protein Engineering*, Oxender, D. L. and Fox, C. F., Eds., A. R. Liss, New York, 1987, 109.
93. **Ellis, R. J.**, Proteins as molecular chaperones, *Nature (London)*, 328, 378, 1987.
94. **Ellis, R. J., van der Vies, S. M., and Hemmingsen, S. M.**, The molecular chaperone concept, *Biochem. Soc. Symp.*, 55, 145, 1989.
95. **Morrison, S. L. and Scharff, M. D.**, Heavy chain-producing variants of a mouse myeloma cell line, *J. Immunol.*, 114, 655, 1975.
96. **Haas, I. G. and Wabl, M.**, Immunoglobulin heavy chain binding protein, *Nature (London)*, 306, 387, 1986.
97. **Bole, D. G., Hendershot, L. M., and Kearney, J. F.**, Posttranslational association of immunoglobulin heavy chain binding protein with nascent heavy chains in nonsecreting and secreting hybridomas, *J. Cell Biol.*, 102, 1558, 1986.
98. **Gething, M.-J., McCammon, K., and Sambrook, J.**, Expression of wild-type and mutant forms of influenza hemagglutinin: the role of folding in intracellular transport, *Cell*, 46, 39, 1986.
99. **Copeland, C. S., Doms, R. W., Bolzau, E. M., Webster, R. G., and Helenius, A.**, Assembly of influenza hemagglutinin trimers and its role in intracellular transport, *J. Cell Biol.*, 103, 1179, 1986.

100. Copeland, C. S., Zimmer, K.-P., Wagner, K. R., Healey, G. A., Mellman, I., and Helenius, A., Folding, trimerization, and transport are sequential events in the biogenesis of influenza virus hemagglutinin, *Cell,* 53, 197, 1988.
101. Doms, R., Keller, D., Helenius, A., and Balch, W., Role for adenosine triphosphate in regulating the assembly and transport of vesicular stomatitis virus trimers, *J. Cell Biol.,* 105, 1957, 1987.
102. Freedman, R., Protein disulfide isomerase: multiple roles in the modification of nascent secretory proteins, *Cell,* 57, 1069, 1989.
103. Lang, K. and Schmid, F., Protein-disulphide isomerase and prolyl isomerase act differently and independently as catalysts of protein folding, *Nature (London),* 331, 453, 1988.
104. Adhya, S. and Gottesman, M., Control of transcription termination, *Annu. Rev. Biochem.,* 47, 967, 1978.
105. Stent, G. S. and Brenner, S., A genetic locus for the regulation of ribonucleic acid synthesis, *Proc. Natl. Acad. Sci. U.S.A.,* 47, 2005, 1961.
106. Cashel, M. and Gallant, J., Two compounds implicated in the function of the RC gene of *Escherichia coli, Nature (London),* 221, 838, 1969.
107. Cashel, M. and Rudd, K. E., The stringent response, in *Escherichia coli and Salmonella typhimurium: Cellular and Molecular Biology,* Vol. 2, Neidhardt, F. C., Ingraham, J. L., Low, K. B., Magasanik, B., Schaechter, M., and Umbarger, H. E., Eds., American Society for Microbiology, Washington, D. C., 1987, 1410.
108. Kingston, R. E. and Chamberlin, M. J., Pausing and attenuation of *in vitro* transcription in the *rrnB* operon of *E. coli, Cell,* 27, 523, 1981.
109. Kassavetis, G. A. and Chamberlin, M. J., Pausing and termination of transcription within the early region of bacteriophage T7 DNA *in vitro, J. Biol. Chem.,* 256, 2777, 1981.
110. Kingston, R. E., Nierman, W. C., and Chamberlin, M. J., A direct effect of guanosine tetraphosphate on pausing of *Escherichia coli* RNA polymerase during RNA chain elongation, *J. Biol. Chem.,* 256, 2787, 1981.
111. Schreiber, G., Metzger, S., Einat, A., Shmuel, R., Cashel, M., and Glaser, G., Characterization of the *relA1* mutation and a comparison of *relA1* with new *relA* null alleles in *Escherichia coli, J. Biol. Chem.,* 266, 3760, 1991.
112. Staden, R., Measurement of the effects that coding for a protein has on a DNA sequence and their use for finding genes, *Nucleic Acids Res.,* 12, 551, 1984.
113. Nomura, M., Bedwell, D. M., Yamagishi, M., Cole, J. R., and Kolb, J. M., *RNA Polymerase and Regulation of RNA Synthesis in Escherichia coli: RNA Polymerase and the Regulation of Transcription,* Reznikoff, W. Z. et al., Elsevier Science Publishing Co., New York, 1987, 137.

Chapter 8

VARIATIONS IN READING THE GENETIC CODE

Jack Parker

TABLE OF CONTENTS

I.	Introduction	192
II.	Types of Errors and Alternatives in Codon Reading	193
III.	Misreading Sense Codons	196
	A. Misacylation	197
	B. Missense Reading	199
	C. Missense Alternatives	206
IV.	Errors Involving Termination	207
	A. Leaky Termination Codons: Readthrough Errors	209
	B. Required Readthrough: Alternatives to Termination	210
	C. What is a Native tRNA?	215
	D. UGA as an Alternative Sense Codon	217
	E. Premature Termination and Ribosome Pausing	217
V.	Ribosomal Frameshifts	220
	A. Frameshift Errors	222
	B. Required Programmed Frameshifts	228
	C. Shift to Terminate — the Case Against Alternative Termination Codons	235
VI.	Control of Error	236
	A. Codon Usage and Codon Context	236
	B. Proofreading, the Stringent Response, and Editing	237
	C. The Balanced Cell	240
VII.	Conclusions	241
	A. What is the Error Level in Cells?	241
	B. How Much Error Can Cells Tolerate?	244

Acknowledgments ... 245

References ... 245

> "Just the facts, Ma'am."
> "... all we want is the facts..."
> —Sergeant Joe Friday, *Dragnet*

I. INTRODUCTION

The genetic code is usually shown as a codon table, where all 64 possible three base sequences are presented along with the amino acid or the termination signal that each encodes. The cell's translational apparatus reads this code as the ribosome moves along the messenger RNA. In order to be translated, the message must also contain an initiation signal which tells the ribosome where and in which reading frame to begin. This initiation signal includes a start codon, but it is not usually limited to simply these three bases. It was demonstrated very early that widely divergent organisms seemed to use the same code, and the very pleasing idea of a universal code was born.[1] Although difficult to decipher, the genetic code did not seem particularly rich as a language. There were only 20 amino acids to correspond to the 61 sense codons, so the code was rich in synonyms, but not much else. This also may have been pleasing, in that one can imagine that such a language can be translated with great ac-curacy, particularly as most of the synonyms are very closely related. Obviously, like all steps in the transfer of genetic information, translation must occur with considerable fidelity. As we shall see, however, it is not yet clear either exactly what the level of fidelity actually is or how much error cells can tolerate.

The deciphering of any code is dependent on the number of encoded messages that are available. Since the late 70s, there has been a tremendous increase in the number of sequenced genes, and in many cases some sequence information has also become available on the proteins they encode. This information has led to two important discoveries about the genetic code. First, some organisms and some organelles use slightly different codes, which seem to be derivatives of the "universal" code.[2-4] Second, even organisms which use the universal code sometimes read it in nonstandard ways, giving the code unexpected subtlety and nuance. This chapter will not concern itself too much with nonuniversal codes. Rather, in addition to the fidelity of codon reading, I will examine the nonstandard or alternative readings which an individual organism can use in deciphering the universal code. As we will see, it is sometimes very difficult to differentiate between an error and an alternative, and some of my distinctions are certainly arbitrary. However, for the most part alternatives will be considered those events which are required for functional gene expression. In some cases, alternative readings seem very much like errors that occur at propitious sites.

II. TYPES OF ERRORS AND ALTERNATIVES IN CODON READING

In this review I will refer to codon misreading, or alternative codon reading, as an event which seems to involve unexpected codon:anticodon interactions. There are a reasonable number of general types of misreading events (or alternatives), and a very large number of specific errors, which could happen during the synthesis of a protein from a "perfect" message. Figure 1 illustrates a few examples of these using a short hypothetical message and its standard, or expected, translational product (protein 1).

The ribosome could begin translation at an incorrect codon either in the correct (0) reading frame or in the other two reading frames (+1 or –1). If this happens in the 0 frame, the protein product will be similar to the expected product but with either an addition or truncation at the amino-terminal end (protein 2 in Figure 1). These alternative products differ in size, but are otherwise closely related to the expected protein. There are many examples of genes which seem to use internal initiation to produce alternative proteins.[5-7] There are also examples of what seem quite clearly to be errors in initiation, and some of these can be detected *in vivo*.[8] Some of the signals involved in the initiation of protein synthesis are discussed in Chapter 11 of this book, it will be unnecessary to deal with them here. However, alternative initiation must sometimes be eliminated as a possibility when searching for other translational errors. This is particularly true for errors involving termination codons, since reinitiation often happens after termination.[9-11]

During elongation there are several different types of errors which may occur. One would be the misreading of a sense codon. Many possible "misreadings" will not actually result in any change in the protein because of the degeneracy of the code (see Chapter 12). However, this chapter will deal only with those codon misreadings which result in an amino acid substitution — a missense error. Proteins with a single such error will usually differ from the expected product only slightly (protein 3 in Figure 1), and therefore one would predict they would be relatively difficult to detect.

Another type of elongation error would be the loss of maintenance of the reading frame. Frameshifted products would be identical to the expected product up to the frameshift, and then differ depending on the direction and extent of the shift and the location of stop codons in the other frames. Since stop codons are usually abundant in other frames, most frameshift products would be truncated. Protein 4 is the result of a ribosome translocating an extra base to read UUU in the +1 frame. Other very interesting types of frameshifting have been detected. Protein 5 represents a +6 frameshift where a single leucine residue is incorporated in apparent response to three codons; in this case both the 5′ and 3′ codons are CUU. This type of frameshift has been termed a

Hypothetical message

Protein	AUG	nnn	nnn	AUG	nnn	nnn	CUU	UCG	CUU	nnn	GUG	Ann	UGA	nnn	...	Type of translational error
1	Met	X	X	Met	X	X	Leu	Ser	Leu	X	Val	X	Term			none
2				Met	X	X	Leu	Ser	Leu	X	Val	X	Term			internal initiation in 0 frame
3	Met	X	X	Met	X	X	Val	Ser	Leu	X	Val	X	Term			misread 1st position of CUU
4	Met	X	X	Met	X	X	Phe	Arg	Y	Y	Term					frameshift into +1 frame at CUU
5	Met	X	X	Met	X	X	-	Leu	-	X	Val	X	Term			+6 frameshift "over" UCG (tRNA hop)
6	Met	X	X	Met	X	X	Leu	Term								premature termination at UCG
7	Met	X	X	Met	X	X	Leu	Ser	Leu	X	Val	X	Trp	Y...		readthrough of UGA

FIGURE 1. A hypothetical messenger RNA and some of its possible protein products. The small letter n in the mRNA represents any base, and the corresponding X in the protein products represents the encoded amino acid. The most 5′ AUG codon is the initiation codon and Term indicates a termination codon. Other designations are from the universal genetic code except as noted.

"hop",[12] and its product can be very similar to the predicted, or standard, protein.

The remaining types of unexpected translational events involve termination. One class of these is premature termination (protein 6 in Figure 1), where protein synthesis terminates at a codon other than a termination codon. Another error involves the insertion of an amino acid in response to a termination codon, a process termed readthrough. Readthrough proteins (protein 7 in Figure 1) would be identical to the standard protein except that they will have carboxy-terminal extensions.

The difficulty in determining whether nonstandard codon reading has taken place is not simply related to the fact that some of the products will be very difficult to detect. It is essential to determine that the event leading to the nonstandard product occurred during translation. Errors in DNA synthesis can usually be ignored because they happen at frequencies many orders of magnitude below detectable translational errors. Much less is known about errors in transcription, but they may play a non-negligible role in generating error-containing protein. In *E. coli*, the *in vivo* error rate seems to vary between 10^{-4} to 10^{-5} mistakes per base incorporation in a context-dependent fashion.[13,14] Slippage of *E. coli* RNA polymerase at long runs of A or T can give rise to very high error levels (greater than 20%) during transcription.[15] Addition of nontemplated G residues by a *programmed* transcriptional slippage or "stuttering" mechanism is used to generate alternate mRNA's in the paramyxoviruses.[16-18]

Even if a particular transcript were a perfect copy of the DNA, this situation can change before the transcript becomes a message. In addition to splicing out introns, a very large number of other programmed sequence-altering events may occur. These include the addition or removal of U's in kinetoplastid mitochondria,[19,20] the insertion of C's in *Physarum*,[21] and the changing of C's to U's in some plant mitochondrial messages.[22,23] The processing of C to U is also used to produce tissue-specific alternative messages from the mammals.[24-27] Presumably such processes are themselves error-prone and may therefore contribute to unexpected translational products. All of the alternative proteins shown in Figure 1 could be generated by various types of RNA editing, and, therefore, great care must be used in interpreting data if one knows only the sequence of the gene and the protein, and not the sequence of the mRNA.

Even so, some of the apparently erroneous or alternative proteins used as examples can be derived by mechanisms other than misreading even from a perfect message. Proteins 2 and 6 could be the result of posttranslational processing at the ends of the protein. Protein 5 could also be the result of a protein splicing mechanism, for which there is now considerable evidence.[28-30] Protein 3 could also be synthesized if the codon:anticodon interation were perfect, but the tRNA had been misacylated. Misacylation could clearly be an important source of mistranslated protein, but it is not the result of unexpected codon:anticodon interaction, and I shall therefore differentiate it from misreading.

In each of the following sections, I will first deal with the unexpected events that occur and attempt to differentiate errors from alternatives. After these sections, I will deal briefly with how cells control error. There have been several recent reviews which have dealt with many of these areas.[31-34] This chapter will emphasize new findings, particularly those dealing with translational alternatives.

III. MISREADING SENSE CODONS

While maintaining the correct frame, a ribosome could misread a sense codon either as another sense codon or as a termination codon. The latter is termed premature termination and will be discussed in the section dealing with termination errors (see Section IV). The former error is termed missense misreading (or often simply misreading), by analogy with missense mutation, and is the subject of this section of the review. These errors were intensively examined in *in vitro* protein synthesizing systems,[35] not because they were considered more interesting than other types of errors, but because they are the only errors which could be detected using simple, repetitive messages. Errors were detected in such systems by looking for the incorporation of an amino acid which should not be encoded by the codons being tested, e.g., the incorporation of leucine using poly-U as message. Almost all the errors observed could be explained by a mechanism which involved a one-base codon:anticodon mismatch. This limits the potential number of different misincorporation events for which testing is needed, but even UUU can be misread as six other amino acids using one-base mismatches. Note also that this analysis presumes that the error is caused by misreading of the codon, not misacylation of the tRNA. These will have to be differentiated to learn the precise nature of the error, but the possibility of misacylation can increase the number of potential errors at a given codon.

The detection of missense misreading (or misacylation) *in vivo* can be difficult, because the aberrant protein will have essentially the same size and amino acid composition as the native protein. In many cases, it will also have normal or nearly normal, activity, since many amino acid substitutions which could occur by a one-base mismatch or by misacylation are conservative. Even if the activity is changed, this may not be observed unless the error occurs in a significant fraction of the protein molecules. Therefore, many missense errors that have been detected *in vivo* involve the misincorporation of an amino acid that normally does not occur in a given protein, or amino acid substitutions that lead to detectable changes in the normal electrophoretic or chromatographic behavior of the protein. Because of these difficulties, most *in vivo* assays for missense errors are both specific and limited, i.e., only a very few of the possible missense errors can be detected.

It is usually possible to differentiate between misreading and misacylation by the use of certain antibiotics, and in *E. coli,* at least, by a variety of mutants. A number of aminoglycoside antibiotics can increase mistranslation events on the ribosome in either eukaryotes or prokaryotes.[36-38] In *E. coli,* some of the mutant alleles of *rpsL,* the gene coding for ribosomal protein S12, can lead to reduction of a variety of translational errors, including misreading, both *in vitro* and *in vivo.*[36] This restrictive phenotype can be antagonized by aminoglycoside antibiotics, and by mutations in the structural genes of other ribosomal proteins, e.g., S4 and S5, which themselves have a Ram (ribosomal ambiguity) phenotype and increase translational errors.[36,39] Additionally, streptomycin seems only to increase the frequency of intrinsic errors, rather than to cause new ones.[40,41] Therefore, if the level of misincorporation is increased or decreased as expected in the mutants or in cells treated with antibiotics, it is likely that misreading and not misacylation is involved. However, studies on other types of errors which must take place on the ribosome (readthrough and frameshifting) indicate that the mutants can have extremely context-dependent effects (see following discussion), so a negative result does not rule out misreading. Apparently, even some of the antibiotics yield context-specific increases in misreading, further complicating analysis.[42] Sequencing the aberrant protein will not always be definitive when other tests have been negative, since amino acids that are structurally related, and therefore might be involved with misacylation, may also have related codons.

Using a combination of *in vivo* and *in vitro* systems, it has been determined that mutants with defects in several proteins involved with translation can affect the level of missense errors caused by codon misreading. These include ribosomal proteins S4, S5, and S12, as previously mentioned, and also L6, L7/L12, and S17.[43-46] Ribosomal protein mutants in other organisms have also been shown to affect translational accuracy.[47] Mutants of elongation factor-Tu (EF-Tu) also lead to increased misreading,[48,49] and a mutant of the equivalent yeast translation factor, EF-1α, may do likewise.[50] In addition there are "relaxed" mutants of *E. coli, relA,* and *relC (rplK),* which lead to higher levels of translational errors than wild-type cells during amino acid starvation.[51,52] The mechanism by which some of these mutants affect error frequency is known, and will be discussed in a later section. For the most part these are mutations affecting the general level of translational errors on the ribosome, i.e., misreading, readthrough, and frameshifting. Mutations affecting misacylation are more restricted in their activity because there are usually 20 different aminoacyl-tRNA synthetases in the cell.

A. MISACYLATION

As a group, the aminoacyl-tRNA synthetases carry out the aminoacylation of tRNA with great accuracy. These enzymes must recognize the correct tRNA (often a number of isoacceptors), and also recognize the correct amino acid. There is much current research being done on "tRNA identity", i.e., what specific features of an individual tRNA molecule are recognized by the cognate aminoacyl-tRNA synthetase (see Chapter 10).[53] The identities and specificities of tRNAs can be altered not only by changing the base sequence, but also by changing base modifications.[54,55] As part of the studies on tRNA identities, synthetic nonsense suppressor tRNAs have been created which can support the incorporation of more than one amino acid at a given codon *in vivo*.[56,57] Mutant initiator tRNAs can also be constructed which allow insertion of amino acids other than methionine at alternative initiation codons *in vivo*, and at least one of these can also be charged by more than one amino acid.[58] In these cases, the tRNA is recognized by more than one synthetase. Most interestingly, there is also a missense suppressor known in *E. coli* which shows a similar lowered fidelity. There are missense mutations in *trpA* which yield lysine codons (AAA/G) and which can be suppressed by mutations in one of the genes encoding tRNA^{Lys}.[59] The mutant tRNA has an altered acceptor stem,[60] and although it is still predominantly aminoacylated with lysine, a low level aminoacylation with alanine accounts for the missense suppressor activity observed.[61]

In the nonsense suppression studies mentioned above, many of the tRNAs were found to be "misacylated" by wild-type lysyl- and glutaminyl-tRNA synthetases of *E. coli*. There are also mutants of glutaminyl-tRNA synthetase known which increase the misacylation of suppressor tRNAs (while maintaining the ability to acylate tRNA^{Gln}).[62,63] When the wild-type glutaminyl-tRNA synthetase is overproduced in *E. coli*, it misacylates a mutant derivative of tRNA^{Tyr} encoded by *supF*.[64] This misacylation can be decreased if the wild-type tRNA^{Gln} is overproduced.[64] This emphasizes the need for maintaining the appropriate ratio of synthetase to tRNA in a cell. We shall return again to the concept of a balanced cell. The mischarging discussed previously involved mutant, suppressor tRNAs. For the most part throughout this review I shall confine myself to the discussion of errors involving native tRNAs, i.e., tRNAs which make up the cell's normal complement. Although aminoacylation takes place with great accuracy, it seems likely that misacylations involving closely related amino acids may occur at frequencies in the range of 4×10^{-4} to 5×10^{-5}.[65,66] However, it is possible that some misacylation errors, specifically cysteine misacylation by isoleucyl-tRNA synthetase, can occur at much higher frequencies.[67]

The first measurements of missense errors *in vivo* involved work on the misincorporation of branched chain amino acids into certain peptides of chicken ovalbumin and rabbit hemoglobin.[68,69] The results could be interpreted as showing that valine was being incorporated at isoleucine codons at a frequency of approximately 3×10^{-4}, involving either misreading (at the first position of the codon) or misacylation. However, it has subsequently been shown that in at least one case valine (codons: GUN) was incorporated at a threonine codon (ACN), almost certainly by misacylation.[70]

However, errors generated by misacylation need not always involve the misincorporation of one of the 20 standard amino acids; many amino acid

Variations in Reading the Genetic Code

analogs can be esterified to tRNA and hence incorporated into protein.[71] Although nonstandard amino acids are often detected in certain proteins, their presence need not be indicative of a high level of misacylation by the relevant aminoacyl-tRNA synthetase,[72] since a very large number of posttranslational modifications can also occur.[73] Proteins with certain amino acid analogs are often degraded at a rapid rate.[74] These can be protected from turnover either if the protein is secreted,[75] or if the protein forms an insoluble inclusion body in the cell.[76,77] This latter observation allowed an extremely interesting demonstration of the fact that metabolism in *E. coli* can become so unbalanced by translational demand that nonprotein amino acids are synthesized and incorporated into protein at high levels.[76] Overproduction of bovine somatotropin, which is a leucine-rich peptide, in *E. coli* leads to an increase in the levels of the leucine biosynthetic enzymes, which are also involved in synthesizing norleucine, a methionine analog. Under these conditions, cells incorporate norleucine into protein in place of methionine at levels up to 14%.[76]

Not all cells or organelles have a full complement of the aminoacyl-tRNA synthetases. Gram-positive cells, archaebacteria, cyanobacteria, chloroplasts, and mitochondria do not have a glutaminyl-tRNA synthetase, but rather first charge tRNA^{Gln} with glutamate and then use an amidotransferase to convert Glu-tRNA^{Gln} to Gln-tRNA^{Gln}.[78] Since an abnormally high level of glutamate misincorporation has not been reported for these systems, it seems likely that, as suggested by Schön et al.,[78] either the free intermediate is discriminated against by some translational factor, or the intermediate is transferred directly to the amidotransferase. It is of some interest that many of the errors and this aminoacylation "alternative" involve glutaminyl-tRNA synthetase. This enzyme and its interactions with tRNA is being intensively studied.[79]

B. MISSENSE MISREADING

Experiments with *in vitro* protein synthesizing systems are a useful guide for predicting the types of errors that may happen *in vivo*, and are also now proving useful for dissecting the mechanisms used by the translational apparatus to limit errors. Only a few *in vivo* misincorporations have been rigorously demonstrated to be caused by misreading, but the number is increasing, and it is clear that misreading happens in organisms other than *E. coli*. Throughout this review we will find examples where investigators intentionally (or unintentionally) increased the *in vivo* error frequency in order to detect or analyze a particular error. The assumption is that the *in vivo* manipulations would, like the *in vitro* manipulations, increase only those errors that normally happen and not cause new errors. There is some evidence that this is true. However, in only a few instances has the actual basal-level *in vivo* error frequency been measured.

As previously mentioned, one way to detect errors is to search for the incorporation of an amino acid residue that is not encoded for by a gene. Since cysteine codons are the least used in *E. coli* genes,[80] and since cysteine is conveniently and relatively specifically labeled by using [^{35}S]-sulfate in the presence of nonradioactive methionine, there have been a number of experiments carried out to look for cysteine misincorporation. In many cases though, little information was obtained regarding the site of misincorporation. Therefore, the measurements do not distinguish between misacylation (which may represent an important avenue of cysteine misincorporation)[67] and misreading. However, some experiments have given a clear indication of both the nature and frequency of the substitution.

Edelmann and Gallant found cysteine misincorporated into flagellin at a level of 6×10^{-4} mol/mol.[81] This level could be increased by streptomycin and neomycin and also by arginine limitation of a *relA* mutant. The latter data would indicate that cysteine is likely being incorporated at one of the arginine codons in *fliC* (10 CGU and 1 CGC), the structural gene for flagellin. This would give an average misreading frequency of 5×10^{-5} per codon in unstarved cells.

Bouadloun et al. looked for cysteine misincorporation in ribosomal proteins S6 and L7/L12, using an assay that depends on the amino acid substitution removing a specific protein cleavage site.[82] In L7/L12 they found cysteine incorporation at CGU at a frequency of 10^{-3}. In S6 they found cysteine incorporation at the tryptophan codon UGG at a frequency of 4×10^{-3}. The misincorporation at UGG was affected predictably by the *rpsD* and *rpsL* alleles,[82] and was greatly reduced in an *miaA* mutant which is missing a modification in tRNA^{Cys}.[83] However, the level of misreading of the CGU codon was much less responsive to all these mutations. It seems most likely that the high frequency of this error, and the inability of various mutations to affect it, is context specific. An error rate of 10^{-3} at the highly used CGU codon cannot be typical of *E. coli*, since it would lead to demonstrable charge heterogeneity of proteins, and this is not observed. This error rate was also not found as the average for the *fliC* gene. It seems likely that cysteine is also incorporated at phenylalanine codons, at least during phenylalanine starvation,[84] and the Cys-tRNA can apparently read UGA codons, at least *in vitro*.[85] Therefore, this tRNA seems to be able to misread at a reasonable frequency making first, second, or third position mismatches with the codon.

In other sections of the review we shall see that studying the leakiness of nonsense or frameshift mutations in suppressor free cells can give valuable information on translational errors. A similar assay has shown that the glycine codon GGC can be misread as serine.[86] A glycine codon was substituted for the AGC codon specifying an essential serine residue in β-lactamase. Although the amount of protein produced from the mutants was the same whether the glycine codon was GGC or GGA, cells containing the former, but not the latter, produced a low level of active enzyme. The activity was increased in *rpsD* mutants, although a restrictive *rpsL* allele had no effect. However, it seems probable that the serine tRNA which normally reads AGU and AGC reads GGC (a first position error) at a frequency of about 10^{-3}.[86] The misreading of

GGA as serine would not be possible by a single base mismatch, and this error was not detected.

Based on *in vitro* data, Woese predicted that most misreading errors occur at the third position of the codon,[35] which can lead to missense errors only in nondegenerate codon families. O'Farrell sought to demonstrate the existence of these errors *in vivo*, by starving *E. coli relA* mutants of amino acids where third position misreading could result in the misincorporation of an amino acid with a different charge.[87] Starvation for histidine resulted in a high level of charge heterogeneity in cellular proteins, as if the histidine codons (CAU and CAC) were being misread as a less basic amino acid, presumably glutamine (codons: CAA and CAG). Starvation for arginine also seemed to give some charge heterogeneity, but this was most likely the result of misincorporation of cysteine, a first position error. We have shown that starvation for asparagine (codons: AAU and AAC) leads to incorporation of lysine (codons: AAA and AAG), also a third position error.[88]

There is now more information on the misreading of asparagine codons as lysine than for any other *in vivo* missense error. The level of this error responds as expected in mutants.[89-91] By sequencing mistranslated MS2 coat protein from unstarved cells, we have demonstrated that this error occurs at a frequency of about 2×10^{-3} at AAU codons, and at a frequency of about 4×10^{-4} at AAC codons.[91] Also, using sequence analysis, we have shown that asparagine starvation and streptomycin addition increases the error level at these codons coordinately,[41,92] and that the difference in the level of misreading between the AAU and AAC is not context dependent.[93] The high level of misreading of AAU codons can be confirmed by quantitating the level of charge heterogeneity in proteins from messages which contain this codon, and may typically be nearer 5×10^{-3}.[91,94] Starvation for asparagine increases the error level at these codons approximately 100-fold in *relA* mutants, so that misincorporation can become the predominant event at AAU codons.[90,92] Asparagine starvation also leads to the same type of charge heterogeneity in mammalian cell cultures.[88,95,96] Although the mistranslated protein has not been sequenced, nor identified in unstarved mammalian cells, if one assumes that the frequency of misreading is linearly related to the severity of starvation, one can calculate that the basal-level of misreading of asparagine codons in these cells is 7×10^{-5},[96] well below the error level in *E. coli*.

As mentioned above, O'Farrell chose histidine starvation on the assumption that glutamine would be misincorporated by a third position misreading error, and the type of charge heterogeneity seemed to confirm this.[87] This starvation-induced heterogeneity is high in *relA* mutants,[87] and reduced in *rpsL* mutants and in *hisT* mutants.[97] The latter are missing a pseudouridine modification found in the anticodon loop of many tRNAs, including tRNA^{His} and the tRNA^{Gln} isoacceptors. On the basis of charge heterogeneity observed in different proteins, it seems possible that CAU is misread more frequently than CAC. Preliminary sequence information seems to support this idea.[98] Starvation for histidine also seems to cause this error in cultured mammalian cells.[88,95,96] In unstarved mammalian cells this error may occur at frequencies of 0.3 to 3.5 × 10^{-4}, depending on the cell type.[95,96] The higher levels are found in cells that have been transformed with simian virus 40.[95]

The AAU and CAU codons are each used less in highly expressed messages than their more accurately read partners, AAC and CAC (see Chapter 3). We therefore predicted that the less frequently used codon UUU might give a higher level of misreading than UUC, and that we could detect differential leucine misincorporation during phenylalanine starvation. Using both wild-type and mutant alleles of the *argI* gene of *E. coli*, we found that we could get high levels of misincorporation of leucine at residue eight of the product whether that codon was UUU or UUC (presumably by third position misreading),[99] but saw no misincorporation at residue three with either codon.[100,101] This situation exemplifies the shortcomings of a typical misreading assay. Although there was clearly a strong context effect, we could not distinguish between whether the context specified accuracy or some error other than leucine misincorporation. Our most recent evidence indicates that there is a ribosomal frameshift which happens near the codon for residue three.[102] It would seem, though, that when the frameshift doesn't happen there is no misreading.

Our hypothesis that codon usage might be an important mechanism used by the cell to control misreading events also led us to search for methionine incorporation as a misreading of a codon rarely used in *E. coli,* the isoleucine codon AUA.[80] We found no detectable incorporation of methionine at a particular AUA codon in unstarved cells under conditions where we could have detected an error frequency as low as 3×10^{-4}.[103] During isoleucine starvation we could detect misreading of this codon, as well as the commonly used AUU codon, at a frequency of 3×10^{-2}. Therefore, it seems likely that the basal-level, third position misreading of these codons is about 10^{-4}. Although this does not rule out the possibility of other errors, it seems most likely that this experiment clearly refutes our hypothesis that misreading happens frequently at rarely used codons.

However, the rarely used arginine codon AGA may be misread at a high frequency. When *E. coli* is utilized to produce high levels of an insulin-like growth factor from a gene construct, the product contains lysine at a frequency of 5 to 12% at the position of three arginine residues encoded by AGA.[104] The same results were seen at positions encoded by AGA in a gene construct encoding atrial peptide III.[77] The error involved could be either misacylation or misreading, but it is difficult to see how the increased usage of the codon could generate misacylation. However, since AGA is normally read by a minor tRNA (see Chapter 2),[105] misreading could be increased by prolonging the ribosome's pause at the A-site. The misreading is apparently very sensitive to the level of usage of the AGA codon, since changing only one of the three AGA codons in the insulin-like gene to the commonly used CGU codon abolished

Variations in Reading the Genetic Code

misreading at all positions.[104] Little information was given concerning the host(s) used in these experiments.

Histidine and tryptophan codons are also not abundant in *E. coli*, and the misincorporation of these amino acids has been detected in proteins which should not contain them. Histidine is found in MS2 coat protein at a level of about 1.4 mmol/mol, and this level can be increased by streptomycin and decreased by a restrictive *rpsL* mutation.[106] Histidine and tryptophan were also found misincorporated at a high frequency in the coat protein of Qβ, but the level of at least the former was not increased by a Ram mutation.[94] Since histidine codons can be misread as glutamine by third position misreading, we sought to determine if starvation for glutamine might increase the reciprocal error, as measured by charge heterogeneity. We have recently shown that this error can only be detected in cells which, in addition to being starved for glutamine, are also overproducing His-tRNA.[103] Therefore, it seems unlikely that the reasonably high basal-level misincorporation of histidine seen in MS2 coat protein is at the six CAA and CAG codons (unless it is context specific), but there remain 29 other codons where a single codon:anticodon mismatch is possible. It is also not known where tryptophan is incorporated in the Qβ coat protein, but the gene does contain cysteine codons, and *E. coli* Trp-tRNA can read cysteine codons *in vitro*.[107] The level of tryptophan misincorporation observed could indicate a very high *in vivo* misreading frequency if the incorporation is by a typical codon:anticodon mismatch.[94] If this is the *in vivo* error, the cysteine ↔ tryptophan system will be the only one where errors happen at a high frequency in either direction. We believe we have generated asparagine for lysine substitutions during lysine starvation, but at a frequency of only 8×10^{-3}, well over an order of magnitude below the reciprocal error.[89]

It seems likely that *Streptomyces* can misread UUA, but the level of misreading has not yet been quantified, nor has the misincorporated amino acid been identified. The *bldA* gene in this organism encodes a minor leucine tRNA which reads the codon UUA,[108] a rarely used codon in these G-C-rich organisms.[80] Several genes containing this codon were cloned into isogenic sets of *S. lividians* and *S. coelicolor* having wild-type and mutant *bldA* genes.[109] Protein synthesis from *E. coli ampC* and *lacZ* was dependent on having the wild-type tRNA, as was synthesis from the *carB* gene from another *Streptomyces* species. If the UUA codons in the *carB* message were changed to the leucine codon CUC by site-directed mutagenesis, then the mutated *carB* gene could function in *bldA* mutants. These experiments would seem to establish that the wild-type *bldA* gene is essential to read UUA and, since the *bldA* mutants are viable, would indicate that no essential genes contain this codon. However, two genes containing UUA codons did continue to function in the *bldA* mutant. The *aad* gene (from an R-plasmid) contains three UUA codons located throughout the gene, and its product confers spectinomycin resistance.[110] Mutants of *bldA* growing on minimal medium were somewhat resistant if they contained the *aad* gene, and fully resistant on rich medium. The *hyg* gene, from another *Streptomyces*, contains a single UUA codon and encodes a protein which confers hygromycin resistance. Once again, *bldA* mutants containing this gene were partially resistant in minimal medium and fully resistant on rich medium. Internal initiation was ruled out in the case of the *hyg* gene by use of site-directed mutagenesis, and can be ruled out in the case of the *aad* gene simply by the location of the UUA codons. Leskiw et al. found that very low levels of the *hyg* gene product can confer high level resistance.[109] The simplest explanation for the results is that the UUA codons are misread by a noncognate tRNA in *Streptomyces*, and that in only a few cases can this error give sufficient amounts of product to be measured.

More is known about misreading in *E. coli* than in any other organism. Although it is possible that misreading frequencies in *E. coli* are higher than in eukaryotes, there is simply very little data from eukaryotes. In Table 1, I have tabulated the data discussed in this section. I have made several assumptions, particularly about tryptophan misincorporation at cysteine codons. However, in view of the fact that these codons can be "misread" as selenocysteine at nearly 100% in a specific context (see also following),[111,112] I believe this assumption is not unwarranted. In cases where the nature of the error is precisely known, the frequency is often at or above 10^{-3}. This is apparently true in general of misreading of AAU, but otherwise is known for a single UGG codon, a single GGC codon, and a single CGU codon. In the latter case, it was argued that the average frequency of cysteine substitution at this codon must be lower. There is some evidence that the histidine codon CAU may be misread at a frequency of about 10^{-3} in most contexts and, also as discussed, it is possible that tryptophan misincorporation is also a relatively frequent error. However, the misreading of AAC codons as lysine is about 4×10^{-4}, and many other errors seem to happen at or below this frequency, although most potential errors (of which there are hundreds) have not been measured.

I believe that the fact that, in many cases, the errors most carefully measured occur at high frequencies has at least two explanations. First, for some of them there was every reason to believe they would happen at high frequency, which was the reason for developing assays to look for them. We chose to look for lysine for asparagine errors because Woese had predicted that third position errors might occur at high frequency,[35] and this particular error was easy to assay. The second reason is that many other errors occur at frequencies too low to be detected or quantified easily. Asparagine for lysine substitution also involves third position errors, and is also easy to assay. However, even during lysine starvation of *relA* mutants, this error happens at a very low frequency. Therefore, although there is undoubtedly an unexpectedly wide spectrum of actual misreading frequencies, it seems likely that the "average" misreading error will be found to happen at a frequency no higher than 4×10^{-4}, the same level as misreading at AAC, and the estimate of global error arrived at by Ellis and Gallant using analysis of charge heterogeneity on two-dimensional gels.[113]

TABLE 1
Missense Misreading of Codons *in vivo* in *Escherichia coli*

Codon	Amino Acid Residue Normal	Amino Acid Residue Substituted	Codon Position Misread	Basal-level Frequency	Comments	Ref.
AAU	Asn	Lys	3rd	5×10^{-3}	Only slight context effects	88-91, 93, 94
AAC	Asn	Lys	3rd	4×10^{-4}	Only slight context effects	88-91, 93, 94
AAA/G	Lys	Asn	3rd	low	8×10^{-3} during lysine starvation, substitution inferred	89
AGA	Arg	Lys	2nd	not known	$\cong 0.1$ during certain conditions, misacylation possible	77, 104
AUA	Ile	Met	3rd	$<3 \times 10^{-4}$	3×10^{-2} during isoleucine starvation	103
AUU	Ile	Met	3rd	low	3×10^{-2} during isoleucine starvation	103
CGU	Arg	Cys	1st	1×10^{-3}	One codon in *rplL*	82, 83
CGU/C	Arg	Cys	1st	5×10^{-5}	Average at 11 codons in *fliC*	81
CAU	His	Gln	3rd	high	>0.1 during histidine starvation, substitution, and basal-level inferred	87, 97, 98
CAC	His	Gln	3rd	low	>0.1 during histidine starvation, substitution, and basal-level inferred	87, 97, 98
CAA/G	Gln	His	3rd	low	only observed during glutamine starvation in His-tRNA overproducing cells	103
GGC	Gly	Ser	1st	1×10^{-3}	missense suppression in *bla* gene	86
UGG	Trp	Cys	3rd	3×10^{-3}	only single site examined, but may be average at all UGG	82, 83
UGU/C	Cys	Trp	3rd	high	substitution and frequency inferred (see text)	94, 106
UGU/C	Cys	Sec	3rd	low	substitution $\cong 100\%$ at specific context	111, 112
UUU	Phe	Leu	3rd	unknown	0.6 during phenylalanine starvation	100, 101
UUC	Phe	Leu	3rd	unknown	0.6 during phenylalanine starvation	100, 101
UUU/C	Phe	Cys	2nd	unknown	seen during phenylalanine starvation	84

C. MISSENSE ALTERNATIVES

There are two situations in nature where it is clear that the sense codons of the universal genetic code have alternative meanings. One of these is in genomes which use certain nonuniversal codes. Although in many cases nonuniversal codes have one or more reassigned termination codons, there are also instances where sense codons are reassigned, e.g., CUG (but not CUU and CUC) as serine in the nuclear genome of *Candida*,[114] or the entire CUN family as threonine in the mitochondria of yeasts.[115-118] These and other changes from the universal code have been recently reviewed.[2,33,119] Such changes are obviously interesting, and offer insight into the evolution of the code. However, it seems most likely that they do not represent one end of a continuum of ambiguity, but rather are examples of the old adage that nature abhors a vacuum. It is thought likely that these divergent codes have evolved through "codon capture".[120] First, through evolutionary pressure, often involving selection for or against G-C content, certain codons became unused, and then, after mutation has caused the loss of a functional cognate tRNA, the codon becomes unassigned. Unassigned codons are free to be captured by mechanisms involving mutations in the structural genes of an appropriate tRNA or an aminoacyl-tRNA synthetase. There would be no intermediate state where a codon had two meanings. Some native tRNAs do have the ability to misread certain codons (at least termination codons) at measurable levels, but it is unclear whether this represents anything other than the optimization of translation allowing some ambiguity in codon reading. The capture of termination codons may also be rather different, since it seems most likely that actual termination sites are not simply a three-base codon.

It is now clear that in most (perhaps all) organisms, certain sense codons can have more than one meaning, depending on codon context. The context is that involved with the initiation of protein synthesis. Codon context is covered extensively in Chapter 11 of this book, and I refer the reader to that chapter and references therein for a discussion of the use of contexts in initiation. What is important to note here is that several sense codons can specify methionine, and that these codons would allow a single codon:anticodon mismatch at the first (UUG and GUG), second (ACG), or third positions (AUU and AUA). However, these alternatives require a particular tRNA and initiation factors, which play an important role in specificity.[121] We shall see another example of context-dependent alternative reading of a codon, in this case UGA, which requires a specific tRNA and a special translational factor.

There is no known example where a sense codon has an alternative meaning within a message, nor do I know of any examples where organisms make use of the reasonable heterogeneity that can be generated by codons which are misread at frequencies approaching 1%. However, the ribosome uses the code considerably more flexibly for every other type of translational event we will consider.

IV. ERRORS INVOLVING TERMINATION

The message signals the end of the reading frame by using a specific termination codon. In the universal code, the codons UAA (ochre), UAG (amber), and UGA (opal or umber) are used as the termination signals, and are therefore usually referred to as stop codons or termination codons. Mutations which generate these codons within a gene are called nonsense mutations. Therefore, the codons themselves are sometimes referred to as nonsense codons, and were given the trivial names based on extragenic suppressors.[122] If no tRNA is available, clearly a codon cannot be read as "sense"; but, as we shall see, termination is not the *de facto* result of having no available specific tRNA. Interestingly, in some genomes, actual unassigned codons seem to exist — codons that are neither read by a tRNA nor recognized as a termination signal.[123,124] Because they do not seem to be read at all, unassigned codons have also been referred to as nonsense codons,[124] making the use of this term somewhat ambiguous. In addition, an unassigned codon will cause a ribosome to stop, so the use of the term "stop codon" also seems inadvisable. Therefore, in this review, codons that signal the end of an open reading frame will be referred to as termination codons.

Any mechanism which allows the cell to avoid the effects of a nonsense mutation would be termed nonsense suppression. Although these can include translational events such as ribosomal frameshifts and internal initiations,[8,9,125] in this section, I will confine myself to the situation where the termination codon is read as an amino acid — a process termed readthrough.

Before discussing how cells subvert the termination process, I shall briefly discuss the mechanism of termination. There are excellent recent reviews on various aspects of translation termination.[126-128]

In order to function in termination, the codons must be "recognized" by specific protein release factors, RF.[128,129] In addition to the RFs, this recognition and/or termination seems to involve ribosomal proteins[128,130,131] and 16S rRNA.[127,130,132,133] It may well be the 16S rRNA that actually "reads" the stop codon.[127,132] The precise mechanism of termination is not known, but in bacteria it appears that when these codons are in the ribosomal A site, the recognition occurs and the nascent protein is hydrolyzed from the peptidyl-tRNA in the ribosomal P site. In bacteria, RF1 is specific for UAG, RF2 is specific for UGA, and both can be used to terminate at UAA. A third bacterial release factor, RF3, may also be involved, but in a non-codon-specific manner. In the eukaryotic cytoplasm, there seems to be only a single release factor which recognizes all three stop codons. In mammals, at least, the RF requires tetranucleotides, e.g., UAAA, for *in vitro* activity.[134,135]

RF1 and RF2 are both essential proteins in bacteria,[136-138] and their genes (*prfA* and *prfB*, respectively) have considerable homology to each other.[139,140] Fascinatingly, the gene for the rabbit RF shows no homology to the bacterial RFs, but has significant homology to the tryptophanyl-tRNA synthetases of bacteria and mitochondria.[135] This could reflect either its mechanism of action or its evolution.[135] The RFs of mitochondria seem to be related to the prokaryotic RFs, in that they can function heterologously *in vitro* under certain conditions.[123] Many genomes use nonuniversal codes in which the number and/or identity of termination codons differ from those of the universal code.[2,33] Therefore, the specificity, and perhaps the number, of RFs present in mitochondria will depend on the code actually used.[123]

If a nonsense mutation is read as an amino acid (or misread, since the interaction is unexpected), and a functional protein is formed, the process is termed nonsense suppression. "Natural" nonsense suppression is the level obtained in cells without an extragenic suppressor. Our understanding of nonsense suppression, termination, and codon reading in general, has been greatly aided by studies using suppressor tRNAs which read termination codons.[141-145] However, mutants with changes in many other components of the translational apparatus can also suppress nonsense mutations. *E. coli* mutants lacking the ms^2i^6A at position 37, immediately 3′ to the anticodon on various tRNAs, have been shown to have decreased levels of nonsense suppression, presumably because the tRNA(s) no longer misreads termination codons.[83,146] Fascinatingly, the absence of this modification can also increase some misreading events.[147] Modification of tRNAs and its implication for codon interaction is covered in Chapter 2 of this book. As might be expected, nonsense suppression in *E. coli* can be increased by mutations in the structural genes for RF1 and RF2.[136-138,148,149] Although the mutations show the expected codon specificity, i.e., mutations in RF2 do not suppress amber mutations, most show additional context specificity. In yeast there is an extrachromosomal cytoplasmic element, the Ψ^+ factor, which increases suppression of some mutations,[143,150] and which may inhibit normal termination.[150] In *E. coli* there are mutants of 16S rRNA which can suppress certain UGA codons,[127,132,133] and, in yeast mitochondria, a mutant of 15S rRNA suppresses UAA codons.[151]

Obviously, readthrough and termination are competitive processes, and the elongation factors and RFs bind at very similar sites on the ribosome.[128,131] As would be expected, either overproduction of the wild-type RF or lowering the expression of the wild-type EF-Tu can reduce suppression.[138,152,153] One would also expect that mutants of EF-Tu may also have altered readthrough efficiency. Some EF-Tu mutants have been found which increase nonsense suppression,[152,154] while others seem to decrease nonsense suppression, at least at some codons.[155] EF-Tu mutants can be quite specific, both as to which codon can be suppressed and to the context of the codon. Mutants of the equivalent yeast protein, EF-1α, which were isolated as frameshift suppressors, also suppress certain nonsense mutations.[50] In yeast several omnipotent suppressors are known which confer a variety of phenotypes on the cell, including increased nonsense suppression.[156] Many members of this class of suppressors seem to be essential proteins that are closely related to EF-1, and are associated with the ribosome.[150,157-160]

Numerous mutants with altered ribosomal proteins also have altered levels of nonsense suppression. Mutations in the genes for L6, S12, and S17 can restrict readthrough,[36,43,44,161] while mutations in the genes for S4, S5, and L7/L12 can increase readthrough.[39,45,46,161-163] Recently, it was reported that ribosomes missing S15 could readthrough one specific opal mutation.[164] The use of restrictive mutants is an important tool to demonstrate that the suppression observed is caused by translation and not by transcription.[13,165]

For the most part I will deal with readthrough involving cells which use the universal code and do not contain mutant suppressor tRNAs. However, most of the studies on "natural" readthrough have of necessity utilized nonsense mutations, and not the termination codons at the end of a cistron. This is important to bear in mind. In the long list of mutants affecting readthrough, only a restrictive *rpsL* (S12) mutation has been found to apparently restrict readthrough of an authentic termination codon,[166] and even in this case the codon is one where readthrough is required to give an essential product.

A. LEAKY TERMINATION CODONS: READTHROUGH ERRORS

As previously mentioned, studies on the suppression of nonsense mutations by various suppressors has shown that different codons, and the same codon in different contexts, are often suppressed at different frequencies. This is also true when looking at the natural leakiness of termination codons. UGA can be misread as tryptophan in *Bacillus subtilis* at frequencies of 1 to 6% depending on the context, and in *Staphylococcus aureus* at a similar frequency.[167] UGA readthrough at or above 1% is also known to occur in *E. coli*,[136,168,169] and probably also involves incorporation of tryptophan.[170,171] In *Salmonella typhimurium*, UGA mutations were found to be suppressed at a frequency of at least 10^{-2} to 10^{-3}.[172] At least in *E. coli*, a much lower level of readthrough was observed at some UGA codons,[173] but in general this codon has a well deserved reputation for leakiness. As we shall see in the next section, some viruses require readthrough of UGA to produce an essential protein. In mammalian cells there is one known example of readthrough of a chromosomal gene's normal termination codon, and this codon is UGA. Geller and Rich observed readthrough of the rabbit β-hemoglobin in an *in vitro* system.[174] The readthrough protein can be detected *in vivo*, and the UGA codon may be read as tryptophan at a frequency of about 0.5%.[174,175] However, a UGA nonsense mutation in the chloramphenicol acetyltransferase gene (*cat*) could not be misread at a measurable frequency, (approximately 10^{-3}), in mammalian cells unless aminoglycosides were added to the culture.[38] Overproduction of wildtype tryptophan tRNA in yeast can suppress a UGA mutation,[176] but it is unclear what the normal level of UGA leakiness is in this organism. Interestingly, some *E. coli* ribosomal protein mutants may affect which amino acid is inserted in response to UGA.[177] Although it is generally assumed that Trp-tRNA is always involved in UGA readthrough, Cys-tRNA may also read this codon.[85]

Natural leakiness of amber and ochre mutations has also been examined in a variety of organisms. In enteric bacteria, most UAG nonsense mutations are suppressed at frequencies of 7×10^{-3} to 1.1×10^{-4},[137,178,179] but several examples of UAG readthrough above 1% have been observed.[173] The readthrough of UAA seems to occur at frequencies from almost 10^{-3} to less than 10^{-5}.[136] In *Bacillus* and *Staphylococcus*, readthrough of UAA is also considerably lower than that of UGA.[167] In *E. coli*, readthrough of UGA usually happens at a higher frequency than readthrough of UAG, and both happen at a higher frequency than readthrough of UAA. This difference is not simply based on context.[169]

Less is known about natural, erroneous suppression of these codons in eukaryotes. Some nonsense mutations in yeast may be suppressed at a frequency of about 10^{-3},[180] and readthrough of UAG can be detected in suppressor-free strains.[181] Overproduction of the appropriate wild-type tRNA^{Gln} increases suppression of UAA and UAG.[182,183] Introducing the termination codons in-frame in a construction involving the *cat* gene showed that readthrough is below 10^{-3} in normal mammalian cells.[38,184]

Organisms have preferences for use of the termination codons, as they do for other codons. From the previous discussion it would seem clear that in bacteria at least, UAA would be the preferred termination codon and that UGA would be very rarely used (unless readthrough was required). Termination codon usage has been compiled for many organisms,[185-188] and is reviewed in Chapter 14 of this book. In *E. coli,* UAA is the preferred codon, as expected — expected not merely from data on leakiness though, but because it is also the only codon at which both RFs function. Unexpectedly, UGA is also used at a considerable frequency and in highly expressed genes, indicating that it can be an efficient terminator. The UAG codon is rarely used, but as previously discussed, its release factor, RF1 (which also reads UAA), is essential. The reasons for the preferred use of UGA over UAG is unclear, but may have to do with RF abundance or efficiency.[186] However, it need not be that UGA is leaky when used at the end of a message, since suppressor tRNAs are not effective at suppressing normal terminators.[189,190] It seems likely that the context of the terminator (or the nonsense mutation) plays an extremely important role in termination. At least some part of this context may involve the RFs, and if it is actually the 16S rRNA that reads the termination codon, then presumably it also plays a role in determining context effects.

B. REQUIRED READTHROUGH: ALTERNATIVES TO TERMINATION

Since most of the work previously discussed has to do with leaky nonsense mutations, this makes it easier to define these readthrough events as errors. Anytime readthrough of an actual termination codon is observed, there is the possibility that the readthrough protein is required by the organism. In this section, we shall discuss events where the readthrough event has been proven to be essential, or where there is a high likelihood that it will be.

The first example of a required readthrough event was found in the RNA coliphage Qβ. It was determined that the UGA terminating its coat protein gene is read as tryptophan about 6.5% of the time, a level which can be increased only marginally in suppressor-containing strains.[94,171,191] This protein is assembled into the viral capsid, and its presence in the virion is required for infectivity.[192] The essential nature of the readthrough protein is supported by the fact that bacterial mutants with restrictive *rpsL* alleles prohibit productive infection by Qβ.[166,193]

Most of the rest of the examples of required readthrough proteins also come from RNA viruses, but RNA viruses of animals and plants. Readthrough in tobacco mosaic virus (TMV) has been extensively characterized. The most 5′ open-reading-frame encodes a 126-kilodalton(kDa) protein and terminates at UAG.[194] Reading through this codon allows synthesis of a 183-kDa polyprotein whose downstream domain is apparently the RNA replicase.[195,196] This 183-kDa protein is found in *in vitro* systems[197] and is also found *in vivo*, where the apparent rate of readthrough can exceed 10% and may vary during the infection cycle.[198]

The sequence surrounding the leaky UAG has been intensively analyzed by site-directed mutagenesis, either to make mutant viruses,[199] or to examine the readthrough activity by a downstream reporter gene.[200,201] As expected, the 183-kDa protein is required for virus replication, and UAA will also allow production of the protein.[199] Interestingly, if the UAG was changed to the tyrosine codon UAU, virus replication was severely reduced, and the progeny of this mutant virus contained a high number of new mutants which now had an in-frame UAA (none were recovered with UAG).[199] It seems that the correct ratio of 183- and 126-kDa proteins is required for optimal replication.

Using the reporter gene fusion constructions in an *in vivo* system, where readthrough was about 5%, it was shown that UAG codons are not simply leaky, but that the context 3′ to the codon is important.[200,201] The important elements of this context are the next six bases, which are normally CAA UUA. They found that either glutamine codon could be used, and that the U residues in this sequence could be replaced by C residues, but the final A was essential. Moving the six base sequence into the +1 frame also abolished readthrough. This could indicate that some part of the sequence, presumably the glutamine codons, must be capable of being translated in-frame, but it is equally possible that it is the sequence itself, perhaps interacting with 18S rRNA, that is important.[201] Skuzeski et al. discuss the fact that it is not possible to distinguish whether the changes increase RF efficiency or decrease tRNA misreading. Unpublished preliminary experiments were reported as indicating that the TMV UAG codon with surrounding bases is not leaky in *E. coli*.[201]

A large number of plant RNA viruses are now known which seem to use readthrough.[201,202] Several of these have a UAG followed by the TMV six base consensus sequence, and all these that were tested allow readthrough in tobacco cell protoplasts when fused to a reporter gene.[200] Other viruses have a different sequence surrounding UAG, and these sequences gave a very low readthrough efficiency (approximately 10^{-4}) in this system.[200] It is possible that there will prove to be a group of different signals, each of which is sufficient in different cell types to induce readthrough. Several viruses have the sequence AAA-UAG-GGG or AAA-UAG-GUA,[200,203,204] but other sequences are also known to be used.[205,206] There is one example from plants, tobacco rattle virus, where the readthrough codon is UGA.[207,208] The sequence surrounding the termination codon is unlike those previously discussed, and it was inefficiently suppressed in the reporter test system.[200]

In many of these RNA viruses, as in TMV, the readthrough protein apparently contains the RNA replicase, indicating that readthrough is essential. However, it is unclear if the readthrough product which gives a replicase extension in turnip yellow mosaic virus,[209] and a similar second potential readthrough product in carnation mottle virus are essential.[205,210] In some cases, though, the readthrough produces a coat protein extension protein (there is no tendency to use a different consensus depending on the type of protein synthesized), as in the bacteriophage Qβ, and it is unclear if all of these are essential. The coat extension of potato leafroll virus (PLRV) appears to be synthesized at about 1% of coat itself and can be found in the virion, where it may play a role in transmission by an insect vector.[211,212] The luteoviruses, of which PLRV is a member, also use other interesting translational alternatives, as will be discussed.

The alphaviruses of animals are RNA viruses whose putative RNA replication enzymes are also encoded as part of a polyprotein. In some of these viruses, including Sindbis virus (SIN), the synthesis of this polyprotein involves the readthrough of an in-frame UGA codon.[213,214] However, other viruses synthesize the polyprotein as an uninterrupted reading frame with an arginine codon rather than with UGA.[214,215] In the case of SIN, at least, the readthrough protein is found in only very small amounts, so that readthrough of UGA need not be particularly efficient.[213,216] Li and Rice have recently constructed an interesting series of mutant derivatives of SIN to test the efficacy of substitutions for the UGA codon.[217] All substitutions tested, both sense and nonsense, yielded infectious mutants. Substitution of UAA gave slower growth rates in most cases (probably indicating it is less efficiently misread). UAG was also less efficient, but only in certain cells and only when the cells were stressed.[217] This fits quite satisfactorily with the relative leakiness of these codons in *E. coli*. Interestingly, mutants with sense codons (serine, arginine, or tryptophan) grew well. A competition between the wild-type and the serine-containing mutant showed that the latter was displaced from the population at 7 to 8% per passage. Analysis of the proteins produced by the cells infected with sense-codon-containing mutant indicates that the polyprotein is overproduced, but processing still takes place. Fascinatingly, mutants with serine or tryptophan codons actually gave larger plaques than wild-type virus (or those with an arginine codon) on some cell lines. Li and Rice suggest that different alphaviruses require different levels of protein for optimal RNA replication, and one way to modulate the amount of protein (either the polyprotein or the replicase) is to use readthrough.[217]

Retroviruses are RNA viruses which encode a reverse transcriptase and replicate through a DNA intermediate which is inserted into the genome of the host cell as a provirus (reviewed in Chapter 1 of this volume).[218] The retroviral genome consists of at least three open-reading-frames (sometimes referred to as genes). These are, in order from 5′ to 3′, *gag*, *pol*, and *env*. The products of these reading frames are processed into a number of proteins: *gag* encodes the structural proteins of the viral core, *pol* the reverse transcriptase and the integrase, and *env* the envelope proteins. A protease is in some cases encoded as part of *gag* and/or *pol*, and in some cases as a separate open-reading-frame called *pro*. The *env* mRNA is generated by splicing, but the other reading frames are read from the genome-length transcript. As we shall see, different retroviruses (and their relatives) have different strategies to accomplish this but some use readthrough of a termination codon at the *gag-pol* junction.

In all cases of readthrough the "termination" codon of *gag* is a UAG,[219-222] read as glutamine,[223,224] at a frequency of 2 to 10% in infected cells.[225] The mechanism of readthrough has been examined in some detail for AKV murine leukemia virus (AKV) and Moloney murine leukemia virus (Mo-MLV). First, unlike the case of the alphaviruses, changing the UAG codon to a sense codon prevents virus replication, possibly by preventing processing.[226,227] Panganiban found that a number of higher eukaryotic cell types could readthrough the UAG as part of a 300 base-pair sequence from the *gag-pol* junction of AKV fused to *lacZ*.[228] The UAG codon was not simply leaky in these cells, since it gave efficient termination in other contexts.[228] Fascinatingly, the same AKV-*lacZ* fusion gave no detectable readthrough in *E. coli*, unless the cells contained a known amber suppressor and then the efficiency of suppression was only 2%. Although all these viruses use UAG, all three termination codons can function *in vivo* in mutant forms of Mo-MLV.[227] In this system, UAA was very inefficient, while in similar constructs UAA seemed to function equivalently to UAG and UGA in a different *in vivo* system.[229] However, it was noticed in this latter study that UAA allowed readthrough less efficiently *in vitro*,[229] so the differences could have to do with the cell lines used. Interestingly, the UAG codon in the *gag-pol* junction in these viruses is in the loop of a stem-loop.[219,220,227,228] Site-directed mutagenesis of nucleotides which were 15 bases 5′ or 3′ of the UAG, but which should destabilize the stem, led to loss of virus production, and mutations which would restabilize the stem were compensating.[227] However, it is possible that these mutations could have affected an alternative, downstream RNA structure.[230]

All the examples we have discussed so far are in RNA viruses. In some cases, these could be examples of strategies used by these very small genomes to overcome difficulties faced in producing proteins from polycistronic messages in eukaryotic cells. However, in other cases, certainly for the coat protein extensions, it seems more likely a convenient method to make small amounts of a protein specifically targeted to a cellular location. If that is so, we would expect to see examples of readthrough in DNA genomes, including that of the host. As I mentioned earlier in the section on readthrough errors, the only known example of a message from a chromosomal gene in eukaryotes which is known to have readthrough of its termination codon is β-hemoglobin, and there is no evidence that this readthrough is required.

In *E. coli* there is one clear example of readthrough of a plasmid encoded gene. Enterotoxigenic strains carry plasmids which encode proteins which form pili and help facilitate colonization of the host. When the genes from the CFA/II system are cloned into *E. coli* K-12, only one of the antigens, CS3, is expressed.[231] Interestingly, the number of proteins required in the synthesis and assembly of CS3 pili was found to exceed the simple coding capacity of the cloned DNA. It was subsequently found that one of the required proteins for the synthesis of CS3 pili is synthesized by readthrough of an in-frame UAG codon at the end of the reading frame for another component of the system.[232] In the absence of readthrough, apparently pilin is synthesized but pili are not produced. The level of readthrough in the native *E. coli* which contains the wild-type plasmid was not measured, but the synthesis of the protein in K-12 was directly related to the serendipitous presence of an amber suppressor.[232]

There are a number of examples of viruses (RNA and DNA) which cannot infect *E. coli* with a restrictive *rpsL* allele,[193,233] but in no case other than Qβ has it been precisely determined why this is so. Interestingly, the virus T7 normally does not replicate in *E. coli* containing an F-plasmid, but can do so if the host has a restrictive *rpsL* allele.[234] The F-plasmid can regain its ability to exclude T7 in such cells if an ochre suppressor is added. Therefore, readthrough of UAA may be involved in the mechanism the F-plasmid normally uses to exclude T7.[234] Although the mechanism is by no means proven, this is the only example of a required readthrough which may normally involve UAA.

The examples mentioned all seem to require the readthrough protein to be made at a fixed, lower level than the normally terminated protein in the host. Obviously, considerably more data will be necessary on the production of the proteins in the natural host before this can be proven, but as yet there seems no reason to doubt the statement. However, it should be possible for a system to exist where changing readthrough efficiency is an important regulatory event. A mechanism like this was proposed to explain the effect of some *rpsL* mutants on regulation of the tryptophan operon.[235] The tryptophan operon leader peptide is terminated by UGA, and it was proposed that Trp-tRNA readthrough of this codon could modulate attenuation. However, changing the termination codon to UAA or UGA did not affect expression.[236] Possibly the *rpsL* mutation altered the elongation rate of the ribosome,[237] thereby uncoupling transcription and translation.

There is one as yet unproven mechanism involving controlled readthrough which seems a very likely possibility, since it involves the regulation of RF1 by readthrough of UAG.[238] In *Salmonella typhimurium*, the RF1 structural gene, *prfA*, is downstream in an operon with *hemA*, a gene involved with δ-aminolevulinic acid synthesis. Interestingly, the *hemA* gene is essential unless the cell has a separate *prfA* gene.[238] The *hemA* gene ends in a UAG codon and a possible readthrough product was observed. Elliot postulates that when RF1 is low, the ribosome reads through the UAG and terminates at an in-frame UGA very close to the *prfA* start codon.[238] The ribosome then reinitiates. In this case, regulated readthrough can couple the translation of these two proteins. We shall see that the regulation of the synthesis of RF2 also involves the regulated avoidance of a termination codon. However, with RF2 the avoidance is not by readthrough but by a programmed frameshift.

The fact that UAG and UGA are involved in almost all cases of required readthrough would seem to argue that there is something special about termination at the UAA codon in all organisms. The additional fact that eukaryotes seem to have only a single RF, would argue that this is not just a matter of RF availability.

C. WHAT IS A NATIVE tRNA?

In several instances in this review, I distinguish between mutant suppressor tRNAs and the native complement of tRNAs, at least one of which must be able to read each termination codon if these codons are leaky. As previously mentioned, several organisms and "all" mitochondria, except those of plants, use genetic codes that are derived from the universal code, but are not identical to it.[3,33] Many of these codes use one or more termination codons as sense codons, for example, UAA and UAG are used as glutamine codons in the nuclear genes of some ciliates,[239-243] other ciliates use UGA as a cysteine codon,[244] and UGA is used as a tryptophan codon in the mycoplasmas.[245-249] Such cells and organelles clearly have tRNAs which efficiently read these codons. However, these organisms and organelles, however numerous they may be, are considered interesting examples of how the universal code can evolve. The nucleus of human cells uses the universal code, as do our favorite model systems, and it is from these that we derive our ideas about the standard complement of tRNA and what is "native", and what is not.

Nowhere is this better exemplified than for *E. coli* K-12. Although the original, unmutagenized strain itself no longer exists, *E. coli* K-12 has been defined as "wild-type" for the basis of genetic experimentation. Obviously if the genome structure, regulation, and metabolism of K-12 had turned out to be very much different from other strains of *E. coli,* from other enterics, or from many other characterized bacteria, it would have been abandoned as an ideal model organism. However, it did not, so much of our understanding of how the genetic code is read, and therefore what is an error and what is an alternative, is based on what complement of tRNA exists in this organism. Since the original K-12 is presumed to have been suppressor minus (Su⁻),[250] we think of having a tRNA which normally reads termination codons reasonably efficiently as an aberration. Certainly, use of Su⁻ strains allows one to study errors, and also to note how the evolutionary process has allowed parasites like Qβ to take advantage of error-prone processes.

But did Qβ evolve in an organism like K-12, and what does "normally read" mean? As was noted by Stringini and Brinkman, wobble (a normal event) differs from misreading only in quantity.[169] Several studies have shown that strain collections of *E. coli* isolated from nature contain reasonable numbers of representatives that suppress nonsense mutations, at least UAG.[251,252] We have seen that synthesis of the CS3 pili of certain enterotoxigenic *E. coli* requires readthrough of a UAG codon.[232] Although this readthrough does occur in Su⁻ lab strains, presumably by misreading the UAG as glutamine (first position wobble), not enough protein is synthesized for the production of pili. Clearly, the naturally occurring enterotoxigenic strains expressing this pili can readthrough this UAG more efficiently, probably because they have a "native" UAG-reading tRNA.[232] It seems unlikely that the UAG codon has become a sense codon in these strains, but takeover of termination codons has been proposed as an important step in the evolution of the universal code.[253] A variety of products are made from this open-reading-frame and it may be essential to maintain them in the correct ratio. *E. coli* may not be able to readthrough UAG efficiently enough without a specific tRNA, and the leakier UGA codon is not used, since the product may require a glutamine residue at this site.[232]

It is also not always clear what tRNA is involved in the readthrough events observed in eukaryotes. In yeast, overproduction of the glutamine tRNA which normally reads CAA can lead to suppression at UAA, and overproduction of the glutamine tRNA which normally reads CAG can lead to suppression of UAG. In each case, these essential tRNAs suppress inefficiently, and by first position wobble.[182,183] A low level of readthrough of UAG can be demonstrated even without tRNA overproduction.[181] At least some mammalian cells seem to have a minor glutamine tRNA with the anticodon UmUG (Um representing 2′-O-methyluridine) which can read UAG as well as CAA in some *in vitro* systems.[254] It is unclear if this particular tRNA is actually involved in retroviral readthrough *in vivo*,[255] but some normal cellular glutamine tRNA must be used, since readthrough can occur in uninfected cells.[228,255]

In the retroviruses, it is known that glutamine is incorporated *in vivo* in response to UAG. Yet in TMV, in spite of the excellent data available upon the components of the readthrough signal, the readthrough protein itself has not been sequenced. Conceivably, glutamine could be incorporated (possibly accounting for the requirement for the prevalence of following glutamine codons in plant virus readthrough), or leucine, since minor leucine-tRNAs are known from mammalian cells which can read the TMV UAG codon.[202] However, the most likely candidates for reading the TMV UAG codon *in vivo* are the major tyrosine tRNAs found in these cells. These tRNAs can read UAG *in vitro* and have the anticodons GΨA (where Ψ is pseudouridine).[256,257] It has been shown that when the anticodon is modified by converting the G to Q (queuosine), the tRNAs no longer read UAG.[256,257] This modification is both tissue specific and developmentally regulated.[258]

However, the evidence is now very strong that UAA and UGA can also be used efficiently *in vivo* in the TMV readthrough context.[199,201] It is very difficult to imagine that all these codons are read efficiently by the same tRNA. This does not mean that the Tyr-tRNA is not used, or that the altered modification patterns are not important. However, it seems that several alternate routes are available for readthrough if necessary in these cells.

In addition to these examples, however, there has been a recent discovery which completely changed our idea about the universal code — that there is a 21st amino acid and that it is incorporated using a special, but native, tRNA.

D. UGA AS AN ALTERNATIVE SENSE CODON

As previously mentioned, alternative codes have been discovered that use UGA to encode either cysteine or tryptophan. However, many organisms which use UGA as a termination codon can also use it to encode selenocysteine. Since incorporation of selenocysteine is very efficient in these messages, the mechanism that permits selenocysteine incorporation must also inhibit termination. Because selenocysteine incorporation is extensively covered in Chapter 9 of this book, and has been recently reviewed,[259] only a few relevant aspects will be mentioned here.

In *E. coli*, the UGA codon encoding selenocysteine is read by a specific tRNA which is aminoacylated with serine, and the serine is then converted to selenocysteine.[260-263] After modification the aminoacylated tRNA is transported to the ribosome by SelB, the product of the *selB* gene.[264] Even with this machinery intact, readthrough incorporation of selenocysteine at actual terminating UGA codons has not been detected.[265,266] The context which allows selenocysteine incorporation seems to include a stem-loop structure downstream from the UGA.[112] When the UGA codon in this special context is replaced by the cysteine codons, UGU or UGC, selenocysteine incorporation is still observed at 10% of the normal level.[111] This very high "error" level can be further increased by removing a restrictive *rpsL* allele in the host, indicating that in this context UGU/C can be misread at almost 100% efficiency.[112] This type of context-specific, efficient, and alternative use of a codon is very similar to the use of alternative start codons in initiation.

E. PREMATURE TERMINATION AND RIBOSOME PAUSING

In addition to reading termination codons as sense, and giving a readthrough product, ribosomes could also terminate at a sense codon. Protein 6 in Figure 1 is the product of premature termination at a UCG codon. One could imagine that the simplest way for this to happen is that the termination mechanism (the RF and other necessary components) responds erroneously to the sense codon, and the nascent peptide is released. Errors from this mechanism have not been identified. If such an "error" happened at a high level, perhaps at an unassigned codon, the organism could evolve an alternative termination codon.[267] Of course, the protein could also be shortened because a frameshifting event had shifted the ribosome into a frame where it encountered a termination codon. Protein 4 in Figure 1 is the result of such an event. However, here the termination is simply an indirect product of the frameshift, and I will deal with frameshifting in a separate section of this review.

There is another translational mechanism which could result in a prematurely terminated protein. The peptidyl-tRNA could simply drop off the ribosome, the peptide being released in the cytoplasm by a peptidyl-tRNA hydrolase. As inefficient and unlikely as such a mechanism might seem, there is considerable support for it. Most of the evidence dealing with "drop-off" has been recently reviewed.[31] However, this topic is crucial to understanding translational fidelity, so some of that analysis relating to *in vivo* events, as well as some new data, will be discussed here.

To actually measure nonribosomal peptidyl-tRNA it is necessary to have a mutant of the peptidyl-tRNA hydrolase. There is a temperature-sensitive mutant of *E. coli* missing this enzymatic activity.[268] At high temperatures most tRNA species quickly become blocked with peptides in the *pth* mutant.[269] This is a phenotype which could also result from defective termination, but the peptidyl-tRNA hydrolase does not seem to be a RF.[270] Measurements of the rate of buildup of peptidyl-tRNA indicate that drop-off could happen *in vivo* at frequencies of between 4×10^{-3} and 1×10^{-4} per elongation event, depending on the particular tRNA.[269] There is also some supporting evidence for drop-off from work on polysome levels in various cells.[31]

If the *pth* mutant affected termination *per se*, then there might be a correlation with tRNAs which are rapidly converted to peptidyl-tRNA and those corresponding to codons next to the termination codon. A comparison of the data of Menninger[269] with that of Brown et al.[188] indicates little support for a role for *pth* in termination. It is true that the tRNA which became blocked the fastest was that for lysine,[269] and the lysine codons are very frequently found before UAA.[188] However, Lys-tRNA also misreads asparagine codons at a high frequency and frameshifts at a high frequency. Menninger has proposed that in the cell, drop-off is coupled to other translational errors as an editing mechanism.[271] We shall discuss this editing hypothesis later.

There is very little evidence that drop-off occurs simply because a ribosome pauses, although it has been postulated as a plausible event at some pause sites.[272] Ribosomes of both prokaryotes and eukaryotes certainly can be shown to stall at various sites on the message, both *in vitro* and *in vivo*,[273-275] but there is no evidence that they release peptidyl-tRNA at these sites. Some of these pause sites are at initiation and termination,[274] and some occur at rarely used codons while others do not.[274,276-278] These instances serve to emphasize the fact that the translation rate is not uniform along the message (see Chapter 7). Possibly the binding of uncharged tRNA at the ribosomal A-site might be a mechanism to cause drop-off.[279] Clearly, ribosomes will stall at unassigned codons, but at least *in vitro* the peptidyl-tRNA is not released.[124] There is as yet no evidence for what a ribosome does at an unassigned codon *in vivo,* but I argue below that it may likely frameshift. In all cases of *in vivo* experimentation, there exists the caveat that any prematurely released product is degraded. In several instances to follow, we will see examples of mRNA secondary structure which seems to participate in either lengthening ribosome pauses or preventing termination. It is likely that there are several mechanisms, including codon usage, codon context, and messenger secondary structure, which can be used by the cell to slow the ribosome, but not apparently to make it drop off.

If drop-off does occur at a frequency of approximately 10^{-3} per elongation event (no matter what its cause), it should leave some evidence, unless all such proteins are very rapidly degraded. A number of experiments have been carried out to measure the success with which a ribosome beginning translation successfully completes the full-length protein, in each case the protein being β-galactosidase or a multimeric derivative. Manley found that over 30% of the enzyme synthesized *in vivo* precipitated by antibody was truncated, apparently because of translational events, not protein degradation.[8] These truncated fragments represented only nine distinct size classes. The frequency at which a ribosome fails to complete a full-length protein, "processivity errors",[280] has also been measured using genes which encoded β-galactosidase as tandem dimers, or even larger constructions (producing a protein of over 6000 residues).[275,280] Tsung et al. found that there was only a 45% probability that a ribosome which had translated to the end of the first monomer would reach the end of the second.[275] They found evidence of distinct ribosomal pause sites along the message, and also found distinct size classes of truncated proteins. However, these investigators used a spectinomycin-resistant mutant, presumably having an altered S5 ribosomal protein, for all their experiments and, as we have seen, some S5 mutants have an increased level of translational errors (Ram phenotype). Jørgensen and Kurland did a similar set of experiments, but using strains with well-characterized ribosomes.[280] They found a 76% probability of completing the synthesis of β-galactosidase. As much as 10% of the processivity errors could have been due to abortive transcriptional termination.[280] Although they do not describe different size classes of truncated proteins, it is clear from their data that these exist. Interestingly, a strain with ribosomes that are hyperaccurate with regard to missense errors has a lower probability of translating full-length protein (much the same as the possible Ram mutant used by Tsung et al.).[275,280] Jørgensen and Kurland propose that a major fraction of the processivity errors is caused by drop-off.[280] As I will discuss later it would seem more likely that frameshifting is the predominant mechanism at work. In any case, their data gives a very accurate measurement of the chances of a ribosome *not* completing the translation of a full-length (perfect) message. If *lacZ* has a typical message, and if averages per codon are meaningful in this case, the ribosome may erroneously terminate, drop off, or frameshift with a probability of about 1.4×10^{-4} per codon.

V. RIBOSOMAL FRAMESHIFTS

Although ribosomes may misread certain sense codons and read through some termination codons at unexpectedly high rates, the resulting protein will still be closely related to the normal product, and in many cases will have normal activity. However, this will not be true if the ribosome shifts reading frame after a correct initiation. Therefore, the *a priori* prediction is that maintenance of the reading frame will be very carefully controlled, and that ribosomal frameshifting may be very infrequent.

The precise nature of a frameshifted product will depend on the location and direction of the frameshift. An apparent four-base translocation will shift the ribosome into the +1 frame (protein 4 in Figure 1), while a two-base translocation shifts the ribosome into the −1 frame; these events will be referred to as a +1 frameshift and a −1 frameshift, respectively. A −2 frameshift will also shift the ribosome into the +1 frame, but the resulting protein will have a different sequence. The ribosome will then continue in the new frame until it reaches a termination codon. Since these codons occur frequently in the alternative frames of most genes, the typical frameshift product of an mRNA will be shorter than the normal product. Unless the frameshift occurred near the 3' end of the gene, the protein will also be inactive. Note that a +1 frameshift at any position in the mRNA depicted in Figure 1 after initiation and before the UGA in the +1 frame will yield a protein of the same size.

As in the previous sections of this review, the discussion of frameshifting will be directed in so far as possible toward *in vivo* events in cells with a normal translational apparatus. However, a considerable amount of information on the maintenance of the reading frame has been gained by studying the extragenic suppressors of frameshift mutations, and the continued isolation of such mutations is essential for understanding the mechanism of frame maintenance.[281] As might be expected, a large number of mutant tRNAs have been isolated which can specifically suppress various frameshift mutations.[282] Many of these tRNAs have altered anticodon loops,[283] but changes outside this region can alter the level of frameshifting,[281] as can specific tRNA modifications.[284,285] Elongation factor G (EF-G) is involved in translocation in bacteria, and recently a mutant with an altered EF-G was isolated which reduces both +1 and −1 frameshifts, but only at specific sites.[286] Interestingly, mutant forms of EF-Tu are known which increase frameshifting,[287,288] as are mutants of the equivalent yeast translational factor.[50] Mutations affecting an essential protein related to this elongation factor can also suppress frameshifts in yeast.[157] We discussed these omnipotent suppressors in the section on readthrough. The wild-type protein is one of a number of similar proteins which seem to be part of the translational apparatus, but whose function is unknown. Mutants in *supK*, which apparently have an altered RF2,[140] can also suppress frameshift mutations.[289] A mutant 15S rRNA found in yeast mitochondria can suppress all types of frameshift mutations studied,[290,291] and a mutation in the *ksgA* gene of

E. coli, which causes the loss of a specific methylation pattern in 16S rRNA, also increases frameshift leakiness.[292] The amount of frameshifting can also be affected by the allele of the genes for ribosomal proteins S4 (*rpsD*) and S12 (*rpsL*), just as we have seen for other types of reading errors.[293]

Some mutants restrict or increase only frameshifts, or only certain frameshifts, but some have a broader range of activity and can affect missense errors or the readthrough of stop codons. Even mutant tRNAs which were selected to suppress missense or nonsense mutations have been found to also suppress frameshift mutations.[294] Clearly, translocation is not divorced from other steps in elongation, but it is unclear whether this has to do with shared or overlapping mechanisms, or "error coupling"; the possibility exists that errors in codon recognition at the A-site can cause a subsequent frameshifting event.[50,295,296]

It is often difficult to detect frameshifting using native messages because there are a very large number of possible frameshift events. In addition, although most such frameshift products will be at least partially homologous to the native protein, so will prematurely terminated proteins and degradation products. Compounding the difficulty of distinguishing these is the fact that many truncated proteins are unstable. For these reasons most studies involve assaying the suppression of frameshift mutations. Even here one must eliminate all the transcription and processing events, and any other possible translational events, that could also lead to a detectable product. This is often difficult, particularly in eukaryotes. Also, by studying frameshift suppression the range of errors that can be detected will be limited to those that can compensate for the original mutational phenotype, i.e., not only will the frameshift have to be at or near the original mutation and to the correct frame, but any additional amino acid(s) inserted must be compatible with activity. Presumably a frameshift mutation can also be compensated by more than one specific ribosomal frameshift. Therefore, the frequencies of suppression measured cannot usually be converted directly into frequencies per codon.

This introduction to frameshifting sounds like a preparation for a statement such as "Therefore, we have little new information on frameshifting." In fact, in the last three or four years a very large amount of such data has been obtained. One reason for this is the exploitation of cleverly designed gene constructs which make it possible to assay for frameshifting at a large number of sequences. Another reason is the surprising finding that a large number of genes require ribosomal frameshifting for expression. As in the previous sections of this review I will divide the discussion of frameshifts into errors and alternatives which will be designated as required, programmed frameshifts. Some of this division is straightforward, but some is much less so. Ribosomal suppression of a frameshift mutation is an error (even though the result is a functional product), and clearly a frameshift which is essential to make an unmutated gene's sole functional product is required. However, much of what we know about "errors" was actually derived from experiments which were designed to dissect the programming signals on the message for the required shifts. Although frameshifting can be a way to express two (or more) functional proteins from a message,[297] in some cases it is not clear if frameshifted products found *in vivo* are anything other than "the ribosome's rubbish".[298] Hopefully, the reader will forgive me if I consign a favorite protein to the trash can. Much of this work has been recently and capably reviewed.[298-301]

A. FRAMESHIFT ERRORS

Early studies showed that both insertion and deletion frameshift mutations could be partially suppressed in wild-type bacteria.[293] The level of suppression varied widely among the different mutations, indicating that frameshifting is not equivalent at all codons. Although the highest level of suppression observed was only 0.06% of wild-type enzyme activity, the collection of mutants that was examined was originally chosen to exclude those that were leaky. As in the case of the other types of ribosomal errors so far discussed, these were reduced by *rpsL* mutations and increased both by a Ram mutation and by streptomycin.

Phenotypic frameshift suppression was also found to be increased by starvation for certain amino acids.[302-304] The increase was specific to particular mutations and to particular amino acid starvations, indicating shifting occurred when the ribosome was paused at particularly shifty codons. As is the case of missense errors increased by amino acid starvation, these increases were sensitive to both the *relA* and *rpsL* alleles. In one case the protein product from such a starvation-induced shift has been sequenced.[305] During lysine limitation the lysine codon AAG in the sequence AAG-C was read in the +1 frame as serine (AGC), presumably by Ser-tRNA$_3$. Although Ser-tRNA$_3$ also reads AGU, frameshifting was not induced significantly at AAG-U indicating that the codon-reading rules may be different for overlapping reading. The level of frameshifting seen without starvation is measurable, but very low (less than 0.01% of wild-type), indicating that it is not normally a frequent event. There are many possible mechanisms to explain different frameshift errors. The simplest explanation is that an incoming aminoacyl-tRNA reads an out-of-frame codon, i.e., an A-site error. This type of frameshift mechanism can be referred to as overlapping reading (the term "offset pairing" has been used to refer to a different mechanism of frameshifting).[306] Protein 4 in Figure 1 would result from this type of error.

There is another example of a starvation-induced frameshift in *E. coli*, but in this case the error was found to occur at a frequency of 50% *in vivo*, and the starvation was apparently induced by the gene construct itself.[307] Spanjaard and van Duin constructed a bicistronic operon in which the upstream gene that coded for MS2 coat protein overlapped slightly with the downstream gene that coded for rat interferon. The latter gene was in the +1 frame with regard to the coat protein gene.[307] The construction was intended to boost synthesis of interferon by coupling its initiation to termination of coat protein. A strong Shine/Dalgarno sequence was placed upstream of the interferon start codon, but in the coat gene, and in such a way that the sequence would be read in-frame as AGG-AGG. However, rather than increasing the production of interferon they found a very high level of a coat-interferon frameshift fusion. They demonstrated that it is essential that the codons be in-frame for the frameshift to occur. The tandem arginine codons AGA-AGA also result in high level frameshifting, but the pair AGG-AGA or AGA-AGG did not.[105] In *E. coli*, both these arginine codons are rarely used and are read by different tRNAs.[105] When the appropriate tRNA is increased (or the level of mRNA production is decreased),[105] the level of frameshifting is reduced or abolished. Apparently after the ribosomes read the first of the tandem rare codons on an abundant message they stall because the pool of the minor tRNA is sequestered by the ribosomal P-site and none is available to read the same codon in the A-site. Presumably, the latter codon is then misread by a noncognate tRNA which shifts or overlapping reading occurs.[306] The proteins produced by frameshifting at these sites (all of which are followed by a U) have not been sequenced, so it is not yet possible to distinguish between these mechanisms. These codons are very rarely found in tandem wild-type *E. coli* genes, but a few examples are known,[105] and it will be interesting to see if the tandem rare codons are involved in programmed frameshifts at such sites.

Overlapping reading has also been demonstrated in an *in vitro* system which uses the MS2 phage genome as message.[308] In this system, Ser-tRNA$_3$ can read the alanine codon GCA as serine (AGC) giving a −1 frameshift.[309] Therefore, this tRNA can do overlapping reading in either the −1 or +1 frames. However, in the MS2 system the preceding base did not have to be an A, although a U is required at position 36 in the tRNA.[310] This may also indicate that base-pairing rules might be different than normal for overlapping reading or, of course, that no more than two base-pairs are required (doublet decoding).[310,311] Frameshifting by this tRNA can also occur at the alanine codons GCU/C, but at a lower frequency. Similarly, a Thr-tRNA which normally reads ACC/U can cause frameshifting at the proline codons CCG/A, but not CCU/C.[329] The level of frameshifting observed can be changed by altering the ratios of tRNAs, and once again only certain tRNAs were found to frameshift.[308,310] Only one of the several frameshift products generated by MS2 in this system has also been found *in vivo*, and the precise nature of the error which produces it has not been established.[308,309]

There were indications that runs of bases could lead to detectable frameshifting errors in normal cells.[312,313] The type of frameshifting mechanism that seems to operate at these repetitive sequences is the tRNA "slip". In this case, the incoming aminoacyl-tRNA apparently correctly reads an in-frame codon, but subsequently slips into a different frame. Like overlapping reading, tRNA slips can occur in either the 5′ or 3′ direction. It must be noted that overlapping reading and tRNA slippage, though very different mechanistically, would result in the same product if the overlapping codon and the normal codon could be read by the same tRNA, as they would in strings of

	...CCA	GUG	AAU	UUU	UAA	A	AGC–lacZ
unshifted		Pro	Val	Asn	Phe	Term	
+1 before window	...Gln	Term					
+1 at GUG	...Pro	Term					
+1 at AAU	...Pro	Val	Ile	Phe	Lys	Ser–β-galactosidase	
+1 at UUU	...Pro	Val	Asn	Phe	Lys	Ser–β-galactosidase	

FIGURE 2. Frameshifting at a synthetic frameshift window. The sequence of the frameshift window is underlined. The ribosome must make a +1 shift in this area in order to synthesize active β-galactosidase from the fused *lacZ* gene. The abbreviation Term indicates termination of protein synthesis.

repetitive bases. Actually much of what we know about tRNA slippage comes from work on required frameshifts, which we shall discuss more fully in the next section. However, slipping can also lead to errors, even though these might be programmed into the message.

Weiss and his colleagues have done extensive analysis of tRNA slipping on strings of repetitive bases by constructing a series of "frameshift windows" and fusing them to *lacZ,* the structural gene for the easily assayed enzyme β-galactosidase.[12,300] A simplified example of a +1 frameshift window is shown in Figure 2. The *lacZ* gene, without a start codon, has been fused to an upstream "gene" but in the +1 frame. The 5' end of this window is not bounded by an initiation codon but by a UGA termination codon in the +1 frame. The 3' end of the frameshift window is the in-frame UAA termination codon. Therefore, to make active enzyme a ribosome must make the +1 frameshift after reaching the valine codon GUG but before reaching the UAA. The exact nature of the frameshift can be determined by sequencing the product(s).

This analysis has shown that Gly-tRNA can do −2, −1, and +1 slips, individually or in combination depending on the sequences used, e.g., at the sequence CGG-GGG-UAU both −2 and +1 slips occur (and −1 slips would not have been detected).[300] Since the shifting was higher using GGG as the in-frame codon, presumably Gly-tRNA$_1$ is involved. Phe-tRNA was also found to slip, but apparently only +1, even when the sequence presents other possibilities.[300] Val-tRNA and Leu-tRNA both do +1 slips, as does Lys-tRNA.[12,300] Interestingly, even though Lys-tRNA reads both AAA and AAG, it slips at least 60-fold more frequently at AAA-A than at AAA-G. As we shall see later, this may be because −1 frameshifts are preferred at AAG, and therefore the two shifts would cancel each other (a quivering tRNA). In all cases of slippage, the amount of shifting depends on the strength of alternate codon:anticodon binding within the strings, but the rules governing these interactions are more relaxed, e.g., Leu-tRNA slips +1 at the sequence CUU-U.[300] The frequency of the frameshifts at these constructs was rather high, usually 1 to 5%, but could be reduced considerably by removing the stop codon 3′ to the shift site, and such sites were called "shifty stops".[12,300] O'Mahony et al. showed in *Salmonella* that mutants of Gly-tRNA$_2$ (reads GGA and GGG) could slip (or doublet decode) at GGA, but the wild-type tRNA did not, and there was no indication of misreading by Gly-tRNA1.[311,314] In their constructs the next codon was a sense codon.

Although termination codons increase frameshifting at shifty stops, it is not clear exactly what the mechanism is. In yeast, frameshifting by suppressor tRNAs is higher in cells with the Ψ^+ factor,[315] which may inhibit normal termination.[150] It is known that increasing the level of the appropriate RF will decrease frameshifting at shifty stops in *E. coli*.[300,316] Therefore, frameshifting is not a consequence of an aborted termination.[300] It could be that peptidyl-tRNA has more opportunity to slip while the ribosome is paused with the termination codon at the A-site,[300] and it is known that ribosomes do apparently pause in such situations.[274] However, increasing RF also restricts both +1 and −1 frameshifting which was increased by mutant forms of EF-Tu.[316] It seems difficult to reconcile this with frameshifting at the P-site. More likely, the mutant elongation factors either allow increased misreading of the termination codon as sense[50] and this tRNA subsequently shifts, or it allows overlapping reading. Sequencing the frameshift products induced by EF-Tu may help distinguish between these mechanisms. Either of the latter two mechanisms would indicate that for a −1 frameshift the P-site tRNA did not slip, it was pushed.

We have recently characterized what appears to be a high-level frameshift error in the *argI* mRNA from *E. coli*.[102] As mentioned in the section on missense errors, we were unable to demonstrate leucine misincorporation at a single position early in the *argI* message during phenylalanine starvation.[100,101] Frameshifting during starvation would obscure misincorporation, so we constructed an appropriate frameshift window fused to *lacZ*. To our surprise we found that in unstarved cells there is a +1 frameshift at a frequency about 5% at the sequence GGG-UUU-UAU to give a frameshifted product with the sequence Gly-Phe-Ile rather than the expected Gly-Phe-Tyr.[102] In the wild-type gene the +1 product would terminate after four more codons, and there are no arginine codons in either frame in this region, making it unlikely that the shift is part of a regulatory mechanism. Although this frameshift is almost certainly programmed by the message, it is also almost certainly an error. Since either or both the Gly- or Phe-tRNA could be involved in the slip, it is possible that a mechanism similar to the simultaneous tRNA slippage found in retroviruses could be involved, but in the opposite direction.

There is another mechanism of frameshifting which also occurs *in vivo*: tRNA hopping.[12] Like tRNA slipping, hopping involves the movement of aminoacyl-tRNA after it has correctly decoded an in-frame codon. However, in the hopping mechanism it is not possible to imagine the tRNA moving step by step along the mRNA, breaking and reforming base-pairs at each step. Rather the tRNA appears to disengage, "take-off" from the original in-frame codon and reengage, and "land" on a downstream codon which it can also read — downstream because there is no evidence that hops ever occur 5' to the take-off site.[12,317] The hops were originally characterized as +2, +5, and +6 shifts between codons normally read by a single tRNA, e.g., from AAC to AAU. Mutant Val-tRNA$_1$ can hop long distances (+9) and at high rates (up to 19%),[317,318] whereas the wild-type tRNAs studied do not normally (but see following) make such long hops, and the frequency is usually 1% or less.[12,317] Investigations of tRNA hopping have disclosed that, although termination codons 3' to the take-off site increase the frequency of hopping, they are not required.[12,317] It was also found that there is flexibility in base-pairing rules at the landing site, i.e., a hop from GUG to CUA by Val-tRNA$_1$, which normally recognizes GUU, GUA, and GUG,[317] a situation similar to that found in tRNA slips. Unexpected flexibility apparently exists at the take-off site, since the hop from GUC to GUU was rather efficient and Val-tRNA$_1$ should not read GUC.[317] Using mutant tRNA it was concluded that the third bases of both the take-off and landing codons were not crucial if the hop was over a stop codon.[317] In addition to Val-tRNA$_1$, hopping has also been detected in *E. coli* with the wild-type Asn-tRNA and a Leu-tRNA (presumably Leu-tRNA$_3$ which reads CUU, although the landing site was CUA).[12,318]

It is unclear how large a contribution tRNA hopping might make to the *in vivo* error burden. From the sequence of *lacZ*,[319] one can see that there are 45 codons in the message which can be read by Val-tRNA$_1$, including 15 where a similar codon occurs as a landing site after a hop over two codons or less. The number of potential hops becomes greater if one allows greater base-pairing flexibility at the landing site, or if potential hops by Asn- and/or Leu-tRNA are included. Of these 15 possible hops, eight are either +3 or +6 and therefore maintain the reading frame of the protein while deleting one or two amino acids. Of the seven which would shift the ribosome into a new frame, only three are less than +7 and none are less than +4. Obviously, none of these involves hops over stop codons and it is very unclear at what frequency any of these potential hops may occur, since the wild-type tRNA makes a +2 hop at the sequence GUGUA at a frequency of only 0.3%.[317]

Gallant and his colleagues found that some *nonsense* mutations can be suppressed in *E. coli* by limitation for certain amino acids.[302,320] One explanation for this is a double frameshift in opposite directions. The first is triggered by the ribosome pausing at a hungry codon upstream from the nonsense mutation, and the second is downstream from the nonsense mutation.[320] However, it may well be that the hungry codon triggers a tRNA hop, and that the tRNA lands in-frame downstream of the nonsense mutation. Some mutant tRNAs can hop over two or three intervening codons,[317] and we will discuss an example (to follow) of a programmed hop over many intervening codons. Whatever the mechanism, this suppression is decreased in *relA$^+$* cells.[302]

The use of frameshift mutations has been an essential part of most ribosomal frameshifting studies. However, the suppression of a typical frameshift mutation involves a shift back to the "correct" reading frame, whereas an actual error would be away from the correct frame. Since codon usage is not random, the +1 and −1 reading frames of a typical message can be much different than the 0 frame.[321] Trifonov has proposed a model in which the reading frame is monitored by 16S rRNA base-pairing to the preferred codon repeat GNN-GNN.[322] Several of the cases previously discussed were certainly shifts from the normal reading frame, e.g., the frameshift in *argI*. They seem unlikely to provide information on the average frequency of frameshifting within a given message and certainly do not test the model. However, the model has been tested for −1 shifts.

Two fusions to *lacZ* were constructed, and in each case the frameshift window was a portion of the *rpoC* gene of *E. coli* containing about 100 codons.[300] In one construct it was fused to test for frameshifting from the 0 to the −1 frame, and in the other to assay the opposite shift. In this segment of the gene GNN codons make up 43% of the 0 frame and 27% of the −1 frame, and the codon usage in the −1 frame is very unlike a typical 0 frame. The level of frameshifting was low in both cases, indicating that there were no shifty codons and that being in the −1 frame does not cause a high level of ribosomal frameshifting. Indeed, the level seen for −1 to 0 was the same as that for a shift from 0 to +1 in a *different* segment of *rpoC*. However, for the comparison within the same DNA sequence, the shift from −1 to 0 was 10-fold higher than the reverse (1.1% of wild-type compared to 0.1%), which is consistent with Trifonov's model. An error frequency of 1% in a 100 codon frameshift window gives an average frequency of 10^{-4} per codon.

Little other information is available about the average level of erroneous frameshifting. Using data from the experiments on processivity errors while synthesizing β-galactosidase in *E. coli*,[8,280] one can calculate that the average frameshifting frequency within *lacZ* could be as high as 1.4×10^{-4}, but only if all the translational errors observed were frameshifting errors. Frameshifting at all sites within a given frameshift window gives products of the same size; therefore, frameshifting easily explains the limited number of size classes of products found. Certainly frameshifting can occur at much lower frequencies[288,293,323] but, as we have seen, it can also occur at much higher frequencies, all apparently depending on the particular tRNAs and the mRNA sequence involved. It seems highly likely that codon usage and context may be used to keep frameshifting at tolerable levels; a frequency of even 10^{-4} could prove costly to the cell when translating messages from genes like *lacZ* and *rpoC* which both encode rather large proteins.[31,34]

The reader may have noticed that none of the erroneous frameshifts in suppressor-free cells referred to previously happen in the eukaryotic cytoplasm. Frameshift mutations certainly exist in eukaryotes, and, at least in yeast, can be suppressed by a variety of mutations.[282] In higher eukaryotes, frameshift mutations can lead to "null" and "silent" phenotypes,[324,325] where the level of leakiness is undetected by immunological techniques and is presumably below

10^{-4}. Therefore, it is possible that eukaryotes have a tighter control on ribosomal frameshifting errors than prokaryotes, although it may well be that more information is needed for all cells. In any case, we shall see that eukaryotes can have messages which cause very high levels of programmed frameshifting.

B. REQUIRED PROGRAMMED FRAMESHIFTS

There have been two recent discoveries about the reading of the genetic code which I consider very remarkable. We have already mentioned one of them — the programmed alternative use of the UGA codon to encode selenocysteine. The second is the discovery that many genes require ribosomal frameshifting in order to make functional products. Like selenocysteine incorporation, these frameshifts seem to be specifically programmed into the mRNA. Unlike selenocysteine incorporation, there is as yet no hint of a requirement for special translational factors, and therefore these programmed frameshifts seem very much like magnified errors.

Most of the known required frameshifts occur in viral genes and, interestingly, most often in eukaryotic viral genes. Ribosomal frameshifting does occur in the prokaryotic virus T7,[313] but it has not yet been demonstrated to be essential. In two other prokaryotic viruses where frameshifting originally seemed to be involved in translational regulation,[326,327] recent evidence indicates that it does not.[328,329] Recently, however, frameshifting has been demonstrated to be required for the bacteriophage λ to produce an essential protein.[330] This frameshifting occurs at a level of almost 4% by a mechanism which may be similar to that used to program shifts in mammalian cells.

The fact that eukaryotic ribosomes prefer to initiate at only one start codon per message would seem to severely restrict the use of messages encoding more than one protein. Many eukaryotic viruses seem to have solved this problem using programmed frameshifting to make polyproteins which are subsequently processed.

We have already discussed the fact that some retroviruses use readthrough as a mechanism to generate a polyprotein from the *gag-pol* genes. Sequencing the protein demonstrated that these were readthrough events and not hops.[223,224] However, most retroviruses sequenced have *gag* and *pol* as out-of-frame, overlapping, and open-reading-frames, and therefore could use frameshifting, but not readthrough, to synthesize the polyprotein(s). Only a few such "transframe" polyproteins have been confirmed by protein sequencing,[331-333] but the evidence is now very strong that −1 frameshifting is programmed into these messages.[145,218,301,333] In some cases there is an intervening *pro* reading frame between *gag* and *pol* which also requires a −1 frameshift.[332,334-336] A ribosome translating the entire RNA from such a virus then must make two programmed −1 reading frames. Much of the work characterizing these frameshifts has involved *in vitro* assay systems. In retroviruses where a single frameshift is required, this shift seems to take place at a frequency of 5 to 10% *in vitro* and *in vivo*.[337-340] The double frameshifts seem to occur at a higher frequency,[332,335,341,342] apparently maintaining the correct ratio between the *gag* product and the reverse transcriptase. In other cases, however, the identification of the transframe products, or even the protein in any form, can be difficult to demonstrate *in vivo,* making quantitation extremely difficult.[336,343-345]

In order to produce a transframe protein, the frameshift must take place within the overlap which is, of course, a frameshift window. These overlaps vary from 13 to 373 bases,[145] and by definition must end in a termination codon in the incoming frame of the ribosome. In a significant number of cases the frameshift may occur at or near the stop codon, indicating that it could be part of the programming signal.[145,333] For the *gag-pol* shift in Rous sarcoma virus (RSV), an *in vitro* system has been used to demonstrate that a –1 shift takes place at the sequence A-AAU-UUA-UAG yielding the amino acid sequence Leu-Ile at the shift site.[333] Apparently, the Leu-tRNA which reads UUA slips –1 to UUU. (The shift also occurs using the leucine codon UUG). Any base change in the U-UU sequence greatly reduces frameshifting, and changing the upstream A's also had an effect, as if the tRNA at the asparagine codon was also slipping, but that the final base-pairing requirements were not as critical. From the sequence requirements for the shift, Jacks and his colleagues developed a model of simultaneous slippage, in which both tRNAs on the ribosome simultaneously slipped to the –1 frame.[333] Changes in the termination codon had no effect, indicating that the slippery sequence was in the heptamer, A-AAU-UUA. Interestingly, changing the leucine codon UUA to UUU increased frameshifting. This indicates both that Phe-tRNA also slips, and that the level of frameshifting is not simply maximized in RSV.

Substitution of similar repetitive heptameric sequences which could be involved in frameshifting in other viruses, A-AAA-AAC and U-UUA-AAC, also allowed –1 frameshifting. Therefore, the tandem slippery codons are sequences in which both tRNAs pair normally in the 0 frame and can form at least two-out-of-three pairs in the –1 frame. These requirements can be most easily met at repetitious sequences, but simple repetition is not enough. The sequence A-AAA-AAA was found to be very inefficient in this system, indicating that Lys-tRNA can slip efficiently at the first codon of a slippery pair, but not at the second codon. Weiss and his colleagues showed that retroviral slippery sequences lead to –1 frameshifting in *E. coli,* also by a simultaneous slippage mechanism.[346] In *E. coli,* –1 frameshifting did occur at the sequence A-AAA-AAA, indicating that Lys-tRNA could shift at both codons. Changing the sequence to A-AAA-AAG increased frameshifting almost 20-fold, indicating that *E. coli* Lys-tRNA shifts readily from AAG. Searching for such slippery sequences in the frameshift windows of mRNAs from genetic elements of higher eukaryotes, where –1 frameshifting is thought to occur, indicates that only a limited number of tRNAs (Asn-tRNA, the UUA-reading Leu-tRNA, and Phe-tRNA) slip at codon 2, and that these and only a few more slip at the first codon.[145] Hatfield and his colleagues have demonstrated that Asn- and Phe-tRNA, which normally have a hypermodification (Q base and Wye-base, respectively) near the anticodon, are undermodified in cells infected with retroviruses.[347] Therefore, as in the case of UAG readthrough in TMV, control of tRNA modification could play a role in regulating programmed frameshifting *in vivo*. The original model of simultaneous slippage envisioned the slippage occurring after A-site binding, but before transpeptidation.[333] Weiss et al. have proposed an alternate model based on the three-site ribosome,[348,349] in which the slippage occurs after transpeptidation and possibly during translocation itself.[346]

Experiments with mouse mammary tumor virus (MMTV) had shown, however, that sequences within the overlap were not sufficient to cause frameshifting.[332] Therefore, since the stop codon did not seem to be involved, the RSV message was analyzed for another programming signal. It had been noticed that immediately downstream of such frameshifting sites there were sequences that had the potential to form stem-loops, and it had been postulated that stem-loops might be a part of the frameshifting mechanism.[332,350] Using site-specific mutagenesis to disrupt or delete the proposed stem, Jacks et al. showed that these sequences were critical and suggested that they may be involved in either giving directionality to the shift or causing the ribosome to pause.[333] In *E. coli*, disruption of potential secondary structure in the retroviral sequences could diminish frameshifting up to sixfold.[346] A computer analysis of several retroviral messages indicates that these stem-loops can be very stable.[351]

A very convincing demonstration of the importance of RNA structure in the frameshifting mechanism has been found in coronaviruses, another type of RNA virus. In these viruses, a −1 frameshift occurs between two open-reading-frames within the *pol* "gene".[352] In the case of the avian virus, infectious bronchitis virus (IBV), the programming signal for this frameshift has been intensively analyzed.[343,353] The frameshift takes place at the sequence U-UUA-AAC by a simultaneous slippage mechanism as in retroviruses. Mutational analysis indicates that in order to function efficiently, the downstream stem-loop must be part of a pseudoknot.[353] Pseudoknots occur when nucleotides within a stem-loop structure form a tertiary interaction with a nearby single-stranded region of the message.[354] For frameshifting to occur at the site in IBV, this structure must be precisely positioned downstream from the slippery sequence.[353] An identical slippery sequence, followed by a potential pseudoknot, from the mammalian coronavirus MHV-A59 supports efficient frameshifting in both *in vitro* and *in vivo* test systems.[352] The frequency of frameshifting in the coronaviruses seems to be between 25 to 40% in most of the systems used, including frog oocytes and HeLa cells.[343,352,353]

However, changes in the potential stem-loop downstream from the *gag-pol* frameshift site of the human immunodeficiency virus, HIV-1, gave less straightforward results about the importance of secondary structure.[355] In addition, −1 frameshifting at a frequency of 5 to 10% has been observed both in a

Variations in Reading the Genetic Code 231

mammalian *in vitro* system and *in vivo* in yeast using a short stretch of the HIV-1 *gag-pol* overlap, which included the slippery codons but excluded any potential stem-loop.[339] Since the HIV-1 frameshift is also not near a termination codon, it is unclear if all retroviral sequences have a uniform requirement for a specific programming element beyond the slippery run.

As previously mentioned, retroviral slippery sequences can function in the yeast *Saccharomyces cerevisiae*.[339] Recently, a yeast genetic element has been found to use −1 frameshifting to encode its RNA-dependent RNA polymerase as a fusion with a coat protein.[356,357] This element is the yeast double-stranded RNA virus L-A. The frameshift occurs *in vivo* at the slippery sequence G-GGU-UUA at a frequency of 1.9%.[357] Construction and analysis of mutations at the site have demonstrated that the frameshift involves simultaneous slippage and a downstream pseudoknot.[357] All tRNAs which one could predict as being capable of slipping on repetitive sequences functioned at varying levels and no correlation with tRNA modification could be noted. Several sequences, e.g., U-UUU-UUA, led to considerably higher levels of frameshifting than the wild-type sequence. In wild-type, the first codon which would be read in the −1 frame is AGG. It is unclear whether this may be significant.

Several yeast retrotransposons are also known to use frameshifting,[358-361] but here the mechanism of frameshifting seems different and involves a +1 rather than a −1 shift.[359,360,362,363] The sequence that is involved in the shift in Ty1 and Ty2 has been found to be only seven nucleotides, CUU-AGG-C, and when placed in a heterologous gene leads to 40% frameshifting.[362] This sequence was extensively analyzed by constructing mutants, and it was shown that any change greatly lowers frameshifting. In yeast, the Leu-tRNA with the anticodon UAG can read all six leucine codons.[364] By a variety of techniques, Belcourt and Farabaugh convincingly demonstrated that after this tRNA has read the in-frame CUU, and the ribosome has advanced to and is paused at the codon AGG, the tRNA slips to the +1 UUA.[362] These experiments ruled out both simultaneous slippage and overlapping reading. The AGG codon in yeast is read by a low abundance tRNA, and overproduction of this tRNA was found to greatly reduce frameshifting at this site.[362,363] Fascinatingly, replacing the AGG codon and limiting the amount of tRNA available for the substitute codon was insufficient to induce frameshifting, indicating that the codon AGG must fill some other as yet unknown function.[362] It is interesting that this +1 frameshift is a shift away from the codon AGG, whereas in the −1 frameshift in the L-A virus of yeast the shift is to AGG.

Recent genetic analysis of the prokaryotic insertion element IS*1* indicates that it synthesizes its transposase as a fusion with a DNA-binding protein encoded by the upstream *insA* gene.[365] The most likely location of the shift is at the sequence A-AAA-AAC, which as we have seen is a typical retroviral slippery sequence. Frameshifting by a similar mechanism may also occur in at least some members of the IS*3* family of prokaryotic insertion sequences.[366,367] Some retrotransposons from *Drosophila* may also use frameshifting,[368] but other such elements in *Drosophila*,[369] higher plants,[370] and lower eukaryotes[371] have one long open-reading-frame and no "ribosome gymnastics"[299] are required.

Based on genome sequence analysis, it seems likely that frameshifting occurs in several plant RNA viruses.[203,204,372-374] Some of these viruses, e.g., potato leafroll virus, also use readthrough of in-frame termination codons in other genes.[204,212] However, it must be emphasized that when genes overlap, other methods besides frameshifting can be used for expression of the downstream gene. In the case of cauliflower mosaic virus and hepatitis B virus, both DNA viruses which encode a reverse transcriptase as part of a large mRNA, it seems internal initiation is used to express the reverse transcriptase.[125,375,376]

The first example of a required frameshift in *E. coli* was the demonstration that the synthesis of RF2 required a +1 frameshift at an in-frame UGA codon early in the message.[139] This UGA codon appears at codon 26 in the 0 frame of the *prfB* message. In order to synthesize the remainder of the 339 residue protein, the ribosome must shift into the +1 frame. The frequency of this frameshift is approximately 30 to 50% under normal conditions.[377,378] The signals which program this shift have been extensively analyzed using mutagenized frameshift windows.[12,379,380] They involve a +1 shift by a Leu-tRNA from CUU to UUU in the sequence CUU-UGA.[12] If the sequence is changed to GGG-U, Gly-tRNA can slip and if changed to GUU-U, Val-tRNA can slip. Weiss et al. have extensively analyzed slippage at a large number of "shifty stops", and the efficiency of frameshifting by a variety of tRNAs depends on the strength of the alternate codon:anticodon interactions within the slippery sequence (here only a single tRNA is involved).[300] The other termination codons can also be used, with UAA being about as efficient as UGA and UAG being less efficient.[12,300] Using a sense codon downstream of the slippery sequence still gives measurable frameshifting. However, another critical element in the programming of the RF2 shift is an upstream Shine/Dalgarno-like sequence, AGGGGG. By mutating this sequence and making compensating changes in the anti-Shine/Dalgarno sequence in 16S rRNA, Weiss et al. have demonstrated that the AGGGGG sequence is interacting with 16S rRNA.[380] This sequence may prolong the pause at the termination codon, give direction to the shift, or both. Its possible role in directionality is illustrated by the fact that, in *E. coli*, it can interfere with −1 shifting at a retroviral sequence if it is located at the proper distance upstream from the slippery sequence.[300] Interestingly, some gene constructs with the RF2 shift site frameshift efficiently in an *in vitro* mammalian translational system,[381] while others do not.[382] Since the mammalian 18S rRNA does not have an anti-Shine/Dalgarno sequence, it should not be able to use this determinant. There must be other sequence determinants, possibly operating over some distance, which can cause ribosomal pausing with some constructs.[382]

In any case, in *E. coli* both the UGA (or UAA) codon and the "Shine/Dalgarno" sequence are necessary for efficient frameshifting. In the previous examples we have discussed, the required frameshift is apparently used to allow the synthesis of a fixed, lower level of the product of the downstream overlapping gene. In the case of the RF2 frameshift it seemed clear that a regulatory circuit was involved, since the frameshift takes place at the termination codon which is recognized by RF2. When RF2 levels in the cell were sufficient, synthesis would be terminated here, but if levels were low, then the ribosome would frameshift. There are now several pieces of information which support this model. First, if RF2 (but not RF1) is added to an *in vitro* translation system, it reduces the frameshift at UGA.[377,378] Second, when using a mutant signal with a UAG codon, an amber suppressor could decrease frameshifting (although an ochre suppressor did not work at UGA).[379] Finally, if the construct is used to produce a mutant RF2, which can bind to ribosomes but not terminate, frameshifting can be increased to 100%.[378]

In the latter case, it is clear why the ribosome does not terminate at UGA: the cell is making defective RF2. The situation is less clear in the normal cell where UGA is apparently an efficient termination codon. Brown et al. have found that termination codons are often preceded by repetitive runs of bases, so it would seem this is not an intrinsically leaky combination,[188] although the sequence around this UGA is not optimal.[378] In addition, although the sequence AGG-GGG seems to be operating by binding to 16S rRNA, and not as codons,[380] they are in fact in-frame and AGG is rarely used and GGG infrequently used in *E. coli*. Donly et al. suggest that they may play a dual role here by both slowing the ribosome before the UGA codon and giving directionality to the shift.[378] Since 16S rRNA seems to be involved both in initiation (as the Shine/Dalgarno) and in termination at UGA, I have previously suggested that these two interactions may be mutually exclusive and that in this case termination is blocked.[31]

The most impressive required ribosomal frameshift is a very long tRNA hop. When *E. coli* ribosomes translate the message from gene 60 of T4, they leap over the intervening 50 nucleotides between the codons for residues 46 and 47 of the protein – a +53 frameshift – at a frequency of 70 to 98%.[383,384] As could be expected, there are several programming requirements involved.[384] The take-off site for the hop is a GGA (Glycine) as is the landing site. Although other codons can be used (GCA, alanine), it is not clear whether first or third position mismatches are allowed.[384] Moving the landing site by shortening the translation gap to 21 nucleotides or lengthening it to 66 reduces the efficiency of the hop. Immediately following the take-off codon is a termination codon which is also necessary. It must be followed by (or be included in) a small stem-loop structure.[383,384] In addition, the upstream nascent peptide is required for hopping. This requirement is for the peptide, or some simple sequence determinant in the peptide, acting in *cis*, and is not simply for the ribosome to transit this portion of the message.[384] This remarkable system promotes frameshifting at nearly 100% in *lacZ* for reasons, both mechanistically and evolutionarily, not well understood. It does not seem to be part of a regulatory mechanism and is not found in closely related bacteriophages.[383]

All of the examples of required frameshifting previously discussed have to do with a ribosomal shift opening up or extending the reading frame, i.e., making reverse transcriptase as part of a *gag-pol* polyprotein as opposed to making only the *gag* protein, or making full-length RF2 rather than a truncated peptide. However, recently, a situation has been uncovered where the ribosome shifts to an alternative reading frame and then terminates.[385-387] The *dnaX* gene encodes two polypeptides, called γ and τ, which are found as subunits of DNA polymerase III, the replicase of *E. coli,* where they may play slightly different roles in replication.[385-387] It has now been shown that γ, the shorter protein, is made by a –1 frameshift at the sequence A-AAA-AAG, which causes termination two codons later at a previously out-of-frame UGA. Although this sequence is followed by a stem-loop, it is not clear if the stem-loop is an important feature of the frameshift site. Certainly, the out-of-frame UGA and sequences downstream from the stem-loop are not.[387] Frameshifting is abolished by deleting the run of A's,[387] and by changing the sequence to G-AAG-AAG.[385] Furthermore, it is greatly diminished if the first lysine codon is changed to an asparagine codon (AAU/C).[386] All these mutations implicate a double-shift as the mechanism. However, since it is possible that the γ subunit is dispensable,[387] this could be simply a programmed, high frequency error at a very unexpected place.

In the *carA* gene of *Pseudomonas* there is another example of what could be a translational event leading to a shorter than expected protein. The product of this gene is missing the residues that should be encoded by codons 5 to 8, although these codons are on the message.[388] The 12 nucleotides which are apparently not translated are an almost perfect (11/12) direct repeat of the preceding 12. Because of the repeat, a tRNA hop from either codon 3 to 7 (both CCA), or from 4 to 8 (both GCC) would give identical proteins. However, there is no other obvious programming signal in the message.[388] Although transcriptional events seem ruled out, it is possible that this is an example of protein splicing.[30]

E. coli also expresses genes form *Neisseria gonorrhea* which have been turned off in their host organism by a DNA frameshifting mechanism.[389] The low level of expression in *E. coli* seems to be the result of ribosomal frameshifting, possibly at a run of A's.[389] It is unclear if this ribosomal frameshifting has significance in *Neisseria*. However, programmed frameshifting seems to be a very widely distributed mechanism of gene expression, and examples of it are being discovered at an accelerating place. Although all known eukaryotic examples are found in viruses, this may well be related to the fact we know much less about eukaryotic genomes. In *E. coli,* frameshifting is required for viability since RF2, if not the γ-subunit, is an essential protein.[138] Therefore, it might be expected that mutations which severely restrict frameshifting will be unobtainable,[286] and certainly the restrictive *rpsL* mutants available should have little effect on this type of frameshift.

The programming for these required frameshifts seems to involve at least two and possibly three components. First, there must be slippery codons (and as a corollary, tRNAs which slip at them). Second, in many cases there is also an apparent requirement for a paused ribosome. This is best illustrated in those cases where the frameshift is induced by a termination codon (which does not terminate) or in examples like the yeast Ty retrotransposons, where shifting takes place at a 0 frame codon calling for a scarce tRNA.[362,363] In cases where secondary structure is involved, it seems likely that the structure of the mRNA may slow the ribosome. Such structures also seem to be involved in slowing the movement of scanning eukaryotic ribosomes for initiation.[390] The third component is one of directionality. In many cases, this may well be determined by the slippery sequence itself. However, in *E. coli,* a single tRNA can be capable of frameshifts in more than one direction at repetitive sequences.[300] The Shine/Dalgarno sequence involved in the RF2 shift in *E. coli* may well be at least in part an independent directional signal.

C. SHIFT TO TERMINATE — THE CASE AGAINST ALTERNATIVE TERMINATION CODONS

There are some vertebrate mitochondrial genomes which have been reported to use AGG and AGA as alternative termination codons.[267] However, in many cases it seems possible that a rather different mechanism of termination is involved, a mechanism involving frameshifting.

The mouse mitochondrial genome does not use the codons AGA or AGG,[391] and no release factor recognizing them was found in rat mitochondria, indicating that in some cases these codons might be unassigned.[123] Human mitochondria use each of these codons once, and in each case at the end of an open reading frame.[392] In the mitochondrial genome of the frog *Xenopus laevis,* AGG is not used but AGA appears one time and also at the end of an ORF.[393] However, in all of these examples the preceding codon ends in U. If a ribosome stalled at an unassigned AGA/G and then shifted into the −1 reading frame, there would be a UAG termination codon in the ribosomal A-site. It could also be that the out-of-frame termination codon is read without ribosome movement.[394] Therefore, the AGA/G codons could be used here as "stop" codons but not as termination codons. As we have seen before, frameshifting by stalled ribosomes certainly does occur.

Unfortunately, there is an example of the apparent use of an alternative termination codon that cannot be explained away by a simple stop-shift-terminate model. The bovine mitochondrial genome uses AGA at only one site, at the end of the ORF encoding apocytochrome b; AGG is not used at all.[395] In this case no possible frameshift can lead to termination at either UAA or UAG (although an unusual editing step of the immediately following tRNA could obviate this problem). The preceding codon is UGA, a "universal" termination codon, but it clearly encodes tryptophan within genes in bovine mitochondria.[396] In the cytoplasm of mammalian cells, UGA can have more than one meaning depending on its context; possibly, it can here also.

A model which only explains three of the four known examples may seem without merit. However, mitochondrial codes are noted mostly for simplification, and the only known RF from vertebrate mitochondria does not recognize AGA or AGG (nor UGA). It would seem more information is necessary before we make the assumption that these codons are alternative termination codons, at least in the sense usually implied.

VI. CONTROL OF ERROR

Many errors in protein synthesis occur at measurable frequencies but, considering the complexity of the process, these frequencies are reasonably low. In the previous sections of this review, we have seen that the level of these errors is under genetic control, and that error level can be related to the message or codon being analyzed. In this section I shall briefly discuss some of the mechanisms by which cells seem to control translational error.

A. CODON USAGE AND CODON CONTEXT

As has been previously discussed, the message context is critical for all the programmed translational alternatives that have been discovered (at least all that have been examined) in genomes using the universal code. In addition, it seems clear that many translational errors are extremely context-dependent although, since so few actual error frequencies have been measured *in vivo*, we have little information on the context involved. Some of the elements involved in certain context-dependent events have been discussed within this review, and context is being dealt with extensively in Chapter 11 of this book. Therefore, in this short section I shall deal only with the possibility that some codons themselves are error-prone.

We have established that AAU is misread more frequently than AAC regardless of context.[93] Dix and Thompson offer evidence from an *in vitro* system that synonymous codons, here UUC vs. UUU and CUC vs. CUU, differ in their accuracy.[397] In each case the codon ending in C is read less accurately. In our studies on the misreading of UUU and UUC *in vivo*, we have not yet been able to identify any difference between these codons which is not related to context.[100,101] This analysis is complicated because more than one type of error can happen at these codons.[102] It seems clear from the data presented in this review that all codons of the UGN family are relatively error-prone in *E. coli*. UGG is misread as cysteine at a high frequency,[82] and UGU/C may be misread as tryptophan at a high frequency. In any case, UGC can substitute extremely well for UGA at a site where UGA is used not as a termination codon, but in its alternative sense as a selenocysteine codon.[111,112] As a nonsense mutation, UGA is very leaky and many required readthrough events involve this codon. (It is also true that UGA is very often reassigned in nonuniversal codes.) There are undoubtedly other examples where codon:anticodon interactions allow a considerable amount of in-frame misreading.

Certainly codons that involve runs of bases are also prone to frameshifting, at least in certain contexts. These codons include the lysine codons AAA and AAG, and as we have seen, the *E. coli* Lys-tRNA also misreads asparagine codons at a high frequency. Frameshifting may occur at a higher frequency at AAG than at AAA,[346] and AAG is used less frequently in *E. coli* than AAA. Based on our data with AAU, which is not commonly used in highly expressed genes, we had originally postulated that rarely used codons might be particularly error-prone. Evidence with AUA refutes this as a general model,[103] as does the *in vitro* study on UUU and UUC.[397] Several examples of errors previously mentioned involved rarely used codons, but in most cases, tRNA supply seemed the controlling mechanism rather than some aspect of the codon:anticodon interaction itself.

It is unclear if the codon usage pattern observed in *E. coli*, or any other organism, reflects selection against error-prone codons. I have argued that the errors that still occur at high frequency, e.g., lysine for asparagine, are those that cause the least damage to the cell.[398] It has been postulated that codon usage (and therefore tRNA abundance) has been balanced in such a way that the most conservative misreading errors are the most likely to occur,[399] or that codons most likely to be misread are simply avoided.[399,400] Although there are no doubt many selective pressures contributing to codon usage (see Chapters 2 and 12),[98,401] it still seems likely that error avoidance is one of them.

B. PROOFREADING, THE STRINGENT RESPONSE, AND EDITING

Some of the accuracy involved in translation will come from an initial discrimination or induced interaction between the substrates and the various components of the translational apparatus. However, the level of accuracy achieved in protein synthesis also involves energy-dependent proofreading reactions.

The aminoacyl-tRNA synthetases appear to have two possible proofreading steps (also called editing), one after activation by ATP, but before attaching to tRNA, and the other after the amino acid is attached to the tRNA.[402,403] In the case of methionyl-tRNA synthetase, the pre-transfer proofreading which leads to rejection of homocysteine leads to a product which can be assayed *in vivo*.[404] It appears that in *E. coli* this reaction occurs about one time for every 100 methionine residues incorporated.[404]

In addition to initial discrimination, proofreading of codon:anticodon interactions also occurs on ribosomes and can be monitored *in vitro* by determining the extra GTP that is hydrolyzed.[99,405-409] Use of such assays has allowed measurements which distinguish between proofreading levels of various noncognate tRNAs, e.g., Leu-tRNA$_2$ is proofread more effectively than Leu-tRNA$_4$ at UUU,[99] and between synonymous codons, e.g., Leu-tRNA$_2$ is proofread more effectively at UUU than UUC.[397] In addition, these assays have allowed analysis of whether the various mutant factors and mutant ribosomes which alter translational error levels affect discrimination or proofreading.

The antibiotic streptomycin primarily reduces proofreading,[406,410] and restrictive, streptomycin-resistant mutants seem to mostly have increased proofreading abilities,[406,411,412] although increased discrimination has also been noted in such mutants.[406] Increased proofreading of noncognate aminoacyl-tRNA results in increased wastage of cognate aminoacyl-tRNA, which is also rejected with a higher frequency. Therefore, increased proofreading can lead to a diminished rate of protein synthesis with increased energy consumption. This can have detrimental effects on the cell, and E. coli with hyperaccurate ribosomes grow more slowly.[237] As might be expected, Ram mutants have decreased proofreading,[39,413] as do error-prone mutants of L7/L12,[45] and a Trp-tRNA with a mutant D-arm which allows efficient reading of UGA codons.[141,142] An error-prone mutant EF-Tu seems to lead to defects in both initial discrimination and proofreading.[48] Use of mutants with altered proofreading may help dissect the mechanisms by which frameshifting errors can happen *in vivo*, and possibly help us learn how required frameshifts can be disguised as normal events.

A very large amount of information is becoming available about the interactions of the rRNAs with ribosomal proteins, tRNA and translational factors, and the involvement of rRNA in protein synthesis.[127,128,414-416] There are a number of rRNA mutants, selected or constructed, which affect antibiotic binding to the ribosome, and there is a strong possibility that many of these will also have altered ribosomal proofreading.[417] It should be possible to test many of these *in vivo* in mutants of E. coli being developed which have a single rRNA operon.[418]

As was previously discussed, there is considerable evidence that the stringent response in wild-type bacteria includes a component which can limit the increase in translational errors seen during amino acid starvation.[51,52] Starving wild-type cells also leads to an increase in errors, but the amount of increase is at least 10-fold less than seen in *relA* mutants.[90] The product of the wild-type *relA* gene, stringent factor, is an enzyme which responds to an uncharged tRNA binding to a codon in the ribosomal A-site by producing guanosine tetraphosphate, ppGpp.[419] This activity is clearly sensitive to the ratio of charged to uncharged tRNA, not just the level of uncharged tRNA.[420] The level of ppGpp itself has been implicated in the control of translational fidelity *in vivo* by using amino acid starvations which do not elicit ppGpp in *relA*$^+$ strains of *Salmonella*,[421] or starving *relA* mutants which have had ppGpp induced by another mechanism.[302] However, I am aware of no data which indicate that the basal-level errors which occur in E. coli respond to the levels of ppGpp. We found that the basal-level misreading of the AAU/C codons as lysine is not affected by starvation for amino acids such as serine or isoleucine in *relA*$^+$

cells.[422] Also Mikkola and Kurland did not see any change in the frequency of UGA readthrough by changing a wild-type cell's growth rate,[423] a situation which changes the level of ppGpp.[424] Any model which accounts for the action of ppGpp on limiting translational errors must take into account the fact that this limitation seems not to be a general error dampening, but one that can occur only at hungry codons.[425]

The experiments of O'Farrell studying recovery from amino acid starvation in strains where ppGpp has different turnover rates indicated that ppGpp itself seemed to be an inhibitor of protein synthesis.[87] However, the effect of ppGpp on translational fidelity does not seem to involve this inhibition increasing the charging level of tRNAs which accept the limiting amino acid, since the charging levels seem to be the same in starved *relA* and *relA+* cells.[426-428] Changing the level of tRNA^{His} in the cell did not seem to affect misreading during histidine starvation,[429] which would seem to rule out models involving the level of uncharged tRNA at the hungry codon.[430] However, uncharged tRNA itself does inhibit protein synthesis, and may play an important regulatory role besides simply inducing ppGpp.[279,419] If the stringent factor is overproduced, ppGpp is synthesized in unstarved cells.[431] One explanation is that the relative abundance of the protein in such cells (about one copy per ribosome) allows ppGpp synthesis in response to momentary pauses at codons,[431] possibly in the absence of uncharged tRNA. Cells stressed by protein overproduction may also have ppGpp synthesis induced.[432] It will be interesting to measure the basal-level of translational errors in such strains.

As was previously mentioned, Menninger has proposed that peptidyl-tRNA drop-off is an editing mechanism that the cell uses to abort the synthesis of error-containing peptides.[271] There are data that show that physiological or genetic manipulations that affect missense errors also affect the level of peptidyl-tRNA in the *pth* mutant.[433-435] The fact that introducing the *miaA* mutation, which decreases certain misreading events, lowers the rate of accumulation of peptidyl-tRNA levels in a *pth* mutant, also supports the editor hypothesis.[436] In addition, Anderson and Menninger have isolated a suppressor of the temperature-sensitive phenotype of the *pth* mutant which maps to a ribosomal protein operon and increases the production of aberrant protein.[437,438] This is the expected phenotype of a mutant blocked in the editing function.

However, there is no information which implicates any specific misreading event with drop-off. Lysine tRNA is the species which becomes blocked by peptide most rapidly in *pth* mutants.[269] We also discussed the evidence that this tRNA frameshifts at a high frequency and misreads both AAU and AAC codons. Yet, although misreading can occur at levels above 10% at AAC codons and 70% at AAU codons in full-length protein, there is no evidence that drop-off also occurs. The possibility that only certain misreading errors lead to drop-off could explain the nonrandom collection of truncated proteins found in the processivity experiments (which can also be explained by frameshifting). However, it is difficult to reconcile this with the finding that a large fraction of almost all tRNA species within the cell become blocked in the *pth* mutant.

C. THE BALANCED CELL

In several instances discussed in this review, the level of various errors could be increased or decreased in a predictable way by changing the cellular concentration of the relevant translational factors. In some cases, this emphasizes the competitive nature of the translational error and the normal event, e.g., the lower frequency of readthrough observed when either the level of RFs is increased or the level of EF-Tu is decreased.[138,152,153] Some changes may reflect error-coupling, e.g., increasing the level of RF can affect the frequency of frameshifting induced by mutant EF-Tu.[316] Changing tRNA levels in the cell can also alter error levels. The frequency of readthrough of specific termination codons is increased in yeast in accord with levels of wild-type tRNA^{Gln}[182,183] and tRNA^{Trp}.[176] Swanson et al. demonstrated that changing the level of a tRNA or its cognate aminoacyl-tRNA synthetase can affect the level of misacylation.[64] Presumably in the cell, most tRNA is either aminoacylated (and being used in protein synthesis) or is sequestered on the cognate synthetase. When the synthetase and its tRNA are not in balance, misacylation can occur, i.e., overproduction of a tRNA could make it available to other synthetases for misacylation.[64] Overproduction of a noncognate tRNA can also lead to changes of codon misreading frequency and, as previously mentioned, tRNA levels might be set within the cell to prevent misreadings which lead to nonconservative amino acid replacements.[103,399]

Amino acid limitation by a variety of means, even exploiting the level of a rare tRNA, can be used by the experimenter, or the cell,[362,363] to increase the error level. Bogosian and his colleagues demonstrated that overproduction of a foreign protein can stress the cell by putting unusual demands on amino acid biosynthesis, leading to biosynthesis and incorporation of amino acid analogs.[76,77] Even the amino acid composition of enriched media can apparently be an important factor in maintaining error level.[77] The stress of protein overproduction can also increase the levels of other errors. For example, in cells overproducing bovine somatotropin, there was increased glutamine incorporation at the UAG codon terminating the message, and an increase in suppression of a *lacZ* frameshift mutation.[77] As previously mentioned, overproduction of foreign protein can induce ppGpp synthesis,[432] as can overproduction of stringent factor.[431] An imbalance in the tetrahydrofolate pools, which might affect tRNA modification levels, has been postulated to explain the observation that thymine-requiring *E. coli* frameshift and readthrough nonsense codons at a higher than normal frequency.[439] Changes in the polyamine pool levels can also alter translational fidelity.[106,440]

Some of these imbalances may not have any particular relevance to the normal cellular state, but even so, the mechanisms by which they exert their effects on translational fidelity promise to be interesting. However, some of the effects may well be of physiological relevance, and it is possible that physiological changes which occur during the growth of a cell, or of a cell culture can be used by the cell to alter translational fidelity for regulatory purposes.[441] The ability to misread UUA in *Streptomyces* is related to the medium used and reading (or misreading) of this codon may be a regulatory event.[109] The lack of the tRNA modification ms²i⁶A can decrease misreading of some codons and increase misreading of others.[83,147] There is the strong possibility that the level of modification is an important regulatory parameter, possibly even involving demodification.[442] Modulation of translational error levels may occur in a number of organisms,[443] but more information is needed before we can tell if any general change in translational fidelity is directly involved in a cellular regulatory circuit.

Clearly, the translational components of the cell are carefully balanced, possibly to optimize accuracy and efficiency.[444] Many of the experiments which have been mentioned in the previous sections, including those of the author, have been performed using cells in a state rather far from normal, and in which it is likely the balance has been disturbed. At the minimum, most *in vivo* measurements of mistranslation made in the last few years have involved cells overproducing one type of protein or another – the very situation which Bogosian et al. demonstrate can lead to increased levels of many different translational errors.[77] It should not go unremarked that the lowest levels of frameshifting observed *in vivo* usually involve studies with either single copies of the gene[293] or low copy number plasmids.[323] Most other studies have used strong promoters driving genes on high copy number plasmids or have been done *in vitro*. At least one reported potential *in vivo* ribosomal frameshift may have occurred as the consequence of an unusually high level of mRNA.[327,329] Almost all measurements of missense misreading have of necessity also been made on abundant proteins, whether produced from cloned genes or from naturally abundant messages. This is not to say that all such measurements are suspect, and certainly most of the errors studied seem clearly to happen in the normal cell. However, it is important to remember the concept of the balanced cell when designing experiments to measure misreading.

VII. CONCLUSIONS

There have been considerable advances in the last few years in research into many aspects of how cells read the genetic code. These advances have been fueled in part by the continuing search for new suppressor mutations, the serendipitous discoveries of interesting genetic organization from DNA sequencing, and the ability to use *in vitro* site-directed mutagenesis and other types of gene construction. This work is continuing at a rapid pace and is likely to yield even more information. It is now clear that ribosomes have more flexibility than we had imagined. But how much closer are we to understanding what the error frequency is in cells and how much error can be tolerated?

A. WHAT IS THE ERROR LEVEL IN CELLS?

Estimates of the likelihood of synthesizing a perfect protein can be made based on the best estimates of each error type. In this review, I have estimated the average missense error as at or below 4×10^{-4} per codon, and the average frameshifting error at about 1×10^{-4} per translocation. There is almost no data on readthrough of normal termination sites, but it is likely to be below 10^{-3}, particularly at UAA. Using estimates from *E. coli* very similar to these, I had previously estimated that in attempting to synthesize 500 copies of a typical protein (300 amino acid residues) from 10 perfect messages (obviously an unwarranted assumption about the accuracy of transcription), and allowing no errors in initiation, only 485 of the polypeptides would be full-length and of these 68 would have a missense error.[31] No new information has become available which would change this estimate appreciably, but I must add the usual caveat that translation may be more accurate in at least some eukaryotes.

Such estimates obviously depend critically on rather sparse data. They also depend on what an "average" gene actually is, since certain errors can occur at very high frequencies. For example, we have evidence that 10% of the ribosomes "erroneously" frameshift on the *argI* message at a single site.[102] Although we have only a small amount of data on the level of missense errors which occur *in vivo,* at least this information seems relevant to normally occurring messages. The situation regarding other errors is less clear. There is a considerable amount of data on required frameshifts, frameshifts at or near in-frame stop codons, and frameshifts which compensate for mutations. Yet the typical message is transcribed from a wild-type gene and doesn't require a ribosomal frameshift to make a functional protein, and the only in-frame termination codon is at the end. There is also remarkably little data on readthrough of normal termination sites at the end of a message. What little we do know indicates that readthrough may happen at a low frequency here, and might possibly be rather different than readthrough of a nonsense mutation. Indeed, given our current understanding of termination codon usage, one interpretation of the data of Lu and Rich is that when an actual termination codon is read by a suppressor tRNA,[190] termination can still take place at a significant frequency. Certainly, it is difficult to envision such a mechanism, but more information is needed on termination at termination sites.

It is difficult to dissect processes that are essential to the cell's survival, or processes that happen at a low frequency. But it should now be possible to look for mutations which "suppress" a normal termination codon, perhaps as part of a gene construct, rather than a nonsense mutation. It is very difficult to quantitate frameshifts within a natural frameshift window, but it is essential to learn what the level of frameshifting is in a typical gene or if ribosomes typically frameshift at simple repetitive sequences. Certainly it is important to learn more about the possibility that ribosomes drop off messages, possibly at certain sites. High levels of frameshifting or drop-off will dramatically lower the overall efficiency of translation, so it is important to have accurate information on their frequency.

One further difficulty with data on missense errors, frameshifts, and drop-offs is the possibility that many mistranslated proteins are so unstable that they would not be detected and, therefore, the data seriously underestimate the actual level of error in translation. "Abnormal" proteins themselves seem to be involved in the triggering mechanism of the heat shock response,[445,446] and several heat shock proteins in both prokaryotes and eukaryotes are involved directly or indirectly in protein degradation.[445,447-451] This extremely interesting area of research obviously cannot be dealt with in detail here, but I will attempt to assess whether this response, and/or the presence of other proteases in the cell, has caused us to seriously underestimate the frequency of translational errors.

There is no evidence that the mistranslated proteins containing the lysine for asparagine errors are unstable,[398] and amino acid starvation, which induces mistranslation to very high frequencies in some cases, is a very poor inducer of the heat-shock response.[98,445,452] Many proteins encoded by genes with missense mutations are also stable,[453,454] and unfolded proteins which induce the heat-shock response need not necessarily be particularly unstable.[446] Certainly there may be types of errors underrepresented because of protein turnover (perhaps cysteine for isoleucine), but there is as yet no evidence for it. Appropriate genes can be constructed by site-directed mutagenesis to answer such questions.[455,456] However, it seems possible that the *in vivo* measurements of missense errors which we have may be fairly accurate (except those having to do with certain amino acid analogs).

However, there is considerable evidence that many truncated proteins (nonsense fragments or puromycin-terminated peptides) are unstable, and can be degraded quite rapidly.[74,457] This could seriously reduce the accuracy of measurements of the frequency of either frameshifts or drop-offs. With frameshifts, however, the great majority of our data comes from frameshift mutation suppression, and therefore, the product will be full-length and stable, unless the amino acids inserted at the frameshift site lead to instability. The measurements of Jørgensen and Kurland,[280] using dimeric forms of β-galactosidase, are remarkably similar to those of Manley,[8] which depended on the stability of fragments, and both are similar to frameshift measurements which measured full-length enzymes.[300] Therefore, even though some of the data do not differentiate between frameshifting and drop-offs, the measurements themselves may be an accurate reflection of the errors leading to truncated protein.

However, if there is a tendency for drop-offs to occur soon after initiation, the products would almost certainly be unstable and extremely difficult to detect. Since as much as 20% of the peptide bonds synthesized in bacteria may be rapidly hydrolyzed,[458] serious consideration must be given to this possibility.[459] Although a majority of peptidyl-tRNA found in the *pth* mutants apparently has greater than seven amino acid residues, about 25% have less than this.[460] Gast et al. report that noncognate tRNAs dissociate more readily from poly-U early in polypeptide synthesis.[461] The data of Chen and Inouye relating to the role of rarely used codons near initiation can also be explained by preferential drop-off within the first 50 to 60 codons, although other explanations are also possible.[272] Our experiments with *argI* may also relate to this. In the wild-type message there seems to be a frameshift that happens at a high frequency at codon four (UUU).[102] However, in-frame translation produces a full-length protein that has no leucine misincorporated at this codon.[100,101] This could mean that errors that do not lead to frameshifting lead to drop-off. The yeast Ty ribosomal frameshifting sequence CUU-AGG-C does not give measurable frequencies of frameshifting unless the CUU is at least four codons downstream from initiation.[360,362] This demonstrates the difference between translation immediately after initiation and later steps in elongation, and could be interpreted to mean that, rather than frameshifting, the peptidyl-tRNA has a tendency to drop off the ribosome at this step.

B. HOW MUCH ERROR CAN CELLS TOLERATE?

The existence of missense suppressor mutations and mutants having ribosomes with a Ram phenotype indicates at least some specific errors can be increased without resulting in cell death, or even a higher death rate.[462] However, there is a mutant *S. typhimurium* with a defective EF-Tu in which there is a direct correlation between a reduction in growth rate and the production of functionally defective proteins.[49] The missense error rate in these cells seems to be 10-fold above normal. *E. coli* can also be grown for hundreds of generations with low levels of streptomycin which slow growth and increase the error level at least 10 times above normal.[463] It is difficult to establish if it is the errors *per se* that decrease the growth rate, and therefore it is not known what level of increase in translation error levels will diminish fitness. Analysis of the populations of wild-type *E. coli* indicates that most loci are polymorphic, i.e., the enzymes encoded are demonstrably different.[464] The interesting analysis of Michaels et al. indicates that a surprising number of amino acid substitutions can still give active thymidylate synthase.[456] Many amino acid substitutions in β-galactosidase also have no demonstrable selective effect.[465] Therefore, it is quite likely that having over 10% of the "average" protein in the cell with an amino acid substitution (and usually a conservative one) is not particularly disadvantageous.

The problem with synthesizing a large protein will not be that it might contain another missense error or two, but simply the decreasing amount of product that will be synthesized because of frameshifting or drop-off. At a combined frequency of 1.4×10^{-4}, we have seen that roughly 86% of measurable β-galactosidase initiations do lead to full-length product (discounting the possibility of a high level of error soon after initiation). The actual amount of full-length β-galactosidase synthesized if the error level rose to 1.4×10^{-3} in Ram mutants would decrease to 25%. (Such errors do not increase appreciably with the mutant EF-Tu.)[49] The efficiency of synthesizing very large proteins could become very low.[275] It will be interesting to see if the messages for large proteins, or the codons for critical residues, are protected against error.

Although I have attempted to cover most of "the facts" as they are known to me, it is clear that much more information is required on error levels (particularly in eukaryotes). However, I hope that it is also clear that organisms can use the genetic code with unexpected flexibility. Some organisms synthesize different rRNA molecules, and therefore, different ribosomes, in a developmentally regulated fashion.[466-468] This opens up the possibility that organisms could program rather dramatic changes in reading the code as a part of their lifecycle.

VIII. ACKNOWLEDGMENTS

I thank my many colleagues who have shared their time, information, ideas, and unpublished data, particularly A. Böck, R. Hendrix, E. Murgola, F. C. Neidhardt, C. Squires, and R. Weiss. I particularly wish to thank J. Baker and E. P. Parker for their help and suggestions, and D. Hatfield for his support and patience. The literature cited was arbitrarily limited, and I apologize for all omissions.

Research in my laboratory has been supported by Public Health Service grant GM25855 from the National Institutes of Health.

REFERENCES

1. **Nirenberg, M., Caskey, T., Marshall, R., Brimacombe, R., Kellogg, D., Doctor, B., Hatfield, D., Levin, J., Rottman, F., Pestka, S., Wilcox, M., and Anderson, F.,** The RNA code and protein synthesis, *Cold Spring Harbor Symp. Quant. Biol.,* 31, 11, 1966.
2. **Caron, F.,** Eukaryotic codes, *Experientia,* 46, 1106, 1990.
3. **Jukes, T. H. and Osawa, S.,** The genetic code in mitochondria and chloroplasts, *Experientia,* 46, 1117, 1990.
4. **Osawa, S., Muto, A., Ohama, T., Andachi, Y., Tanaka, R., and Yamao, F.,** Prokaryotic genetic code, *Experientia,* 46, 1097, 1990.
5. **Plumbridge, J. A., Deville, F., Sacerdot, C., Petersen, H. U., Cenatiempo, Y., Cozzone, A., Grunberg-Manago, M., and Hershey, J. W. B.,** Two translational initiation sites in the *infB* gene are used to express initiation factor IF2α and IF2β in *Escherichia coli, EMBO J.,* 4, 223, 1985.
6. **Curran, J. and Kolakofsky, D.,** Ribosomal initiation from an ACG codon in the Sendai virus P/C mRNA, *EMBO J.,* 7, 245, 1988.
7. **Gupta, K. C. and Patwardhan, S.,** ACG, the initiator codon for a Sendai virus protein, *J. Biol. Chem.,* 263, 8553, 1988.
8. **Manley, J. L.,** Synthesis and degradation of termination and premature-termination fragments of β-galactosidase *in vitro* and *in vivo, J. Mol. Biol.,* 125, 407, 1978.
9. **Files, J. G., Weber, K., and Miller, J. H.,** Translational reinitiation: reinitiation of *lac* repressor fragments at three internal sites early in the *lac i* gene of *Escherichia coli, Proc. Natl. Acad. Sci. U.S.A.,* 71, 667, 1974.
10. **Peabody, D. S. and Berg, P.,** Termination – reinitiation occurs in the translation of mammalian cell mRNAs, *Mol. Cell. Biol.,* 6, 2695, 1986.
11. **Adhin, M. R. and van Duin, J.,** Scanning model for translational reinitiation in eubacteria, *J. Mol. Biol.,* 213, 811, 1990.
12. **Weiss, R. B., Dunn, D. M., Atkins, J. F., and Gesteland, R. F.,** Slippery runs, shifty stops, backward steps, and forward hops: –2, –1, +1, +2, +5 and +6 ribosomal frameshifting, *Cold Spring Harbor Symp. Quant. Biol.,* 52, 687, 1987.

13. **Rosenberger, R. F. and Hilton, J.**, The frequency of transcriptional and translational errors at nonsense codons in the *lacZ* gene of *Escherichia coli*, *Mol. Gen. Genet.*, 191, 207, 1983.
14. **Blank, A., Gallant, J. A., Burgess, R. R., and Loeb, L. A.**, An RNA polymerase mutant with reduced accuracy of chain elongation, *Biochemistry*, 25, 5920, 1986.
15. **Wagner, L. A., Weiss, R. B., Driscoll, R., Dunn, D. S., and Gesteland, R. F.**, Transcriptional slippage occurs during elongation at runs of adenine or thymine in *Escherichia coli*, *Nucleic Acids Res.*, 18, 3529, 1990.
16. **Thomas, S. M., Lamb, R. A., and Paterson, R. G.**, Two mRNAs that differ by two nontemplated nucleotides encode the amino coterminal proteins P and V of the paramyxovirus SV5, *Cell*, 54, 891, 1988.
17. **Cattaneo, R., Kaelin, K., Baczko, K., and Billeter, M. A.**, Measles virus editing provides an additional cysteine-rich protein, *Cell*, 56, 759, 1989.
18. **Vidal, S., Curran, J., and Kolakofsky, D.**, A stuttering model for paramyxovirus P mRNA editing, *EMBO J.*, 9, 2017, 1990.
19. **Benne, R.**, RNA editing in trypansomes: is there a message?, *Trends Genet.*, 6, 177, 1990.
20. **Feagin, J. E.**, RNA editing in kinetoplastid mitochondria, *J. Biol. Chem.*, 265, 19373, 1990.
21. **Mahendran, R., Spottswood, M. R., and Miller, D. L.**, RNA editing by cytidine insertion in mitochondria of *Physarum polycephalum*, *Nature (London)*, 349, 434, 1991.
22. **Covello, P. S. and Gray, M. W.**, RNA editing in plant mitochondria, *Nature (London)*, 341, 662, 1989.
23. **Gualberto, J. M., Lamattina, L., Bonnard, G., Weil, J.-H., and Grienenberger, J.-M.**, RNA editing in wheat mitochondria results in the conservation of protein sequences, *Nature (London)*, 341, 660, 1989.
24. **Chen, S.-H., Habib, G., Yang, C.-Y., Gu, Z.-W., Lee, B. R., Weng, S.-A., Silberman, S. R., Cai, S.-J., Deslypere, J. P., Rosseneu, M., Gotto, A. M., Jr., Li, W.-H., and Chan, L.**, Apolipoprotein B-48 is the product of a messenger RNA with an organ-specific in-frame stop codon, *Science*, 238, 363, 1987.
25. **Powell, L. M., Wallis, S. C., Pease, R. J., Edwards, Y. H., Knott, T. J., and Scott, J.**, A novel form of tissue-specific RNA processing produces apolipoprotein-B48 in intestine, *Cell*, 50, 831, 1987.
26. **Davidson, N. O., Powell, L. M., Wallis, S. C., and Scott, J.**, Thyroid hormone modulates the introduction of a stop codon in rat liver apolipoprotein B messenger RNA, *J. Biol. Chem.*, 263, 13,482, 1988.
27. **Wintz, H. and Hanson, M. R.**, A termination codon is created by RNA editing in the petunia mitochondrial *atp*9 gene transcript, *Curr. Genet.*, 19, 61, 1991.
28. **Carrington, D. M., Auffret, A., and Hanke, D. E.**, Polypeptide ligation occurs during post-translational modification of concanavalin A, *Nature (London)*, 313, 64, 1985.
29. **Bowles, D. J., Marcus, S. E., Pappin, D. J. C., Findlay, J. B. C., Eliopoulos, E., Maycox, P. R., and Burgess, J.**, Posttranslational processing of concanavalin A precursors in jackbean cotyledons, *J. Cell Biol.*, 102, 1284, 1986.
30. **Kane, P. M., Yamashiro, C. T., Wolczyk, D. F., Neff, N., Goebl, M., and Stevens, T. H.**, Protein splicing converts the yeast *TFP1* gene product to the 69-kD subunit of the vacuolar H^+ - adenosine triphosphatase, *Science*, 250, 651, 1990.
31. **Parker, J.**, Errors and alternatives in reading the universal genetic code, *Microbiol. Rev.*, 53, 273, 1989.
32. **Valle, R. P. C. and Morch, M.-D.**, Stop making sense, or regulation at the level of termination in eukaryotic protein synthesis, *FEBS Lett.*, 235, 1, 1988.
33. **Jukes, T.H.**, Genetic code 1990. Outlook, *Experientia*, 46, 1149, 1990.
34. **Kurland, C. G., Jörgensen, F., Richter, A., Ehrenberg, M., Bilgin, N., and Rojas, A.-M.**, Through the accuracy window, in *The Ribosome: Structure, Function & Evolution*, Hill, W. E., Dahlberg, A., Garrett, R. A., Moore, P. B., Schlessinger, D., and Warner, J. R., Eds., American Society for Microbiology, Washington, D.C., 1990, chap. 44.

35. Woese, C. R., The present status of the genetic code, *Prog. Nucleic Acids Res. Mol. Biol.*, 7, 107, 1967.
36. Gorini, L., Streptomycin and misreading of the genetic code, in *Ribosomes*, Nomura, M., Tissières, A. and Lengyel, P., Eds., Cold Spring Harbor Laboratory, 1974, 791.
37. Grant, C. M., Firoozan, M., and Tuite, M. F., Mistranslation induces the heat-shock response in the yeast *Saccharomyces cerevisiae*, *Mol. Microbiol.*, 3, 215, 1989.
38. Martin, R., Mogg, A. E., Heywood, L. A., Nitschke, L., and Burke, J. F., Aminoglycoside suppression at UAG, UAA and UGA codons in *Escherichia coli* and human tissue culture cells, *Mol. Gen. Genet.*, 217, 411, 1989.
39. Andersson, D. I., Andersson, S. G. E., and Kurland, C. G., Functional interactions between mutated forms of ribosomal proteins S4, S5 and S12, *Biochimie*, 68, 705, 1986.
40. Davies, J., Streptomycin and the genetic code, *Cold Spring Harbor Symp. Quant. Biol.*, 31, 665, 1966.
41. Johnston, T. C. and Parker, J., Streptomycin-induced, third-position misreading of the genetic code, *J. Mol. Biol.*, 181, 313, 1985.
42. Phoenix, P., Gravel, M., Herrington, M. B., and Brakier-Gingras, L., Neomycin is more efficient than streptomycin in suppressing frameshift mutations, *Can. J. Genet. Cytol.*, 27, 776, 1985.
43. Bollen, A., Cabezón, T., de Wilde, M., Villarroel, R., and Herzog, A., Alteration of ribosomal protein S17 by mutation linked to neamine resistance in *Escherichia coli*. I. General properties of *neaA* mutants, *J. Mol. Biol.*, 99, 795, 1975.
44. Kühberger, R., Piepersberg, W., Petzet, A., Buckel, P., and Böck, A., Alteration of ribosomal protein L6 in gentamicin-resistant strains of *Escherichia coli*. Effects on fidelity of protein synthesis, *Biochemistry*, 18, 187, 1979.
45. Kirsebom, L. A. and Isaksson, L. A., Involvement of ribosomal protein L7/L12 in control of translational accuracy, *Proc. Natl. Acad. Sci. U.S.A.*, 82, 717, 1985.
46. Kirsebom, L. A., Amons, R., and Isaksson, L. A., Primary structures of mutationally altered ribosomal protein L7/L12 and their effects on cellular growth and translational accuracy, *Eur. J. Biochem.*, 156, 669, 1986.
47. Dequard-Chablat, M., Coppin-Raynal, E., Picard-Bennoun, M., and Madjar, J. J., At least seven ribosomal proteins are involved in the control of translational accuracy in a eukaryotic organism, *J. Mol. Biol.*, 190, 167, 1986.
48. Tapio, S. and Kurland, C. G., Mutant EF-Tu increases missense error in vitro, *Mol. Gen. Genet.*, 205, 186, 1986.
49. Hughes, D., Error-prone EF-Tu reduces *in vivo* enzyme activity and cellular growth rate, *Mol. Microbiol.*, 5, 623, 1991.
50. Sandbaken, M. G. and Culbertson, M. R., Mutations in elongation factor EF-1α affect the frequency of frameshifting and amino acid misincorporation in *Saccharomyces cerevisiae*, *Genetics*, 120, 923, 1988.
51. Gallant, J. A., Stringent control in *E. coli*, *Ann. Rev. Genet.*, 13, 393, 1979.
52. Cashel, M. and Rudd, K. E., The stringent response, in *Escherichia coli and Salmonella typhimurium: Cellular and Molecular Biology*, Vol. 2, Neidhardt, F. C., Ingraham, J. L., Low, K. B., Magasanik, B., Schaechter, M., and Umbarger, H. E., Eds., American Society for Microbiology, Washington, D.C., 1987, chap. 87.
53. Normanly, J. and Abelson, J., tRNA identity, *Ann. Rev. Biochem.*, 58, 1029, 1989.
54. Muramatsu, T., Nishikawa, K., Nemoto, F., Kuchino, Y., Nishimura, S., Miyazawa, T., and Yokoyama, S., Codon and amino acid specificities of a transfer RNA are both converted by a single post-transcriptional modification, *Nature (London)*, 336, 179, 1988.
55. Perret, V., Garcia, A., Grosjean, H., Ebel, J.-P., Florentz, C., and Giegé, R., Relaxation of a transfer RNA specificity by removal of modified nucleotides, *Nature (London)*, 344, 787, 1990.
56. McClain, W. H. and Foss, K., Changing the identity of a tRNA by introducing a G-U wobble pair near the 3′ acceptor end, *Science*, 240, 793, 1988.

57. **Normanly, J., Kleina, L. G., Masson, J.-M., Abelson, J., and Miller, J. H.,** Construction of *Escherichia coli* amber suppressor tRNA genes. III. Determination of tRNA specificity, *J. Mol. Biol.,* 213, 719, 1990.
58. **Pallanck, L. and Schulman, L. H.,** Anticodon-dependent aminoacylation of a noncognate tRNA with isoleucine, valine, and phenylalanine *in vivo, Proc. Natl. Acad. Sci. U.S.A.,* 88, 3872, 1991.
59. **Murgola, E. J. and Pagel, F. T.,** Suppressors of lysine codons may be misacylated lysine tRNAs, *J. Bacteriol.,* 156, 917, 1983.
60. **Prather, N. E., Murgola, E. J., and Mims, B. H.,** Nucleotide substitution in the amino acid acceptor stem of lysine transfer RNA causes missense suppression, *J. Mol. Biol.,* 172, 177, 1984.
61. **Murgola, E. J.,** Personal communication, 1991.
62. **Inokuchi, H., Hoben, P., Yamao, F., Ozeki, H., and Söll, D.,** Transfer RNA mischarging mediated by a mutant *Escherichia coli* glutaminyl-tRNA synthetase, *Proc. Natl. Acad. Sci. U.S.A.,* 81, 5076, 1984.
63. **Perona, J. J., Swanson, R. N., Rould, M. A., Steitz, T. A., and Söll, D.,** Structural basis for misaminoacylation by mutant *E. coli* glutaminyl-tRNA synthetase enzymes, *Science,* 246, 1152, 1989.
64. **Swanson, R., Hoben, P., Sumner-Smith, M., Uemura, H., Watson, L., and Söll, D.,** Accuracy of in vivo aminoacylation requires proper balance of tRNA and aminoacyl-tRNA synthetase, *Science,* 242, 1548, 1988.
65. **Lin, S. X., Blatzinger, M., and Remy, P.,** Fast kinetic study of yeast phenylalanyl-tRNA synthetase: role of tRNA^{Phe} in the discrimination between tyrosine and phenylalanine, *Biochemistry,* 23, 4109, 1984.
66. **Okamoto, M. and Savageau, M. A.,** Integrated function of a kinetic proofreading mechanism: steady-state analysis testing internal consistency of data obtained *in vivo* and *in vitro* and predicting parameter values, *Biochemistry,* 23, 1701, 1984.
67. **Freist, W., Sternbach, H., and Cramer, F.,** Isoleucyl-tRNA synthetase from baker's yeast and from *Escherichia coli* MRE 600: discrimination of 20 amino acids in aminoacylation of tRNA^{Ile}-C-C-A, *Eur. J. Biochem.,* 173, 27, 1988.
68. **Loftfield, R. B.,** The frequency of errors in protein biosynthesis, *Biochem. J.,* 89, 82, 1963.
69. **Loftfield, R. B. and Vanderjagt, D.,** The frequency of errors in protein biosynthesis, *Biochem. J.,* 128, 1353, 1972.
70. **Coons, S. F., Smith, L. F., and Loftfield, R. B.,** The nature of amino acid errors in *in vivo* biosynthesis of rabbit hemoglobin, *Fed. Proc. Fed. Am. Soc. Exp. Biol.,* 38, 328, 1979.
71. **Wilson, M. J. and Hatfield, D. L.,** Incorporation of modified amino acids into proteins *in vivo, Biochim. Biophys. Acta,* 781, 205, 1984.
72. **Momand, J. A. and Clark, S.,** The fidelity of protein synthesis: can mischarging by aspartyl-tRNA^{Asp} synthetase lead to the formation of isoaspartyl residues in proteins?, *Biochim. Biophys. Acta,* 1040, 153, 1990.
73. **Wold, F.,** *In vivo* chemical modification of proteins (post-translational modification), *Ann. Rev. Biochem.,* 50, 783, 1981.
74. **Goldberg, A. L. and Goff, S. A.,** The selective degradation of abnormal proteins in bacteria, in *Maximizing Gene Expression,* Reznikoff, W. and Gold, L., Eds., Butterworth, Boston, 1986, chap. 9.
75. **Koide, H., Yokoyama, S., Kawai, G., Ha, J.-M., Oka, T., Kawai, S., Miyake, T., Fuwa, T., and Miyazawa, T.,** Biosynthesis of a protein containing a nonprotein amino acid by *Escherichia coli:* L-2-aminohexanoic acid at position 21 in human epidermal growth factor, *Proc. Natl. Acad. Sci. U.S.A.,* 85, 6237, 1988.
76. **Bogosian, G., Violand, B. N., Dorward-King, E. J., Workman, W. E., Jung, P. E., and Kane, J. F.,** Biosynthesis and incorporation into protein of norleucine by *Escherichia coli, J. Biol. Chem.,* 264, 531, 1989.

77. Bogosian, G., Violand, B. N., Jung, P. E., and Kane, J. F., Effect of protein overexpression on mistranslation in *Escherichia coli*, in *The Ribosome: Structure, Function & Evolution*, Hill, W. E., Dahlberg, A., Garrett, R. A., Moore, P. B., Schlessinger, D., and Warner, J. R., Eds., American Society for Microbiology, Washington, D.C., 1990, chap. 48.
78. **Schön, A., Kannangara, C. G., Gough, S., and Söll, D.**, Protein biosynthesis in organelles requires misaminoacylation of tRNA, *Nature (London)*, 331, 187, 1988.
79. **Söll, D.**, The accuracy of aminoacylation – ensuring the fidelity of the genetic code, *Experientia*, 46, 1089, 1990.
80. **Wada, K.-N., Aota, S.-I., Tsuchiya, R., Ishibashi, F., Gojobori, T., and Ikemura, T.**, Codon usage tabulated from the GenBank genetic sequence data, *Nucleic Acids Res.*, 18, (Suppl.), 2367, 1990.
81. **Edelmann, P. and Gallant, J.**, Mistranslation in *E. coli*, *Cell*, 10, 131, 1977.
82. **Bouadloun, F., Donner, D., and Kurland, C. G.**, Codon-specific missense errors *in vivo*, *EMBO J.*, 2, 1351, 1983.
83. **Bouadloun, F., Srichaiyo, T., Isaksson, L. A., and Björk, G. R.**, Influence of modification next to the anticodon in the tRNA on codon context sensitivity of translational suppression and accuracy, *J. Bacteriol.*, 166, 1022, 1986.
84. **Laughrea, M., Latulippe, J., Filion, A.-M., and Boulet, L.**, Mistranslation in twelve *Escherichia coli* ribosomal proteins: cysteine misincorporation at neutral amino acid residues other than tryptophan, *Eur. J. Biochem.*, 169, 59, 1987.
85. **Caskey, C. T., Beaudet, A., and Nirenberg, M.**, RNA codons and protein synthesis. XV. Dissimilar responses of mammalian and bacterial transfer RNA fractions to messenger RNA codons, *J. Mol. Biol.*, 37, 99, 1968.
86. **Toth, M. J., Murgola, E. J., and Schimmel, P.**, Evidence for a unique first position codon-anticodon mismatch *in vivo*, *J. Mol. Biol.*, 201, 451, 1988.
87. **O'Farrell, P. H.**, The suppression of defective translation by ppGpp and its role in the stringent response, *Cell*, 14, 545, 1978.
88. **Parker, J., Pollard, J. W., Friesen, J. D., and Stanners, C. P.**, Stuttering: high-level mistranslation in animal and bacterial cells, *Proc. Natl. Acad. Sci. U.S.A.*, 75, 1091, 1978.
89. **Parker, J. and Friesen, J. D.**, "Two out of three" codon reading leading to mistranslation in vivo, *Mol. Gen. Genet.*, 177, 439, 1980.
90. **Parker, J., Johnston, T. C., and Borgia, P. T.**, Mistranslation in cells infected with the bacteriophage MS2: direct evidence of Lys for Asn substitution, *Mol. Gen. Genet.*, 180, 275, 1980.
91. **Parker, J., Johnston, T. C., Borgia, P. T., Holtz, G., Remaut, E., and Fiers, W.**, Codon usage and mistranslation: *in vivo* basal level misreading of the MS2 coat protein message, *J. Biol. Chem.*, 258, 10007, 1983.
92. **Johnston, T. C., Borgia, P. T., and Parker, J.**, Codon specificity of starvation-induced misreading, *Mol. Gen. Genet.*, 195, 459, 1984.
93. **Precup, J. and Parker, J.**, Missense misreading of asparagine codons as a function of codon identity and context, *J. Biol. Chem.*, 262, 11351, 1987.
94. **Khazaie, K., Buchanan, J. H., and Rosenberger, R. F.**, The accuracy of Qβ RNA translation I. Errors during the synthesis of Qβ proteins by intact *Escherichia coli* cells, *Eur. J. Biochem.*, 144, 485, 1984.
95. **Pollard, J. W., Harley, C. B., Chamberlain, J. W., Goldstein, S., and Stanners, C. P.**, Is transformation associated with an increased error frequency in mammalian cells?, *J. Biol. Chem.*, 257, 5977, 1982.
96. **Harley, C. B., Pollard, J. W., Stanners, C. P., and Goldstein, S.**, Model for messenger RNA translation during amino acid starvation applied to the calculation of protein synthetic error rates, *J. Biol. Chem.*, 256, 10786, 1981.
97. **Parker, J.**, Specific mistranslation in *hisT* mutants of *Escherichia coli*, *Mol. Gen. Genet.*, 187, 405, 1982.

98. **Parker, J.,** unpublished results, 1991.
99. **Ruusala, T., Ehrenberg, M., and Kurland, C. G.,** Is there proofreading during polypeptide synthesis?, *EMBO J.,* 1, 741, 1982.
100. **Parker, J. and Precup, J.,** Mistranslation during phenylalanine starvation, *Mol. Gen. Genet.,* 204, 70, 1986.
101. **Precup, J., Ulrich, A. K., Roopnarine, O., and Parker, J.,** Context specific misreading of phenylalanine codons, *Mol. Gen. Genet.,* 218, 397, 1989.
102. **Fu, C. and Parker, J.,** unpublished data, 1991.
103. **Ulrich, A. K., Li, L.-Y., and Parker, J.,** Codon usage, transfer RNA availability and mistranslation in amino acid starved bacteria, *Biochim. Biophys. Acta,* 1089, 362, 1991.
104. **Seetharam, R., Heeren, R. A., Wong, E. Y., Braford, S. R., Klein, B. K., Aykent, S., Kotts, C. E., Mathis, K. J., Bishop, B. F., Jennings, M. J., Smith, C. E., and Siegel, N. R.,** Mistranslation in IGF-1 during over-expression of the protein in *Escherichia coli* using a synthetic gene containing low frequency codons, *Biochem. Biophys. Res. Commun.,* 155, 518, 1988.
105. **Spanjaard, R. A., Chen, K., Walker, J. R., and van Duin, J.,** Frameshift suppression at tandem AGA and AGG codons by cloned tRNA genes: assigning a codon to *argU* tRNA and T4 tRNA^{Arg}, *Nucleic Acids Res.,* 18, 5031, 1990.
106. **McMurry, L. M. and Algranati, I. D.,** Effect of polyamines on translation fidelity *in vivo, Eur. J. Biochem.,* 155, 383, 1986.
107. **Buckingham, R. H. and Kurland, C. G.,** Codon specificity of UGA suppressor tRNA^{Trp} from *Escherichia coli, Proc. Natl. Acad. Sci. U.S.A.,* 74, 5496, 1977.
108. **Lawlor, E. J., Baylis, H. A., and Chater, K. F.,** Pleiotropic morphological and antibiotic deficiencies result from mutations in a gene encoding a tRNA-like product in *Streptomyces coelicolor* A3(2), *Genes Dev.,* 1, 1305, 1987.
109. **Leskiw, B. K., Lawlor, E. J., Fernandez-Abalos, J. M., and Chater, K. F.,** TTA codons in some genes prevent their expression in a class of developmental, antibiotic-negative, *Streptomyces* mutants, *Proc. Natl. Acad. Sci. U.S.A.,* 88, 2461, 1991.
110. **Hollingshead, S. and Vapnek, D.,** Nucleotide sequence of a gene encoding a streptomycin/spectinomycin adenyltransferase, *Plasmid,* 13, 17, 1985.
111. **Zinoni, F., Birkmann, A., Leinfelder, W., and Böck, A.,** Cotranslational insertion of selenocysteine into formate dehydrogenase from *Escherichia coli* directed by a UGA codon, *Proc. Natl. Acad. Sci. U.S.A.,* 84, 3156, 1987.
112. **Zinoni, F., Heider, J., and Böck, A.,** Features of the formate dehydrogenase mRNA necessary for decoding of the UGA codon as selenocysteine, *Proc. Natl. Acad. Sci. U.S.A.,* 87, 4660, 1990.
113. **Ellis, N. and Gallant, J.,** An estimate of the global error frequency in translation, *Mol. Gen. Genet.,* 188, 169, 1982.
114. **Kawaguchi, Y., Honda, H., Taniguchi-Morimura, J., and Iwasaki, S.,** The codon CUG is read as serine in an asporogenic yeast *Candida cylindracea, Nature (London),* 341, 164, 1989.
115. **Li, M. and Tzagoloff, A.,** Assembly of the mitochondrial membrane system: sequences of yeast mitochondrial valine and an unusual threonine tRNA gene, *Cell,* 18, 47, 1979.
116. **Sibler, A.-P., Dirheimer, G., and Martin, R. P.,** Nucleotide sequence of a yeast mitochondrial threonine-tRNA able to decode the C-U-N leucine codons, *FEBS Lett.,* 132, 344, 1981.
117. **Clark-Walker, G. D., McArthur, C. R., and Sriprakash, K. S.,** Location of transcriptional control signals and transfer RNA sequences in *Torulopsis glabrata* mitochondrial DNA, *EMBO J.,* 4, 465, 1985.
118. **Osawa, S., Collins, D., Ohama, T., Jukes, T. H., and Watanabe, K.,** Evolution of the mitochondrial genetic code. III. Reassignment of CUN codons from leucine to threonine during evolution of yeast mitochondria, *J. Mol. Evol.,* 30, 322, 1990.
119. **Fox, T. D.,** Natural variation in the genetic code, *Ann. Rev. Genet.,* 21, 67, 1987.

120. Osawa, S. and Jukes, T. H., Codon reassignment (codon capture) in evolution, *J. Mol. Evol.*, 28, 271, 1989.
121. Hartz, D., McPheeters, D. S., and Gold, L., Selection of the initiator tRNA by *Escherichia coli* initiation factors, *Genes Dev.*, 3, 1899, 1989.
122. Brenner, S., Stretton, A. O. W., and Kaplan, S., Genetic code: the 'nonsense' triplets for chain termination and their suppression, *Nature (London)*, 206, 994, 1965.
123. Lee, C. C., Timms, K. M., Trotman, C. N. A., and Tate, W. P., Isolation of a rat mitochondrial release factor: accomodation of the changed genetic code for termination, *J. Biol. Chem.*, 262, 3548, 1987.
124. Oba, T., Andachi, Y., Muto, A., and Osawa, S., CGG: an unassigned or nonsense codon in *Mycoplasma capricolum*, *Proc. Natl. Acad. Sci. U.S.A.*, 88, 921, 1991.
125. Chang, L.-J., Pryciak, P., Ganem, D., and Varmus, H. E., Biosynthesis of the reverse transcriptase of hepatitis B viruses involves *de novo* translational initiation not ribosomal frameshifting, *Nature (London)*, 337, 364, 1989.
126. Craigen, W. J., Lee, C. C., and Caskey, C. T., Recent advances in peptide chain termination, *Mol. Microbiol.*, 4, 861, 1990.
127. Murgola, E. J., Dahlberg, A. E., Hijazi, K. A., and Tiedeman, A. A., rRNA and codon recognition: the rRNA-mRNA base-pairing model of peptide chain termination, in *The Ribosome: Structure, Function & Evolution*, Hill, W. E., Dahlberg, A., Garrett, R. A., Moore, P. B., Schlessinger, D., and Warner, J. R., Eds., American Society for Microbiology, Washington, D.C., 1990, chap. 33.
128. Tate, W. P., Brown, C. M., and Kastner, B., Codon recognition by the polypeptide release factor, in T*he Ribosome: Structure, Function & Evolution*, Hill, W. E., Dahlberg, A., Garrett, R. A., Moore, P. B., Schlessinger, D., and Warner, J. R., Eds., American Society for Microbiology, Washington, D.C., 1990, chap. 32.
129. Craigen, W. J. and Caskey, C. T., The function, structure and regulation of *E. coli* peptide chain release factors, *Biochimie*, 69, 1031, 1987.
130. Lang, A., Friemert, C., and Gassen, H. G., On the role of the termination factor RF-2 and the 16S RNA in protein synthesis, *Eur. J. Biochem.*, 180, 547, 1989.
131. Tate, W. P., Kastner, B., Edgar, C. D., McCaughan, K. K., Timms, K. M., Trotman, C. N. A., Stoffler-Meilicke, M., Stoffler, G., Nag, B., and Traut, R. R., The ribosomal domain of the bacterial release factors: the carboxyl-terminal domain of the dimer of *Escherichia coli* ribosomal protein L7/L12 located in the body of the ribosome is important for release factor interaction, *Eur. J. Biochem.*, 187, 543, 1990.
132. Murgola, E. J., Hijazi, K. A., Göringer, H. U., and Dahlberg, A. E., Mutant 16S ribosomal RNA: a codon-specific translational suppressor, *Proc. Natl. Acad. Sci. U.S.A.*, 85, 4162, 1988.
133. Hänfler, A., Kleuvers, B., and Göringer, H. U., The involvement of base 1054 in 16S rRNA for UGA stop codon dependent translation termination, *Nucleic Acids Res.*, 18, 5625, 1990.
134. Beaudet, A. L. and Caskey, C. T., Mammalian peptide chain termination. II. Codon specificity and GTPase activity of release factor, *Proc. Natl. Acad. Sci. U.S.A.*, 68, 619, 1971.
135. Lee, C. C., Craigen, W. J., Muzny, D. M., Harlow, E., and Caskey, C. T., Cloning and expression of a mammalian peptide chain release factor with sequence similarity to tryptophanyl-tRNA synthetases, *Proc. Natl. Acad. Sci. U.S.A.*, 87, 3508, 1990.
136. Rydén, S. M. and Isaksson, L. A., A temperature-sensitive mutant of *Escherichia coli* that shows enhanced misreading of UAG/A and increased efficiency for some tRNA nonsense suppressors, *Mol. Gen. Genet.*, 193, 38, 1984.
137. Ryden, M., Murphy, J., Martin, R., Isaksson, L., and Gallant, J., Mapping and complementation studies of the gene for release factor 1, *J. Bacteriol.*, 168, 1066, 1986.

138. **Kawakami, K., Inada, T., and Nakamura, Y.,** Conditionally lethal and recessive UGA-suppressor mutations in the *prfB* gene encoding peptide chain release factor 2 of *Escherichia coli, J. Bacteriol.,* 170, 5378, 1988.
139. **Craigen, W. J., Cook, R. G., Tate, W. P., and Caskey, C. T.,** Bacterial peptide chain release factors: conserved primary structure and possible frameshift regulation of release factor 2, *Proc. Natl. Acad. Sci. U.S.A.,* 82, 3616, 1985.
140. **Kawakami, K., Jönsson, Y., Björk, G. R., Ikeda, H., and Nakamura, Y.,** Chromosomal location and structure of the operon encoding peptide-chain-release factor 2 of *Escherichia coli, Proc. Natl. Acad. Sci. U.S.A.,* 85, 5620, 1988.
141. **Smith, D. and Yarus, M.,** Transfer RNA structure and coding specificity. I. Evidence that a D-arm mutation reduces tRNA dissociation from the ribosome, *J. Mol. Biol.,* 206, 489, 1989.
142. **Smith, D. and Yarus, M.,** Transfer RNA structure and coding specificity. II. A D-arm tertiary interaction that restricts coding range, *J. Mol. Biol.,* 206, 503, 1989.
143. **Sherman, F.,** Suppression in the yeast *Saccharomyces cerevisiae*, in *The Molecular Biology of the Yeast Saccharomyces: Metabolism and Gene Expression*, Strathern, J. N., Jones, E. W., and Broach, J. R., Eds., Cold Spring Harbor Laboratory, New York, 1982, 463.
144. **Eggertsson, G. and Söll, D.,** Transfer ribonucleic acid-mediated suppression of termination codons in *Escherichia coli, Microbiol. Rev.,* 52, 354, 1988.
145. **Hatfield, D. L., Smith, D. W. E., Lee, B. L., Worland, P. J., and Oroszlan, S.,** Structure and function of suppressor tRNAs in higher eukaryotes, *CRC Rev. Biochem. Mol. Biol.,* 25, 71, 1990.
146. **Petrullo, L. A., Gallagher, P. J., and Elseviers, D.,** The role of 2-methylthio-N6-isopentyladenosine in readthrough and suppression of nonsense codons in *Escherichia coli, Mol. Gen. Genet.,* 190, 289, 1983.
147. **Wilson, R. K. and Roe, B. A.,** Presence of the hypermodified nucleotide N^6-(Δ^2-isopentenyl)-2-methylthioadenosine prevents codon misreading by *Escherichia coli* phenylalanyl-transfer RNA, *Proc. Natl. Acad. Sci. U.S.A.,* 86, 409, 1989.
148. **Chang, Z., Inokuchi, H., and Ozeki, H.,** Novel UGA-suppressors in *Escherichia coli* K-12, *Jpn. J. Genet.,* 65, 71, 1990.
149. **Wu, E.-D., Inokuchi, H., and Ozeki, H.,** Identification of the mutations in the *prfB* gene of *Escherichia coli* K12, which confer UGA suppressor activity, *Jpn. J. Genet.,* 65, 115, 1990.
150. **Tuite, M. F.,** Protein synthesis, in *The Yeasts*, Vol. 3, 2nd ed., Rose, A. H. and Harrison, J. S., Eds., Academic Press, New York, 1989, chap. 5.
151. **Shen, Z. and Fox, T. D.,** Substitution of an invariant nucleotide at the base of the highly conserved '530-loop' of 15S rRNA causes suppression of yeast mitochondrial ochre mutations, *Nucleic Acids Res.,* 17, 4535, 1989.
152. **Vijgenboom, E., Vink, T., Kraal, B., and Bosch, L.,** Mutants of the elongation factor EF-Tu, a new class of nonsense suppressors, *EMBO J.,* 4, 1049, 1985.
153. **Martin, R., Weiner, M., and Gallant, J.,** Effects of release factor context at UAA codons in *Escherichia coli, J. Bacteriol.,* 170, 4714, 1988.
154. **Hughes, D.,** Mutant forms of *tufA* and *tufB* independently suppress nonsense mutations, *J. Mol. Biol.,* 197, 611, 1987.
155. **Tapio, S. and Isaksson, L. A.,** Antisuppression by mutations in elongation factor Tu, *Eur. J. Biochem.,* 188, 339, 1990.
156. **Wakem, L. P. and Sherman, F.,** Isolation and characterization of omnipotent suppressors in the yeast *Saccharomyces cerevisiae, Genetics,* 124, 515, 1990.
157. **Wilson, P. G. and Culbertson, M. R.,** SUF12 suppressor protein of yeast: a fusion protein related to the EF-1 family of elongation factors, *J. Mol. Biol.,* 199, 559, 1988.
158. **Himmelfarb, H. J., Maicas, E., and Friesen, J. D.,** Isolation of the *SUP45* omnipotent suppressor gene of *Saccharomyces cerevisiae* and characterization of its gene product, *Mol. Cell. Biol.,* 5, 816, 1985.

159. Kushnirov, V. V., Ter-Avanesyan, M. D., Telckov, M. V., Surguchov, A. P., Smirnov, V. N., and Inge-Vechtomov, S. G., Nucleotide sequence of the *SUP2 (SUP35)* gene of *Saccharomyces cerevisiae*, *Gene*, 66, 45, 1988.
160. Breining, P. and Piepersberg, W., Yeast omnipotent suppressor *SUP1 (SUP45)*: nucleotide sequence of the wild type and a mutant gene, *Nucleic Acids Res.*, 14, 5187, 1986.
161. Gorini, L., Ribosomal discrimination of tRNAs, *Nature (London) New Biol.*, 234, 261, 1971.
162. Olsson, M. O. and Isaksson, L. A., Analysis of *rpsD* mutations in *Escherichia coli*. I. Comparison of mutants with various alterations in ribosomal protein S4, *Mol. Gen. Genet.*, 169, 251, 1979.
163. Piepersberg, W., Böck, A., and Wittman, H. G., Effect of different mutations in ribosomal protein S5 of *Escherichia coli* on translational fidelity, *Mol. Gen. Genet.*, 140, 91, 1975.
164. Yano, R. and Yura, T., Suppression of the *Escherichia coli rpoH* opal mutation by ribosomes lacking S15 protein, *J. Bacteriol.*, 171, 1712, 1989.
165. Rosenberger, R. F. and Foskett, G., An estimate of the frequency of in vivo transcriptional errors at a nonsense codon in *Escherichia coli*, *Mol. Gen. Genet.*, 183, 561, 1981.
166. Engelberg-Kulka, H., Dekel, L., and Israeli-Reches, M., Streptomycin resistant *Escherichia coli* mutant temperature sensitive for the production of Qβ-infective particles, *J. Virol.*, 21, 1, 1977.
167. Lovett, P. S., Ambulos, N. P., Jr., Mulbry, W., Noguchi, N., and Rogers, E. J., UGA can be decoded as tryptophan at low efficiency in *Bacillus subtilis*, *J. Bacteriol.*, 173, 1810, 1991.
168. Sambrook, J. F., Fan, D. P., and Brenner, S., A strong suppressor specific for UGA, *Nature (London)*, 214, 452, 1967.
169. Stringini, P. and Brickman, E., Analysis of specific misreading in *Escherichia coli*, *J. Mol. Biol.*, 75, 659, 1973.
170. Hirsh, D. and Gold, L., Translation of the UGA triplet *in vitro* by tryptophan transfer RNA's, *J. Mol. Biol.*, 58, 459, 1971.
171. Weiner, A. M. and Weber, K., A single UGA codon functions as a natural termination signal in the coliphage Qβ coat protein cistron, *J. Mol. Biol.*, 80, 837, 1973.
172. Roth, J. R., UGA nonsense mutations in *Salmonella typhimurium*, *J. Bacteriol.*, 102, 467, 1970.
173. Miller, J. H. and Albertini, A. M., Effects of surrounding sequence on the suppression of nonsense codons, *J. Mol. Biol.*, 164, 59, 1983.
174. Geller, A. I. and Rich, A., A UGA termination suppression tRNA^{Trp} active in rabbit reticulocytes, *Nature (London)*, 283, 41, 1980.
175. Hatfield, D., Thogeirsson, S. S., Copeland, T. D., Oroszlan, S., and Bustin, M., Immunopurification of the suppressor tRNA dependent rabbit β-globin readthrough protein, *Biochemistry*, 27, 1179, 1988.
176. Kim, D., Raymond, G. J., Clark, S. D., Vranka, J. A., and Johnson, J. D., Yeast tRNA^{Trp} genes with anticodons corresponding to UAA and UGA nonsense codons, *Nucleic Acids Res.*, 18, 4215, 1990.
177. Olsson, M. O. and Isaksson, L. A., Analysis of *rpsD* mutations in *Escherichia coli*. IV. Accumulation of minor forms of protein S7(K) in ribosomes of *rpsD* mutant strains due to translational read-through, *Mol. Gen. Genet.*, 177, 485, 1980.
178. Garen, A. and Siddiqi, O., Suppression of mutations in the alkaline phophatse structural cistron of *E. coli*, *Proc. Natl. Acad. Sci. U.S.A.*, 48, 1121, 1962.
179. Bossi, L., Context effects: translation of UAG codon by suppressor tRNA is affected by the sequence following UAG in the message, *J. Mol. Biol.*, 164, 73, 1983.
180. Manney, T. R., Evidence for chain termination by super-suppressible mutants in yeast, *Genetics*, 60, 719, 1968.

181. **Weiss, W. A., Edelman, I., Culbertson, M. R., and Friedberg, E. C.,** Physiological levels of normal tRNA$_{CAG}^{Gln}$ can effect partial suppression of amber mutations in the yeast *Saccharomyces cerevisiae, Proc. Natl. Acad. Sci. U.S.A.,* 84, 8031, 1987.
182. **Pure, G. A., Robinson, G. W., Naumovski, L., and Friedberg, E. C.,** Partial suppression of an ochre mutation in *Saccharomyces cerevisiae* by multicopy plasmids containing a normal yeast tRNA^{Gln} gene, *J. Mol. Biol.,* 183, 31, 1985.
183. **Lin, J. P., Aker, M., Sitney, K. C., and Mortimer, R. K.,** First position wobble in codon-anticodon pairing: amber suppression by a yeast glutamine tRNA, *Gene,* 49, 383, 1986.
184. **Capone, J. P., Sedivy, J. M., Sharp, P. A., and RajBhandary, U. L.,** Introduction of UAG, UAA, and UGA nonsense mutations at a specific site in the *Escherichia coli* chloramphenicol acetyltransferase gene: use in measurement of amber, ochre, and opal suppression in mammalian cells, *Mol. Cell. Biol.,* 6, 3059, 1986.
185. **Kohli, J. and Grosjean, H.,** Usage of the three termination codons: compilation and analysis of the known eukaryotic and prokaryotic translation termination sequences, *Mol. Gen. Genet.,* 182, 430, 1981.
186. **Sharp, P. M. and Bulmer, M.,** Selective differences among translation termination codons, *Gene,* 63, 141, 1988.
187. **Angenon, G., Van Montagu, M., and Depicker, A.,** Analysis of the stop codon context in plant nuclear genes, *FEBS Lett.,* 271, 144, 1990.
188. **Brown, C. M., Stockwell, P. A., Trotman, C. N. A., and Tate, W. P.,** The signal for the termination of protein synthesis in procaryotes, *Nucleic Acids Res.,* 18, 2079, 1990.
189. **Bienz, M., Kubli, E., Kohli, J., de Henau, S., Huez, G., Marbaix, G., and Grosjean, H.,** Usage of the three termination codons in a single eukaryotic cell, the *Xenopus laevis* oocyte, *Nucleic Acids Res.,* 9, 3835, 1981.
190. **Lu, P. and Rich, A.,** The nature of the polypeptide chain termination signal, *J. Mol. Biol.,* 58, 513, 1971.
191. **Weiner, A. M. and Weber, K.,** Natural read-through at the UGA termination signal of Qβ coat protein cistron, *Nature (London) New Biol.,* 234, 206, 1971.
192. **Hofstetter, H., Monstein, H.-J., and Weissman, C.,** The readthrough protein A$_1$ is essential for the formation of viable Qβ particles, *Biochim. Biophys. Acta,* 374, 238, 1974.
193. **Engelberg-Kulka, H., Dekel, L., Israeli-Reches, M., and Belfort, M.,** The requirement of nonsense suppression for the development of several phages, *Mol. Gen. Genet.,* 170, 155, 1979.
194. **Goelet, P., Lomonossoff, G. P., Butler, P. J. G., Akam, M. E., Gait, M. J., and Karn, J.,** Nucleotide sequence of tobacco mosaic virus RNA, *Proc. Natl. Acad. Sci. U.S.A.,* 79, 5818, 1982.
195. **Haseloff, J., Goelet, P., Zimmern, D., Ahlquist, P., Dasgupta, R., and Kaesberg, P.,** Striking similarities in amino acid sequence among nonstructural proteins encoded by RNA viruses that have dissimilar genomic organization, *Proc. Natl. Acad. Sci. U.S.A.,* 81, 4358, 1984.
196. **Kamer, G. and Argos, P.,** Primary structural comparison of RNA-dependent polymerases from plant, animal and bacterial viruses, *Nucleic Acids Res.,* 12, 7269, 1984.
197. **Pelham, H. R. B.,** Leaky UAG termination codon in tobacco mosaic virus RNA, *Nature (London),* 272, 469, 1978.
198. **Paterson, R. and Knight, C. A.,** Protein synthesis in tobacco protoplasts infected with tobacco mosaic virus, *Virology,* 64, 10, 1975.
199. **Ishikawa, M., Meshi, T., Motoyoshi, F., Takamatsu, N., and Okada, Y.,** *In vitro* mutagenesis of the putative replicase genes of tobacco mosaic virus, *Nucleic Acids Res.,* 14, 8291, 1986.
200. **Skuzeski, J. M., Nichols, L. M., and Gesteland, R. F.,** Analysis of leaky viral translation termination codons *in vivo* by transient expression of improved β-glucuronidase vectors, *Plant Mol. Biol.,* 15, 65, 1990.

201. **Skuzeski, J. M., Nichols, L. M., Gesteland, R. F., and Atkins, J. F.,** The signal for a leaky UAG stop codon in several plant viruses includes the two downstream codons, *J. Mol. Biol.,* 218, 365, 1991.
202. **Valle, R. P. C., Morch, M.-D., and Haenni, A.-L.,** Novel amber suppressor tRNAs of mammalian origin, *EMBO J.,* 6, 3049, 1987.
203. **Mayo, M. A., Robinson, D. J., Jolly, C. A., and Hyman, L.,** Nucleotide sequence of potato leafroll luteovirus RNA, *J. Gen. Virol.,* 70, 1037, 1989.
204. **van der Wilk, F., Huisman, M. J., Cornelissen, B. J. C., Huttinga, H., and Goldbach, R.,** Nucleotide sequence and organization of potato leafroll virus genomic RNA, *FEBS Lett.,* 245, 51, 1989.
205. **Guilley, H., Carrington, J. C., Balàzs, E., Jonard, G., Richards, K., and Morris, T. J.,** Nucleotide sequence and genome organization of carnation mottle virus RNA, *Nucleic Acids Res.,* 13, 6663, 1985.
206. **Carrington, J. C., Heaton, L. A., Zuidema, D., Hillman, B. I., and Morris, T. J.,** The genome structure of turnip crinkle virus, *Virology,* 170, 219, 1989.
207. **Pelham, H. R. B.,** Translation of tobacco rattle virus RNAs *in vitro*: four proteins from three RNAs, *Virology,* 97, 256, 1979.
208. **Hamilton, W. D. O., Boccara, M., Robinson, D. J., and Baulcombe, D. C.,** The complete nucleotide sequence of tobacco rattle virus RNA-1, *J. Gen. Virol.,* 68, 2563, 1987.
209. **Morch, M.-D., Boyer, J.-C., and Haenni, A.-L.,** Overlapping open reading frames revealed by complete nucleotide sequencing of turnip yellow mosaic virus genomic RNA, *Nucleic Acids Res.,* 16, 6157, 1988.
210. **Harbison, S.-A., Davies, J. W., and Wilson, T. M. A.,** Expression of high molecular weight polypeptides by carnation mottle virus RNA, *J. Gen. Virol.,* 66, 2597, 1985.
211. **Tacke, E., Prüfer, D., Salamini, F., and Rohde, W.,** Characterization of a potato leafroll luteovirus subgenomic RNA: differential expression by internal translation initiation and UAG suppression, *J. Gen. Virol.,* 71, 2265, 1990.
212. **Bahner, I., Lamb, J., Mayo, M. A., and Hay, R. T.,** Expression of the genome of potato leafroll virus: readthrough of the coat protein termination codon *in vivo, J. Gen. Virol.,* 71, 2251, 1990.
213. **Strauss, E. G., Rice, C. M., and Strauss, J. H.,** Sequence coding for the alphavirus nonstructural proteins is interrupted by an opal termination codon, *Proc. Natl. Acad. Sci. U.S.A.,* 80, 5271, 1983.
214. **Strauss, E. G., Levinson, R., Rice, C. M., Dalrymple, J., and Strauss, J. H.,** Nonstructural proteins nsP3 and nsP4 of Ross River and O'Nyong-nyong viruses: sequence and comparison with those of other alphaviruses, *Virology,* 164, 265, 1988.
215. **Takkinen, K.,** Complete nucleotide sequence of the nonstructural protein genes of Semliki Forest virus, *Nucleic Acids Res.,* 14, 5667, 1986.
216. **Lopez, S., Bell, J. R., Strauss, E. G., and Strauss, J. H.,** The nonstructural proteins of Sindbis virus as studied with an antibody specific for the C terminus of the nonstructural readthrough polyprotein, *Virology,* 141, 235, 1985.
217. **Li, G. and Rice, C. M.,** Mutagenesis of the in-frame opal termination codon preceding nsP4 of Sindbis virus: studies on translational readthrough and its effect on virus replication, *J. Virol.,* 63, 1326, 1989.
218. **Varmus, H.,** Retroviruses, *Science,* 240, 1427, 1988.
219. **Shinnick, T. M., Lerner, R. A., and Sutcliffe, J. G.,** Nucleotide sequence of Moloney murine leukaemia virus, *Nature (London),* 293, 543, 1981.
220. **Herr, W.,** Nucleotide sequence of AKV murine leukemia virus, *J. Virol.,* 49, 471, 1984.
221. **Tamura, T.-A.,** Provirus of M7 baboon endogenous virus: nucleotide sequence of the *gag-pol* region, *J. Virol.,* 47, 137, 1983.
222. **Philipson, L., Andersson, P., Olshevsky, U., Weinberg, R., Baltimore, D., and Gesteland, R.,** Translation of MuLV and MSV RNAs in nuclease-treated reticulocyte extracts: enhancement of the *gag-pol* polypeptide with yeast suppressor tRNA, *Cell,* 13, 189, 1978.

223. **Yoshinaka, Y., Katoh, I., Copeland, T. D., and Oroszlan, S.,** Murine leukemia virus protease is encoded by the *gag-pol* gene and is synthesized through suppression of an amber termination codon, *Proc. Natl. Acad. Sci. U.S.A.,* 82, 1618, 1985.
224. **Yoshinaka, Y., Katoh, I., Copeland, T. D., and Oroszlan, S.,** Translational readthrough of an amber termination codon during synthesis of feline leukemia virus protease, *J. Virol.,* 55, 870, 1985.
225. **Jamjoom, G. A., Naso, R. B., and Arlinghaus, R. B.,** Further characterization of intracellular precursor polyproteins of Rauscher leukemia virus, *Virology,* 78, 11, 1977.
226. **Felsenstein, K. M. and Goff, S. P.,** Expression of the *gag-pol* fusion protein of Moloney murine leukemia virus without *gag* protein does not induce virion formation or proteolytic processing, *J. Virol.,* 62, 2179, 1988.
227. **Jones, D. S., Nemoto, F., Kuchino, Y., Masuda, M., Yoshikura, H., and Nishimura, S.,** The effect of specific mutations at and around the *gag-pol* gene junction of Moloney murine leukemia virus, *Nucleic Acids Res.,* 17, 5933, 1989.
228. **Panganiban, A. T.,** Retroviral *gag* gene amber codon suppression is caused by an intrinsic *cis*-acting component of the viral mRNA, *J. Virol.,* 62, 3574, 1988.
229. **Feng, Y.-X., Levin, J. G., Hatfield, D. L., Schaefer, T. S., Gorelick, R. J., and Rein, A.,** Suppression of UAA and UGA termination codons in mutant murine leukemia viruses, *J. Virol.,* 63, 2870, 1989.
230. **ten Dam, E. B., Pleij, C. W. A., and Bosch, L.,** RNA pseudoknots: translational frameshifting and readthrough on viral RNAs, *Virus Genes,* 4, 121, 1990.
231. **Manning, P. A., Timmis, K. N., and Stevenson, G.,** Colonization factor antigen II (CFA/II) of enterotoxigenic *Escherichia coli*: molecular cloning of the CS3 determinant, *Mol. Gen. Genet.,* 200, 322, 1985.
232. **Jalajakumari, M. B., Thomas, C. J., Halter, R., and Manning, P. A.,** Genes for biosynthesis and assembly of CS3 pili of CFA/II enterotoxigenic *Escherichia coli*: novel regulation of pilus production by bypassing an amber codon, *Mol. Microbiol.,* 3, 1685, 1989.
233. **Chakrabarti, S. and Gorini, L.,** Growth of bacteriophages MS2 and T7 on streptomycin-resistant mutants of E*scherichia coli, J. Bacteriol.,* 121, 670, 1975.
234. **Krüger, D. H. and Bickle, T. A.,** Abortive infection of *Escherichia coli* F$^+$ cells by bacteriophage T7 requires ribosomal misreading, *J. Mol. Biol.,* 194, 349, 1987.
235. **Engelberg-Kulka, H., Dekel, L., and Israeli-Reches, M.,** Regulation of the *Escherichia coli* tryptophan operon by readthrough of UGA termination codons, *Biochem. Biophys. Res. Comm.,* 98, 1008, 1981.
236. **Roesser, J. R. and Yanofsky, C.,** Ribosome release modulates basal level expression of the *trp* operon of *Escherichia coli, J. Biol. Chem.,* 263, 14251, 1988.
237. **Andersson, D. I., van Verseveld, H. W., Stouthamer, A. H., and Kurland, C. G.,** Suboptimal growth with hyper-accurate ribosomes, *Arch. Microbiol.,* 144, 96, 1986.
238. **Elliot, T.,** Cloning, genetic characterization, and nucleotide sequence of the *hemA-prfA* operon of *Salmonella typhimurium, J. Bacteriol.,* 171, 3948, 1989.
239. **Caron, F. and Meyer, E.,** Does *Paramecium primaurelia* use a different genetic code in its macronucleus?, *Nature (London),* 314, 185, 1985.
240. **Preer, J. R., Jr., Preer, L. B., Rudman, B. M., and Barnett, A. J.,** Deviation from the universal code shown by the gene for surface protein 51A in *Paramecium, Nature (London),* 314, 188, 1985.
241. **Horowitz, S. and Gorovsky, M. A.,** An unusual genetic code in the nuclear genes of *Tetrahymena, Proc. Natl. Acad. Sci. U.S.A.,* 82, 2452, 1985.
242. **Helftenbein, E.,** Nucleotide sequence of a macronuclear DNA molecule coding for α-tubulin from the ciliate *Stylonychia lemnae*. Special codon usage: TAA is not a translation termination codon, *Nucleic Acids Res.,* 13, 415, 1985.
243. **Hanyu, N., Kuchino, Y., and Nishimura, S.,** Dramatic events in ciliate evolution: alteration of UAA and UAG termination codons to glutamine codons due to anticodon mutations in two *Tetrahymena* tRNAs^{Gln}, *EMBO J.,* 5, 1307, 1986.

244. **Meyer, F., Schmidt, H. J., Plümper, E., Hasilik, A., Mersmann, G., Meyer, H. E., Engström, Å, and Heckmann, K.,** UGA is translated as cysteine in pheromone 3 of *Euplotes octocarinatus, Proc. Natl. Acad. Sci. U.S.A.,* 88, 3758, 1991.
245. **Yamao, F., Muto, A., Kawauchi, Y., Iwami, M., Iwagami, S., Azumi, Y., and Osawa, S.,** UGA is read as tryptophan in *Mycoplasma capricolum, Proc. Natl. Acad. Sci. U.S.A.,* 82, 2306, 1985.
246. **Dudler, R., Schmidhauser, C., Parish, R. W., Wettenhall, R. E. H., and Schmidt, T.,** A mycoplasma high-affinity transport system and the *in vitro* invasiveness of mouse sarcoma cells, *EMBO J.,* 7, 3963, 1988.
247. **Inamine, J. M., Denny, T. P., Loechel, S., Schaper, U., Huang, C.-H., Bott, K. F., and Hu, P.-C.,** Nucleotide sequence of the P1 attachment-protein gene of *Mycoplasma pneumoniae, Gene,* 64, 217, 1988.
248. **Andachi, Y., Yamao, F., Muto, A., and Osawa, S.,** Codon recognition patterns as deduced from sequences of the complete set of transfer RNA species in *Mycoplasma capricolum*: resemblance to mitochondria, *J. Mol. Biol.,* 209, 37, 1989.
249. **Blanchard, A.,** *Ureaplasma urealyticum* urease genes; use of a UGA tryptophan codon, *Mol. Microbiol.,* 4, 669, 1990.
250. **Bachmann, B. J.,** Derivations and genotypes of some mutant derivatives of *Escherichia coli* K-12, in *Escherichia coli and Salmonella typhimurium. Cellular and Molecular Biology,* Vol. 2, Neidhardt, F. C., Ingraham, J. L., Low, K. B., Magasanik, B., Schaechter, M., and Umbarger, H. E., Eds., American Society for Microbiology, Washington, D.C., 1987, chap. 72.
251. **Marshall, B. and Levy, S. B.,** Prevalence of amber suppressor-containing coliforms in the natural environment, *Nature (London),* 286, 524, 1980.
252. **Robeson, J. P., Goldschmidt, R. M., and Curtiss, R., III,** Potential of *Escherichia coli* isolated from nature to propogate cloning vectors, *Nature (London),* 283, 104, 1980.
253. **Lehman, N. and Jukes, T. H.,** Genetic code development by stop codon takeover, *J. Theor. Biol.,* 135, 203, 1988.
254. **Kuchino, Y., Beier, H., Akita, N., and Nishimura, S.,** Natural UAG suppressor glutamine tRNA is elevated in mouse cells infected with Moloney murine leukemia virus, *Proc. Natl. Acad. Sci. U.S.A.,* 84, 2668, 1987.
255. **Feng, Y.-X., Hatfield, D. L., Rein, A., and Levin, J. G.,** Translational readthrough of the murine leukemia virus *gag* gene amber codon does not require virus-induced alteration of tRNA, *J. Virol.,* 63, 2405, 1989.
256. **Bienz, M. and Kubli, E.,** Wild-type tRNA$_G^{Tyr}$ reads the TMV RNA stop codon, but Q base-modified tRNA$_Q^{Tyr}$ does not, *Nature (London),* 294, 188, 1981.
257. **Beier, H., Barciszewska, M., Krupp, G., Mitnacht, R., and Gross, H. J.,** UAG readthrough during TMV RNA translation: isolation and sequence of two tRNAs^{Tyr} with suppressor activity from tobacco plants, *EMBO J.,* 3, 351, 1984.
258. **Beier, H., Barciszewska, M., and Sickinger, H.-D.,** The molecular basis for the differential translation of TMV RNA in tobacco protoplasts and wheat germ extracts, *EMBO J.,* 3, 1091, 1984.
259. **Böck, A., Forchhammer, K., Heider, J., Leinfelder, W., Sawers, G., Veprek, B., and Zinoni, F.,** Selenocysteine: the 21st amino acid, *Mol. Microbiol.,* 5, 515, 1991.
260. **Leinfelder, W., Zehelein, E., Mandrand-Berthelot, M.-A., and Böck, A.,** Gene for a novel tRNA species that accepts L-serine and cotranslationally inserts selenocysteine, *Nature (London),* 331, 723, 1988.
261. **Schön, A., Böck, A., Ott, G., Sprinzl, M., and Söll, D.,** The selenocysteine-inserting opal suppressor serine tRNA from *E. coli* is highly unusual in structure and modification, *Nucleic Acids Res.,* 17, 7159, 1989.
262. **Leinfelder, W., Forchhammer, K., Veprek, B., Zehelein, E., and Böck, A.,** *In vitro* synthesis of selenocysteinyl-tRNA$_{UCA}$ from seryl-tRNA$_{UCA}$: involvement and characterization of the *selD* gene product, *Proc. Natl. Acad. Sci. U.S.A.,* 87, 543, 1990.

263. **Forchhammer, K. and Böck, A.,** Selenocysteine synthase from *Escherichia coli*: analysis of the reaction sequence, *J. Biol. Chem.*, 266, 6324, 1991.
264. **Forchhammer, K., Leinfelder, W., and Böck, A.,** Identification of a novel translation factor necessary for the incorporation of selenocysteine into protein, *Nature (London)*, 342, 453, 1989.
265. **Cox, J. C., Edwards, E. S., and DeMoss, J. A.,** Resolution of distinct selenium-containing formate dehydogenases from *Escherichia coli, J. Bacteriol.*, 145, 1317, 1981.
266. **Pecher, A., Zinoni, F., and Böck, A.,** The seleno-polypeptide of formic dehydrogenase (formate hydrogen-lyase linked) from *Escherichia coli*: genetic analysis, *Arch. Microbiol.*, 141, 359, 1985.
267. **Osawa, S., Ohama, T., Jukes, T. H., and Watanabe, K.,** Evolution of the mitochondrial genetic code. I. Origin of AGR serine and stop codons in metazoan mitochondria, *J. Mol. Evol.*, 29, 202, 1989.
268. **Atherly, A. G. and Menninger, J. R.,** Mutant *E. coli* strain with temperature sensitive peptidyl-transfer RNA hydrolase, *Nature (London) New Biol.*, 240, 245, 1972.
269. **Menninger, J. R.,** The accumulation as peptidyl-transfer RNA of isoaccepting transfer RNA families in *Escherichia coli* with temperature-sensitive peptidyl-transfer RNA hydrolase, *J. Biol. Chem.*, 253, 6808, 1978.
270. **Menninger, J. R., Walker, C., Tan, P. F., and Atherly, A. G.,** Studies on the metabolic role of peptidyl-tRNA hydrolase. I. Properties of a mutant *E. coli* with a temperature-sensitive peptidyl-tRNA hydrolase, *Mol. Gen. Genet.*, 121, 307, 1973.
271. **Menninger, J. R.,** Computer simulation of ribosome editing, *J. Mol. Biol.*, 171, 383, 1983.
272. **Chen, G.-F. T. and Inouye, M.,** Suppression of the negative effect of minor arginine codons on gene expression; preferential usage of minor codons within the first 25 codons of the *Escherichia coli* genes, *Nucleic Acids Res.*, 18, 1465, 1990.
273. **Varenne, S., Buc, J., Lloubes, R., and Lazdunski, C.,** Translation is a non-uniform process: effect of tRNA availability on the rate of elongation of nascent polypeptide chains, *J. Mol. Biol.*, 180, 549, 1984.
274. **Wolin, S. L. and Walter, P.,** Ribosome pausing and stacking during translation of a eukaryotic mRNA, *EMBO J.*, 7, 3559, 1988.
275. **Tsung, K., Inouye, S., and Inouye, M.,** Factors affecting the efficiency of protein synthesis in *Escherichia coli*: production of a polypeptide of more than 6000 amino acid residues, *J. Biol. Chem.*, 264, 4428, 1989.
276. **Misra, R. and Reeves, P.,** Intermediates in the synthesis of TolC protein include an incomplete peptide stalled at a rare Arg codon, *Eur. J. Biochem.*, 152, 151, 1985.
277. **Bonekamp, F. and Jensen, K. F.,** The AGG codon is translated slowly in *E. coli* even at very low expression levels, *Nucleic Acids Res.*, 16, 3013, 1988.
278. **Lindhout, P., Neeleman, L., Van Tol, H., and Van Vloten-Doting, L.,** Ribosomes are stalled during *in vitro* translation of alfalfa mosaic virus RNA 1, *Eur. J. Biochem.*, 152, 625, 1985.
279. **Rojiani, M. V., Jakubowski, H., and Goldman, E.,** Relationship between protein synthesis and concentrations of charged and uncharged tRNA^{Trp} in *Escherichia coli, Proc. Natl. Acad. Sci. U.S.A.*, 87, 1511, 1990.
280. **Jørgensen, F. and Kurland, C. G.,** Processivity errors of gene expression in *Escherichia coli, J. Mol. Biol.*, 215, 511, 1990.
281. **Murgola, E. J.,** Suppression and the code: beyond codons and anticodons, *Experientia*, 46, 1134, 1990.
282. **Culbertson, M. R., Leeds, P., Sandbaken, M. G., and Wilson, P. G.,** Frameshift suppression, in *The Ribosome: Structure, Function & Evolution*, Hill, W. E., Dahlberg, A., Garrett, R. A., Moore, P. B., Schlessinger, D., and Warner, J. R., Eds., American Society for Microbiology, Washington, D.C., 1990, chap. 49.
283. **Curran, J. F. and Yarus, M.,** Reading frame selection and transfer RNA anticodon loop stacking, *Science*, 238, 1545, 1987.

284. **Björk, G. R., Wikström, P. M., and Byström, A. S.**, Prevention of translational frameshifting by the modified nucleoside 1-methylguanosine, *Science*, 244, 986, 1989.
285. **Hagervall, T. G., Ericson, J. U., Esberg, K. B., Ji-nong, L., and Björk, G. R.**, Role of tRNA modification in translational fidelity, *Biochim. Biophys. Acta*, 1050, 263, 1990.
286. **Dahlfors, A. A. R. and Kurland, C. G.**, Novel mutants of elongation factor G, *J. Mol. Biol.*, 215, 549, 1990.
287. **Hughes, D., Atkins, J. F., and Thompson, S.**, Mutants of elongation factor Tu promote ribosomal frameshifting and nonsense readthrough, *EMBO J.*, 6, 4235, 1987.
288. **Vijgenboom, E. and Bosch, L.**, Translational frameshifts induced by mutant species of the polypeptide chain elongation factor Tu of *Escherichia coli*, *J. Biol. Chem.*, 264, 13012, 1989.
289. **Atkins, J. F. and Ryce, S.**, UGA and non-triplet suppressor reading of the genetic code, *Nature (London)*, 249, 527, 1974.
290. **Weiss-Brummer, B., Sakai, H., and Kaudewitz, F.**, A mitochondrial frameshift-suppressor (+) of the yeast *S. cerevisiae* maps in the mitochondrial 15S rRNA locus, *Curr. Genet.*, 11, 295, 1987.
291. **Weiss-Brummer, B., Sakai, H., and Magerl-Brenner, M.**, At least two nuclear-encoded factors are involved together with a mitochondrial factor (MF1) in spontaneous mitochondrial frameshift-suppression of the yeast *S. cerevisiae*, *Curr. Genet.*, 12, 387, 1987.
292. **van Buul, C. P. J. J., Visser, W., and van Knippenberg, P. H.**, Increased translational fidelity caused by the antibiotic kasugamycin and ribosomal ambiguity in mutants harboring the *ksgA* gene, *FEBS Lett.*, 177, 119, 1984.
293. **Atkins, J. F., Elseviers, D., and Gorini, L.**, Low activity of β-galactosidase in frameshift mutants of *Escherichia coli*, *Proc. Natl. Acad. Sci. U.S.A.*, 69, 1192, 1972.
294. **Tucker, S. D., Murgola, E. J., and Pagel, F. T.**, Missense and nonsense suppressors can correct frameshift mutations, *Biochimie*, 71, 729, 1989.
295. **Kurland, C. G.**, Reading frame errors on ribosomes, in *Nonsense Mutations and tRNA Suppressors*, Celis, J. E. and Smith, J. D., Eds., Academic Press, New York, 1979, 97.
296. **Kurland, C. G. and Gallant, J. A.**, The secret life of the ribosome, in *Accuracy in Molecular Processes*, Kirkwood, T. B. L., Rosenberger, R. F., and Galas, D. J., Eds., Chapman and Hall, New York, 1986, 127.
297. **Cattaneo, R.**, How 'hidden' reading frames are expressed, *Trends Biochem. Sci.*, 14, 165, 1989.
298. **Weiss, R., Dunn, D., Atkins, J., and Gesteland, R.**, The ribosome's rubbish, in *The Ribosome: Structure, Function & Evolution*, Hill, W. E., Dahlberg, A., Garrett, R. A., Moore, P. B., Schlessinger, D., and Warner, J. R., Eds., American Society for Microbiology, Washington, D.C., 1990, chap. 46.
299. **Atkins, J. F., Weiss, R. B., and Gesteland, R. F.**, Ribosome gymnastics – degree of difficulty 9.5, style 10.0, *Cell*, 62, 413, 1990.
300. **Weiss, R. B., Dunn, D. M., Atkins, J. F., and Gesteland, R. F.**, Ribosomal frameshifting from –2 to +50 nucleotides, *Prog. Nucleic Acid Res. Mol. Biol.*, 39, 159, 1990.
301. **Hatfield, D. and Oroszlan, S.**, The where, what and how of ribosomal frameshifting in retroviral protein synthesis, *Trends Biochem. Sci.*, 15, 186, 1990.
302. **Gallant, J. and Foley, D.**, On the causes and prevention of mistranslation, in *Ribosomes: Structure, Function, and Genetics*, Chambliss, G., Craven, G. R., Davies, J., Davis, K., Kahan, L., and Nomura, M., Eds., University Park Press, Baltimore, 1980, 615.
303. **Weiss, R. and Gallant, J.**, Mechanism of ribosome frameshifting during translation of the genetic code, *Nature (London)*, 302, 389, 1983.
304. **Weiss, R. B. and Gallant, J. A.**, Frameshift suppression in aminoacyl-tRNA limited cells, *Genetics*, 112, 727, 1986.
305. **Weiss, R., Lindsley, D., Falahee, B., and Gallant, J.**, On the mechanism of ribosomal frameshifting at hungry codons, *J. Mol. Biol.*, 203, 403, 1988.
306. **Weiss, R. B.**, Molecular model of ribosome frameshifting, *Proc. Natl. Acad. Sci. U.S.A.*, 81, 5797, 1984.

307. **Spanjaard, R. A. and van Duin, J.,** Translation of the sequence AGG-AGG yields 50% ribosomal frameshift, *Proc. Natl. Acad. Sci. U.S.A.,* 85, 7967, 1988.
308. **Atkins, J. F., Gesteland, R. F., Reid, B. R., and Anderson, C. W.,** Normal tRNAs promote ribosomal frameshifting, *Cell,* 18, 1119, 1979.
309. **Dayhuff, T.J., Atkins, J. F., and Gesteland, R. F.,** Characterization of ribosomal frames hift events by protein sequence analysis, *J. Biol. Chem.,* 261, 7491, 1986.
310. **Bruce, A. G., Atkins, J. F., and Gesteland, R. F.,** tRNA anticodon replacement experiments show that ribosomal frameshifting can be caused by doublet decoding, *Proc. Natl. Acad. Sci. U.S.A.,* 83, 5062, 1986.
311. **O'Mahony, D. J., Hughes, D., Thompson, S., and Atkins, J. F.,** Suppression of a -1 frameshift mutation by a recessive tRNA suppressor which causes doublet decoding, *J. Bacteriol.,* 171, 3824, 1989.
312. **Atkins, J. F., Nichols, B. P., and Thompson, S.,** The nucleotide sequence of the first externally suppressible -1 frameshift mutant, and of some nearby leaky frameshift mutants, *EMBO J.,* 2, 1345, 1983.
313. **Dunn, J. J. and Studier, F. W.,** Complete nucleotide sequence of bacteriophage T7 DNA and the locations of T7 genetic elements, *J. Mol. Biol.,* 166, 477, 1983.
314. **O'Mahony, D. J., Mims, B. H., Thompson, S., Murgola, E. J., and Atkins, J. F.,** Glycine tRNA mutants with normal anticodon loop size cause -1 frameshifting, *Proc. Natl. Acad. Sci. U.S.A.,* 86, 7979, 1989.
315. **Culbertson, M. R., Charnas, L., Johnson, M. T., and Fink, G. R.,** Frameshifts and frameshift suppressors in *Saccharomyces cerevisiae, Genetics,* 86, 745, 1977.
316. **Aulin, M. R. and Hughes, D.,** Overproduction of release factor reduces spontaneous frameshifting and frameshift suppression by mutant elongation factor Tu, *J. Bacteriol.,* 172, 6721, 1990.
317. **O'Connor, M., Gesteland, R. F., and Atkins, J. F.,** tRNA hopping: enhancement by an expanded anticodon, *EMBO J.,* 8, 4315, 1989.
318. **Falahee, M. B., Weiss, R. B., O'Connor, M., Doonan, S., Gesteland, R. F., and Atkins, J. F.,** Mutants of translational components that alter reading frame by two steps forward or one step back, *J. Biol. Chem.,* 263, 18099, 1988.
319. **Kalnins, A., Otto, K., Rüther, U., and Müller-Hill, B.,** Sequence of the *lacZ* gene of *Escherichia coli, EMBO J.,* 2, 593, 1983.
320. **Gallant, J., Erlich, H., Weiss, R., Palmer, L., and Nyari, L.,** Nonsense suppression in aminoacyl-tRNA limited cells, *Mol. Gen. Genet.,* 186, 221, 1982.
321. **Gribskov, M., Devereux, J., and Burgess, R. R.,** The codon preference plot: graphic analysis of protein coding sequences and prediction of gene expression, *Nucleic Acids Res.,* 12, 539, 1984.
322. **Trifonov, E. N.,** Translation framing code and frame-monitoring mechanism as suggested by the analysis of mRNA and 16S rRNA nucleotide sequences, *J. Mol. Biol.,* 194, 643, 1987.
323. **Curran, J. F. and Yarus, M.,** Base substitutions in the tRNA anticodon arm do not degrade the accuracy of reading frame maintenance, *Proc. Natl. Acad. Sci. U.S.A.,* 83, 6538, 1986.
324. **Sifers, R. N., Brashears-Macatee, S., Kidd, V. J., Muensch, H., and Woo, S. L. C.,** A frameshift mutation results in a truncated α_1-antitrypsin that is retained within the rough endoplasmic reticulum, *J. Biol. Chem.,* 263, 7330, 1988.
325. **Nogueira, C. P., McGuire, M. C., Graeser, C., Bartels, C. F., Arpagaus, M., Ven der Spek, A. F. L., Lightstone, H., Lockridge, O., and La Du, B. N.,** Identification of a frameshift mutation responsible for the silent phenotype of human serum cholinesterase, Gly 117 (GG$\underline{T}$→GG$\underline{AG}$), *Am. J. Hum. Genet.,* 46, 934, 1990.
326. **Kastelein, R. A., Remaut, E., Fiers, W., and van Duin, J.,** Lysis gene expression of RNA phage MS2 depends on a frameshift during translation of the overlapping coat protein gene, *Nature (London),* 295, 35, 1982.

Variations in Reading the Genetic Code 261

327. **Buckley, K. J. and Hayashi, M.,** Role of premature translational termination in the regulation of expression of the φX174 lysis gene, *J. Mol. Biol.,* 198, 599, 1987.
328. **Berkhout, B., Schmidt, B. F., van Strein, A., van Boom, J., van Westrenen, J., and van Duin, J.,** Lysis gene of bacteriophage MS2 is actually by translation termination at the overlapping coat protein, *J. Mol. Biol.,* 195, 517, 1987.
329. **Bläsi, U., Nam, K., Lubitz, W., and Young, R.,** Translational efficiency of φX174 lysis gene E is unaffected by upstream translation of the overlapping gene D reading frame, *J. Bacteriol.,* 172, 5617, 1990.
330. **Levin, M., Hendrix, R., and Casjeans, S.,** personal communication, 1991.
331. **Hizi, A., Henderson, L. E., Copeland, T. D., Sowder, R. C., Hixson, C. V., and Oroszlan, S.,** Characterization of mouse mammary tumor virus *gag-pro* gene products and the ribosomal frameshift site by protein sequencing, *Proc. Natl. Acad. Sci. U.S.A.,* 84, 7041, 1987.
332. **Jacks, T., Townsley, K., Varmus, H. E., and Majors, J.,** Two efficient ribosomal frameshifting events are required for synthesis of mouse mammary tumor virus *gag*-related polyproteins, *Proc. Natl. Acad. Sci. U.S.A.,* 84, 4298, 1987.
333. **Jacks, T., Madhani, H. D., Masiarz, F. R., and Varmus, H. E.,** Signals for ribosomal frameshifting in the Rous sarcoma virus *gag-pol* region, *Cell,* 55, 447, 1988.
334. **Hiramatsu, K., Nishida, J., Naito, A., and Yoshikura, H.,** Molecular cloning of the closed circular provirus of human T cell leukaemia virus type I: a new open reading frame in the *gag-pol* region, *J. Gen. Virol.,* 68, 213, 1987.
335. **Moore, R., Dixon, M., Smith, R., Peters, G., and Dickson, C.,** Complete nucleotide sequence of a milk-transmitted mouse mammary tumor virus: two frameshift suppression events are required for translation of *gag* and *pol, J. Virol.,* 61, 480, 1987.
336. **Nam, S. H., Kidokoro, M., Shida, H., and Hatanaka, M.,** Processing of *gag* precursor polyprotein of human T-cell leukemia virus type I by virus-encoded protease, *J. Virol.,* 62, 3718, 1988.
337. **Jacks, T. and Varmus, H. E.,** Expression of the Rous sarcoma virus *pol* gene by ribosomal frameshifting, *Science,* 230, 1237, 1985.
338. **Jacks, T., Power, M. D., Masiarz, F. R., Luciw, P. A., Barr, P. J., and Varmus, H. E.,** Characterization of ribosomal frameshifting in HIV-1 *gag-pol* expression, *Nature (London),* 331, 280, 1988.
339. **Wilson, W., Braddock, M., Adams, S. E., Rathjen, P. D., Kingsman, S. M., and Kingsman, A. J.,** HIV expression strategies: ribosomal frameshifting is directed by a short sequence in both mammalian and yeast systems, *Cell,* 55, 1159, 1988.
340. **Oppermann, H., Bishop, J. M., Varmus, H. E., and Levintow, L.,** A joint product of the genes *gag* and *pol* of avian sarcoma virus: a possible precursor of reverse transcriptase, *Cell,* 12, 993, 1977.
341. **Anderson, S. J., Naso, R. B., Davis, J., and Bowen, J. M.,** Polyprotein precursors to mouse mammary tumor virus proteins, *J. Virol.,* 32, 507, 1979.
342. **Dickson, C. and Atterwill, M.,** Composition, arrangement and cleavage of the mouse mammary tumor virus polyprotein precursor Pr77^{gag} and p110^{gag}, *Cell,* 17, 1003, 1979.
343. **Brierley, I., Boursnell, M. E. G., Binns, M. M., Bilimoria, B., Blok, V. C., Brown, T. D. K., and Inglis, S. C.,** An efficient ribosomal frameshifting signal in the polymerase-encoding region of the coronavirus IBV, *EMBO J.,* 6, 3779, 1987.
344. **Mietz, J. A., Grossman, Z., Lueders, K. K., and Kuff, E. L.,** Nucleotide sequence of a complete mouse intracisternal A-particle genome: relationship to known aspects of particle assembly and function, *J. Virol.,* 61, 3020, 1987.
345. **Kuff, E. L. and Fewell, J. W.,** Intracisternal A-particle gene expression in normal mouse thymus tissue: gene products and strain-related variability, *Mol. Cell. Biol.,* 5, 474, 1985.
346. **Weiss, R. B., Dunn, D. M., Shuh, M., Atkins, J. F., and Gesteland, R. F.,** *E. coli* ribosomes re-phase on retroviral frameshift signals ranging from 2 to 50 percent, *New Biologist,* 1, 159, 1989.

347. **Hatfield, D., Feng, Y.-X., Lee, B. J., Rein, A., Levin, J. G., and Oroszlan, S.,** Chromatographic analysis of the aminoacyl-tRNAs which are required for translation of codons at and around the ribosomal frameshift sites of HIV, HTLV-1 and BLV, *Virology,* 173, 736, 1989.
348. **Moazed, D. and Noller, H. F.,** Intermediate states in the movement of transfer RNA in the ribosome, *Nature (London),* 342, 142, 1989.
349. **Rheinberger, H.-J., Sternbach, H., and Nierhaus, K. H.,** Codon-anticodon interaction at the ribosomal E site, *J. Biol. Chem.,* 261, 9140, 1986.
350. **Rice, N. R., Stephens, R. M., Burny, A., and Gilden, R. V.,** The *gag* and *pol* genes of bovine leukemia virus: nucleotide sequence and analysis, *Virology,* 142, 357, 1985.
351. **Le, S.-Y., Chen, J.-H., and Maizel, J. V.,** Thermodynamic stability and statistical significance of potential stem-loop structures situated at the frameshift sites of retroviruses, *Nucleic Acids Res.,* 17, 6143, 1989.
352. **Bredenbeek, P. J., Pachuk, C. J., Noten, A. F. H., Charité, J., Luytjes, W., Weiss, S. R., and Spaan, W. J. M.,** The primary structure and expression of the second open reading frame of the polymerase gene of the coronavirus MHV-A59; a highly conserved polymerase is expressed by an efficient ribosomal frameshifting mechanism, *Nucleic Acids Res.,* 18, 1825, 1990.
353. **Brierley, I., Digard, P., and Inglis, S. C.,** Characterization of an efficient coronavirus ribosomal frameshifting signal: requirement for an RNA pseudoknot, *Cell,* 57, 537, 1989.
354. **Pleij, C. W. A.,** Pseudoknots: a new motif in the RNA game, *Trends Biochem. Sci.,* 15, 143, 1990.
355. **Madhani, H. D., Jacks, T., and Varmus, H. E.,** Signals for expression of HIV *pol* gene by ribosomal frameshifting, in *The Control of HIV Gene Expression,* Franza, R., Cullen, B., and Wong-Staal, F., Eds., Cold Spring Harbor Laboratory, New York, 1988, 119.
356. **Icho, T. and Wickner, R. B.,** The double-stranded RNA genome of yeast virus L-A encodes its own putative RNA polymerase by fusing two open reading frames, *J. Biol. Chem.,* 264, 6716, 1989.
357. **Dinman, J. D., Icho, T., and Wickner, R. B.,** A −1 ribosomal frameshift in a double-stranded RNA virus of yeast forms a *gag-pol* fusion protein, *Proc. Natl. Acad. Sci. U.S.A.,* 88, 174, 1991.
358. **Clare, J. and Farabaugh, P.,** Nucleotide sequence of a yeast Ty element: evidence for an unusual mechanism of gene expresssion, *Proc. Natl. Acad. Sci. U.S.A.,* 82, 2829, 1985.
359. **Wilson, W., Malim, M. H., Mellor, J., Kingsman, A. J., and Kingsman, S. M.,** Expression strategies of the yeast retrotransposon Ty: a short sequence directs ribosomal frameshifting, *Nucleic Acids Res.,* 14, 7001, 1986.
360. **Clare, J. J., Belcourt, M., and Farabaugh, P. J.,** Efficient translational frameshifting occurs within a conserved sequence of the overlap between the two genes of a yeast Ty1 transposon, *Proc. Natl. Acad. Sci. U.S.A.,* 85, 6816, 1988.
361. **Hansen, L. J., Chalker, D. L., and Sandmeyer, S. B.,** Ty3, a yeast retrotransposon associated with tRNA genes, has homology to animal retroviruses, *Mol. Cell. Biol.,* 8, 5245, 1988.
362. **Belcourt, M. F. and Farabaugh, P. J.,** Ribosomal frameshifting in the yeast retrot ransposon Ty: tRNAs induce slippage on a 7 nucleotide minimal site, *Cell,* 62, 339, 1990.
363. **Xu, H. and Boeke, J. D.,** Host genes that influence transposition in yeast: the abundance of a rare tRNA regulates Ty1 transposition frequency, *Proc. Natl. Acad. Sci. U.S.A.,* 87, 8360, 1990.
364. **Weissenbach, J., Dirheimer, G., Falcoff, R., Sanceau, J., and Falcoff, E.,** Yeast tRNA^{Leu} (anticodon U-A-G) translates all six leucine codons in extracts from interferon treated cells, *FEBS Lett.,* 82, 71, 1977.
365. **Sekine, Y. and Ohtsubo, E.,** Frameshifting is required for production of the transposase encoded by insertion sequence 1, *Proc. Natl. Acad. Sci. U.S.A.,* 86, 4609, 1989.

366. Fayet, O., Ramond, P., Polard, P., Prère, M. F., and Chandler, M., Functional similarities between retroviruses and the IS3 family of bacterial insertion sequences? *Mol. Microbiol.,* 4, 1771, 1990.
367. Prère, M.-F., Chandler, M., and Fayet, O., Transposition in *Shigella dysenteriae*: isolation and analysis of IS911, a new member of the IS3 group of insertion sequences, *J. Bacteriol.,* 172, 4090, 1990.
368. Sagio, K., Kugimiya, W., Matsuo, Y., Inouye, S., Yoshioka, K., and Yuki, S., Identification of the coding sequence for a reverse transcriptase-like enzyme in a transposable genetic element in *Drosophila melanogaster, Nature (London),* 312, 659, 1984.
369. Mount, S. M. and Rubin, G. M., Complete nucleotide sequence of the *Drosophila* transposable element copia: homology between copia and retroviral proteins, *Mol. Cell. Biol.,* 5, 1630, 1985.
370. Voytas, D. F. and Ausubel, F. M., A copia-like transposable element family in *Arabidopsis thaliana, Nature (London),* 336, 242, 1988.
371. Levin, H. L., Weaver, D.C., and Boeke, J. D., Two related families of retrotransposons from *Schizosaccharomyces pombe, Mol. Cell. Biol.,* 10, 6791, 1990.
372. Miller, W. A., Waterhouse, P. M., and Gerlach, W. L., Sequence and organization of barley yellow dwarf virus genomic RNA, *Nucleic Acids Res.,* 16, 6097, 1988.
373. Veidt, I., Lot, H., Leiser, M., Scheidecker, D., Guilley, H., Richards, K., and Jonard, G., Nucleotide sequence of beet western yellows virus RNA, *Nucleic Acids Res.,* 16, 9917, 1988.
374. Xiong, Z. and Lommel, S. A., The complete nucleotide sequence and genome organization of red clover necrotic mosaic virus RNA-1, V*irology,* 171, 543, 1989.
375. Schlicht, H.-J., Radziwill, G., and Schaller, H., Synthesis and encapsidation of duck hepatitis B virus reverse transcriptase do not require formation of core-polymerase fusion proteins, *Cell,* 56, 85, 1989.
376. Schultze, M., Hohn, T., and Jiricny, J., The reverse transcriptase gene of cauliflower mosaic virus is translated separately from the capsid gene, *EMBO J.,* 9, 1177, 1990.
377. Craigen, W. J. and Caskey, C. T., Expression of peptide chain release factor 2 requires high-efficiency frameshift, *Nature (London),* 322, 273, 1986.
378. Donly, B. C., Edgar, C. D., Adamski, F. M., and Tate, W. P., Frameshift autoregulation in the gene for *Escherichia coli* release factor 2: partly functional mutants result in frameshift enhancement, *Nucleic Acids Res.,* 18, 6517, 1990.
379. Curran, J. F. and Yarus, M., Use of tRNA suppressors to probe regulation of E*scherichia coli* release factor 2, *J. Mol. Biol.,* 203, 75, 1988.
380. Weiss, R. B., Dunn, D. M., Dahlberg, A. E., Atkins, J. F., and Gesteland, R. F., Reading frame switch caused by base-pair formation between the 3' end of 16S rRNA and the mRNA during elongation of protein synthesis in E*scherichia coli, EMBO J.,* 7, 1503, 1988.
381. Williams, J. M., Donly, B. C., Brown, C. M., Adamski, F. M., Trotman, C. N. A., and Tate, W. P., Frameshifting in the synthesis of *Escherichia coli* polypeptide chain release factor two on eukaryotic ribosomes, *Eur. J. Biochem.,* 186, 515, 1989.
382. Donly, C., Williams, J., Richardson, C., and Tate, W., Frameshifting at the internal stop codon within the mRNA for bacterial release factor-2 on eukaryotic ribosomes, *Biochim. Biophys. Acta,* 1050, 283, 1990.
383. Huang, W. M., Ao, S.-Z., Casjens, S., Orlandi, R., Zeikus, R., Weiss, R., Winge, D., and Fang, M., A persistent untranslated sequence within bacteriophage T4 DNA topisomerase gene 60, *Science,* 239, 1005, 1988.
384. Weiss, R. B., Huang, W. M., and Dunn, D. M., A nascent peptide is required for ribosomal bypass of the coding gap in bacteriophage T4 gene *60, Cell,* 62, 117, 1990.
385. Blinkowa, A. L. and Walker, J. R., Programmed ribosomal frameshifting generates the *Escherichia coli* DNA polymerase III γ subunit from within the τ subunit reading frame, *Nucleic Acids Res.,* 18, 1725, 1990.

386. **Flower, A. M. and McHenry, C. S.**, The γ subunit of DNA polymerase III holoenzyme of*Escherichia coli* is produced by ribosomal frameshifting, *Proc. Natl. Acad. Sci. U.S.A.*, 87, 3713, 1990.
387. **Tsuchihashi, Z. and Kornberg, A.**, Translational frameshifting generates the γ subunit of DNA polymerase III holoenzyme, *Proc. Natl. Acad. Sci. U.S.A.*, 87, 2516, 1990.
388. **Wong, S. C. and Abdelal, A. T.**, Unorthodox expression of an enzyme: evidence for an untranslated region within c*arA* from *Pseudomonas aeruginosa*, *J. Bacteriol.*, 172, 630, 1990.
389. **Belland, R. J., Morrison, S. G., van der Ley, P., and Swanson, J.**, Expression and phase variation of gonococcal P.II genes in *Escherichia coli* involves ribosomal frameshifting and slipped-strand mispairing, *Mol. Microbiol.*, 3, 777, 1989.
390. **Kozak, M.**, Downstream secondary structure facilitates recognition of initiator codons by eukaryotic ribosomes, *Proc. Natl. Acad. Sci. U.S.A.*, 87, 8301, 1990.
391. **Bibb, M. J., Van Etten, R. A., Wright, C. T., Walberg, M. W., and Clayton, D. A.**, Sequence and gene organization of mouse mitochondrial DNA, *Cell*, 26, 167, 1981.
392. **Anderson, S., Bankier, A. T., Barrell, B. G., de Bruijn, M. H. L., Coulson, A. R., Drouin, J., Eperon, I. C., Nierlich, D. P., Roe, B. A., Sanger, F., Schreier, P. H., Smith, A. J. H., Staden, R., and Young, I. G.**, Sequence and organization of the human mitochondrial genome, *Nature (London)*, 290, 457, 1981.
393. **Roe, B. A., Ma, D.-P., Wilson, R. K., and Wong, J. F.-H.**, The complete nucleotide sequence of the *Xenopus laevis* mitochondrial genome, *J. Biol. Chem.*, 260, 9759, 1985.
394. **Buckingham, K., Chung, D.-G., Neilson, T., and Ganoza, M. C.**, Recognition of translational termination signals, B*iochim. Biophys. Acta*, 909, 92, 1987.
395. **Anderson, S., de Bruijn, M. H. L., Coulson, A. R., Eperon, I. C., Sanger, F., and Young, I. G.**, Complete sequence of bovine mitochondrial DNA: conserved features of the mammalian mitochondrial genome, *J. Mol. Biol.*, 156, 683, 1982.
396. **Young, I. G. and Anderson, S.**, The genetic code in bovine mitochondria: sequence of genes for the cytochrome oxidase subunit II and two tRNAs, *Gene*, 12, 257, 1980.
397. **Dix, D. B. and Thompson, R. C.**, Codon choice and gene expression: synonymous codons differ in translational accuracy, *Proc. Natl. Acad. Sci. U.S.A.*, 86, 6888, 1989.
398. **Parker, J.**, Mistranslated protein in *Escherichia coli*, *J. Biol. Chem.*, 256, 9770, 1981.
399. **McPherson, D. T.**, Codon preference reflects mistranslational constraints: a proposal, *Nucleic Acids Res.*, 16, 4111, 1988.
400. **Kato, M.**, Codon discrimination due to presence of abundant non-cognate competitive tRNA, *J. Theor. Biol.*, 142, 35, 1990.
401. **Andersson, S. G. E. and Kurland, C. G.**, Codon preferences in free-living microorganisms, M*icrobiol. Rev.*, 54, 198, 1990.
402. **Fersht, A. R.**, The charging of tRNA, in A*ccuracy of Molecular Processes*, Kirkwood, T. B. L., Rosenberger, R. F., and Galas, D. J., Eds., Chapman & Hall, New York, 1986, 67.
403. **Freist, W.**, Mechanisms of aminoacyl-tRNA synthetases: a critical consideration of recent results, *Biochemistry*, 28, 6787, 1989.
404. **Jakubowski, H.**, Proofreading *in vivo*: editing of homocysteine by methionyl-tRNA synthetase in *Escherichia coli*, *Proc. Natl. Acad. Sci. U.S.A.*, 87, 4504, 1990.
405. **Thompson, R. C. and Stone, P. J.**, Proofreading of the codon-anticodon interaction on ribosomes, *Proc. Natl. Acad. Sci. U.S.A.*, 74, 198, 1977.
406. **Yates, J. L.**, Role of ribosomal protein S12 in discrimination of aminoacyl-tRNA, *J. Biol. Chem.*, 254, 11550, 1979.
407. **Thompson, R. C., Dix, D. B., Gerson, R. B., and Karim, A. M.**, A GTPase reaction accompanying the rejection of Leu-tRNA$_2$ by UUU-programmed ribosomes, *J. Biol. Chem.*, 256, 81, 1981.
408. **Ehrenberg, M. and Kurland, C. G.**, Measurements of translational kinetic parameters, *Meth. Enzymol.*, 164, 611, 1988.
409. **Thompson, R. C.**, EFTu provides an internal kinetic standard for translational accuracy, *Trends Biochem. Sci.*, 13, 91, 1988.

410. **Ruusala, T. and Kurland, C. G.,** Streptomycin preferentially perturbs ribosomal proofreading, M*ol. Gen. Genet.,* 198, 100, 1984.
411. **Bohman, K., Ruusala, T., Jelenc, P. C., and Kurland, C. G.,** Kinetic impairment of restrictive streptomycin-resistant ribosomes, *Mol. Gen. Genet.,* 198, 90, 1984.
412. **Ruusala, T., Andersson, D., Ehrenberg, M., and Kurland, C. G.,** Hyper-accurate ribosomes inhibit growth, *EMBO J.,* 3, 2575, 1984.
413. **Andersson, D. I. and Kurland, C. G.,** Ram ribosomes are defective proofreaders, *Mol. Gen. Genet.,* 191, 378, 1983.
414. **Noller, H. F., Moazed, D., Stern, S., Powers, T., Allen, P. N., Robertson, J. M., Weiser, B., and Triman, K.,** Structure of rRNA and its functional interactions in translation, in *The Ribosome: Structure, Function & Evolution,* Hill, W. E., Dahlberg, A., Garrett, R. A., Moore, P. B., Schlessinger, D., and Warner, J. R., Eds., American Society for Microbiology, Washington, D.C., 1990, chap. 3.
415. **Tapprich, W. E., Göringer, H. U., De Stasio, E., Prescott, C., and Dahlberg, A. E.,** Studies of ribosome function by mutagenesis of *Escherichia coli* rRNA, in *The Ribosome: Structure, Function & Evolution,* Hill, W. E., Dahlberg, A., Garrett, R. A., Moore, P. B., Schlessinger, D., and Warner, J. R., Eds., American Society for Microbiology, Washington, D.C., 1990, chap. 16.
416. **Zimmermann, R. A., Thomas, C. L., and Wower, J.,** Structure and function of rRNA in the decoding domain and at the peptidyltransferase center, in *The Ribosome: Structure, Function & Evolution,* Hill, W. E., Dahlberg, A., Garrett, R. A., Moore, P. B., Schlessinger, D., and Warner, J. R., Eds., American Society for Microbiology, Washington, D.C., 1990, chap. 26.
417. **Cundliffe, E.,** Recognition sites for antibiotics within rRNA, in *The Ribosome: Structure, Function & Evolution,* Hill, W. E., Dahlberg, A., Garrett, R. A., Moore, P. B., Schlessinger, D., and Warner, J. R., Eds., American Society for Microbiology, Washington, D.C., 1990, chap. 41.
418. **Condon, C., Squires, C., and Squires, C.,** personal communication, 1991.
419. **Goldman, E. and Jakubowski, H.,** Uncharged tRNA, protein synthesis, and the bacterial stringent response, *Mol. Microbiol.,* 4, 2035, 1990.
420. **Rojiani, M. V., Jakubowski, H., and Goldman, E.,** Effect of variation of charged and uncharged tRNA^{Trp} levels on ppGpp synthesis in *Escherichia coli, J. Bacteriol.,* 171, 6493, 1989.
421. **Nègre, D., Cortay, J.-C., Donini, P., and Cozzone, A. J.,** Relationship between guanosine tetraphosphate and accuracy of translation in *Salmonella typhimurium, Biochemistry,* 28, 1814, 1989.
422. **Parker, J. and Holtz, G.,** Control of basal-level codon misreading in E*scherichia coli,* Biochem. Biophys. Res. Comm., 121, 487, 1984.
423. **Mikkola, R. and Kurland, C. G.,** Media dependence of translational mutant phenotypes, F*EMS Microbiol. Lett.,* 56, 265, 1988.
424. **Sarubbi, E., Rudd, K. E., and Cashel, M.,** Basal ppGpp level adjustment shown by new *spoT* mutants affect steady state growth rates and *rrnA* ribosomal promoter regulation in *Escherichia coli, Mol. Gen. Genet.,* 213, 214, 1988.
425. **Ninio, J.,** Fine tuning of ribosomal accuracy, *FEBS Lett.,* 196, 1, 1986.
426. **Böck, A., Faiman, L. E., and Neidhardt, F. C.,** Biochemical and genetic characterization of a mutant of *Escherichia coli* with a temperature-sensitive valyl ribonucleic acid synthetase, *J. Bacteriol.,* 92, 1076, 1966.
427. **Piepersberg, W., Geyl, D., Buckel, P., and Böck, A.,** Studies on the coordination of tRNA-charging and polypeptide synthesis activity in E*scherichia coli,* in *Regulation of Macromolecular Synthesis by Low Molecular Weight Mediators,* Koch, G. and Richter, D., Eds., Academic Press, New York, 1979, 39.
428. **Yegian, C. D. and Stent, G. S.,** An unusual condition of leucine transfer RNA appearing during leucine starvation of *Escherichia coli, J. Mol. Biol.,* 39, 45, 1969.

429. Ulrich, A. K. and Parker, J., Strains overproducing tRNA for histidine, M*ol. Gen. Genet.*, 205, 540, 1986.
430. Gallant, J. A., Uncharged tRNA error damping model, *FEBS Lett.*, 206, 185, 1986.
431. Schreiber, G., Metzger, S., Aizenman, E., Roza, S., Cashel, M., and Glaser, G., Overexpression of the *relA* gene in *Escherichia coli, J. Biol. Chem.*, 266, 3760, 1991.
432. Brinkmann, U., Mattes, R. E., and Buckel, P., High-level expression of recombinant genes in *Escherichia coli* is dependent on the availability of the *dnaY* gene product, *Gene*, 85, 109, 1989.
433. Caplan, A. B. and Menninger, J. R., Tests of the ribosomal editing hypothesis: amino acid starvation differentially enhances the dissociation of peptidyl-tRNA from the ribosome, *J. Mol. Biol.*, 134, 621, 1979.
434. Caplan, A. B. and Menninger, J. R., Dissociation of peptidyl-tRNA from ribosomes is perturbed by streptomycin and by *strA* mutations, *Mol. Gen. Genet.*, 194, 534, 1984.
435. Menninger, J. R., Caplan, A. B., Gingrich, P. K. E., and Atherly, A. G., Tests of the ribosome editor hypothesis. II. Relaxed (*relA*) and stringent (*relA+*) E. coli differ in rates of dissociation of peptidyl-tRNA from ribosomes, *Mol. Gen. Genet.*, 190, 215, 1983.
436. Petrullo, L. A. and Elseviers, D., Effect of a 2-methylthio-N6-isopentenyladenosine deficiency on peptidyl-tRNA release in *Escherichia coli, J. Bacteriol.*, 165, 608, 1986.
437. Anderson, R. P. and Menninger, J. R., Tests of the ribosome editor hypothesis. III. A mutant *Escherichia coli* with a defective ribosome editor, *Mol. Gen. Genet.*, 209, 313, 1987.
438. Anderson, R. P. and Menninger, J. R., Genetic mapping of a ribosome editor mutation in *E. coli, J. Cell Biol.*, 107, 548a, 1988.
439. Herrington, M. B., Kohli, A., and Faraci, M., Frameshift suppression by *thyA* mutants of *Escherichia coli* K-12, *Genetics*, 114, 705, 1986.
440. Tabor, H. and Tabor, C. W., Polyamine requirement for efficient translation of amber codons *in vivo, Proc. Natl. Acad. Sci. U.S.A.*, 79, 7087, 1982.
441. Engelberg-Kulka, H. and Schoulaker-Schwarz, R., Stop is not the end: physiological implications of translational readthrough, *J. Theor. Biol.*, 131, 477, 1988.
442. Connolly, D. M. and Winkler, M. E., Structure of *Escherichia coli* K-12 *miaA* and characterization of the mutator phenotype caused by *miaA* insertion mutations, *J. Bacteriol.*, 173, 1711, 1991.
443. Bulté, L. and Bennoun, P., Translational accuracy and sexual differentiation in C*hlamydomonas reinhardtii, Curr. Genet.*, 18, 155, 1990.
444. Ehrenberg, M. and Kurland, C. G., Costs of accuracy determined by a maximal growth rate constraint, *Q. Rev. Biophys.*, 17, 45, 1984.
445. Goff, S. A. and Goldberg, A. L., Production of abnormal proteins in E. coli stimulates transcription of *lon* and other heat shock genes, *Cell*, 41, 587, 1985.
446. Parsell, D. A. and Sauer, R. T., Induction of a heat shock-like response by unfolded protein in *Escherichia coli*: dependence on protein level not protein degradation, *Genes Dev.*, 3, 1226, 1989.
447. Ciechanover, A., Finley, D., and Varshavsky, A., Ubiquitin dependence of selective protein degradation demonstrated in the mammalian cell cycle mutant ts85, *Cell*, 37, 57, 1984.
448. Bond, U. and Schlesinger, M. J., Ubiquitin is a heat shock protein in chicken embryo fibroblasts, *Mol. Cell. Biol.*, 5, 949, 1985.
449. Straus, D. B., Walter, W. A., and Gross, C. A., *Escherichia coli* heat shock gene mutants are defective in proteolysis, *Genes Dev.*, 2, 1851, 1988.
450. Lipinska, B., Zylicz, M., and Georgopoulos, C., The HtrA (DegP) protein, essential for *Escherichia coli* survival at high temperatures, is an endopeptidase, *J. Bacteriol.*, 172, 1791, 1990.
451. Seufert, W. and Jentsch, S., Ubiquitin-conjugating enzymes UBC4 and UBC5 mediate selective degradation of short-lived and abnormal proteins, *EMBO J.*, 9, 543, 1990.

452. **VanBogelen, R. A., Kelley, P. M., and Neidhardt, F. C.**, Differential induction of heat shock, SOS, and oxidation stress regulons and accumulation of nucleotides in *Escherichia coli*, *J. Bacteriol.*, 169, 26, 1987.
453. **Betz, J. L.**, Cloning and characterization of several dominant-negative and tight-binding mutants of *lac* repressor, *Gene*, 42, 283, 1986.
454. **Zipser, D. and Bhavsar, P.**, Missense mutations in the *lacZ* gene that result in degradation of β-galactosidase structural protein, *J. Bacteriol.*, 127, 1538, 1976.
455. **Parsell, D. A. and Sauer, R. T.**, The structural stability of a protein is an important determinant of its proteolytic susceptibility in *Escherichia coli*, *J. Biol. Chem.*, 264, 7590, 1989.
456. **Michaels, M. L., Kim, C. W., Matthews, D. A., and Miller, J. H.**, *Escherichia coli* thymidylate synthase: amino acid substitutions by suppression of amber nonsense mutations, *Proc. Natl. Acad. Sci. U.S.A.*, 87, 3957, 1990.
457. **Miller, C. G.**, Protein degradation and proteolytic modification, in *Escherichia coli and Salmonella typhimurium: Cellular and Molecular Biology*, Vol. 1, Neidhardt, F. C., Ingraham, J. L., Low, K. B., Magasanik, B., Schaechter, M., and Umbarger, H. E., Eds., American Society for Microbiology, Washington, D.C., 1987, chap. 44.
458. **Yen, C. L., Green, L., and Miller, C. G.**, Peptide accumulation during growth of peptidase deficient mutants, *J. Mol. Biol.*, 143, 35, 1980.
459. **Ninio, J.**, Kinetic devices in protein synthesis, DNA replication, and mismatch repair, *Cold Spring Harbor Symp. Quant. Biol.*, 52, 639, 1987.
460. **Menninger, J. R.**, Peptidyl transfer RNA dissociates during protein synthesis from ribosomes of *Escherichia coli*, *J. Biol. Chem.*, 251, 3392, 1976.
461. **Gast, F.-U., Peters, F., and Pingoud, A.**, The role of translocation in ribosomal accuracy: translocation rates for cognate and noncognate aminoacyl- and peptidyl-tRNAs on *Escherichia coli* ribosomes, *J. Biol. Chem.*, 262, 11920, 1987.
462. **Jörgensen, F. and Kurland, C. G.**, Death rates of bacterial mutants, *FEMS Microbiol. Lett.*, 40, 43, 1987.
463. **Gallant, J. and Palmer, L.**, Error propogation in viable cells, M*ech. Ageing Dev.*, 10, 27, 1979.
464. **Hartl, D. L. and Dykhuizen, D. E.**, The population genetics of *Escherichia coli*, *Ann. Rev. Genet.*, 18, 31, 1984.
465. **Dean, A. M., Dykhuizen, D. E., and Hartl, D. L.**, Fitness effects of amino acid replacements in the β-galactosidase of *Escherichia coli*, *Mol. Biol. Evol.*, 5, 469, 1988.
466. **Gunderson, J. H., Sogin, M. L., Wollett, G., Hollingdale, M., de la Cruz, V. F., Waters, A. P., and McCutchan, T. F.**, Structurally distinct, stage-specific ribosomes occur in *Plasmodium, Science*, 238, 933, 1987.
467. **Waters, A. P., Syin, C., and McCutchan, T. F.**, Developmental regulation of stage-specific ribosome populations in *Plasmodium, Nature (London)*, 342, 438, 1989.
468. **Zhu, J., Waters, A. P., Appiah, A., McCutchan, T. F., Lal, A. A., and Hollingdale, M. R.**, Stage-specific ribosomal RNA expression switches during sporozoite invasion of hepatocytes, *J. Biol. Chem.*, 265, 12740, 1990.

Chapter 9

SELENOCYSTEINE, A NEW ADDITION TO THE UNIVERSAL GENETIC CODE

Dolph L. Hatfield, In Soon Choi, Byeong J. Lee, and Jae-Eon Jung

TABLE OF CONTENTS

I. Introduction 270

II. UGA Specifies Selenocysteine 270

III. Selenocysteyl-tRNA[Ser]Sec 271

IV. Selenocysteine Biosynthesis and Incorporation of Selenocysteine into Protein 272

V. Selenocysteyl-tRNA is Ubiquitous in Nature 273

References 275

I. INTRODUCTION

The use of selenium in the metabolism of organisms was reported almost 40 years ago. Pinset showed in 1954 that selenium was required for active formate dehydrogenase in procaryotes[1] and Schwarz and Foltz subsequently showed that this element was required in the diet of mammals to prevent liver necrosis.[2] Selenium has also been implicated in cancer and in heart and other muscle disorders in mammals.[3] This element was initially shown to exist in glutathione peroxidase[4] and the selenium containing amino acid in protein was subsequently identified as selenocysteine.[5] Selenocysteine has now been shown to occur in a variety of proteins from a large number of organisms.[6-8]

Interestingly, selenocysteine, whose codon is UGA, is now recognized as the 21st naturally occurring amino acid in protein.[9] The corresponding selenocysteyl-tRNA and/or selenocysteyl-tRNA gene is known to be widespread, if not ubiquitous, in nature[10-13] suggesting that selenocysteine belongs in the universal genetic code.[12,13] The assignment of this amino acid to the universal genetic code and several features of the corresponding selenocysteyl-tRNA that donates selenocysteine to protein are the subjects of this chapter.

II. UGA SPECIFIES SELENOCYSTEINE

The gene encoding glutathione peroxidase in the genome of mice[14] and the gene encoding formate dehydrogenase in the genome of *E. coli*[15] have been sequenced and both genes were found to contain a TGA codon in their open reading frames. The TGA codon corresponded to a selenocysteine residue in the respective gene products of both the mammalian[16] and *E. coli*[15] enzymes. The glutathione peroxidase gene was also sequenced from human[17,18] and bovine[17] sources and found to have a TGA codon at the same position as that reported for the mouse gene. Furthermore, a number of other selenocysteine-containing proteins in which the selenocysteine moiety is encoded by TGA have been characterized in bacteria,[6,7] including that encoded by the glycine reductase gene in *Clostridium,* recently described by Garcia and Stadtman. In addition to glutathione peroxidase, two additional selenoproteins have recently been characterized in mammals. Their gene sequences reveal that the corresponding selenocysteine moieties are also encoded by TGA codons. One of these proteins is selenoprotein P[19] whose function is unknown and the other is iodothyronine (Type I) deiodinase[20] whose function is to catalyze the deiodination of the prohormone L-thyroxine to the biologically active thyroid hormone 3,3′,5-triiodothyronine. Interestingly, the rat selenoprotein P gene contains 10 TGA codons[21] which encode selenocysteine residues in the gene product.[19]

Following the initial studies which reported that the selenocysteine moiety in glutathione peroxidase[14] and in formate dehydrogenase[15] was encoded by TGA, a question arose as to whether selenocysteine was incorporated directly into protein or was generated by posttranslational modification of the nacent

Selenocysteine, a New Addition to the Universal Genetic Code

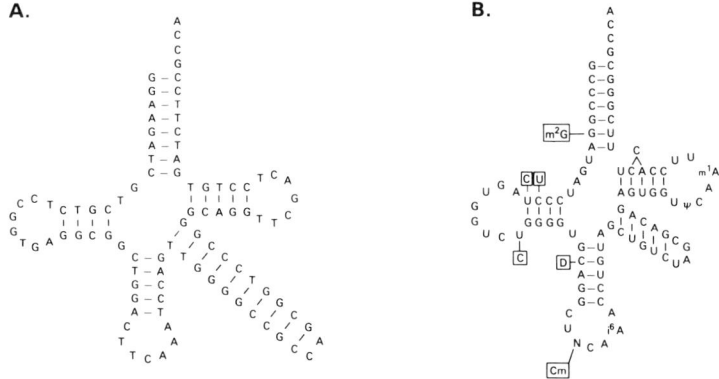

FIGURE 1. Primary structures of (A) selenocysteine tRNA[Ser]Sec from *E. coli* and (B) selenocysteine tRNA[Ser]Sec isoacceptors from bovine liver. The bovine isoacceptors, designated NCA and CmCA on the basis of their anticodon sequences, differ from each other by five pyrimidine transitions and a methylG as shown by the bases in boxes.

polypeptide. We now know that selenocysteine is incorporated directly into protein by the selenocysteyl-tRNA which decodes specific UGA codons.[9]

III. SELENOCYSTEYL-tRNA[SER]SEC

Selenocysteine tRNAs have been characterized from *E. coli*[22] and mammalian sources.[23] In mammals, selenocysteine tRNA was originally identified as a minor seryl-tRNA that recognized UGA in a ribosomal binding assay.[24] It was subsequently shown to suppress UGA in protein biosynthesis[25,26] and to form phosphoseryl-tRNA.[26,27] In addition, two isoacceptors were observed in mammalian tissues[28] and both isoacceptors have been sequenced from bovine liver.[25,26] These isoacceptors have been designated CmCA and NCA based on the structures of their anticodons. These tRNAs have also been sequenced from HeLa and mouse cells.[29] In *E. coli*, the selenocysteine tRNA is encoded by a gene designated *selC*.[30] The corresponding gene product, which is aminoacylated with serine,[30] has been sequenced.[31] Since the mammalian and *E. coli* selenocysteine isoacceptors are aminoacylated with serine and selenocysteine is then synthesized on the RNA molecule, this tRNA is designated as selenocysteine tRNA[Ser]Sec. The primary structures of the bovine[25,26] and *E. coli*[31] tRNAs are shown in Figure 1. They have numerous unique features which set them apart from all other tRNAs. For example, both are longer than other tRNAs; the *E. coli* isoacceptor has 95 nucleotides and the mammalian one 90 nucleotides. Transfer RNAs from both sources are highly undermodified as compared to other tRNAs; the *E. coli* isoacceptor contains four modified nucleotides[31] and the bovine CmCA and NCA isoacceptors have five and six modified nucleotides, respectively.[25,26] The *E. coli* tRNA has an eight base pair aminoacyl acceptor stem[30,31] as compared to seven in other tRNAs and several of the invariant positions in other tRNAs do indeed vary in *E. coli* tRNA[Ser]Sec.[31] The longer acceptor stem and the long extra loop are crucial to the incorporation of selenocysteine into protein from selenocysteyl-tRNA[Ser]Sec in *E. coli*.[32] The mammalian tRNAs have two extra nucleotides between the universal U at position 8 and the A at position 14, and contain an extra unpaired nucleotide within the stem of loop IV.[25,26] Most certainly, some of the unique features in mammalian tRNA[Ser]Sec have a role in the specific incorporation of selenocysteine into protein in response to specific UGA codons.

The gene for selenocysteine tRNA has been shown to occur in single copy in the genomes of several mammals, including humans,[33] rabbits,[34] and cattle,[35] and also in the genomes of chickens,[36] *Xenopus, Drosophila,* and *C. elegans*.[11] The gene has been mapped to human chromosome 19,[37] and it is closely linked on chromosome 19 to apolipoprotein C-II.[38] Transcription of the selenocysteine tRNA gene in higher vertebrates is controlled by multiple upstream regulatory elements[39,40] and expression of tRNA[Ser]Sec occurs by an unique pathway in that it is the only known tRNA which begins transcription at the first nucleotide within the gene.[41]

The fact that the vertebrate selenocysteine tRNA gene occurs in single copy in the genomes of mammals and birds,[33-37] but yet exists in two different forms which differ from each other by several pyrimidine transitions (see Figure 1)[24] suggests that these isoacceptors arise by RNA editing.[35] Furthermore, it appears that the editing process is responsive to selenium, since the presence or absence of selenium in the media of mammalian (HL60 and rat mammary tumor) cells grown in culture affect the intracellular levels and distributions of these two isoacceptors.[42] For example, the level of the selenocysteine tRNA[Ser]Sec population increased about 20% and the distributions of the two isoacceptors shifted to an enrichment of the CmCA isoacceptor in cells grown in the presence of selenium. In addition, similar alterations in the levels and distributions of the selenocysteine tRNA[Ser]Sec isoacceptors occurred in various tissues of rats on diets with, as compared to those without, selenium.[43] We are presently in the process of studying the RNA editing process.

IV. SELENOCYSTEINE BIOSYNTHESIS AND INCORPORATION OF SELENOCYSTEINE INTO PROTEIN

Incubation of *E. coli*[22] and mammalian cells[23] in the presence of [^{75}Se]selenite resulted in the intracellular synthesis of [^{75}Se]selenocysteyl-tRNA[Ser]Sec. Since the selenocysteine tRNAs from these organisms are first aminoacylated with serine and since the corresponding selenocysteyl-tRNAs were shown to occur *in vivo*,[22,23] the biosynthesic pathway of selenocysteine must then occur on the intact tRNA. Indeed, in *E. coli*, two genes, designated *selA* and *selD*, encode proteins which participate in the biosynthesis of selenocysteyl-tRNA[Ser]Sec from seryl-tRNA$^{[Ser]Sec}$.^{22,44} The SELA protein interacts with seryl-tRNA$^{[Ser]Sec}$ and promotes the 2,3-elimination of water from serine. This reaction generates an aminoacrylyl-tRNA$^{[Ser]Sec}$ intermediate. The SELD protein reduces selenium and donates active selenium (the precise state of the active donor is not known) to the enzyme bound aminoacrylyl-tRNA$^{[Ser]Sec}$ intermediate. Selenocysteyl-tRNA$^{[Ser]Sec}$ is then released from SELA.^{9,22,44} In mammalian systems, seryl-tRNA$^{[Ser]Sec}$ is converted to phosphoseryl-tRNA$^{[Ser]Sec}$ [26] by a kinase which has been partially purified from rabbit liver.[27] The proposed phosphoseryl-tRNA$^{[Ser]Sec}$ intermediate in the biosynthesis of selenocysteine has been shown to occur *in vivo* in mammalian cells.[23] However, direct proof that phosphoserine is the intermediate between serine and selenocysteine is still lacking.

In *E. coli*, the product of the *selB* gene is a translation factor specific for selenocysteyl-tRNA$^{[Ser]Sec}$ which serves a similar role as EF-Tu for all other tRNAs.[45,46] Furthermore, selenocysteyl-tRNA$^{[Ser]Sec}$ has been shown to interact poorly with EF-Tu.[47] SELB presumably also plays a role in recognizing UGA codons specific for selenocysteine and not the UGA codons which dictate termination. In addition, approximately 40 bases, which occur immediately downstream of the selenocysteine UGA codon in formate dehydrogenase and which are capable of forming a stem-loop structure, are critical to this codon being read as selenocysteine.[48] In the mammalian Type I deiodinase, about 200 bases within the 3′-untranslated region are essential for insertion of selenocysteine into the nascent polypeptide in response to the selenocysteine codon.[49] Replacement of this region with the corresponding 3′-flank of the mammalian glutathione peroxidase gene (although there is little sequence homology between these two regions) also supports selenocysteine incorporation into Type I deiodinase.[49] Both 3′ untranslated regions in the Type I deiodinase and in glutathione peroxidase genes are capable of forming stem-loop structures suggesting the requirement of a downstream RNA secondary structure in the use of UGA as a selenocysteine codon in mammalian systems. It is of interest to note, however, that selenoprotein P mRNA contains 9 UGA codons in its terminal 122 codewords and the possible role (if any) of downstream secondary structure(s) in decoding each (carboxy terminal) UGA as selenocysteine remains to be established.[21]

V. SELENOCYSTEYL-tRNA IS UBIQUITOUS IN NATURE

Upon identification of a selenocysteyl-tRNA which decodes UGA in *E. coli*[22] and in mammalian cells,[23] a question arose as to how widespread is this tRNA in nature. The selenocysteine tRNA gene was found to be widespread in the subkingdom, Eubacteria,[10] and the selenocysteine tRNA gene and/or its gene product were (was) found to be ubiquitous in the animal kingdom.[11] However, all living organisms may be placed into one of five kingdoms, e.g., see Margulis and Schwartz[50]:

1. Monera (with its two subkingdoms, Eubacteria and Archaebacteria)
2. Protists
3. Plants
4. Animals
5. Fungi

Even though the selenocysteine tRNA gene appears to be ubiquitous in the Eubacteria subkingdom[10] and in the animal kingdom,[11] organisms from the other kingdoms must be examined for the presence of a selenocysteine tRNA that decodes UGA before it can indeed be concluded that selenocysteine should be assigned to this codon in the universal genetic code.

A study was undertaken to determine how widespread the selenocysteyl-tRNA is in nature.[12,13] The assay employed was a simple one which had been used to identify selenocysteyl-tRNA in mammalian cells.[23] Cells (from a given organism) were grown in culture and administered [^{75}Se]selenite, the resulting selenium-labeled tRNA isolated and characterized for its ability to decode UGA and to contain selenocysteine as the aminoacyl moiety. Two quite diverse microorganisms, a diatom, *Thalassiosira*, and a ciliate, *Tetrahymena*, which are both protists, were examined and found to contain selenocysteyl-tRNAs that decode UGA.[12] In addition, *Beta vulgaris*, a plant, and *Gliocladium virens*, a fungi, were examined and found to contain UGA-decoding selenocysteyl-tRNAs.[13] Thus, all five kingdoms contain organisms that have selenocysteyl-tRNAs decoding UGA. It seems reasonable, therefore, that the universal genetic code should be expanded to include selenocysteine which is assigned the codon UGA (see Figure 2). The dual function of AUG as a codon for the initiation of protein synthesis and as a codon for methionine at internal positions of proteins (see Figure 2) has been known since the code was established[51,52] and shown to be universal.[51,53] The use of UGA as a codon for selenocysteine, in addition to its known role in dictating termination, however, is the first addition to the universal genetic code since it was deciphered in the mid 1960s.

5' Base \ Middle Base	U	C	A	G	Middle Base / 3' Base
U	Phenylalanine	Serine	Tyrosine	Cysteine	U
	Phenylalanine	Serine	Tyrosine	Cysteine	C
	Leucine	Serine	Terminator	Selenocysteine / Terminator	A
	Leucine	Serine	Terminator	Tryptophan	G
C	Leucine	Proline	Histidine	Arginine	U
	Leucine	Proline	Histidine	Arginine	C
	Leucine	Proline	Glutamine	Arginine	A
	Leucine	Proline	Glutamine	Arginine	G
A	Isoleucine	Threonine	Asparagine	Serine	U
	Isoleucine	Threonine	Asparagine	Serine	C
	Isoleucine	Threonine	Lysine	Arginine	A
	Methionine / Initiator	Threonine	Lysine	Arginine	G
G	Valine	Alanine	Aspartic acid	Glycine	U
	Valine	Alanine	Aspartic acid	Glycine	C
	Valine	Alanine	Glutamic acid	Glycine	A
	Valine	Alanine	Glutamic acid	Glycine	G

FIGURE 2. The universal genetic code, 1992.

REFERENCES

1. **Pinset, J.,** The need of selenite and molybdate in the formation of formic acid dehydrogenase by members of the coliaerogenes group of bacteria, *Biochem. J.,* 57, 10, 1954.
2. **Schwarz, K. and Foltz, C. M.,** Selenium as an integral part of factor 3 against dietary necrotic liver degeneration, *J. Am. Chem. Soc.,* 79, 3292, 1957.
3. **Burk, R. F.,** Molecular biology of selenium with implications for its metabolism, *FASEB,* 5, 2274, 1991.
4. **Flohé, L., Günzler, W. A., and Schock, H. H.,** Glutathione peroxidase: a selenoenzyme, *FEBS Letts.,* 32, 132, 1973.
5. **Cone, J. E., Del Rio, R. M., Davis, J. N., and Stadtman, T. C.,** Chemical characterization of the selenoprotein component of clostridial glycine reductase: identification of selenocysteine as the organoselenium moiety, *Proc. Natl. Acad. Sci. U.S.A.,* 73, 2659, 1976.

6. **Stadtman, T. C.,** Selenium biochemistry, *Annu. Rev. Biochem.,* 59, 111, 1990.
7. **Stadtman, T. C.,** Biosynthesis and function of selenocysteine-containing enzymes, *J. Biol. Chem.,* 266, 16257, 1991.
8. **Garcia, G. and Stadtman, T. C.,** Selenoprotein A component of the glycine reductase complex from *Clostridium purinolyticum:* nucleotide sequence of the gene shows that selenocysteine is encoded by UGA, *J. Bacteriol.,* 173, 2093, 1991.
9. **Böck, A., Forchhammer, K., Heider, J., Leinfelder, W., Sawers, G., Veprek, B., and Zinoni, F.,** Selenocysteine: the 21st amino acid, *Mol. Microbiol.,* 5, 515, 1991.
10. **Heider, J., Leinfelder, W., and Böck, A.,** Occurrence and functional compatibility within Enterobacteriaceae of a tRNA species which inserts selenocysteine into protein, *Nucleic Acids Res.,* 17, 2529, 1989.
11. **Lee, B. J., Rajagopalan, M., Kim, Y. S., You, K.-H., Jacobson, K. B., and Hatfield, D.,** Selenocysteine tRNA$^{[Ser]Sec}$ gene is ubiquitous within the animal kingdom, *Mol. Cell. Biol.,* 10, 1940, 1990.
12. **Hatfield, D. L., Lee, B. J., Price, N. M., and Stadtman, T. C.,** Selenocysteyl-tRNA occurs in the diatom *Thalassiosira* and in the ciliate *Tetrahymena, Mol. Microbiol.,* 5, 1183, 1991.
13. **Hatfield, D. L., Choi, I. S., Mischke, S., and Owens, L. D.,** Selenocysteyl-tRNAs recognize UGA in *Beta vulgaris,* a higher plant, and in *Gliocladium virens,* a filamentous fungi, *Biochem. Biophys. Res. Commun.,* 184, 254, 1992.
14. **Chambers, I., Frampton, J., Goldfarb, P., Affara, N., McBain, W., and Harrison, P. R.,** The structure of the mouse glutathione peroxidase gene: the selenocysteine in the active site is encoded by the "termination" codon, TGA, *EMBO J.,* 5, 1221, 1986.
15. **Zinoni, F., Birkmann, A., Stadtman, T. C., and Böck, A.,** Nucleotide sequence and expression of the selenocysteine-containing polypeptide of formate dehydrogenase (formate-hydrogen-lysate-linked) from *Escherichia coli, Proc. Natl. Acad. Sci. U.S.A.,* 84, 3156, 1986.
16. **Günzler, W. A., Steffens, G. J., Grossmann, A., Kim, S.-M., Ötting, F., Wendel, A., and Flohé, L.,** The amino-acid sequence of bovine glutathione peroxidase, *Hoppe-Seyler's Z. Physiol. Chem.,* 365, 195, 1984.
17. **Mullenbach, G. T., Tabrizi, A., Irvine, B. D., Bell, G. I., Tainer, G. I., and Hallewell, R.,** Selenocysteine's mechanism of incorporation and evolution revealed in cDNAs of three glutathione peroxidases, *Protein Eng.,* 2, 239, 1988.
18. **Sukenaka, Y., Ishida, K., Takeda, T., and Takagi, K.,** cDNA sequence coding for human glutathione peroxidase, *Nucleic Acids Res.,* 15, 7178, 1987.
19. **Read, R., Bellew, T., Yang, J.-G., Hill, K. E., Palmer, I. S., and Burk, R. F.,** Selenium and amino acid composition of selenoprotein P, the major selenoprotein in rat serum, *J. Biol. Chem.,* 265, 17899, 1990.
20. **Berry, M. J., Banu, L., and Larsen, P. R.,** Type I iodothyronine deiodinase is a selenocysteine-containing enzyme, *Nature (London),* 349, 438, 1991.
21. **Hill, K. E., Lloyd, R. S., Yang, J.-G., Read, R., and Burk, R. F.,** The cDNA for rat selenoprotein P contains ten TGA codons in the open reading frame, *J. Biol. Chem.,* 266, 10050, 1991.
22. **Leinfelder, W., Stadtman, T. C., and Böck, A.,** Occurrence *in vivo* of selenocysteyl-tRNA$^{Ser}_{UCA}$ in *Escherichia coli, J. Biol. Chem.,* 264, 9720, 1989.
23. **Lee, B. J., Worland, P. J., Davis, J. N., Stadtman, T. C., and Hatfield, D. L.,** Identification of a selenocysteyl-tRNA^{Ser} in mammalian cells that recognizes the nonsense codon, UGA, *J. Biol. Chem.,* 264, 9724, 1989.
24. **Hatfield, D., Smith, D. W. E., Lee, B. J., Worland, P. J., and Oroszlan, S.,** Structure and function of suppressor tRNAs in higher eucaryotes, *Crit. Rev. Biochem. Mol. Biol.,* 25, 71, 1990.
25. **Diamond, A., Dudock, B., and Hatfield, D.,** Structure and properties of a bovine liver UGA suppressor serine tRNA with a tryptophan anticodon, *Cell,* 497, 1981.

26. **Hatfield, D., Diamond, A., and Dudock, B.,** Opal suppressor serine tRNAs from bovine liver form phosphoseryl-tRNA, *Proc. Natl. Acad. Sci. U.S.A.,* 79, 6215, 1982.
27. **Mizutani, T. and Hashimoto, A.,** Purification and properties of suppressor seryl-tRNA: ATP phosphotransferase from bovine liver, *FEBS Lett.,* 169, 319, 1984.
28. **Hatfield, D.,** Recognition of nonsense codons in mammalian cells, *Proc. Natl. Acad. Sci. U.S.A.,* 69, 3014, 1972.
29. **Kato, N., Hoshino, H., and Harada, F.,** Minor serine tRNA containing anticodon NCA(C4 RNA) from human and mouse cells, *Biochem. Int.,* 7, 635, 1983.
30. **Zinoni, F., Birkmann, A., Leinfelder, W., and Böck, A.,** Cotranslational insertion of selenocysteine into formate dehydrogenase from *Escherichia coli* directed by a UGA codon, *Proc. Natl. Acad. Sci. U.S.A.,* 84, 3156, 1987.
31. **Schön, A., Böck, A., Ott, G., Sprinzl, M., and Söll, D.,** The selenocysteine-inserting opal suppressor tRNA from *E. coli* is highly unusual in structure and modification, *Nucleic Acids Res.,* 17, 7159, 1989.
32. **Baron, C., Heider, J., and Böck, A.,** Mutagenesis of *selC*, the gene for the selenocysteine-inserting tRNA-species in *E. coli*: effects on *in vivo* function, *Nucleic Acids Res.,* 18, 6761, 1990.
33. **O'Neill, V., Eden, F. C., Pratt, D., and Hatfield, D.,** A human opal suppressor tRNA gene and pseudogene, *J. Biol. Chem.,* 260, 2501, 1985.
34. **Pratt, K., Eden, F. C., You, K. H., O'Neill, V. A., and Hatfield, D.,** Conserved sequences in both coding and 5' flanking regions of mammalian opal suppressor tRNA genes, *Nucleic Acids Res.,* 13, 4765, 1985.
35. **Diamond, A. M., Montero-Puerner, Y., Lee, B. J., and Hatfield, D.,** Selenocysteine inserting tRNAs are likely generated by tRNA editing, *Nucleic Acids Res.,* 18, 6727, 1990.
36. **Hatfield, D., Dudock, B., and Eden, F. C.,** Characterization and nucleotide sequence of a chicken gene encoding an opal suppressor tRNA and its flanking DNA segments, *Proc. Natl. Acad. Sci. U.S.A.,* 80, 4940, 1983.
37. **McBride, O. W., Rajagopalan, M., and Hatfield, D.,** Opal suppressor phosphoserine gene and pseudogene are located on human chromosomes 19 and 22, respectively, *J. Biol. Chem.,* 262, 11163, 1987.
38. **Mitchell, A., Bale, A. E., Lee, B. J., Hatfield, D., Harley, H., Rudle, S., Fan, Y. S., Fukushima, Y., Shows, T. B., and McBride, O. W.,** Regional localization of the selenocysteine tRNA gene (TRSP) on human chromosome 19, *Cytogenet. Cell Genet.,* in press.
39. **Lee, B. J., Kang, S. G., and Hatfield, D.,** Transcription of *Xenopus* selenocysteine tRNA^{Ser} (formerly designated opal suppressor phorphoserine tRNA) gene is directed by multiple 5'-extragenic regulatory elements, *J. Biol. Chem.,* 264, 9696, 1989.
40. **Carbon, P. and Krol, A.,** Transcription of the *Xenopus laevis* selenocysteine tRNA$^{(Ser)Sec}$ gene: a system that combines an internal B box and upstream elements also found in U6 snRNA genes, *EMBO J.,* 10, 599, 1991.
41. **Lee, B. J., de la Pena, P., Tobian, J. A., Zasloff, M., and Hatfield, D.,** Unique pathway of expression of an opal suppressor phosphoserine tRNA, *Proc. Natl. Acad. Sci. U.S.A.,* 84, 6384, 1987.
42. **Hatfield, D., Lee, B. J., Hampton, L., and Diamond, A. M.,** Selenium induces changes in the selenocysteine tRNA$^{[Ser]Sec}$ population in mammalian cells, *Nucleic Acids Res.,* 19, 939, 1991.
43. **Hatfield, D., Choi, I. S., Hill, K. E., Burk, R. F., and Diamond, A. M.,** (unpublished data).
44. **Leinfelder, W., Forchhammer, K., Veprek, B., Zehelein, E., and Böck, A.,** *In vitro* synthesis of selenocysteinyl-tRNA$_{UCA}$ from seryl-tRNA$_{UCA}$: involvement and characterization of the *selD* gene product, *Proc. Natl. Acad. Sci. U.S.A.,* 87, 543, 1990.
45. **Forchhammer, K., Leinfelder, W., and Böck, A.,** Identification of a novel translation factor necessary for the incorporation of selenocysteine into protein, *Nature (London),* 342, 453, 1989.

46. **Forchhammer, K., Rücknagel, K.-P., and Böck, A.**, Purification and biochemical characterization of SELB, a translation factor involved in selenoprotein synthesis, *J. Biol. Chem.*, 265, 9346, 1990.
47. **Förster, C., Ott, G., Forchhammer, K., and Sprinzl, M.**, Interaction of a selenocysteine-incorporating tRNA with elongation factor Tu from *E. coli, Nucleic Acids Res.*, 18, 487, 1990.
48. **Zinoni, F., Heider, J., and Böck, A.**, Features of the formate dehydrogenase mRNA necessary for decoding of the UGA codon as selenocysteine, *Proc. Natl. Acad. Sci. U.S.A.*, 87, 4660, 1990.
49. **Berry, M. J., Banu, L., Chen, Y., Mandel, S. J., Kieffer, J. D., Harney, J. W., and Larsen, P. R.**, Recognition of UGA as a selenocysteine codon in Type I deiodinase requires sequences in the 3' untranslated region, *Nature (London)*, 353, 273, 1991.
50. **Margulis, L. and Schwartz, K. V.**, *Five Kingdoms, An Illustrated Guide to the Phyla of Life on Earth*, 2nd ed., W. H. Freeman, San Francisco, 1988.
51. **Nirenberg, M., Caskey, T., Marshall, R., Brimacombe, R., Kellog, D., Doctor, B., Hatfield, D., Levin, J., Rottman, F., Pestka, S., Wilcox, M., and Anderson, F.**, The RNA code and protein synthesis, *Cold Spring Harbor Symp. Quant. Biol.*, 31, 11, 1966.
52. **Khorana, G. H., Büchi, H., Ghosh, H., Gupta, N., Jacob, T. M., Kössel, H., Morgan, R., Narang, S. A., Ohtuska, E., and Wells, R. D.**, Polynucleotide synthesis and the genetic code, *Cold Spring Harbor Symp. Quant. Biol.*, 31, 39, 1966.
53. **Marshall, R. E., Caskey, C. T., and Nirenberg, M.**, Fine structure of RNA codewords recognized by bacterial, amphibian and mammalian transfer RNA, *Science* 155, 820, 1967.

Chapter 10

tRNA DISCRIMINATION IN AMINOACYLATION

Leo Pallanck and LaDonne H. Schulman

TABLE OF CONTENTS

I. Introduction 280

II. tRNA Identity Assays 280

III. The Anticodon and Acceptor Arm are Major Recognition
     Sites in tRNAs 283

IV. The Role of the Anticodon 284
    A. *In Vitro* Studies 284
    B. *In Vivo* Studies 285
    C. *E. coli* Methionine tRNA 289
    D. Valine tRNAs 292

V. Role of the Acceptor Stem and "Discriminator" Base
   at Position 73 293
    A. Alanine tRNAs 293
    B. *E. coli* Histidine tRNA 297
    C. Other *E. coli* tRNAs 297

VI. The *E. coli* Glutamine tRNA-Glutamine Synthetase
    Complex 299

VII. Recognition of other tRNA Domains 301
    A. tRNAs Containing a Large Variable Arm 301
    B. *E. coli* Arginine tRNA 303
    C. Phenylalanine tRNAs 304
    D. Yeast Aspartate tRNAs 306

VIII. The Role of Modified Bases 308
    A. *E. coli* Isoleucine tRNAs 308
    B. Recognition by Yeast Arginine tRNA Synthetase 308

IX. Conclusions 309

References 312

I. INTRODUCTION

The recognition of a tRNA by its aminoacyl-tRNA synthetase is a classic example of the specificity often encountered in biology. Each of the 20 aminoacyl-tRNA synthetases in a cell must distinguish its own set of isoacceptor tRNAs from the many noncognate tRNAs and efficiently catalyze the covalent attachment of the correct amino acid to the 3' end of only these species. Ultimately, the fate of the cell rests on this interaction, as there are no subsequent proofreading steps in protein synthesis whereby the amino acid is matched against the anticodon to ensure that the proper amino acid is inserted in response to a given codon. How a synthetase is able to select its tRNA substrates from a pool of noncognate species sharing similar tertiary structure[1-3] has been the focus of over 20 years of research. Recent technical refinements in the types of assays used to study this interaction have contributed a wealth of new information to this field, allowing the identification of nucleotides conferring a particular amino acid acceptor identity for a number of tRNAs.[4-8] The goal of this article will be to try to summarize these more recent developments in tRNA recognition. A related and equally fascinating subject, namely the study of aminoacyl-tRNA synthetases, will not be covered in this review. However, several recent reviews are suggested.[9,10]

II. tRNA IDENTITY ASSAYS

Central to the study of tRNA aminoacylation is the identification of the set of nucleotides in a given tRNA allowing efficient aminoacylation exclusively by the cognate synthetase. In essence, there are two components to this tRNA specificity problem. The first component involves determinants in a tRNA which allow efficient recognition and aminoacylation by its cognate synthetase, nucleotides referred to as "recognition elements". The second component to the specificity problem is the set of nucleotides in a tRNA which protect it from aminoacylation by a noncognate synthetase, or rather the "negative elements". These two components together define the amino acid acceptor specificity, or "identity" of a tRNA, and the recognition elements together with the negative elements are known as the "identity elements" of a tRNA.[11]

There have typically been both *in vitro* and *in vivo* approaches to the study of tRNA identity. Each method has its own particular strengths and weaknesses. However, the two approaches yield different kinds of information, and so, results obtained *in vitro* complement those obtained *in vivo*. In early studies, tRNA substrates for *in vitro* analysis were obtained following mutagenesis and expression of the mutant tRNAs *in vivo*.[12-15] Unfortunately, many tRNA mutants are expressed poorly, processed incorrectly, or have altered patterns of base modification, limiting the usefulness of this approach. Currently, the favored *in vitro* method for tRNA identity studies involves T7 RNA polymerase runoff transcription.[16-18] Sampson and Uhlenbeck have placed a synthetic tRNA gene immediately adjacent to a bacteriophage T7 RNA polymerase late promoter so that the first nucleotide transcribed constitutes the 5' end of the tRNA. Runoff transcription of this DNA *in vitro* with T7 RNA polymerase yields a transcript with the proper 5' and 3' ends. Because T7 RNA polymerase prefers to initiate transcription with a purine, tRNAs beginning with pyrimidines have been synthesized with a 5' leader sequence followed by posttranscriptional processing to the mature size with RNAse P.[19] Thus, this method allows one to create any desired mutation in a tRNA and then to recover that tRNA in mg quantities for study. tRNAs made by runoff transcription, however, lack the modified bases typically found in tRNAs made *in vivo*. Fortunately, these transcripts devoid of base modifications have generally been efficient substrates for aminoacyl-tRNA synthetases, allowing quantitative comparisons of the effects of base changes at specific sites on recognition by purified synthetases.

While much can be learned about tRNA recognition from *in vitro* experiments, the complete set of tRNA identity elements can only be derived from *in vivo* studies in which synthetases specific for all 20 amino acids compete for the tRNA substrate under normal physiological conditions. Some of the earliest information on tRNA identity in fact derives from *in vivo* study. In 1972, two groups working independently, used a genetic approach to try to identify mutations in an amber suppressor derivative of *E. coli* tRNA^{Tyr} which would alter its specificity.[20,21] The selection involved suppression of *E. coli* auxotrophic mutants with amber codons at sites where insertion of tyrosine would not produce a biologically active protein. A number of different mutants were obtained, but in each case the acceptor identity was converted to Gln.[22-25] Additional identity changes were probably not obtained because of the requirement for multiple base changes in the tRNA. With the advent of oligonucleotide synthesizers, however, a large number of tRNA sequence alterations could be created that would otherwise be difficult to achieve using genetics. Armed with this new technique, Abelson, Miller, and co-workers returned to this amber suppression assay, but instead of selecting for tRNA mutations they have synthesized them.[11,26,27] Amber suppressor derivatives of 13 different *E. coli* tRNAs were constructed and cloned under the control of the strong constitutive *E. coli lpp* promoter. In this system the translational efficiency of the tRNA is determined by assaying its ability to suppress an amber codon in a reporter gene, whereas its specificity is determined by sequencing the suppressed protein and identifying the amino acid inserted at the position corresponding to the amber codon. *E. coli* dihydrofolate reductase (DHFR) with an amber mutation at position 10 was the reporter protein used in these studies mainly because of its ease of purification. Position 10 in DHFR is in a loop on the exterior of the protein away from the catalytic and substrate binding sites and is therefore thought to tolerate a wide variety of amino acid substitutions. Using this type of approach a wealth of information on tRNA identity in *E. coli* has been obtained.[11,28-37] However, despite the gains afforded by this system of study there are some important limitations as well.

Perhaps the primary deficiency of the amber suppression (and opal and ochre suppression) approach is that it requires altering the anticodon of the tRNA to something complementary to a stop codon. If the anticodon contains no identity elements this poses no problems. However, a great deal of *in vitro* as well as *in vivo* data suggest that this is not the case for most tRNAs. Thus, by simply changing the anticodon of a tRNA, its original acceptor identity may be altered or perhaps weakened to the point that the effects of additional mutations introduced into the suppressor tRNA will be given undue importance. The ideal *in vivo* system would be one in which mutations could be introduced into a test tRNA without having to alter its natural anticodon. However, a potential problem with this type of assay is that the test tRNA would probably be quite toxic to the cell if it had an acceptor identity different from its decoding capacity. Such a tRNA would insert the wrong amino acid in response to a given codon. To get around such obstacles, the *E. coli* initiator tRNA (tRNA^{fMet}) has been used as the test tRNA in recent studies.[38-43] The features of the initiator tRNA making it attractive for these analyses are that there is only one kind of initiator tRNA (Met) and it is the only tRNA able to initiate translation. By altering the anticodon of the initiator tRNA, one can exploit its singular capacity to initiate protein synthesis by constructing a complementary initiation codon in a reporter gene. Only in the presence of the initiator tRNA with the complementary anticodon will the reporter protein be made, further, and most importantly, since no other tRNA in the cell will be able to initiate translation from a nonmethionine codon, the amino terminal residue is uniquely inserted by the test tRNA. Purification of the reporter protein (as in the suppression assay, DHFR) and sequencing to identify the amino terminal residue(s), indirectly allow determination of the test tRNA's identity. An attractive feature of this "initiation assay" is that these mutant initiator tRNAs are not toxic to cells, irrespective of their acceptor identities, since they are unable to participate in polypeptide chain elongation.[44,45]

The *in vivo* initiation assay has been used to study the role of the anticodon as well as to examine the effects of additional changes outside of the anticodon on aminoacylation specificity.[39-43] The *in vivo* initiation assay has also proven useful in locating important features of the initiator tRNA allowing it to act in initiation.[38,46,47] What remains to be seen, however, is whether this assay will prove useful for all amino acids, and to what extent changes can be introduced into the initiator tRNA before it loses its capacity to initiate protein synthesis.

Although the *in vivo* initiation assay overcomes one of the major problems of the nonsense suppression approach, it is still subject to many of the same limitations and considerations. Like the suppression assay, the initiation assay is an indirect measure of aminoacylation. Thus, neither suppression efficiency nor efficiency of initiation can be directly correlated with aminoacylation. In essence, efficiency not only reflects how good a substrate a tRNA is for an aminoacyl-tRNA synthetase, but additionally, how well it is synthesized, processed and modified, as well as how efficiently it interacts with the translational machinery. Therefore, it is not possible to assess from *in vivo* data how good a substrate is, but only what it is a substrate of. Nonetheless, *in vivo* studies can and do provide valuable information on tRNA identity assuming that they are carried out properly and supported by *in vitro* data. Perhaps the primary consideration in performing such *in vivo* studies is to ensure that the tRNA is expressed at physiological levels. Theoretical,[48] as well as empirical observations,[33,49,50] have shown that simple overexpression of a tRNA can result in mischarging by synthetases which do not recognize the tRNA with high efficiency *in vitro*. Also genetic screens using cells which require a specific amino acid inserted at a nonsense codon for growth should be considered a minimal test of tRNA identity. These assays are extremely sensitive and will yield a signal even if the suppressor tRNA inserts a very tiny amount of that amino acid (in some cases below the level detected by sequencing). Furthermore, these types of screens do not rule out the possibility of insertion of more than one amino acid by a given suppressor.

III. THE ANTICODON AND ACCEPTOR ARM ARE MAJOR RECOGNITION SITES IN tRNAS

Sufficient work has now been carried out, in *E. coli* at least, to begin to see a general pattern for tRNA recognition. The overwhelming majority of results point to nucleotides of the anticodon, in addition to nucleotides in the distal part of the acceptor arm, as playing the largest role in recognition by aminoacyl-tRNA synthetases. The relative importance of these two widely spaced tRNA domains varies greatly amongst different tRNA species. At one extreme, the anticodon appears to contain all the major recognition elements for the methionine tRNAs from *E. coli*. At the other extreme, the acceptor arm alone constitutes the major recognition site for the *E. coli* alanine tRNAs. Between these two extremes lie the majority of tRNAs which harbor nucleotides important for their identities in both these tRNA domains. This conservation of the major recognition elements of tRNAs in the same locations may play an important functional role in discrimination, and therefore, in identity. By overlapping the recognition sites of tRNAs, nucleotides which form positive contacts with the cognate synthetase may make negative contacts with a noncognate synthetase requiring different recognition nucleotides at the same location. Thus, the recognition elements of a tRNA may simultaneously define some, or perhaps most, of its identity elements. However, the delineation of the entire set of identity elements for a given tRNA could prove much more difficult because many regions of the tRNA playing no role in recognition could potentially serve to block aminoacylation by competing synthetases. Thus, the number of potential identity elements is very large (perhaps most of the nucleotides in a tRNA). To date, most investigations have focused on elucidation of the recognition elements of tRNAs, in which great progress has been made, as will be seen in the ensuing discussion.

IV. THE ROLE OF THE ANTICODON

Since the anticodon specifies the decoding capacity of a tRNA, it would seem a logical site to specify tRNA acceptor identity as well. The anticodon is also one of the only domains where no two tRNAs will have the same sequence, and so, could act as a site of discrimination by aminoacyl-tRNA synthetases. However, despite the fact that these observations were first pointed out in 1964,[51] the role of the anticodon in aminoacylation was not generally accepted until fairly recently. One of the primary arguments against the anticodon as a site of identity initially came from studies of nonsense suppressor derivatives of Gln, Leu, Ser, and Tyr tRNAs, which had lost their wild-type anticodons and yet maintained their original identities. Nevertheless, one of the first amber suppressor derivatives studied, su^{+7} tRNA^{Trp}, which had undergone a CCA→CUA change in its anticodon, was found to insert both Trp and Gln at amber codons *in vivo*.[52] Further characterization of this tRNA *in vitro* showed that aminoacylation by tryptophanyl-tRNA synthetase (TrpRS) was decreased, while the affinity of GlnRS towards the tRNA had increased relative to wild-type tRNA^{Trp}.[53,54] Similarly, introducing anticodon base changes into *E. coli* tRNA^{fMet},[55] tRNA^{Arg},[56] tRNA^{Gly},[14,15] and yeast tRNA^{Val}[55] resulted in dramatic effects on aminoacylation by their cognate synthetases *in vitro*. These early results were just the tip of the iceberg, in terms of the evidence supporting the anticodon's role in aminoacylation, and have been nicely reviewed by Kisselev.[58] The remainder of this discussion will focus on data acquired mainly since 1985, with an emphasis on results obtained in *E. coli*.

A. *IN VITRO* STUDIES

In vitro tRNA recognition studies are typically carried out by measuring the rate of aminoacylation of tRNA derivatives with base changes at sites under investigation. The relative specificity of a synthetase for different tRNA substrates is evaluated by comparing k_{cat}/K_m, the "specificity constant" for each tRNA, where k_{cat} is the turnover number of the enzyme and K_m is the Michaelis constant for the tRNA substrate.[59] The ratio of specificity constants for aminoacylation of cognate vs noncognate tRNAs by *E. coli* synthetases is generally between 10^5 and 10^8. Interestingly, discrimination by yeast aminoacyl-tRNA synthetases *in vitro* appears less stringent as the ratio of aminoacylation of cognate vs. noncognate tRNAs by yeast synthetases is in the range of 10^4 to 10^6. Thus, when analyzing kinetic data obtained in studies of yeast aminoacyl-tRNA synthetase recognition, one must keep such matters in mind so as not to underestimate the effect of mutational analysis.

Using this sort of approach, significant reductions in the kinetics of aminoacylation following single nucleotide replacements in the anticodon have been observed for *E. coli* Arg, Gln, Gly, Ile, Met, Phe, Thr, Trp, Tyr, and Val tRNAs (Table 1, and references therein). The magnitude of the defect varies from one tRNA to the next, possibly reflecting the relative importance of the anticodon in recognition of the tRNA. MetRS and ThrRS are also known to protect the anticodons of their cognate tRNAs from chemical modification and nuclease digestion,[60,61] as expected from direct interaction with this site. At least five yeast (Asp, Met, Phe, Tyr, and Val) and two mammalian (bovine Trp and human Phe) tRNAs also contain recognition elements in their anticodons (Table 1, and references therein). For three of the yeast tRNAs, Asp, Met, and Val, single nucleotide substitutions in the anticodon have large effects on recognition by the cognate synthetases (k_{cat}/K_m decreases of $>10^2$). Studies carried out with yeast tRNA^{Phe} and tRNA^{Tyr} indicate that individual base changes in the anticodons of these tRNAs have more modest effects on recognition by the cognate synthetases. Nevertheless, the sum of the individual contributions of the three nucleotides represents the major part of the recognition profile of yeast tRNA^{Phe}.[16,62,63] *In vitro* experiments carried out with bovine tRNA^{Trp} and human tRNA^{Phe} have shown that single anticodon nucleotide substitutions can result in severe reductions in aminoacylation with the cognate synthetases. Footprinting experiments performed with bovine TrpRS and tRNA^{Trp} indicated protection of the anticodon domain by the enzyme,[64] consistent with the kinetic data showing this to be a site of recognition in the tRNA. Thus, the anticodon's important role in recognition has been conserved in at least some higher eukaryotes as well.

Although a decrease in aminoacylation upon nucleotide substitution may indicate that a recognition element has been eliminated, other interpretations are possible from negative results, i.e., affects on folding of the tRNA. The most convincing way to demonstrate that an anticodon dictates the aminoacylation of a tRNA is to replace the anticodon of one tRNA with that of another and show that the amino acid acceptor identity of the tRNA coincides with the anticodon. In this fashion, substituting the anticodons of an *E. coli* Arg,[65] Met,[66] Thr,[67] and Val[66] tRNA into a noncognate tRNA background results in increases of aminoacylation of four to six orders of magnitude by the cognate synthetases. Similar studies were also carried out with yeast tRNA^{Phe} and tRNA^{Tyr}, resulting in anticodon-dependent aminoacylation, but to a less dramatic extent.[68,69] The fact that several of the anticodon swap experiments resulted in tRNAs aminoacylating less well than the wild-type counterpart suggests that additional recognition elements lie outside the anticodon of these tRNAs.

B. *IN VIVO* STUDIES

In vivo studies have also provided strong support for the important role played by the anticodon. Abelson and Miller have attempted to construct a complete collection of amber suppressor tRNAs in *E. coli* and have examined their acceptor identities in the suppression assay described earlier.[26,27,70,71] While many of the amber suppressor tRNAs function efficiently and insert the correct amino acid, a large proportion of them insert the incorrect amino acid, mixtures of different amino acids, or fail to function at all as suppressors.

TABLE 1
Role of the Anticodon in Recognition of tRNAs

Organism	tRNA	Anticodon recognition elements	In vivo(a) and/or in vitro(b) data	Effect of single base changes in vitro	Ref.
Ec	Ala	None	b	No effect	78
Sc	Ala	None	b	No effect	79,80,81
Ec	Arg	C_{35}	a, b	Large loss of activity	36,56,65
Ec	Asn	$U_{35}U_{36}$[c]	—		93
Ec	Asp	—[a]	a		
Sc	Asp	$G_{34}U_{35}C_{36}$	b	k_{cat}/K_m reduced 20 to 500-fold; mainly k_{cat}	143
Ec	Cys	$G_{34}C_{35}A_{36}$	a		41
Ec	Gln	$Py_{34}U_{35}G_{36}$	b	k_{cat}/K_m reduced 70 to 3×10^4-fold; 35>34, 36; $k_{cat}>K_m$	37,121[b]
Ec	Glu	—[a]	a		
Ec	Gly	$C_{35}C_{36}$	a, b	Reduced rate	14,15,43, 115,116, 118
All	His	Unknown			
Ec	Ile	$L/G_{34}A_{35}U_{36}$	a, b	Reduced rate	40,86,151
Ec	Leu	Unknown			
Ec	Lys	U_{35}[c]	a		36,70,71
Ec	Met	$C_{34}A_{35}U_{36}$	a, b	k_{cat}/K_m reduced 10 to 10^5-fold; 34>36>35; K_m and k_{cat}	55,66,84[b]
Sc	Met	$C_{34}A_{35}U_{36}$	b		91
Ec	Phe	$G_{34}A_{35}A_{36}$	a, b	k_{cat}/K_m reduced 10 to >10^3-fold; 34,35>36	39,40,138
Sc	Phe	$G_{34}A_{35}A_{36}$	b	k_{cat}/K_m reduced 2 to 13-fold; 34,35>36	62,139
H	Phe	$G_{34}A_{35}A_{36}$	b	k_{cat}/K_m reduced 10 to >10^3-fold; 34,35>36	141
All	Pro	Unknown			
Ec	Ser	None[d]			
Ec	Thr	$G_{35}U_{36}$	a, b	Large loss of activity	67,74–77
Ec	Trp	$C_{34}C_{35}A_{36}$	a, b	$C_{35}\rightarrow U$ k_{cat}/K_m reduced 400-fold; $K_m>k_{cat}$	42,52,53, 155
Bov	Trp	C_{34}[c]	b		153
Ec	Tyr	$Q_{34}U_{35}A_{36}$	a, b	$Q_{34}\rightarrow C$ k_{cat}/K_m reduced 20-fold; $U_{35}\rightarrow G$ k_{cat}/K_m reduced 200-fold	33,43,73
Sc	Tyr	$G_{34}\Psi_{35}$[c]	b	k_{cat}/K_m reduced 7 to 14-fold	69,154
Ec	Val	$A_{35}C_{36}$	a, b	k_{cat}/K_m reduced 10^2 to >10^3-fold; 35>36; K_m and k_{cat}	39,40, 66,94

TABLE 1 (continued)
Role of the Anticodon in Recognition of tRNAs

Organism	tRNA	Anticodon recognition elements	In vivo(a) and/or in vitro(b) data	Effect of single base changes in vitro	Ref.
Sc	Val	$A_{35}C_{36}$	b	Large loss of activity	57,98
W	Val	$A_{35}C_{36}{}^e$	b	k_{cat}/K_m reduced 20 to >10^3- fold; 35>36; K_m and k_{cat}	99

L = Lysidine
Q = Queuosine
Ec = *Escherichia coli*
Sc = *Saccharomyces cerevisiae*

All = All organisms tested
H = Human
Bov = Bovine
W = Wheat germ

a Anticodon is required for in vivo identity;[71] data on recognition are unclear.
b See text for additional references.
c Additional anticodon recognition sites may exist.
d Based on the fact that serine isoacceptor tRNAs contain base changes at all three positions of the anticodon.
e Data obtained using the tRNA-like structure from TYMV RNA.

TABLE 2
E. coli Amber Suppressor tRNAs Classified According to their Amino Acid Acceptor Identities[a]

CLASS I identity preserved	CLASS II mischarged with Gln	CLASS III mischarged with Lys
Ala2	GluA	Arg
Cys	Gly2	AspM
Gly1	Ile1	Ile2
HisA	Met (f)	Met (m)
Leu	Trp (su$_7^+$)	Thr2
Lys		Val
Phe		
ProH		
Gln (su$_2^+$)		
Ser (su$_1^+$)		
Tyr (su$_3^+$)		

[a]Data from Reference 71 and additional references therein.

Those tRNAs which functioned as suppressors were divided into three classes (see Table 2): Class I suppressors are those that accept the correct amino acid; Class II encompasses those which are mischarged by glutamine; and Class III are the set of suppressors mischarged by lysine. Five of the tRNAs, Asp, Ile_1, Ile_2, Met_m and Val were found to have lost their original acceptor identities altogether, whereas another four, Arg, Gly_2, Glu, and Thr_2 were found to insert mixtures of either Gln or Lys in addition to their cognate amino acids.

Though it might be tempting to conclude that the Class I suppressor tRNAs which are not mischarged *in vivo* lack important identity elements in their anticodons, to do so would clearly be incorrect. The Class I tRNAs might maintain their charging specificities because of additional identity elements lying outside the anticodon, and/or because of the presence of negative determinants blocking the GlnRS and LysRS so that the cognate synthetase does not have a strong competitor for its substrate, and finally, because of the conservation of nucleotides important for recognition in the amber anticodon. Perhaps the best example of the third point is provided by the large number of suppressors mischarged by Gln and Lys. The amber anticodon CUA shares two out of three nucleotides with a Gln anticodon CUG and one with a Lys anticodon UUU. Thus, the most likely explanation of the Class II and III suppressors is that GlnRS and LysRS strongly recognize nucleotides of the anticodon which are conserved in the amber anticodon, a notion strongly supported by additional *in vivo* and *in vitro* data.[36,37,53,72] Likewise, many of the other Class I tRNAs might also recognize nucleotides conserved in the amber anticodon. Use of the *in vivo* initiation assay described earlier supports this notion.

The introduction of anticodons corresponding to a Cys (GCA), Gln (UUG), Ile (GAU), Phe (GAA), Val (GAC), Trp (CCA), or Tyr (GUA) tRNA into the *E. coli* initiator tRNA resulted in the insertion of the corresponding amino acids at the amino terminus of DHFR initiated with the complementary initiation codon.[39-43] This shows that the anticodons of these tRNAs contain sufficiently strong identity elements to allow aminoacylation of a noncognate tRNA. Four of these, Gln, Cys, Phe, and Tyr, fall into the Class I group of amber suppressor tRNAs (Table 2). Interestingly, all of these tRNAs have nucleotides in their anticodons that are conserved in the amber anticodon which might explain their retention of identity, as previously discussed for Gln. Nucleotide A_{36} (Figure 1) in the Phe anticodon GAA_{36} has been shown to play an important role in allowing aminoacylation with Phe, since $tRNA^{fMet}$ derivatives with the closely related GAU_{36} and GAC_{36} anticodons are not substrates for PheRS *in vivo*.[40] Likewise, nucleotides A_{36} and U_{35} in the Cys and Tyr anticodons, respectively, are conserved in the amber anticodon and have also been found to play an essential role in allowing mischarging of $tRNA^{fMet}$ derivatives with Cys and Tyr.[41,43] U_{35} in the Tyr anticodon is also known to constitute a TyrRS recognition element from studies carried out *in vitro*.[73] Thus, $tRNA^{Phe}$, $tRNA^{Cys}$, and $tRNA^{Tyr}$ are additional examples of Class I tRNAs which preserve their identity at least partly because of anticodon nucleotides conserved in the amber anticodon. Based on results obtained with the *in vivo* initiation assay the synthetases specific for $tRNA^{Ile}$, $tRNA^{Tre}$, and $tRNA^{Val}$, also strongly recognize nucleotides in the anticodon, rationalizing the entry of these tRNAs as Class II and III suppressors (Table 2). Likewise, genetic experiments have

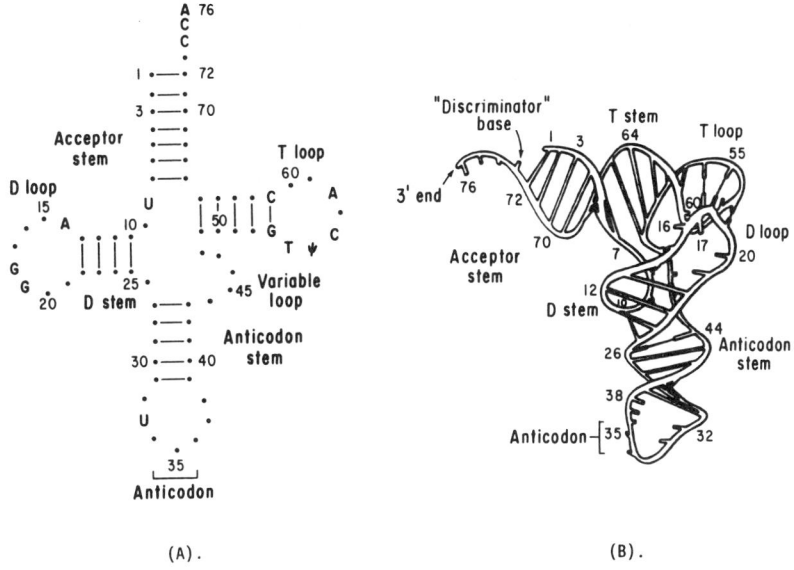

FIGURE 1. (A) Cloverleaf structure of a typical Class 1 tRNA, numbered according to Sprinzl et al.[110] Bases conserved in all *E. coli* tRNAs are included. (B) Three-dimensional structure of yeast tRNA^{Phe}.[1,2] (From Schulman, L. H., *Prog. Nucleic Acid Res. Mol. Biol.*, 41, 24, 1991. With permission.)

shown that ThrRS recognizes the anticodon of its cognate tRNA.[74-77]

Although additional members of the Class I suppressor tRNAs may turn out to have important recognition elements in their anticodons (see Table 1), there are several members that clearly do not. A derivative of tRNA^{Ala}, with base changes at all three anticodon positions, has been characterized *in vitro* and found to lack detectable anticodon dependent aminoacylation.[78] Likewise, the five tRNA^{Ser} isoacceptors in *E. coli* do not contain a single conserved nucleotide at any of the three anticodon positions. In addition, no functional groups are conserved in the bases found at any anticodon position making it unlikely that recognition elements lie in this domain. So, at the least, tRNA^{Ala} and tRNA^{Ser} are two members of a class of tRNAs lacking important elements in their anticodons. Studies carried out with yeast tRNA^{Ala} indicate that it also falls into this class of tRNAs, as the tRNA containing a noncognate anticodon is efficiently recognized by yeast AlaRS.[79-81] One might imagine that the anticodons found in this class of tRNAs contain nucleotide combinations not strongly recognized by any of the noncognate synthetases. So these anticodons could be functioning, not as recognition elements to recruit the cognate synthetases, but rather, as identity elements in blocking noncognate synthetases.

C. *E. COLI* METHIONINE tRNA

Perhaps the best example of a tRNA with important elements in its anticodon is offered by methionine tRNA. An extensive catalogue of mutations at all three anticodon nucleotides has been made in *E. coli* tRNA^{fMet} and tRNA$^{Met}_m$

and these have been assayed *in vitro* with purified MetRS for aminoacylation activity.[55,65-67,72] The results of this analysis have shown that all three anticodon bases are important for MetRS recognition, with the largest effects seen upon replacement of the wobble base, C_{34} (Table 1).[84,85] Although this sort of approach illustrates the importance of the anticodon in tRNA^{Met}, the results of anticodon "swap" experiments, where the Met anticodon was transplanted into a noncognate tRNA and then assayed for Met acceptance, provides even more dramatic evidence for the significance of the Met anticodon. Transfer of the Met anticodon into tRNA$^{Ile}_2$ *E. coli* tRNA^{Ile},[86] tRNA$^{Ile}_1$,[87] tRNA^{Phe},[88] tRNA^{Trp},[89] or tRNA^{Val} [66] background confers efficient mischarging by MetRS upon the chimeras. In fact, tRNA^{Val} with the Met CAU anticodon is nearly as good a substrate as the wild-type Met tRNA. Primary sequence comparisons between tRNAs which are efficient substrates of MetRS show that, aside from nucleotides conserved in all *E. coli* tRNAs, only the three anticodon bases are common to all. The anticodon of the yeast Met tRNA has also been found to contain nucleotides recognized by the yeast MetRS, based on genetic studies as well as direct *in vitro* analysis.[90,91]

While the anticodon is clearly the dominant identity element in tRNA^{fMet}, additional sites outside of the anticodon appear to modulate the affinity of MetRS. For example, MetRS has a 15,000-fold higher k_{cat}/K_m for tRNA^{Trp} with a Met anticodon {tRNA$^{Trp}_{CAU}$} than for wild-type tRNA^{Trp}; nonetheless, it is still 50-fold below that of wild-type tRNA^{Met}.[89] Altering the discriminator base (N_{73}) in tRNA$^{Trp}_{CAU}$ from G_{73} to the A_{73} found in tRNA^{Met} increases k_{cat}/K_m by a factor of ten. Coupling this change to an additional change of G_3C_{70} to the Met sequence C_3G_{70} raises the activity of the tRNA to that of authentic tRNA^{Met}.

Other studies aimed at the role of the discriminator base in tRNA^{Met} identity have suggested that A_{73} is not a Met recognition element.[92] Substitutions at this site in tRNA^{Met} have very small effects on V_{max} (two- to threefold), in the order $A_{73} > U_{73} > C_{73}, G_{73}$. Furthermore, bacteriophage T5 encodes a tRNA^{fMet} containing U_{73} which is thought to be a substrate of MetRS *in vivo*. Since A and U have no functional groups in common, it seems unlikely that MetRS makes positive contacts at this site. The larger effect of the change in the 73 position on Met identity in tRNA^{Trp} is apparently a context effect of the Trp acceptor stem. Similarly, the effect of the 3-70 position is dependent on the tRNA context because tRNA$^{Val}_{CAU}$ with C_3G_{70} [66] and tRNA^{fMet} altered to A_3U_{70} [46] are both efficiently aminoacylated by MetRS. Thus, the only significant recognition elements for Met lie in the anticodon (Figure 2A). Because of the strong tendency of MetRS to aminoacylate substrates with methionine nucleotides in the anticodon, particularly C_{34}, there are likely to be negative identity elements for MetRS outside the anticodons of noncognate tRNAs which share some of these bases. Candidates of this class of tRNA include tRNA$^{Arg}_{CCU}$, tRNA$^{Arg}_{CCG}$, and tRNA$^{Leu}_{CAG}$.

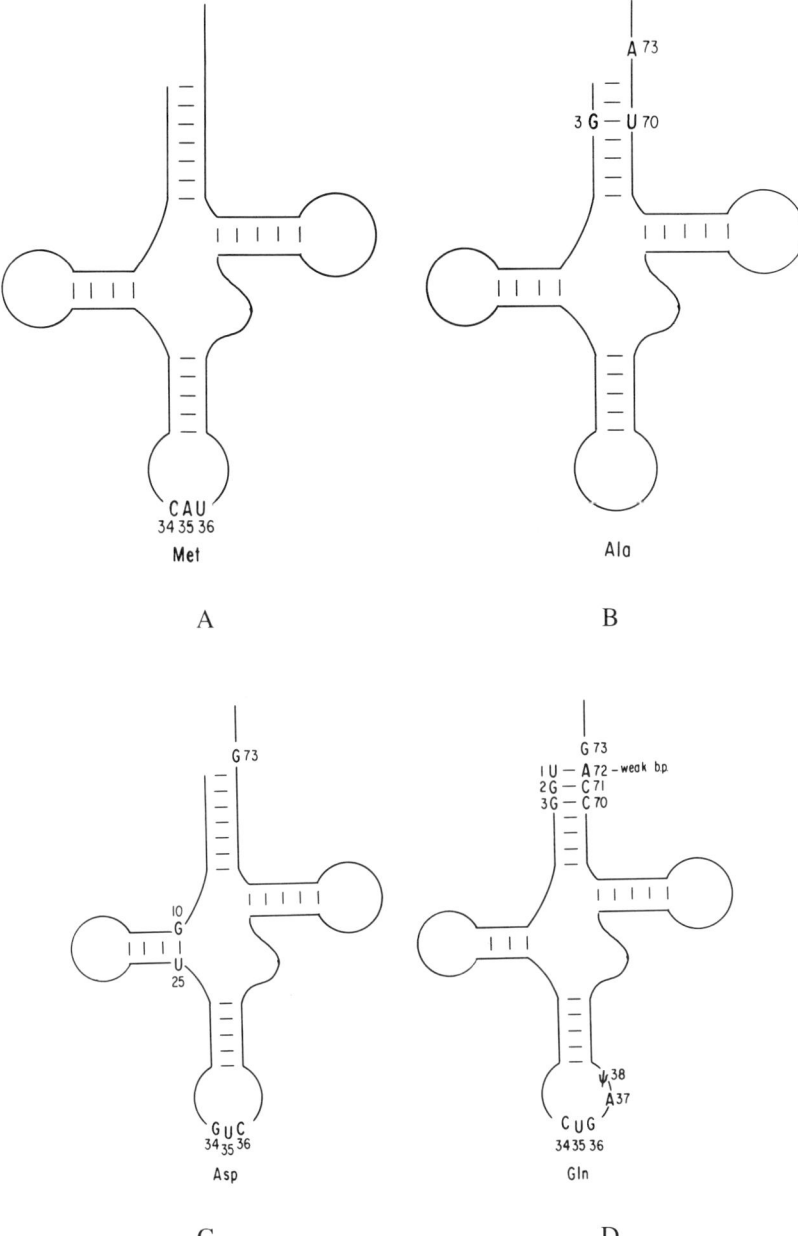

FIGURE 2. The major recognition elements of (A) *E. coli* tRNA[Met], (B) *E. coli* tRNA[Ala], (C) yeast tRNA[Asp], and (D) *E. coli* tRNA[Gln].

D. VALINE tRNAs

The anticodon has also been found to make a strong contribution to recognition of tRNAs and tRNA-like structures by *E. coli*, yeast, and wheat germ valyl-tRNA synthetases (Table 1). The first direct evidence of this in *E. coli* came from *in vitro* work in which the anticodon of *E. coli* tRNA$_1$^{Val}($U_{34}A_{35}C_{36}$) was transferred to tRNA$_m^{Met}$ allowing a 10^4-fold increase in aminoacylation of the hybrid tRNA with valine.[66] Aminoacylation studies carried out with a tRNA$_m^{Met}$ (anticodon UAU) derivative established the importance of C_{36} for Val recognition, because this tRNA was aminoacylated with Val only slightly better than wild-type tRNA$_m^{Met}$ (CAU).[66] *In vivo* experiments in *E. coli* using the initiation assay have illustrated the importance of both A_{35} and C_{36} for aminoacylation by ValRS.[40] In support of this, recent *in vitro* experiments have found that substitutions at positions 35 and 36 result in a large loss of aminoacylation activity with ValRS, with the greatest effect seen at A_{35}.[94] The nucleotide in the wobble position is not conserved in the three *E. coli* Val isoacceptors and *in vitro* experiments have shown that it is not a site of recognition by ValRS.[94]

Early experiments carried out with a yeast tRNA^{Val}(anticodon $I_{34}A_{35}C_{36}$) indicated that position 36 (C) in the anticodon is an important recognition element in this tRNA as well. Substituting U_{36} for C_{36} in the anticodon resulted in loss of detectable aminoacylation of the tRNA by the yeast ValRS.[57] A large part of the information on tRNA^{Val} recognition in eukaryotes comes from studies of the tRNA-like structure at the 3' end of turnip yellow mosaic virus (TYMV) RNA, which is efficiently and specifically aminoacylated with Val by ValRS from bacteria,[95] yeast,[96] and plants.[97] An 86-nucleotide segment of the 3' end of the TYMV RNA has been used in studies of recognition by yeast and wheat germ ValRS. Mutations introduced into the middle position of the sequence corresponding to the tRNA$_{CAC}^{Val}$ anticodon greatly reduced the efficiency of aminoacylation by the yeast enzyme.[98] Thus, the yeast ValRS, like *E. coli* ValRS, recognizes the anticodon bases corresponding to both A_{35} and C_{36} in tRNA^{Val}. The effects of the mutations at the middle and third positions of the "anticodon" of TYMV RNA on recognition by wheat germ ValRS varied depending on the substitution, with roughly 10^3-fold reductions in aminoacylation upon changing the middle position and from 16 to 300-fold reductions associated with substitutions in the third position.[99] As in *E. coli*, the wobble nucleotide in wheat germ tRNA^{Val} appears to play no role in recognition by wheat germ ValRS, since substitutions at the equivalent position of TYMV RNA have no effect on aminoacylation.

Additional *in vitro* work suggests that the discriminator base may also be important to tRNA^{Val} identity. Mutations introduced at position 73 using transcripts of *E. coli* tRNA^{Val} resulted in up to 10^3-fold reductions in aminoacylation with *E. coli* ValRS.[94] Studies using tRNA fragments as competitive inhibitors of aminoacylation by yeast ValRS, or mutants of TYMV RNA containing base changes at the site corresponding to position 73 in tRNA^{Val}, indicate that the contribution of the discriminator base to Val identity may be a conserved feature.[98]

V. ROLE OF THE ACCEPTOR STEM AND "DISCRIMINATOR" BASE AT POSITION 73

If, prior to the study of tRNA identity, one were to guess which sites of the tRNA might be involved in base-specific interactions with a synthetase, perhaps the anticodon, for reasons outlined earlier, and the acceptor stem, because of its proximity to the catalytic site, would be suspect. This has in fact largely proven true because, aside from the anticodon, most tRNA identity elements have been found to cluster in the acceptor stem and the fourth base from the 3' end of the tRNA. The latter, a single-stranded site in all but E. coli tRNA^{His}, has been coined the "discriminator" base because of its hypothesized role in dividing tRNAs into four distinct recognition groups.[100] Although some of the original ideas about the discriminator base appear to be incorrect, much recent evidence indicates that it does play an important role in tRNA identity. Base pairs in the acceptor stem have been found to contain recognition elements for Ala, Gln, Gly, His, and Ser, whereas the discriminator site is important for recognition of at least twelve E. coli tRNAs (Table 3). tRNAs with recognition elements lying in the acceptor stem and discriminator base have also been postulated to be the earliest to evolve, since they conserve the recognition and amino acid attachment sites in the same domain.[101]

A. ALANINE tRNAs

tRNA^{Ala} represents the premier example of a tRNA with important recognition elements in the acceptor stem. The identity of tRNA^{Ala} has been found to lie almost exclusively in a single G_3U_{70} base pair, based on results carried out both *in vitro*[32,33,78] and *in vivo*.[28,32] The importance of this site was suggested from comparative sequence analysis which found that tRNA^{Ala} alone contains this base pair at the 3-70 position.[102,103] Construction of amber suppressor derivatives of tRNA^{Ala} which replace the G_3U_{70} sequence with G:C, A:U, or U:G base pairs, results in tRNAs with dramatically decreased suppression efficiency. Additionally, introducing the G_3U_{70} sequence into amber suppressor derivatives of tRNA^{Cys}, tRNA^{Lys}, or tRNA^{Phe}, resulted in tRNAs capable of inserting Ala into protein, indicating the importance of this site to tRNA^{Ala} identity.

Data generated with an extensive set of tRNA^{Ala} derivatives, which include mutations introduced into the anticodon stem and loop, the D-stem and loop, the TΨC-stem, and the acceptor stem and discriminator base, indicate that the majority of bases in the tRNA are dispensable for alanine accepting activity.[32] Further proof of this comes from the fact that a minihelix consisting of the acceptor stem and T-stem and loop of tRNA^{Ala} is aminoacylated with kinetics only fivefold below that of authentic tRNA^{Ala}.[104] A microhelix of tRNA^{Ala}

TABLE 3
Role of the Discriminator Base in Recognition of *E. coli* tRNAs

	tRNA	Important site	*In vivo*(a) and/or *in vitro*(b)	Effect of base changes *in vitro*	Ref.
A_{73}	Ala	+	b	Reduced rate of transfer; G>U>C	108,109
	Arg	+	a		36
	Ile	+	b	Large loss of activity	151
	Lys	+	a		36
	Met	−	b	≤threefold effect on k_{cat}/K_m	92
	Phe	−	b	≤threefold effect on k_{cat}/K_m	138
	Tyr	+	a, b	k_{cat}/K_m reduced 8 to 40-fold; G>U>C	23,73
	Val	+	b	k_{cat}/K_m reduced 4 to 200-fold; U>>G	94
G_{73}	Asp	+	b	k_{cat}/K_m reduced 200 to >700-fold;C>A>U; K_m and k_{cat}	119
	Gln	+	a, b	k_{cat}/K_m reduced 2 to 10^3-fold;U>C>>A; K_m and k_{cat}	20,21, 23,37
	Trp	+	a, b		42, 155
U_{73}	Cys	+	a		41
	Gly	+	a, b	Large loss of activity	43,106,115
C_{73}	His	+	b	k_{cat}/K_m reduced 10 to >10^4-fold; G>>U>A; mainly k_{cat}	106,113, 114

consisting of only the acceptor stem and a seven membered loop is also a substrate of AlaRS, but in this case with a 50-fold reduction in specificity. This modest reduction of aminoacylation of the microhelix by AlaRS suggests that, though the microhelix contains the major recognition determinants in tRNA^{Ala}, there is a loss of contacts which somewhat improve both binding and aminoacylation. Footprinting experiments performed with the tRNA^{Ala}-AlaRS complex show that the enzyme protects only phosphates in the acceptor stem and on the 3' side of residues 64-70.[105] Nucleotides at positions 16, 17, 20, and 60 have also been suggested to play a role in tRNA^{Ala} identity,[28] however, these bases are not conserved in *B. mori* and human alanine tRNAs which are known to be efficient substrates of the *E. coli* AlaRS,[34] suggesting they play minor roles.

In vitro analysis with a set of tRNA^{Ala} mutants supports the idea that the G_3U_{70} base pair in tRNA^{Ala} is directly recognized by AlaRS. Introducing A_3U_{70}, U_3G_{70}, or G_3C_{70} sequences into tRNA^{Ala} prevents aminoacylation, even in the presence of high concentrations of AlaRS. In addition, delivering the G_3U_{70} determinants into a number of different substrates allows efficient aminoacylation with Ala *in vitro*, strongly suggesting that the G_3U_{70} base pair is the major recognition determinant. More recent work with tRNA^{Ala} microhelices indicates that the 2-71 position also modulates AlaRS recognition, since substitutions at this site reduce aminoacylation with Ala.[106] The observation that a G_3U_{70} base pair can act as an tRNA^{Ala} recognition element, whereas a G_3C_{70} base pair combination cannot, prompted further study of the specific role of this determinant. In a normal G:C base pair, the two-amino group of guanine is involved in hydrogen bonding interaction with the oxygen at position 2 (O-2) of cytosine. However, NMR studies indicate that the two-amino group of G_3 in the minor groove of the tRNA helix is not involved in base pairing interaction with U_{70}. The minor groove of A-form RNA is very wide and shallow and is therefore readily accessible to amino acid side chains which could make discriminating contacts with the tRNA. To test the possibility that AlaRS makes base-specific contact with the two-amino group of G_3, oligoribonucleotide substrates were synthesized containing base analogs which would either maintain the two-amino group or delete it without disturbing base pairing interactions.[107] RNA duplex substrates which substitute the guanine of the G_3U_{70} base pair with inosine or adenine, bases lacking the two-amino group of a guanine nucleotide, were found to be inactive in aminoacylation with AlaRS. Additionally, substituting two-amino purine (2-AP) at position 3 resulted in a duplex with the two-amino group of 2-AP involved in base pairing interaction with the O-2 of uracil. This duplex was also found to be devoid of alanine acceptor activity, as would be expected if an unpaired two-amino group is required for recognition. Only the RNA duplex with the G:U base pair was an efficient substrate of AlaRS, leading the authors to conclude that an unpaired two-amino group of G_3 is the determinant specifically recognized. In contrast to this, McClain et al have found that substitutions of the 3-70 position in tRNA$^{Ala}_{CUA}$, by "wobble" base pairs which contain neither G_3 nor U_{70}, still allow alanine insertion into protein with low efficiency.[30] Furthermore, two sets of mispaired bases, G_3A_{70} and A_3C_{70} (in addition to G_3U_{70}) allowed insertion of Ala into protein by tRNA$^{Lys}_{CUA}$. Based on this, the authors concluded that a "helix irregularity", owing to the G_3U_{70} "wobble" base pair, is the motif recognized by AlaRS rather than a specific sequence at this site. A compromise between these two sets of conflicting data might be arranged by suggesting that unique conformational effects, as well as base specific contacts, allow recognition by AlaRS. In the absence of the guanine two-amino group of G_3, AlaRS may still possess weak affinity for substrates with helical irregularities in the acceptor stem, particularly if the tRNA substrate is overproduced several 100-fold and there is no strongly competing synthetase.

Several of the suppressor tRNAs to which the G_3U_{70} sequence was transferred failed to insert Ala into protein (tRNA$^{Gly}_2$ and tRNA^{Tyr}),[30,33] or were found to insert Ala along with the cognate amino acid (tRNA^{Phe}).[30,32] Mutations introduced at additional sites in the Phe suppressor allowed complete conversion to Ala identity.[30] The role of these mutations was not investigated *in vitro* however, so it remains unclear whether the effect was to reduce the affinity of PheRS or to increase the affinity of AlaRS. With the tRNA^{Tyr} suppressor, *in vitro* kinetic studies showed that it was only a tenfold poorer substrate for AlaRS than wild-type tRNA^{Ala}, and it was suggested that the lack of aminoacylation with Ala *in vivo* might be due to competition from the TyrRS. Overproduction of AlaRS from a plasmid carrying the *alaS* gene resulted in Ala insertion by the Tyr suppressor, confirming this hypothesis.[33] tRNA$_2^{Gly}$(CUA) with the G_3U_{70} sequence, which is charged with GLN was not extensively characterized, but will most likely be found to contain negative elements for AlaRS resulting in an inability of AlaRS unable to compete with GlnRS. There may be additional structural features present in tRNA^{Ala} modulating efficient aminoacylation by AlaRS which are absent in noncognate tRNAs that might otherwise have some affinity for AlaRS, as has been proposed.[28,107a]

Of six different tRNAs to which the alanine G_3U_{70} sequence has been transferred and analyzed *in vitro*, tRNA$_{CUA}^{Cys}$ is by far the worst Ala acceptor.[32] Work with a tRNA^{Cys} minihelix containing the G_3U_{70} sequence suggested that the defect lies in the acceptor stem or TΨC stem, since it also proved to be a poorer substrate of AlaRS than other tRNA minihelices tested.[108] Comparison of the sequences of the six tRNAs indicated that the discriminator base might be the cause of the reduced aminoacylation of tRNA^{Cys}, since it contains U at position 73 whereas all the others have A. Introduction of an A in place of the U at position 73 of the Cys minihelix restored aminoacylation efficiency to very near the level of the Ala minihelix. Further work showed that while any nucleotide at position 73 will allow aminoacylation by Ala, providing the G_3U_{70} base pair is present, the efficiency is substantially reduced when G, C, or U occupy this site (Table 3). Schimmel et al have explored the catalytic step where discrimination of the N73 nucleotide takes place.[109] Since the effect on aminoacylation is mainly on k_{cat} the authors tested whether pre- or posttransfer hydrolysis editing was responsible for the rate reduction, or alternatively, whether transfer of the activated aminoacyl-adenylate to the tRNA was impaired. Results obtained from single turnover charging experiments indicated that transfer of the amino acid to the bound tRNA was greatly reduced for the N73-substituted substrates. This represented the first demonstration of nucleotide substitutions in a tRNA affecting the transition state of the aminoacylation reaction. In addition to the G_3U_{70} sequence, therefore, the discriminator base also makes a contribution to Ala recognition (Figure 2).

Conservation of the alanine G_3U_{70} sequence in higher organisms[110] suggests that the function of this important element has been maintained in evolution. Early studies carried out *in vitro* with RNA duplex substrates derived from yeast tRNA^{Ala} established that the acceptor stem alone is sufficient for aminoacylation by the yeast AlaRS.[111] In addition, recent data indicates that the G_3U_{70} sequence in human and *B. mori* tRNA^{Ala} are required for aminoacylation by their respective synthetases.[34]

B. *E. COLI* HISTIDINE tRNA

tRNA^{His} is unique in being the only *E. coli* tRNA with an eight base-pair acceptor stem and only three unpaired 3' nucleotides (Sel-Cys tRNA has an eight base pair acceptor stem but four unpaired 3' nucleotides).[112] The difference lies in the fact that the His tRNA contains an extra 5' nucleotide which can base-pair with the discriminator base at position 73. Experiments conducted *in vitro* with tRNA^{His} transcripts lacking the extra 5' nucleotide (denoted the -1 position), or with substrates in which the G-1:C$_{73}$ base pair has been altered to A-1:U$_{73}$, resulted in reductions of aminoacylation of greater than three orders of magnitude.[113] C$_{73}$ is also unique to tRNA^{His}[110] and has been found to play an important role in His tRNA recognition whether or not G-1 is present (Table 3).[113] The major effect of substitutions at these sites is on k$_{cat}$, indicating that this part of the tRNA plays a role in the proper positioning of the 3' end during the catalytic step. Additional evidence for the function of the -1-73 position in *E. coli* tRNA^{His} recognition comes from studies carried out with a tRNA^{His} microhelix which was found to be a substrate of the HisRS only when the G-1:C$_{73}$ base pair is present, regardless of the neighboring sequence context.[106,114] Because this base pair was also found to block aminoacylation by several noncognate synthetases *in vitro*, it apparently serves a dual function in establishing tRNA^{His} identity. Additional work showed that the 2-71 and 3-70 sequences were important in modulating the acceptor activity of the minihelix, with the wild-type His sequences optimal.[106]

C. OTHER *E. COLI* tRNAs

Aside from the synthetases mentioned above, there are at least twelve other *E. coli* synthetases which discriminate on the basis of acceptor stem or N73 nucleotides (Table 3). One of the better characterized of these is tRNA^{Gly}. A microhelix based on the sequence of tRNA^{Gly} has been constructed and found to be a good substrate of GlyRS, indicating that important sites of recognition lie in the acceptor stem and/or discriminator base.[106] Results of tRNA^{Gly} charging profiles of a large number of microhelix mutants found that C$_2$G$_{71}$ and the discriminator base U$_{73}$ are the two primary recognition determinants in the acceptor stem, with the 3-70 position playing an additional minor role. *In vivo* results using the initiation and suppression assays largely support these contentions.[43,115] However, one must not make the assumption that, because the tRNA^{Gly} microhelix is a substrate of GlyRS, the presence of additional identity determinants is ruled out. In fact, a good deal of evidence suggests that the anticodon plays a significant role in recognition by GlyRS.[41,43,115-118] Aminoacylation studies carried out with a large number of permutations of the tRNA^{Gly}, tRNA^{Ala}, and tRNA^{His} minihelices indicate that an overlap of specificity-determining nucleotides, where positive elements for one synthetase are negative elements for the others, is important in maintaining fidelity in this class of synthetases.[106]

The discriminator base has also been found to play a role in the aminoacylation of a number of other tRNAs. *E. coli* tRNA^{Asp}, which encodes G at position 73, has been shown to fall into this class since substitution to other nucleotides can effect aminoacylation by a factor of 200 or more (Table 3).[120] In contrast to the case of tRNA^{His}, the consequence of nucleotide replacement at this site is split between K_m and k_{cat}, possibly indicating that both binding and catalysis are perturbed. Nucleotide substitutions at this position with the least severe effects (U>A) conserve functional groups with G, suggesting that direct contacts are made by the enzyme. Similar findings have also been made with yeast tRNA^{Asp} (see following). *E. coli* TyrRS also appears to recognize determinants in the discriminator position of tRNA^{Tyr}, since conversion of A_{73} to G_{73} reduces both tyrosine insertion of a tRNA^{Tyr} suppressor *in vivo* and hinders aminoacylation *in vitro* (Table 3).[23,73] Modeling studies of the interaction of *B. stear.* TyrRS and tRNA^{Tyr} has led to the proposal that protein-nucleic acid contacts are made at the discriminator site to ensure proper positioning of the 3' end of the tRNA at the active site.[120]

Use of the *in vivo* initiation assay has also proven useful in studies aimed at the role of the discriminator base. For example, the *E. coli* initiator tRNA with a Cys anticodon alone and A_{73} {tRNA$^{fMet}_{GCA}$} is only weakly aminoacylated with Cys *in vivo*. However, introduction of the Cys discriminator base, U_{73}, resulted in greater than 90% Cys insertion into protein.[41] Altering the discriminator base to C and G results in loss of Cys acceptor activity by tRNA$^{fMet}_{GCA}$ mutants. Similarly, a tRNA^{fMet} derivative with the Trp (CCA) anticodon inserts 40% Trp and 60% Met when it exists in an A_{73} background.[42] The Met inserted by tRNA$^{fMet}_{CCA}$ probably arises from weak mischarging by MetRS because of the strong tRNA^{Met} recognition element C_{34} in the Trp anticodon. However, altering the tRNA^{fMet} A_{73} discriminator base to the G_{73} sequence found in tRNA^{Trp}, results in quantitative Trp insertion into protein.[42] The loss of Met acceptor activity of the G_{73} vs. the A_{73} derivative likely indicates superior competition by TrpRS due to its increased affinity for the tRNA. Additionally, substitutions of A, U, and C at the discriminator site of a Trp opal suppressor tRNA results in dramatic decreases in suppression, further supporting the conclusion that the discriminator base is an important identity element in tRNA^{Trp}.[42]

Studies carried out with a Lys amber suppressor suggest that the discriminator base is also important for aminoacylation with Lys. Replacing A_{73} with G_{73} results in a decrease of suppression efficiency by 1/3 and insertion of Gln in addition to Lys.[36] The effect of this substitution most likely reflects reduced aminoacylation with Lys rather than increased aminoacylation with Gln, based on *in vitro* studies carried out with tRNA^{Gln} transcripts (see following). Substituting U_{73} into the discriminator site of tRNA$^{Lys}_{CUA}$ reduced the suppression efficiency even more dramatically, further indicating that aminoacylation by LysRS is impaired in substrates lacking the wild-type Lys discriminator nucleotide.[36]

VI. THE *E. COLI* GLUTAMINE tRNA-GLUTAMINE SYNTHETASE COMPLEX

One of the most important recent developments in the tRNA identity field is the solution of the tRNA^{Gln}-GlnRS-ATP co-crystal structure to 2.8Å resolution.[121] The X-ray crystal structure now provides a clear explanation of over a decade's worth of biochemical data and also illustrates several unexpected features of tRNA recognition. This achievement was made possible by the earlier cloning and sequencing of GlnRS and the development of efficient methods of overproduction of the synthetase and tRNA$_2^{Gln}$(CUG) isoacceptor.[122-124] GlnRS functions as a 63-Kd monomer of 553 amino acids arranged in an elongated structure made up of four domains. The enzyme contacts the inside surface of the tRNA^{Gln} L-shaped structure extending from the anticodon all the way to the 3′ end of the acceptor stem. As discussed earlier, a large body of evidence suggests that nucleotides in the anticodon of tRNA^{Gln} are strongly recognized by GlnRS, making the observation that the anticodon adopts an unusual conformation while bound to GlnRS somewhat less surprising.[125] Nucleotides C_{34} and G_{36} are unstacked and splayed out, while U_{35} is stacked under A_{37} and buried in a tight fitting protein pocket. The important role played by nucleotide U_{35} is illustrated by the large number of protein contacts made with this base. These strong interactions, involving several charged amino acid residues, reveal why U is found in all the known GlnRS substrates, since no other nucleotide could make the same protein contacts. In addition to the interactions made with U_{35} further contacts are observed with the other two anticodon bases. Modeling studies suggest that the protein pocket which contains C_{34} could also form base-specific contacts with U, the nucleotide occupying this site in the other tRNA^{Gln} isoacceptor species, by making small protein conformational adjustments; but this pocket could not accommodate a purine.[125] However, the protein contacts made with G_{36} appear to be specific to this base, and no other nucleotide would efficiently substitute for guanine at this site. Recent *in vitro* studies carried out with tRNA^{Gln} transcripts containing mutations at this site have confirmed this point.[37]

In addition to the anticodon base requirements of tRNA^{Gln}, anticodon loop nucleotide substitutions at positions 37 and 38 result in severe reductions in aminoacylation by GlnRS *in vitro*.[37] Mutations were introduced at these sites based on examination of the co-crystal structure which showed protein interactions with A_{37} and Ψ_{38}. The effect of alterations at these sites may be due to direct loss of contacts with the enzyme. However, the crystal structure indicates that the anticodon loop nucleotides U_{32} and U_{33} are involved in non-Watson-Crick base pairs with their cross-loop counterparts at positions 37 and 38, respectively. Thus, the effect of substitutions at positions 37 and 38 may also lie in the fact that the nucleotides introduced at these sites are unable to form the same interactions, impairing adoption of the proper anticodon loop conformation for aminoacylation by GlnRS. Substitutions at positions 33 and 34, nucleotides which do not contact GlnRS in the crystal structure, also result in decreased aminoacylation *in vivo*, lending weight to this argument.[126]

GlnRS has also been found to prefer substrates with certain acceptor stem features in addition to nucleotides in the anticodon.[20-25] The first clue to this came from the results of genetic studies carried out with the tRNA^{Tyr} amber suppressor mentioned earlier. A mutation which replaced the A_{73} discriminator base of tRNA^{Tyr} with G_{73}, a nucleotide found in the two tRNA^{Gln} isoacceptor species, resulted in mischarging with Gln. Interestingly, the $G_{73} \rightarrow A_{73}$ change in a tRNA^{Gln} transcript resulted in very little effect on aminoacylation with Gln *in vitro*.[37] Furthermore, only three of the five amber suppressor tRNAs mischarged with Gln contain the G_{73} sequence. *In vitro* work carried out with tRNA^{Tyr} indicates that the effect of the mutation was more to decrease the competition from TyrRS, rather than to increase the affinity of GlnRS (Table 3). Introducing pyrimidines at the discriminator site of tRNA^{Gln}, however, resulted in sizable reductions in aminoacylation with Gln *in vitro*.

Additional mutations of the tRNA^{Tyr} suppressor toward Gln identity were found at the 1-72 position. The mutations isolated did not necessarily correspond to the tRNA^{Gln} sequence at this site, but were always found to convert the G_1C_{72} base pair of tRNA$_{CUA}^{Tyr}$ to a weaker, or mismatched base pair, indicating no sequence or base pairing requirement at this site for recognition by GlnRS. An amber suppressor of tRNA^{fMet} containing a C_1A_{72} mismatch in the acceptor stem was also found to be a good substrate of GlnRS, consistent with these earlier observations.[72] In fact, conversion of the C_1A_{72} mismatch of tRNA^{fMet} to a U_1A_{72} base pair decreases the efficiency of aminoacylation by GlnRS.[45]

Examination of the co-crystal structure now provides an explanation of this biochemical data.[121] While bound to GlnRS, the 3' end of tRNA^{Gln} forms a hairpin structure, with the 1-72 base pair broken and the 3' single-stranded bases looped back towards the anticodon, rather than away from it. This structure is stabilized by a number of protein-RNA contacts as well as an RNA-RNA interaction between the N_2 amino group of G_{73} and a phosphate oxygen of A_{72}. The two distal 3' bases, A_{76} and C_{75}, are stacked together with G_{73}, whereas C_{74} is flipped out of the stack to interact with amino acid side chains of GlnRS. Modeling studies aimed to address the role of the discriminator site suggest that an A_{73} substitution would not seriously disturb the observed base stacking interactions, while introducing a pyrimidine would severely hinder the stacking effect with C_{75}, a view consistent with aminoacylation studies. Also in keeping with the biochemical data, no sequence specific protein-nucleic acid interactions occur between GlnRS and the nucleotides in the 1-72 position, indicating that the only requirement at this position is an easily disrupted base pair. Mutations have also been made at the 2-71 and 3-70 position of tRNA^{Gln} to study the role of these sites in tRNA^{Gln} identity. A:C mismatches, A:U base pairs, and G:U wobble base pairs were inserted at these sites in tRNA^{Gln} transcripts and used to measure the kinetics of aminoacylation by GlnRS.[37] All of the mutations introduced resulted in reductions in aminoacylation of from two to four orders of magnitude, indicating that these bases are important for Gln identity. The co-crystal complex shows that the side chain of Asp235 of GlnRS contacts the N2 amino group of G_3 and an amide carbonyl of Pro181 forms a hydrogen bond with the N2 amino group of G_2, with another G_2-protein interaction occurring through a water molecule. Based on the crystal structure therefore, the 2-71 and 3-70 positions in tRNA^{Gln} are sites of base-specific contact by GlnRS. Additional support for the significance of these protein-RNA interactions is provided by a mutant form of GlnRS in which the Asp235 is replaced by Asn.[127-129] This enzyme is able to aminoacylate an A_3U_{70} tRNA^{Gln} species far more efficiently than the wild-type enzyme, and also exhibits a more relaxed substrate specificity based on its ability to mischarge a number of noncognate tRNAs. The results of biochemical analysis of the 2-71 and 3-70 position, however, indicate that conformational effects as well as base-specific contacts are important at these sites, because G_2U_{71} or G_3U_{70} substitutions which conserve the guanine two-amino group are nonetheless greatly reduced in aminoacylation.

In addition to the nucleotides found important in the anticodon and acceptor stem, bases in the D-stem and loop may also play a role in Gln identity. Contacts with the enzyme are observed at G_{10} and C_{16} in the crystal structure.[121] Substitutions at these sites have not yet been made in tRNA^{Gln}, however, and, of the suppressors mischarged with Gln, only tRNA^{fMet} contains C_{16}, so the importance of this interaction is presently unclear. Amongst the set of suppressor tRNAs mischarged with Gln, only tRNA^{Trp} has all of the acceptor stem sequences preferred by GlnRS as well as G_{10} (but not C_{16}) in the D-stem. This shows that significant charging by GlnRS can occur in the absence of tRNA^{Gln} identity elements in the acceptor stem and D-loop *in vivo*. Thus, the high rate of mischarging of tRNA amber mutants clearly reflects the combination of weak recognition by the cognate synthetase coupled with moderate recognition by GlnRS, largely through nucleotides conserved in the amber and tRNA^{Gln} anticodon. Figure 2. summarizes the known major tRNA^{Gln} recognition elements.

VII. RECOGNITION OF OTHER tRNA DOMAINS

A. tRNAs CONTAINING A LARGE VARIABLE ARM

The first "identity swap" experiment, in which a limited number of nucleotides from one tRNA were transferred to a noncognate tRNA to give it the charging identity of the former, was carried out with *E. coli* tRNA^{Ser}.[11] In this work, tRNA$^{Leu}_5$ containing the amber anticodon was used as the acceptor of the tRNA^{Ser} nucleotides and a selection was utilized whereby insertion of Ser into an amber codon of the β-lactamase gene was required for ampicillin resistant growth. Twelve nucleotide changes of tRNA$_5^{Leu}$(CUA) were found to convert it to Ser acceptance based on both β-lactamase activity and sequence analysis of amber DHFR. Subsequent studies have found that of these 12, only 8 (not counting the anticodon changes) are required for conversion of tRNA^{Leu} to tRNA^{Ser} identity.[4] Six of the eight are found to lie in the acceptor arm and four of these, C$_{72}$ (tRNA^{Leu} has a G$_1$U$_{72}$ base pair, so only the 72 position was replaced), G$_2$C$_{71}$, and G$_{73}$, are conserved in all five *E. coli* serine isoacceptors, as well as bacteriophage T4 tRNA^{Ser}. Despite the fact that either U:A or A:U, are found in the 3-70 position of the six Ser accepting tRNAs, only A$_3$U$_{70}$ was found to specify Ser in the tRNA^{Leu} suppressor. *In vivo* studies with a tRNA^{Ser} suppressor, indicate that the 1-72 and 3-70 positions are important in preventing Gln mischarging, since substitutions at these two sites lead to Gln insertion by the suppressor.[35] The functional significance of this for wild-type tRNA^{Ser} is uncertain though, because it is very unlikely that GlnRS would recognize a tRNA^{Ser} with its natural anticodon. Based on recent *in vitro* studies, the 2-71 position makes the largest acceptor stem contributions to tRNA^{Ser} identity.[130,131] An R:Y pair at the 4-69 position also favors interaction with SerRS and is a conserved feature of all the *E. coli* serine accepting tRNAs. However, this feature is not absolutely essential for tRNA^{Ser} identity because the Ser accepting tRNA^{Leu} suppressor has a C$_4$G$_{69}$ base pair at this position.

Aside from the mutations required in the acceptor stem only one additional change was needed to convert tRNA$_5^{Leu}$(CUA) into a Ser accepting tRNA. This change resulted in replacement of the tRNA^{Leu} U$_{11}$A$_{24}$ sequence in the D-stem with C$_{11}$G$_{24}$. The C$_{11}$G$_{24}$ sequence is conserved in all *E. coli* tRNA^{Ser} isoacceptors as well as the T4 Ser tRNA, but is not found in the Ser-accepting selenocysteine tRNA. The *E. coli* selenocysteine tRNA is aminoacylated by SerRS with kinetics 100-fold below that of the other Ser-accepting tRNAs,[132,133] followed by enzymatic conversion of Ser to selenocysteine.[134] Aside from the 11-24 sequence, there are several additional features of the selenocysteine tRNA distinguishing it from the other serine accepting tRNAs, including, an eight base pair acceptor stem, a four base pair D-stem, and the lack of several nucleotides conserved in the D-stem and loop of all other *E. coli* tRNAs. Thus, it is difficult to assign the decreased aminoacylation of this tRNA to a particular set of nucleotides, since any number of these deviations may be responsible.

tRNA$_5^{Leu}$ was used in these identity swap studies because it shares a number of characteristics with tRNA^{Ser}, including the presence of a long variable arm. Because there are only three tRNA isoacceptor families in *E. coli* which possess this feature, it would seem a logical site of discrimination. Studies carried out with a set of tRNA^{Ser} mutants have substantiated this suspicion and indicate that the variable arm plays a role in tRNA^{Ser} identity by adopting an orientation favorable for interaction with SerRS.[73] Phosphorothioate footprinting experiments conducted with tRNA^{Ser} and SerRS indicate that the synthetase makes direct contacts with nucleotides in the variable arm, consistent with the function of this site as a recognition element in tRNA^{Ser}.[135] Only a very limited number of sites outside the acceptor stem and variable arm are conserved in the six Ser-accepting tRNAs, suggesting that these domains contain the major tRNA^{Ser} recognition elements.

Adding or deleting nucleotides in the variable arm of *E. coli* tRNA^{Tyr} transcripts caused reductions in aminoacylation by TyrRS of greater than 10^2.[73] Thus, the variable arm of tRNA^{Tyr}, like tRNA^{Ser}, contributes to recognition by its cognate synthetase. In addition to the variable arm, the anticodon and discriminator base were found to contain recognition determinants for TyrRS, with substitutions at U_{35} resulting in defects similar in magnitude to those obtained from mutations introduced into the variable arm. Alterations of the discriminator base had smaller effects on aminoacylation (Table 3). *E. coli* tRNA^{Leu} has not been extensively characterized, so it remains to be seen whether this tRNA, like the others, harbors important recognition determinants in its variable arm. Less explored, but also possible in this class of tRNAs is that the variable arm functions as an identity element in preventing aminoacylation by noncognate synthetases by steric hindrance.

Recent *in vivo* results indicate that a large variable arm may function as an identity element in eukaryotic tRNAs as well.[136] Expression of an *E. coli* tRNA^{Tyr} amber suppressor in yeast results in mischarging with Leu based on the sequence of suppressed protein. The yeast tRNA^{Tyr} differs from its *E. coli* counterpart in that it does not contain a long variable arm, and no cross-species charging of the tyrosine tRNAs is observed *in vitro*. Yeast tRNA^{Leu} is a member of the class of tRNAs possessing large variable arms, suggesting that the cognate tRNA^{Leu} and *E. coli* tRNA^{Tyr} may both be recognized by the yeast LeuRS at least partly as a consequence of this feature.

B. *E. COLI* ARGININE tRNA

Early biochemical data suggested an important role in synthetase recognition for the middle nucleotide in the anticodon of *E. coli* Arg tRNAs, C_{35}.[56] More recently, this site, and A_{20} in the D loop, were identified as potential sites of discrimination by ArgRS from a computer assisted analysis of *E. coli* tRNAs designed to identify unique nucleotide combinations.[102] To investigate the role of A_{20}, a nucleotide found exclusively in *E. coli* arginine isoacceptors, amber suppressor derivatives were constructed with or without A_{20} and analyzed for Arg insertion at amber codons of DHFR.[29,71] An amber suppressor derivative of the major tRNA^{Arg} inserted a mixture of Arg and Lys, but upon deletion of A_{20}, or replacement with U, the suppressor inserted only a small trace of Arg. Even more convincingly, substitution of U_{20} with A_{20} in an amber suppressor derivative of tRNA^{Phe} (with an additional $U_{59} \rightarrow A_{59}$ substitution to allow proper processing of the tRNA) resulted in the insertion of mainly Arg. Nucleotides 16, 17, 20, 59, and 60 reside in close proximity to one another in the tRNA tertiary structure to form a potential region of interaction with the synthetase known as the variable pocket. It has been postulated that, because of the accessibility of this region and the fact that the number and identity of nucleotides making it up are different in different tRNAs, this tRNA domain might serve as a site of discrimination amongst synthetases.[137] The evidence obtained for A_{20} in tRNA^{Arg} is consistent with this notion.

The other potential tRNA^{Arg} recognition element, identified in the computer analysis and from biochemical data as a potential contributor to tRNA^{Arg} identity, C_{35} in the anticodon, has also been explored in detail. *In vivo* experiments have shown that the amber suppressor derivative of the major tRNA^{Arg}, lacking C_{35}, inserts mainly Lys, while an opal suppressor derivative with C_{35} inserts mainly Arg.[29,36] *In vitro* analysis of tRNA transcripts have indicated that C_{35} is the primary recognition element in the conversion of a tRNA^{Met} to a tRNA^{Arg}.[65] Introduction of an Arg anticodon into tRNA$_m^{Met}$ resulted in a 40,000-fold increase in specificity for aminoacylation by ArgRS compared to tRNA$_m^{Met}$ with the Met anticodon, while inserting the Arg A_{20} sequence into tRNA$_m^{Met}$ resulted in only 1000-fold improvement in aminoacylation compared with the substrate lacking this change. Both changes, when present together, resulted in charging very near to that observed for wild-type tRNA^{Arg}. In contrast to this, *in vivo* work carried out with opal and amber suppressor tRNA^{Arg} derivatives suggest that A_{20}, rather than C_{35} is the main tRNA^{Arg} identity element.[36] As these two studies were carried out in different tRNA contexts using different experimental systems, it is difficult to directly compare the results. It is clear from both studies, however, that C_{35} and A_{20} together make the primary contributions to Arg identity.

In addition to the previously mentioned, the discriminator base in tRNA^{Arg} has also recently been implicated as a site of tRNA^{Arg} identity.[36] Substitutions of the discriminator base of an opal suppressor tRNA^{Arg} from A_{73} to C_{73} or U_{73} resulted in notable reductions in suppression efficiency. Significantly, alteration of the discriminator base to G_{73}, the nucleotide found in two of the four arginine isoacceptors, had little effect on suppression. This result raises the possibility that both G_{73} and A_{73} are recognized equally well by ArgRS, or alternatively, that C_{73} and U_{73} are negative elements for ArgRS. The resolution of these possibilities, as well as determination of the kinetic contribution of the discriminator base to tRNA^{Arg} identity will have to await more direct inquiry.

C. PHENYLALANINE tRNAs

In vitro studies have now been carried out with an extensive set of mutants of *E. coli*,[138] yeast,[62,63,139,140] and human[141] Phe tRNAs, allowing a detailed comparison of the tRNA recognition profiles for one type of amino acid acceptor in three widely divergent species. In addition, a number of *in vivo* experiments have been performed to study tRNA^{Phe} identity in *E. coli*.[31,39,40] The results of *in vitro* inquiry, as well as the *in vivo* studies in *E. coli*, indicate that the anticodon is the primary recognition site in the Phe tRNAs of all three organisms, with reductions in aminoacylation of greater than three orders of magnitude associated with substitutions at this site in *E. coli* and human tRNAs, and smaller effects in yeast tRNA. Amongst the three anticodon bases, G_{34} and A_{35} provide large contributions of similar magnitude for recognition,

tRNA Discrimination in Aminoacylation 305

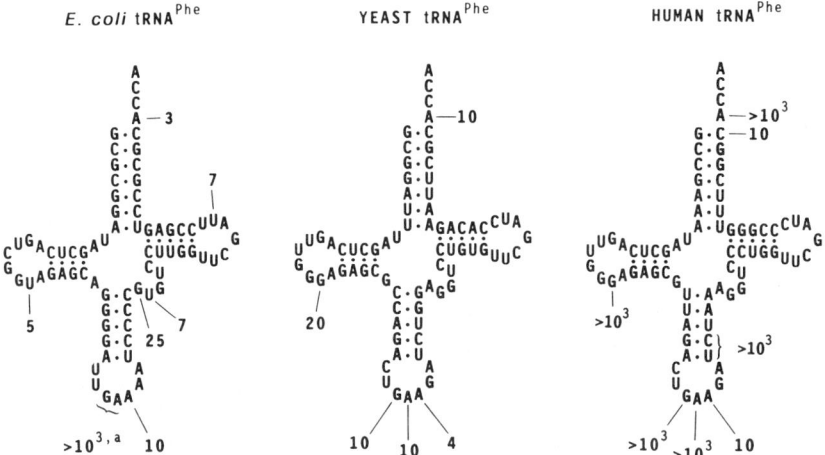

FIGURE 3. The kinetic effect of nucleotide substitutions of *E. coli*, yeast, and human tRNA^{Phe} on aminoacylation by the cognate synthetase.[16,62,63,138,139,141] The numbers shown refer to the fold reduction in k_{cat}/K_m upon substitution at that site. Where variable effects of aminoacylation were observed, kinetic data for the nucleotide substitution causing the largest defect are shown.

with A_{36} playing a smaller role (Figure 3 and Table 1). Although the contribution of the anticodon to yeast tRNA^{Phe} identity appears to be less than that of the other two, its overall importance relative to other sites in the yeast tRNA is quite similar to that of the *E. coli* and human tRNAs. Interestingly, the magnitude of discrimination by the human enzyme seems to more closely resemble *E. coli* than yeast, indicating that the diminished discrimination observed in yeast is not necessarily a general trait of eukaryotes.

In vivo experiments carried out with an *E. coli* suppressor tRNA^{Phe} have indicated that U_{20}, $G_{27}C_{43}$, $G_{28}C_{42}$, G_{44}, U_{45}, U_{59}, U_{60}, and A_{73} also contribute to tRNA^{Phe} identity.[31] Many of these sites have also been studied *in vitro*, in particular positions 20, 44, 45, 59, and 73, and individually contribute only modestly to tRNA^{Phe} recognition in *E. coli*.[138] This indicates many of the nucleotides identified from the *in vivo* studies are not tRNA^{Phe} recognition elements, but rather nucleotides keeping noncognate synthetases at bay. Nonetheless, these, or other elements outside of the anticodon, clearly have some importance to tRNA^{Phe} identity in *E. coli* since the tRNA^{Phe} amber suppressor retains its identity[26,31] despite the loss of the two most important tRNA^{Phe} recognition elements identified, G_{34} and A_{35}. Taken together, the *in vitro* and *in vivo* results suggest that the retention of tRNA^{Phe} identity in the absence of the primary anticodon elements results from the sum of a large number of small contributions, including that provided from A_{36} which is conserved in the amber anticodon. In addition, the suppressor tRNA^{Phe} may be a poor substrate for GlnRS and LysRS.

Several of the sites outside the anticodon found to contribute to *E. coli* tRNA^{Phe} identity have also been explored in the yeast and human Phe tRNAs. Of these, only positions 20 and 73 contribute to recognition by the yeast and human enzymes. G_{20} is unique to phenylalanine tRNAs among the known yeast and human cytoplasmic species and has been found to be a major recognition site for the PheRS of both organisms. Substituting U_{20} for G_{20} in the human tRNA^{Phe} results in more than a 1000-fold reduction in aminoacylation by human PheRS. Substitutions of the G at position 20 of the yeast counterpart result in smaller effects on aminoacylation by yeast PheRS, but show that this site contributes a large fraction of the total recognition of yeast tRNA^{Phe} (Figure 3). Substitution of position 20 in *E. coli* tRNA^{Phe} results in, at most, a fivefold reduction in aminoacylation by the cognate synthetase and represents only a tiny fraction of the total recognition profile for the *E. coli* tRNA (Figure 3). Native *E. coli* tRNA^{Phe} contains a dihydrouridine residue at position 20, a modified base found at this site in many other *E. coli* tRNAs. Thus, both the importance and identity of the base at position 20 are conserved in the eukaryotes, yeast and human, and neither are conserved in the prokaryote *E. coli*.

The role of the discriminator nucleotide, A_{73} in all three Phe tRNAs, has also been explored and found to vary in importance in *E. coli*, yeast, and human. In *E. coli*, the discriminator base plays essentially no role at all in recognition (Table 3). In yeast, on the other hand, this is an important recognition site, and in human, position 73 plays a very large role (Figure 3). Again, recognition by the eukaryotic synthetases appear more similar to each other than the prokaryote is to either of the eukaryotes.

Studies with the human PheRS have shown that changes at the 1-72 base pair also have a small effect on aminoacylation. Alteration of two base pairs adjacent to the anticodon loop (positions 30-40 and 31-39) have a significant effect on recognition by human PheRS. At present, it is not clear whether the latter is a direct effect on recognition or whether base substitutions in the anticodon stem indirectly alter recognition of important nucleotides in the anticodon itself. Interestingly, although all species of tRNA^{Phe} contain the same primary sequence in this region, only the human PheRS is sensitive to the changes in these base pairs (Figure 3), emphasizing that significant differences exist in the mode of interaction of these evolutionarily diverse enzymes with their substrates. Nevertheless, it is possible that, like yeast tRNA^{Phe}, the major recognition sites for human tRNA^{Phe} reside in the anticodon bases $G_{34}A_{35}A_{36}$ and nucleotides G_{20} and A_{73}, with the 1-72 position playing a minor role. In addition, it appears that the most important feature of *E. coli*, yeast, and human tRNA^{Phe} recognition is conserved, in that the primary recognition element, the anticodon, is the same for all three. Whether the similar magnitude of discrimination by the human and *E. coli* enzymes is specific to tRNA^{Phe} or is a more general phenomenon remains to be seen.

D. YEAST ASPARTATE tRNA

The high resolution X-ray crystal structure of the yeast tRNA^{Asp} and AspRS complex has recently been solved, revealing contacts between the protein and the anticodon loop and stem as well as the distal portion of the acceptor stem.[144]

A comprehensive set of yeast tRNA^{Asp} mutants have been constructed by *in vitro* transcription and assayed with yeast AspRS to identify the nucleotides in these domains important for aminoacylation.[143] In addition, single-stranded nucleotides protected in earlier footprinting experiments,[144,145] G:U base pairs,[28,32] and nucleotides identified by computer analysis[146] as correlated with Asp identity were tested. Each position in the wild-type anticodon of tRNA$^{Asp}_{GUC}$ was mutated in turn to all possible nucleotide combinations. Depending on the nucleotide substitution, aminoacylation was found to be reduced from 19 to greater than 500-fold. Interestingly, all nine of the mutations increased Km by a similar degree of from four to less than sevenfold. Thus, the variable effect of the mutations was manifest in the value of k_{cat}. The greatest defects in aminoacylation were observed when G_{34} was replaced with C (400-fold reduction in aminoacylation) and U35 with A (530-fold reduction in aminoacylation). However, C_{36} is also an important nucleotide for AspRS recognition, since alteration to A or G results in equal or larger defects in aminoacylation than are observed in two out of three of the mutations introduced at position 34 or 35. Base changes made at position 37 on the 3′ side of the anticodon resulted in at most threefold effects on aminoacylation, indicating that the defects observed from anticodon substitutions are specific to the three anticodon nucleotides. As in the case of tRNA^{Gln}, the anticodon of tRNA^{Asp} undergoes a large conformational change on binding to the protein.[142]

In addition to the anticodon, mutations were also created in the D-stem, anticodon stem, acceptor stem, and discriminator base. The most profound effects on aminoacylation were associated with mutations introduced at the discriminator site, followed by the 10-25 base pair of the D-stem. Changing the discriminator base from G_{73} to U, A, or C decreased aminoacylation by 36, 160, and 200-fold, respectively. The effects of these mutations were, like the anticodon, split between K_m and k_{cat}, with all three substitutions resulting in increases of 4 to 6-fold in K_m. In contrast, the mutations created at the 10-25 position of the D-stem specifically affected K_m and had little effect on k_{cat}. Converting $G_{10}U_{25}$ to an A:U or G:C base pair increased K_m by 28 and 5-fold, respectively. So it appears that alterations at the 10-25 position in yeast tRNA^{Asp} disrupt binding to AspRS without affecting catalysis. The $G_{10}U_{25}$ base pair may be important for the proper folding of the tRNA, for creating a specific local conformation in the D-stem complementary to the binding domain of AspRS, or may directly provide functional groups for interaction with the enzyme. The crystal structure of the tRNA^{Asp}-AspRS complex suggests a possible contact with the D-stem and ethylnitrosourea footprinting of the complex shows that phosphates around G_{10}, but not U_{25}, are protected by the enzyme.[142,144]

None of the other tRNA^{Asp} mutants affected aminoacylation by more than threefold, indicating that the major tRNA^{Asp} recognition elements lie in the three bases of the anticodon, the discriminator base at position 73, and the $G_{10}U_{25}$ base pair in the D-stem. Substituting these nucleotides into yeast tRNA^{Phe} (not including G_{34} and G_{10}, which are already present in yeast tRNA^{Phe})

allowed aminoacylation of the chimera by AspRS with kinetics only 12-fold below that of tRNA^{Asp}, a value in close agreement with the above conclusion. Figure 1C illustrates the nucleotides in tRNA^{Asp} responsible for recognition by AspRS *in vitro*.

VIII. THE ROLE OF MODIFIED BASES

A. *E. COLI* ISOLEUCINE tRNAs

A minor species of *E. coli* Ile tRNA, tRNA$_2^{Ile}$, which reads the codon AUA[147] was found to contain a modified cytosine in the "wobble" base of the anticodon.[148] The gene for tRNA$_2^{Ile}$ encodes the methionine anticodon CAU. However, the cytosine is posttranscriptionally modified by covalent addition of a lysine residue to position 2 of the pyrimidine ring to create the base lysidine. This base modification is required for recognition by IleRS, in addition to bestowing on the tRNA its AUA decoding capacity.[86] Without the modification, this tRNA is an excellent Met acceptor, but is poorly aminoacylated with Ile. Following the modification, the tRNA is no longer a substrate of MetRS. Thus, the modification simultaneously changes the acceptor identity and decoding capacity of the tRNA. The fact that the AUA decoding tRNAs of bacteriophage T4 and spinach chloroplasts also receive a posttranscriptional modification to a "wobble" cytosine suggests that this mechanism may be conserved.[149,150]

The major Ile tRNA, tRNA$_1^{Ile}$, contains the anticodon GAU which has been shown to allow *in vivo* aminoacylation by Ile when transplanted into the *E. coli* methionine initiator tRNA.[40] The chemical structures of guanine and lysidine are quite different from one another, however, the imine tautomer of lysidine shares a number of functional groups with guanine and one or more of these may be the motif recognized by IleRS.[151]

B. RECOGNITION BY YEAST ARGININE tRNA SYNTHETASE

Works carried out with the yeast ArgRS and transcripts of unmodified yeast tRNA^{Asp} suggest that base modifications play a role in helping to maintain the fidelity of aminoacylation in yeast.[152] The yeast tRNA^{Asp} was found to be a 500-fold better substrate of ArgRS, when devoid of the normal complement of base modifications. In fact, the kinetics of aminoacylation of this tRNA were found to lie only 20-fold below that of authentic tRNA^{Arg}. Further study of the tRNA^{Asp} transcript found that it was not a substrate of the yeast HisRS, PheRS, or ValRS indicating that the mischarging was specific to ArgRS. The negative effect of the tRNA^{Asp} base modifications on charging by ArgRS is mainly on k_{cat}, because native tRNA^{Asp}, native tRNA^{Arg}, and the tRNA^{Asp} transcript lacking the modifications have similar K_m's for aminoacylation by ArgRS. Comparing the extent of base modifications in tRNA^{Asp} and tRNA^{Arg} indicates that, at most, three modifications are responsible for the observed effects (Ψ_{13}, Ψ_{32}, and m^{1}G$_{37}$). The unmodified tRNA has been found to assume a different tertiary structure than the fully-modified species, possibly allowing the tRNA to bind to the noncognate synthetase in a manner that leads to catalysis.

IX. CONCLUSIONS

The past few years have seen an explosion of new information added to the tRNA identity field. Table 4 summarizes the major recognition elements of several tRNAs which have been particularly well characterized. For some of these, the illustrated recognition elements are likely the only significant ones in the entire tRNA. The time is now not far off when the complete recognition set of *E. coli* tRNAs will be available. Thus, comprehensive models for discrimination amongst tRNAs by the 20 aminoacyl-tRNA synthetases will shortly begin to emerge for testing. Perhaps the major challenge remaining will be to work out a clearer picture of identity elements preventing aminoacylation by competing synthetases. The primary limitation to this task is the lack of a direct assay of aminoacylation *in vivo*.

To date, very little information on aminoacylation exists at all for higher eukaryotes, and appropriate *in vivo* assays are presently lacking for investigation of tRNA identity in these organisms. Our understanding of eukaryotic discrimination would be greatly accelerated if many of the features of prokaryotes are found to be conserved in eukaryotes. Much more work must be done before any firm conclusions can be made on this matter. However, the recently acquired data on tRNA^{Ala}, tRNA^{Phe}, and tRNA^{Val} recognition in higher eukaryotes suggests that at least some features of prokaryotic recognition mechanisms have been retained.

The resolution of the crystal structures of the *E. coli* tRNA^{Gln}-GlnRS and yeast tRNA^{Asp}-AspRS complexes represent major breakthroughs in illustrating how synthetases discriminate between substrates at the molecular level. It is also quite satisfying to see the ultimate resolution of years of biochemical data (in the case of Gln) in the crystal structure complex in a readily interpretable form. The X-ray structures also indicate that many surprises may lie ahead, since a tRNA can apparently bind to its synthetase with a conformation quite different than that in the free state. Because the free protein structures have not been determined, it is possible that RNA binding induces analogous conformational alterations in them as well. The next few years promise to yield important insights on the mechanism of protein-RNA interaction as recognition studies shift from locating recognition elements to identifying particular functional groups or structural features responsible for recognition, and as more high resolution structures emerge. Thus, comprehension of the biological machinery responsible for making sense of the most fundamental cellular information, the genetic code, appears within grasp.

TABLE 4
A Summary of Known Recognition Elements for some tRNAs[a]

Organism	tRNA	Site	Location	Proposed role	Ref.
Ec	Ala	G_2C_{71}	Acceptor stem	Modulates activity	106
		G_3U_{70}	Acceptor stem	Recognition	28,32[b]
		A_{73}	Discriminator base	Transition state for transfer	c
Ec	Arg	A_{20}	D loop	Recognition of "variable pocket"	29
		C_{35}	Anticodon	Recognition	d
Sc	Asp	$G_{10}U_{25}$	D-stem	Possibly conformational	143
		$G_{34}U_{35}C_{36}$	Anticodon	Recognition	143
		G_{73}	Discriminator base	Recognition	143
Ec	Cys	$G_{34}C_{35}A_{36}$	Anticodon	Recognition	41
		U_{73}	Discriminator base	Recognition	41
Ec	Gln	Weak 1-72 base pair	Acceptor stem	Facilitates conformational change at 3′ terminus	121[b]
		G_2C_{71}	Acceptor stem	Recognition	37,121,127
		G_3C_{70}	Acceptor stem	Recognition	37,121,127
		$U/C_{34}U_{35}G_{36}$	Anticodon	Recognition	d
		A_{37}	Anticodon loop	Possibly conformational	37,121
		Ψ_{38}	Anticodon loop	Possibly conformational	37,121
		G_{73}	Discriminator base	Facilitates conformational change at 3′ terminus	c
Ec	Gly	C_2G_{71}	Acceptor stem	Recognition	106
		G_3C_{70}	Acceptor stem	Modulates activity	106
		$C_{35}C_{36}$	Anticodon	Recognition	d
		U_{73}	Discriminator base	Recognition	c

tRNA Discrimination in Aminoacylation

Organism	aa	Nucleotides	Location	Function	Refs
Ec	His	$G_{-1}C_{73}$	Acceptor stem	Recognition	106,113,114
		U_2A_{71}	Acceptor stem	Modulates activity	106
		G_3C_{70}	Acceptor stem	Modulates activity	106
Ec	Ile	$L/G_{34}A_{35}U_{36}$	Anticodon	Recognition	d
		A_{37}	Anticodon loop	Recognition	151
		A_{73}	Discriminator base	Recognition	c
Ec	Met	$C_{34}A_{35}U_{36}$	Anticodon	Recognition	16,63
Sc	Phe	G_{20}	D loop	Recognition	d
		$G_{34}A_{35}A_{36}$	Anticodon	Recognition	c
		A_{73}	Discriminator base	Recognition	
Ec	Ser	G_2C_{71}	Acceptor stem	Recognition	130,131
		Variable arm		Recognition or conformation	73
Ec	Tyr	$Q_{34}U_{35}A_{36}$	Anticodon	Recognition	d
		Variable arm		Recognition or conformation	73
		A_{73}	Discriminator base	Recognition	c
Ec	Trp	$C_{34}C_{35}A_{36}$	Anticodon	Recognition	d
		G_{73}	Discriminator base	Recognition	c
Ec, Sc	Val	$A_{35}C_{36}$	Anticodon	Recognition	d
		A_{73}	Discriminator base	Recognition	c

^a Additional as yet unidentified recognition elements may also be present.
^b See text for additional references.
^c See Table 3 for additional references.
^c See Table 3 for additional references.
^d See Table 1 for additional references.

REFERENCES

1. **Kim, S.-H., Suddath, F. L., Quigley, G. J. McPherson, A., Sussman, J. L., Wang, A. H. J., Seeman, N. C., and Rich, A.,** Three-dimensional tertiary structure of yeast phenylalanine transfer RNA, *Science,* 185, 435, 1974.
2. **Robertes, J. D., Ladner, J. E., Finch, J. T., Rhodes, D., Brown, R. S., Clark, B. F. C., and Klug, A.,** Structure of yeast phenylalanine tRNA at 3 Å resolution, *Nature (London),* 250, 546, 1974.
3. **Westhof, E., Dumas, P., and Moras, D.,** Crystallographic refinement of yeast aspartic acid transfer RNA, *J. Mol. Biol.,* 184, 119, 1985.
4. **Normanly, J. and Abelson, J.,** tRNA identity, *Ann. Rev. Biochem.,* 58, 1029, 1989.
5. **Schimmel, P.,** Parameters for the molecular recognition of transfer RNAs, *Biochemistry,* 28, 2747, 1989.
6. **Schulman, L. H. and Abelson, J.,** Recent excitement in understanding transfer RNA identity, *Science,* 240, 1591, 1988.
7. **Yarus, M.,** tRNA identity: a hair of the dogma that bit us, *Cell,* 55, 739, 1988.
8. **RajBhandary, U. L.,** Genetic code. Modified bases and aminoacylation, *Nature (London),* 336, 112, 1988.
9. **Schimmel, P.,** Aminoacyl-tRNA synthetases: general scheme of structure-function relationships in the polypeptides and recognition of transfer RNAs, *Ann. Rev. Biochem.,* 56, 125, 1987.
10. **Schulman, L. H.,** Recognition of tRNAs by aminoacyl-tRNA synthetases, *Prog. Nucleic Acid Res. Mol. Biol.,* 41, 23, 1991.
11. **Normanly, J., Ogden, R. C., Horvath, S. J., and Abelson, J.,** Changing the identity of a transfer RNA, *Nature (London),* 321, 213, 1986.
12. **Smith, J. D.,** *Nonsense Mutations and tRNA Suppressors,* Academic Press, New York, 1979, 109.
13. **Ozeki, H., Inokuchi, H., Yamao, F., Kodaira, M., Sakano, T., Ikemura, T., and Schimmel, P. R.,** *Transfer RNA: Biological Aspects,* Söll, D., Abelson, J., and Schimmel, P. R., Eds., Cold Spring Harbor Laboratory, New York, 1980, 341.
14. **Squires, C. and Carbon, J.,** Normal and mutant glycine transfer RNAs, *Nature (London) New Biol.,* 233, 274, 1971.
15. **Roberts, J. W. and Carbon, J.,** Molecular mechanism for missense suppression in *E. coli, Nature (London),* 250, 412, 1974.
16. **Sampson, J. R. and Uhlenbeck, O. C.,** Biochemical and physical characterization of an unmodified yeast phenylalanine transfer RNA *in vitro, Proc. Natl. Acad. Sci. U.S.A.,* 85, 1033, 1988.
17. **Samualsson, T., Boren, T., Johanson, T.-I., and Lustig, F.,** Properties of a transfer RNA lacking modified nucleosides, *J. Biol. Chem.,* 263, 13692, 1988.
18. **Hall, K. B., Sampson, J. R., Uhlenbeck, O. C., and Redfield, A. G.,** Structure of an unmodified tRNA molecule, *Biochemistry,* 28, 5794, 1989.
19. **Altman, S., Guerrier-Takada, C., Frankfort, H. M., and Robertson, H. D.,** *Nucleases,* Linn, S. and Roberts, R., Eds., Cold Spring Harbor Laboratory, New York, 1982, 243.
20. **Shimura, Y., Aono, A., Ozeki, H., Sabrabhai, A., Lamfrom, H., and Abelson, J.,** Mutant tyrosine tRNA of altered amino acid specificity, *FEBS Lett.,* 22, 144, 1972.
21. **Hooper, M. L., Russell, R., and Smith, J. D.,** Mischarging in mutant tyrosine transfer RNAs, *FEBS Lett.,* 22, 149, 1972.
22. **Smith, J. D. and Celis, J. E.,** Mutant tyrosine transfer RNA that can be charged with glutamine, *Nature (London) New Biol.,* 243, 66, 1973.
23. **Celis, J. E., Hooper, M. L., and Smith, J. D.,** Amino acid acceptor stem of *E. coli* suppressor tRNA^{Tyr} is a site of synthetase recognition, *Nature (London) New Biol.,* 244, 261, 1973.
24. **Ghysen, A. and Celis, J. E.,** Mischarging single and double mutants of *Escherichia coli sup3* tyrosine transfer RNA, *J. Mol. Biol.,* 83, 333, 1974.

25. **Inokuchi, H., Celis, J. E., and Smith, J. D.**, Mutant tyrosine transfer ribonucleic acids of *Escherichia coli:* construction by recombination of a double mutant A1G82 chargeable with glutamine, *J. Mol. Biol.,* 85, 187, 1974.
26. **Normanly, J., Masson, J.-M., Kleina, L. G., Abelson, J., and Miller, J. H.**, Construction of two *Escherichia coli* amber suppressor genes: $tRNA^{Phe}_{CUA}$ and $tRNA^{Cys}_{CUA}$, *Proc. Natl. Acad. Sci. U.S.A.,* 83, 6548, 1986.
27. **Masson, J.-M. and Miller, J. H.**, Expression of synthetic suppressor tRNA genes under the control of a synthetic promoter, *Gene,* 47, 179, 1986.
28. **McClain, W. H. and Foss, K.**, Changing the identity of a tRNA by introducing a G-U wobble pair near the 3' acceptor end, *Science,* 240, 793, 1988.
29. **McClain, W. H. and Foss, K.**, Changing the acceptor identity of a transfer RNA by altering nucleotides in a variable pocket, *Science,* 241, 1804, 1988.
30. **McClain, W. H., Chen, Y.-M., Foss, K., and Schneider, J.**, Association of transfer RNA acceptor identity with a helical irregularity, *Science,* 242, 1681, 1988.
31. **McClain, W. H. and Foss, K.**, Nucleotides that contribute to the identity of *Escherichia coli* $tRNA^{Phe}$, *J. Mol. Biol.,* 202, 697, 1988.
32. **Hou, Y.-M. and Schimmel, P.**, A simple structural feature is a major determinant of the identity of a transfer RNA, *Nature (London),* 333, 140, 1988.
33. **Hou, Y.-M. and Schimmel, P.**, Modeling with *in vitro* kinetic parameters for the elaboration of transfer RNA identity *in vivo, Biochemistry,* 28, 4942, 1989.
34. **Hou, Y.-M. and Schimmel, P.**, Evidence that a major determinant for the identity of a transfer RNA is conserved in evolution, *Biochemistry,* 28, 6800, 1989.
35. **Rogers, M.J. and Söll, D.**, Discrimination between glutaminyl-tRNA synthetase and seryl-tRNA synthetase involves nucleotides in the acceptor helix of tRNA, *Proc. Natl. Acad. Sci. U.S.A.,* 85, 6627, 1988.
36. **McClain, W. H., Foss, K., Jenkins, R. A., and Schneider, J.**, Nucleotides that determine *Escherichia coli* $tRNA^{Arg}$ and $tRNA^{Lys}$ identities revealed by analyses of mutant opal and amber suppressor tRNAs, *Proc. Natl. Acad. Sci. U.S.A.,* 87, 9260, 1990.
37. **Jahn, M., Rogers, M. J., and Söll, D.**, Anticodon and acceptor stem nucleotides in $tRNA^{Gln}$ are major recognition elements for *E. coli* glutaminyl-tRNA synthetase, *Nature (London),* 352, 258, 1991.
38. **Varshney, U. and RajBhandary, U. L.**, Initiation of protein synthesis from a termination codon, *Proc. Natl. Acad. Sci. U.S.A.,* 87, 1586, 1990.
39. **Chattapadhyay, R., Pelka, H., and Schulman, L. H.**, Initiation of *in vivo* protein synthesis with non-methionine amino acids, *Biochemistry,* 29, 4263, 1990.
40. **Pallanck, L. and Schulman, L. H.**, Anticodon-dependent aminoacylation of a noncognate tRNA with isoleucine, valine, and phenylalanine *in vivo, Proc. Natl. Acad. Sci. U.S.A.,* 88, 3872, 1991.
41. **Pallanck, L., Li, S., and Schulman, L. H.**, The anticodon and discriminator base are major determinants of cysteine tRNA identity *in vivo, J. Biol. Chem.,* in press, 1992.
42. **Pak, M., Pallanck, L., and Schulman, L. H.**, Conversion of a methionine initiator tRNA into a tryptophan-inserting elongator tRNA *in vivo, Biochemistry,* in press, 1992.
43. **Pallanck, L. and Schulman. L. H.**, unpublished data, 1992.
44. **Seong, B. L. and RajBhandary, U. L.**, Mutants of *Escherichia coli* formylmethionine tRNA: a single base change enables initiator tRNA to act as an elongator *in vitro, Proc. Natl. Acad. Sci. U.S.A.,* 84, 8859, 1987.
45. **Seong, B. L., Lee, C.-P., and RajBhandary, U. L.**, Suppression of amber codons *in vivo* as evidence that mutants derived from *Escherichia coli,* initiator tRNA can act at the step of elongation in protein synthesis, *J. Biol. Chem.,* 264, 6504, 1989.
46. **Lee, C. P., Seong, B. L., and RajBhandary, U. L.**, Structural and sequence elements important for recognition of *Escherichia coli* formylmethionine tRNA by methionyl-tRNA transformylase are clustered in the acceptor stem, *J. Biol. Chem.,* 266, 18012, 1991.
47. **Varshney, U., Lee, C. P., Seong, B. L., and RajBhandary, U. L.**, Mutants of initiator tRNA that function both as initiators and elongators, *J. Biol. Chem.,* 266, 18018, 1991.

48. **Yarus, M.,** Intrinsic precision of aminoacyl-tRNA synthesis enhanced through parallel systems of ligands, *Nature (London) New Biol.,* 239, 106, 1972.
49. **Yarus, M., Cline, S. W., Wier, P., Breeden, L., and Thompson, R. C.,** Actions of the anticodon arm in translation on the phenotypes of RNA mutants, *J. Mol. Biol.,* 192, 235, 1986.
50. **Swanson, R., Hoben, P., Sumner-Smith, M., Uemura, H., Watson, L., and Söll, D.,** Accuracy of *in vivo* aminoacylation requires proper balance of tRNA and aminoacyl-tRNA synthetase, *Science,* 242, 1548, 1988.
51. **Fralova, L. and Kisselev, L. L.,** *Biokhimiya,* 29, 1177, 1964.
52. **Yaniv, M., Folk, W. R., Berg, P., and Soll, L.,** A single mutational modification of a tryptophan-specific transfer RNA permits aminoacylation by glutamine and translation of the codon UAG, *J. Mol. Biol.,* 86, 245, 1974.
53. **Yarus, M., Knowlton, R. G., and Soll, L.,** *Nucleic Acid-Protein Recognition,* Vogel, H., Eds., Academic Press, New York, 1977, 391.
54. **Knowlton, R. G., Soll, L., and Yarus, M.,** Dual specificity of *su+7* tRNA. Evidence for translational discrimination, *J. Mol. Biol.,* 139, 705, 1980.
55. **Schulman, L. H. and Goddard, J. P.,** Loss of methionine acceptor activity resulting from a base change in the anticodon of *Escherichia coli* formylmethionine transfer ribonucleic acid, *J. Biol. Chem.,* 248, 1341, 1973.
56. **Chakraburtty, K.,** Effect of sodium bisulfite modification on the arginine acceptance of *E. coli* tRNA^{Arg}, *Nucleic Acids Res.,* 2, 1793, 1975.
57. **Chambers, R. W., Aoyasi, S., Furukawa, Y., Zawadzka, H., and Bhanot, O.,** Inactivation of valine acceptor activity by a C→U missense change in the anticodon of yeast valine transfer ribonucleic acid, *J. Biol. Chem.,* 248, 5549, 1973.
58. **Kisselev, L. L.,** The role of the anticodon in recognition of tRNA by aminoacyl-tRNA synthetases, *Prog. Nucleic Acid Res. Mol. Biol.,* 32, 237, 1985.
59. **Fersht, A.,** *Enzyme Structure and Mechanism,* W. H. Freeman, New York, 1985, chap. 3.
60. **Pelka, H. and Schulman, L. H.,** Study of the interaction of *Escherichia coli* methionyl-tRNA synthetase with tRNA^{fMet} using chemical and enzymatic probes, *Biochemistry,* 25, 4450, 1986.
61. **Theobald, A., Springer, M., Grunberg-Manago, M., Ebel, J.-P., and Giegé, R.,** Tertiary structure of *Escherichia coli* tRNA$_3^{Thr}$ in solution and interaction of this tRNA with the cognate threonyl-tRNA synthetase, *Eur. J. Biochem.,* 175, 511, 1974.
62. **Bruce, A. G. and Uhlenbeck, O. C.,** Specific interaction of anticodon loop residues with yeast phenylalanyl-tRNA synthetase, *Biochemistry,* 21, 3921, 1982.
63. **Sampson, J. R., DiRenzo, A. B., Behlen, L. S., and Uhlenbeck, O. C.,** Nucleotides in yeast tRNA^{Phe} required for the specific recognition by its cognate synthetase, *Science,* 243, 1363, 1989.
64. **Garrett, M., LaBouesse, B., Litvak, S., Romby, P., Ebel, J.-P., and Giegé, R.,** Tertiary structure of animal tRNA^{Trp} in solution and interaction of tRNA^{Trp} with tryptophanyl-tRNA synthetase, *Eur. J. Biochem.,* 138, 67, 1984.
65. **Schulman, L. H. and Pelka, H.,** The anticodon contains a major element of the identity of arginine transfer RNAs, *Science,* 246, 1595, 1989.
66. **Schulman, L. H. and Pelka, H.,** Anticodon switching changes the identity of methionine and valine transfer RNAs, *Science,* 242, 765, 1988.
67. **Schulman, L. H. and Pelka, H.,** An anticodon change switches the identity of *E. coli* tRNA$_m^{Met}$ from methionine to threonine, *Nucleic Acids Res.,* 18, 285, 1990.
68. **Bare, L. A. and Uhlenbeck, O. C.,** Aminoacylation of anticodon loop substituted yeast tyrosine transfer RNA, *Biochemistry,* 24, 2354, 1985.
69. **Uhlenbeck, O. C.,** Structure and function of RNA, *Chemica Scripta,* 26B, 97, 1986.
70. **Kleina, L. G., Masson, J.-M., Normanly, J., Abelson, J., and Miller, J. H.,** Construction of *Escherichia coli* amber suppressor tRNA genes. II. Synthesis of additional tRNA genes and improvement of suppressor efficiency, *J. Mol. Biol.,* 213, 705, 1990.

71. Normanly, J., Kleina, L. G., Masson, J.-M., Abelson, J., and Miller, J. H., Construction of *Escherichia coli* amber suppressor tRNA genes. III. Determination of tRNA specificity, *J. Mol. Biol.*, 213, 719, 1990.
72. Schulman, L. H. and Pelka, H., *In vitro* conversion of a methionine to a glutamine-acceptor tRNA, *Biochemistry*, 24, 7309, 1985.
73. Himeno, H., Hasegawa, T., Ueda, T., Watanabe, K., and Shimizu, M., Conversion of aminoacylation specificity from tRNA^{Tyr} to tRNA^{Ser} *in vitro*, *Nucleic Acids Res.*, 18, 6815, 1990.
74. Springer, M., Graffe, M., Dondon, J., and Grunberg-Manago, M., tRNA-like structures and gene regulation at the translational level: a case of molecular mimicry in *Escherichia coli*, *EMBO J.*, 8, 2417, 1989.
75. Springer, M., Graffe, M., Butler, J. S., and Grunberg-Manago, M., Genetic definition of the translational operator of the threonine-tRNA ligase gene in *Escherichia coli*, *Proc. Natl. Acad. Sci. U.S.A.*, 83, 4384, 1986.
76. Moine, H., Romby, P., Springer, M., Grunberg-Manago, M., Ebel, J.-P., Ehresmann, C., and Ehresmann, B., Messenger RNA structure and gene regulation at the translational level in *Escherichia coli*: the case of threonine:tRNA^{Thr} ligase, *Proc. Natl. Acad. Sci. U.S.A.*, 85, 7892, 1988.
77. Springer, M., Graffe, M., Dondon, J., Grunberg-Manago, M., Romby, P., Ehresmann, B., Ehresmann, C., and Ebel, J.-P., Translational control in *E. coli:* the case of threonyl-tRNA synthetase, *Bioscience Rep.*, 8, 619, 1988.
78. Park, S. J., Hou, Y.-M., and Schimmel, P., A single base pair affects binding and catalytic parameters in the molecular recognition of a transfer RNA, *Biochemistry*, 28, 2740, 1989.
79. Chambers, R. W., Bhanot, O. S., Aoyasi, S., Furukawa, Y., and Zawadzka, H., Effects of C→U transitions on the adaptor function of tRNA$^{Ala}_{lab}$ and tRNA$^{Val}_1$, *Fed. Proc.*, 33, 1422, 1974.
80. Qiu, M. S., Jin, Y. X., Li, W. Q., Bao, J. R., and Wang, D., Biological function of modified nucleotides in tRNA molecules-synthesis and biological activity of the analogues of yeast alanyl-tRNA with I$_{34}$ replaced by A$_{34}$ or G$_{34}$, *Sci. Sin. [B]*, 31, 695, 1988.
81. Jin, Y. X., Qiu, M. S., Li, W. Q., Zeng, K. Q., Wang, D., Bao, J., Gong, P., Wu, R., and Wang, D., Effect of the anticodon loop size of yeast alanyl tRNA on its biological activity, *Anal. Biochem.*, 161, 453, 1987.
82. Stern, L. and Schulman, L. H., Role of anticodon bases in aminoacylation of *E. coli* methionine tRNAs, *J. Biol. Chem.*, 252, 6403, 1977.
83. Schulman L. H. and Pelka, H., Structural requirements for aminoacylation of *E. coli* formylmethionine tRNA, *Biochemistry*, 16, 4256, 1977.
84. Schulman, L. H. and Pelka, H., Anticodon loop size and sequence requirements for recognition of formylmethionine tRNA by methionyl-tRNA synthetase, *Proc. Natl. Acad. Sci. U.S.A.*, 80, 6755, 1983.
85. Ghosh, G., Pela, H., and Schulman, L. H., Identification of the tRNA anticodon recognition site of *Escherichia coli* methionyl-tRNA synthetase, *Biochemistry*, 29, 2220, 1990.
86. Muramatsu, T., Nishikawa, K., Nemoto, F., Kuchino, Y., Nishimura, S., Miyazawa, T., and Yokoyama, S., Codon and amino acid specificities of a transfer RNA are both converted by a single post-transcriptional modification, *Nature (London)*, 336, 179, 1988.
87. Meinnel, T., Mechulam, Y., LeCorre, D., Panvert, M., Blanquet, S., and Fayat, G., Selection of suppressor methionyl-tRNA synthetases: mapping the tRNA anticodon binding site, *Proc. Natl. Acad. Sci. U.S.A.*, 88, 291, 1991.
88. Bahramian, B., Lee, C. P., and RajBhandary, U. L., unpublished data, 1991.
89. Schulman, L. H. and Pelka, H., unpublished data, 1991.
90. Cigan, A. M., Feng, L., and Donahue, T. F., tRNA$^{Met}_i$ functions in directing the scanning ribosome to the start site of translation, *Science*, 242, 93, 1988.

91. **Fasiolo, F., Despons, L., Laforêt, M., Ebel, J. P., and Walter, Ph.,** Anticodon mediated tRNA recognition of yeast methionyl-tRNA synthetase requires at least two distinct helices in the C-terminal domain, (abstract), 14th Int. tRNA Workshop, Rydzyna, Poland, May 4–9, 1991.
92. **Uemura, H., Imai, M., Ohtsuka, E., Ikehara, M., and Söll, D.,** *E. coli* initiator tRNA analogs with different nucleotides in the discriminator base position, *Nucleic Acids Res.*, 10, 6531, 1982.
93. **Li, S. and Schulman, L. H.,** unpublished data, 1992.
94. **Horowitz, J., Chu, W.-C., Feiz, V., and Derrick, W. B.,** Recognition of *E. coli* valine tRNA by its cognate synthetase, (Abstract), 14th Int. tRNA Workshop, Rydzyna, Poland, May 4 to 9, 1991.
95. **Pinck, M., Yot, P., Chapeville, F., and Duranton, H. M.,** A new principle of RNA folding based on pseudoknotting, *Nucleic Acids Res.*, 13, 1717, 1985.
96. **Giegé, R., Briand, J. P., Mengual, R., Ebel, J. P., and Hirth, L.,** Valylation of the two RNA components of turnip yellow mosaic virus and specificity of the tRNA aminoacylation reaction, *Eur. J. Biochem.*, 84, 251, 1978.
97. **Dreher, T. W., Florentz, C., and Giegé, R.,** Valylation of tRNA-like transcripts from cloned cDNA of turnip yellow mosaic virus RNA demonstrate that the L-shaped region at the 3′ end of the viral RNA is not sufficient for optimal aminoacylation, *Biochimie,* 70, 1719, 1988.
98. **Florentz, C., Dreher, T. W., Rudinger, J., and Geigé, R.,** Specific valylation identity of turnip yellow mosaic virus RNA by yeast valyl-tRNA synthetase is directed by the anticodon in a kinetic rather than affinity-based discrimination, *Eur. J. Biochem.,* 195, 229, 1991.
99. **Tsai, C.-H. and Dreher, T. W.,** Turnip yellow mosaic virus RNAs with anticodon loop substitutions that result in decreased valylation fail to replicate efficiently, *J. Virol.*, 65, 3060, 1991.
100. **Crothers, D. M., Seno, T., and Söll, D. G.,** Is there a discriminator site in transfer RNA?, *Proc. Natl. Acad. Sci. U.S.A.,* 69, 3063, 1972.
101. **deDuve, C.,** The second genetic code, *Nature (London),* 333, 117, 1988.
102. **Atilgan, T., Nicholas, H. B., Jr., and McClain, W. H.,** A statistical method for correlating tRNA sequence with amino acid specificity, *Nucleic Acids Res.,* 14, 375, 1986.
103. **McClain, W. H. and Nicholas, H. B., Jr.,** Differences between transfer RNA molecules, *J. Mol. Biol.,* 194, 635, 1987.
104. **Francklyn, C. and Schimmel, P.,** Aminoacylation of RNA minihelices with alanine, *Nature (London),* 337, 478, 1989.
105. **Park, S. J. and Schimmel, P.,** Evidence for interaction of an aminoacyl-transfer RNA synthetase with a region important for the identity of its cognate transfer RNA, *J. Biol. Chem.,* 263, 16527, 1988.
106. **Francklyn, C., Shi, J.-P., and Schimmel, P.,** Overlapping nucleotide determinants for specific aminoacylation of RNA microhelices, *Science,* 255, 1121, 1991.
107. **Musier-Forsyth, K., Usman, N., Scaringe, S., Doudna, J., Green, R., and Schimmel, P.,** Specificity for aminoacylation of an RNA helix: an unpaired, exocyclic amino group in the minor groove, *Science,* 253, 784, 1991.
107a. **McClain, W. H., Foss, K., Jenkins, R. A., and Schneider, J.,** Four sites in the acceptor helix and one site in the variable pocket of tRNA^{Ala} determine the molecule's acceptor identity, *Proc. Natl. Acad. Sci. U.S.A.,* 88, 9272, 1991.
108. **Shi, J.-P., Francklyn, C., Hill, K., and Schimmel, P.,** A nucleotide that enhances the charging of RNA minihelix sequence variants with alanine, *Biochemistry,* 29, 3626, 1990.
109. **Shi, J.-P. and Schimmel, P.,** Aminoacylation of alanine minihelices, *J. Biol. Chem.,* 266, 2705, 1991.
110. **Sprinzl, M., Hartmann, T., Weber, J., Blank, J., and Zeidler, R.,** Compilation of tRNA sequences and sequences of tRNA genes, *Nucleic Acids Res.,* 17, r1, 1989.
111. **Imura, N., Weiss, G. B., and Chambers, R. W.,** Reconstitution of alanine acceptor activity from fragments of yeast tRNA$_2^{Ala}$, *Nature (London),* 222, 1147, 1969.

112. Singer, C. E. and Smith, G. R., Histidine regulation in *Salmonella typhimurium, J. Biol. Chem.,* 247, 2989, 1972.
113. Himeno, H., Hasegawa, T., Ueda, T., Watanabe, K., Miura, K., and Shimizu, M., Role of the extra G-C pair at the end of the acceptor stem of tRNA^{His} in aminoacylation, *Nucleic Acids Res.,* 17, 7855, 1989.
114. Francklyn, C. and Schimmel, P., Enzymatic aminoacylation of an eight-base-pair microhelix with histidine, *Proc. Natl. Acad. Sci. U.S.A.,* 87, 8655, 1990.
115. McClain, W. H., Foss, K., Jenkins, R. A., and Schneider, J., Rapid determination of nucleotides that define tRNA^{Gly} acceptor identity, *Proc. Natl. Acad. Sci. U.S.A.,* 88, 6147, 1991.
116. Hill, C. W., Combriato, G., and Dolph, W., Three different missense suppressor mutations affecting the tRNA$^{Gly}_{GGG}$ species of *Escherichia coli, J. Bact.,* 117, 351, 1974.
117. Riddle, D. L. and Carbon, J., Frameshift suppression: a nucleotide addition in the anticodon of a glycine transfer RNA, *Nature (London) New Biol.,* 242, 230, 1973.
118. Sabban, E. L. and Bhanot, O., The effect of bisulfite-induced C→U transitions on aminoacylation of *Escherichia coli* glycine tRNA, *J. Biol. Chem.,* 257, 4796, 1982.
119. Hasegawa, T., Himeno, H., Ishikura, H., and Shimizu, M., Discriminator base of tRNA^{Asp} is involved in amino acid acceptor activity, *Biochem. Biophys. Res. Comm.,* 163, 1534, 1989.
120. Labouze, E. and Bedouelle, H., Structural and kinetic bases for the recognition of tRNA^{Tyr} by tyrosyl-tRNA synthetase *J. Mol. Biol.,* 205, 729, 1989.
121. Rould, M. A., Perona, J. J., Söll, D., and Steitz, T. A., Structure of *E. coli* glutaminyl-tRNA synthetase complexed with tRNA^{Gln} and ATP at 2.8Å resolution, *Science,* 246, 1135, 1989.
122. Yamao, F., Inokuchi, H., Cheung, A., Ozeki, H., and Söll, D., *Escherichia coli* glutaminyl-tRNA synthetase, *J. Biol. Chem.,* 257, 11639, 1982.
123. Hoben, P., Royal, N., Cheung, A., Yamao, F., Biemann, K., and Söll, D., *Escherichia coli* glutaminyl-tRNA synthetase, *J. Biol. Chem.,* 257, 11644, 1982.
124. Perona, J. J., Swanson, R., Steitz, T. A., and Söll, D., Overproduction and purification of *Escherichia coli* tRNA$^{Gln}_2$ and its use in crystallization of the glutaminyl-tRNA synthetase-tRNA^{Gln} complex, *J. Mol. Biol.,* 202, 121, 1988.
125. Rould, M. A., Perona, J. J., and Steitz, T. A., Structural basis of anticodon loop recognition by glutaminyl-tRNA synthetase, *Nature (London),* 352, 213, 1991.
126. Yarus, M., Cline, S. W., Wier, P., Breeden. L., and Thompson, R. C., Actions of the anticodon arm in translation on the phenotypes of RNA mutants, *J. Mol. Biol.,* 192, 235, 1986.
127. Perona, J. J., Swanson, R. N., Rould, M. A., Steitz, T. A., and Söll, D., Structural basis for misaminoacylation by mutant *E. coli* glutaminyl-tRNA synthetase enzymes, *Science,* 246, 1152, 1989.
128. Hoben, P., Ph.D. thesis, Yale University, New Haven, CT, 1984.
129. Inokuchi, H., Hoben, P., Yamao, F., Ozeki, H., and Söll, D., Transfer RNA mischarging mediated by a mutant *Escherichia coli* glutaminyl-tRNA synthetase, *Proc. Natl. Acad. Sci. U.S.A.,* 81, 5076, 1984.
130. Saks, P., Sampson, J. R., and Abelson, J., unpublished data, 1991.
131. Shimizu, M., personal communication, 1991.
132. Leinfelder, W., Zehelein, E., Mandrand-Berthelot, M., and Böck, A., Gene for a novel tRNA species that accepts L-serine and cotranslationally inserts selenocysteine, *Nature (London),* 331, 723, 1988.
133. Böck, A., personal communication, 1991.
134. Leinfelder, W., Forchhammer, K., Veprek, B., Zehelein, E., and Böck, A., *In vitro* synthesis of selenocysteinyl-tRNA(UCA) from seryl-tRNA(UCA): involvement and characterization of the *selD* gene product, *Proc. Natl. Acad. Sci. U.S.A.,* 87, 543, 1990.
135. Schatz, D., Leberman, R., and Eckstein, F., Interaction of *Escherichia coli* tRNA^{Ser} with its cognate aminoacyl-tRNA synthetase as determined by footprinting with phosphorothioate containing tRNA transcripts, *Proc. Natl. Acad. Sci. U.S.A.,* 88, 6132, 1991.

136. **Edwards, H., Trézéguet, V., and Schimmel, P.,** An *Escherichia coli* tyrosine transfer RNA is a leucine-specific transfer RNA in the yeast *Saccharomyces cerevisiae, Proc. Natl. Acad. Sci. U.S.A.,* 88, 1153, 1991.
137. **Ladner, J. E., Jack, A., Robertus, J. D., Brown, R. S., Rhodes, D., Clark, B. F. C., and Klug, A.,** Structure of yeast phenylalanine transfer RNA at 2.5 Å resolution, *Proc. Natl. Acad. Sci. U.S.A.,* 72, 4414, 1975.
138. **Tinkle Peterson, E. F. and Uhlenbeck, O. C.,** unpublished data, 1991.
139. **Bruce, A. G. and Uhlenbeck, O. C.,** Enzymatic replacement of the anticodon of yeast phenylalanine transfer ribonuceic acid, *Biochemistry,* 21, 855, 1982.
140. **Sampson, J. R., DiRenzo, A. B., Behlen, L. S., and Uhlenbeck, O. C.,** Role of the tertiary nucleotides in the interaction of yeast phenylalanine tRNA with its cognate synthetase, *Biochemistry,* 29, 2523, 1990.
141. **Nazarenko, I. A., Tinkle Peterson, E., Zakharova, O. D., Lavrik, O. I., and Uhlenbeck, O. C.,** Recognition nucleotides for human phenylalanyl-tRNA synthetase, *Nucleic Acids Res.,* 20, 475, 1992.
142. **Ruff, M., Krishnaswamy, S., Boeglin, M., Poterszman, A., Mitschler, A., Podjarny, A., Rees, B., Thierry, J. C., and Moras, D.,** Class II aminoacyl transfer RNA synthetases: crystal structure of yeast aspartyl-tRNA synthetase complexed with tRNA^{Asp}, *Science,* 252, 1682, 1991.
143. **Pütz, J., Puglisi, J. D., Florentz, C., and Giegé, R.,** Identity elements for specific aminoacylation of yeast tRNA^{Asp} by cognate aspartyl-tRNA synthetase, *Science,* 252, 1696, 1991.
144. **Romby, P., Moras, D., Bergdoll, M., Dumas, P., Vlassov, V. V., Westhof, E., Ebel, J. P., and Giegé, R.,** Yeast tRNA^{Asp} tertiary structure in solution and areas of interaction of the tRNA with aspartyl-tRNA synthetase. A comparative study of the yeast phenylalanine system by phosphate alkylation experiments with ethylnitrosourea, *J. Mol. Biol.,* 184, 455, 1985.
145. **Garcia, A., Giegé, R., and Behr, J.-P.,** New photoactivatable structural and affinity probes of RNAs: Specific features and applications for mapping of spermine binding sites in yeast tRNA^{Asp} and interactions of this tRNA with yeast aspartyl-tRNA synthetase, *Nucleic Acids Res.,* 18, 89, 1990.
146. **Nicholas, H. B. and McClain, W. H.,** An algorithm for discriminating sequences and its application to yeast transfer RNA, *Cabios,* 3, 177, 1987.
147. **Harada, F. and Nishimura, S.,** Purification and characterization of AUA specific isoleucine transfer ribonucleic acid from *Escherichia coli* B, *Biochemistry,* 13, 300, 1974.
148. **Muramatsu, T., Yokoyama, S., Horie, N., Matsuda, A., Ueda, T., Yamaizumi, Z., Kuchino, Y., Nishimura, S., and Miyazawa, T.,** A novel lysine-substituted nucleoside in the first position of the anticodon of minor isoleucine tRNA from *Escherichia coli, J. Biol. Chem.,* 263, 9261, 1988.
149. **Guthrie, C. and McClain, W. H.,** Rare transfer ribonucleic acid essential for phage growth. Nucleotide sequence comparison of normal and mutant T4 isoleucine-accepting transfer ribonucleic acid, *Biochemistry,* 18, 3786, 1979.
150. **Francis, M. A. and Dudock, B. S.,** Nucleotide sequence of a spinach chloroplast isoleucine tRNA, *J. Biol. Chem.,* 257, 11195, 1982.
151. **Yokoyama, S.,** personal communication, 1991.
152. **Perret, V., Garcia, A., Grosjean, H., Ebel, J.-P., Florentz, C., and Giegé, R.,** Relaxation of a transfer RNA specificity by removal of modified nucleotides *Nature (London),* 344, 787, 1990.
153. **Scheinker, V., Beresten, S. F., Mashkova, T. D., Mazo, A. M., and Kisselev, L. L.,** Role of exposed cytosine residues in aminoacylation activity of tRNA^{Trp}, *FEBS Lett.,* 132, 349, 1981.
154. **Bare, L. A. and Uhlenbeck, O. C.,** Specific substitutions into the anticodon loop of yeast tyrosine transfer RNA, *Biochemistry,* 25, 5825, 1986.
155. **Himeno, H., Hasegawa, T., Asahara, H., Tamura, K., and Shimizu, M.,** Identity determinants of *E. coli* tryptophan tRNA, *Nucleic Acids Res.,* 19, 6379, 1991.

Chapter 11

THE TRANSLATIONAL CONTEXT EFFECT

Michael Yarus and James Curran

TABLE OF CONTENTS

I.	Introduction	320
II.	Mechanisms of Context Action	320
III.	The Existence of Context Constraints	321
IV.	A Summary of Statistical Context Investigations	323
V.	Clues to the Context Mechanism?	325
VI.	Study of Translation using Suppression	325
VII.	Context Effects on tRNA-Mediated Nonsense Suppression	327
VIII.	UAG Suppression Reflects 3' Neighbor Effects on both tRNA and Release Factor	327
IX.	A Biochemical Demonstration of a 3' Neighbor Effect on tRNA Selection	328
X.	Context Biases at Natural Terminators	330
XI.	Context Sensitive Action by Release Factors	330
XII.	Wobble-Wobble Context Interactions are Observable *In Vivo*	331
XIII.	A Mechanism for Wobble Interaction	333
XIV.	tRNA Modification Affects Context Sensitivity	334
XV.	A Context that Determines the Meaning of the UGA Codon	335
XVI.	A Downstream Sequence Determines the Meaning of the UAG Codon	336
XVII.	Effects of Context on Initiation	337

XVIII. Context Effects on Missense Errors 338

XIX. Context and Frameshifting 338

XX. Possible Consequences for the Context Structure of Messages 341

XXI. Context and Expression 342

XXII. Summary and Prospects 346

Acknowledgments 347

Appendices 348

References 360

I. INTRODUCTION

A protein can be encoded by a prodigious number of nucleotide sequences. For example, an amino acid can be specified on average by three codons, so the number of isocoding nucleotide sequences of an average (300 codon) gene is $\approx 3^{300}$, or about 10^{142}, a truly astronomic number about equal to the number of subatomic particles in the universe, *squared* (give or take a few orders of magnitude). Because many amino acids in proteins can be varied without loss of function, the number of nucleotide sequences that could encode the same function is yet greater, and recedes even further from intuitive grasp.

Would all possible mRNA sequences specifying the same protein be equally functional in translation? The answer surely is no; this means that evolution, even in the small fraction of possible mRNAs that it can explore, utilizes one of many kindred sequences that encode the same gene product. In this chapter, we discuss one aspect of choice between kindred sequences — the evidence for effects of context (nearby nucleotides) on translation of a given codon. Emphasis is on the better-defined situation in prokaryotes, with comments on eukaryotes. Reviews of the effects of codon usage on gene expression, a substantially overlapping topic, can be found in Chapters 3, 4, and 13 of this volume. Other recent comments on codon choice are to be found in Andersson and Kurland[1] and on codon context effects in Buckingham.[2]

II. MECHANISMS OF CONTEXT ACTION

By definition, a context effect is a functional interaction between distinct sites in an mRNA; namely, a codon and its neighboring nucleotides. Therefore, the interaction is either *direct* (e.g., by stacking, base-pairing, or tertiary interactions between mRNA nucleotides), or *indirect,* through other molecules. While indirect interaction can be circuitous, involving any element in the cell, the most likely indirect context mechanisms involve the tRNA, tRNA-tRNA contacts, or interactions mediated by the ribosome during decoding.

Context may have immediate translational effects; for example, on the frequency of missense, frameshift, and chain loss errors. Context may also be subtle, triggering a chain of events that ultimately affect expression by another route. For example, differences in translational rate alter ribosomal distribution, which may conceivably lead to effects on mRNA synthesis and decay (Appendix I). In this review, we introduce evidence for the existence of the context effect first, that is, evidence for translationally relevant context bias in mRNA sequences. Later, we will address the role of context bias in the expression of genes.

III. THE EXISTENCE OF CONTEXT CONSTRAINTS

Structural genes show evidence of nonrandom codon contexts. Some of this nonrandomness seems attributable to translation rather than to other effects on sequence, such as sequence-specific mutation rates, or constraints on amino acid sequence.

Lipman and Wilbur[3] first detected context in mRNA sequences; they calculated the information in the three successive nucleotide doublets that overlap codons (the 1-2, 2-3, and 3-1 doublets). The third dinucleotide (3-1) juxtaposes the wobble position of one codon with the first nucleotide of a second codon, and Lipman and Wilbur conclude that such juxtapositions were nonrandom for eukaryotes, and most likely nonrandom for their selection of *E. coli* and *S. typhimurium* genes. In earlier studies by Nussinov,[4] this doublet had been implicitly considered to be random.

Yarus and Folley[5] analyzed a selection of *E. coli* genes with known high or low levels of expression, asking if the sequences of the five codons to either side of specified sense codons were randomly selected. Sequence bias flanked codons on both sides; the single 5′ neighboring nucleotide, and the 3′ neighboring triplet were significantly nonrandom in weakly expressed genes, with nucleotide choice becoming random rather sharply past these limits. In highly expressed genes, unexpectedly, only the 3′ neighbor of a codon appears distinctly nonrandom. Thus, context bias is more significant, and spread to more positions, in weakly expressed genes. Both frequently used codons and rarely used codons showed context preferences. Yarus and Folley[5] then examined sequences around codons of specified structure, for example, the sequences around all codons that have G in the wobble position. The overall nonrandomness in weakly expressed genes appeared most strongly as correlation with the wobble position of the test codon, that is, there is a tendency for a codon that has a wobble purine to be preceded and followed by another codon with a wobble purine, and similarly for pyrimidine wobbles (a correlated 3-3 doublet).

There was also a smaller, but very significant, tendency for wobble purines to be followed by the pyrimidine C in the first position of the next codon, and wobble pyrimidines to be followed by G in the first position of the next codon (3-1 doublet). The second position of the following codon was also somewhat correlated with the preceding wobble (3-2 doublet) in weakly expressed genes, but this was not much greater than in a set of computed sequences with the same overall base composition (corrected for bias by position within codons), same codon selection, and same amino acid neighbor frequencies as the real genes. Taking into account all departures from expectation, they suggested that weakly expressed genes might prefer the patterns Y RRY or R YYR. Such preferences appeared to have a translational origin, because they were correlated with expression, but particularly because the strongest effect, the correlation between wobble positions in weakly expressed genes, had the unique triplet periodicity of translation.

Shpaer[6] analyzed *E. coli* genes of high and low expression, looking particularly at the nucleotides neighboring the codon on the 3' and 5' sides (the 3-1 doublets). Again, expression was an important variable, but both neighbors were nonrandom, and in both expression classes.

Gouy[7] studied a larger group of eubacterial genes (181 sequences), and utilized a sophisticated set of reference sequences that preserved both the amino acid sequence and codon usage of the real genes, but whose messages were otherwise randomly ordered. Gouy found that bacterial codons have nonrandom 5' neighboring nucleotides and nonrandom triplets as 3' neighbors. In particular, neighboring wobble positions (3-3 doublets) in presumptive weakly expressed genes (little bias in codon choice) tend to be purines or pyrimidines (R NNR and Y NNY), and purine-pyrimidine and pyrimidine-purine adjacent wobbles are avoided (R NNY and Y NNR). These tendencies are much less significant in highly expressed genes. This 3-3 correlation agrees with that suggested by Yarus and Folley,[5] but Gouy's analysis suggests that the 3-1 tendency is a subset of that previously suggested; G CNN preferred, C CNN and G GNN avoided. A particularly useful control showed that genes have a tendency to correlate wobbles that varies with the potential variance of the underlying codon set. Thus, because highly expressed genes use only a subset of the total codon set, they may not be able to choose neighboring codons as freely, obscuring context preferences. Analysis of yeast and human genes showed no wobble correlation (3-3 doublet), but did find a wobble-first position (3-1 doublet) bias that was similar in both organisms. However, this 3-1 bias is similar to doublet bias in untranslated regions, and Gouy argues that such bias is probably not translational.

Gutman and Hatfield[8] analyzed all adjacent codon pairs in 237 *E. coli* genes, and found many cases of overrepresentation and underrepresentation among neighboring codons, even after correction for the tendency toward certain pairs of neighboring amino acids. Codons tend to both prefer and avoid specific neighboring codons and biases for or against a specific neighbor occur either

3' or 5' of a given codon. These directional effects have a short range, because they are very much smaller or do not exist if nonadjacent codons are analyzed. Neighbor preference is found to exist in both highly and weakly expressed genes, but can be qualitatively distinguished by expression level. Though only part of the raw data is shown, the greatest deviations from expectation among the cases shown are principally wobble correlations like those described earlier.

Hanai and Wada[9] specifically reinvestigated the question of context preferences in weakly expressed genes. On one hand, they find that 3-1 doublets occur within coding regions in frequencies correlated to frequencies in noncoding regions. This is interpreted as evidence that the bias in the 3-1 doublet is not translational, but perhaps a result of sequence-biased mutation pressure (e.g., AT -> GC mutations may occur more readily than the reverse), as suggested earlier by Shields and Sharp[10]. In contrast, in their set of 14 genes with low frequencies of favored codons, they not only find highly-significant 3-3 doublet bias, but also that this is best described as a preference for purines or pyrimidines at adjacent wobble positions. Even more striking is the fact that all 16 sequences of four nucleotides (Y YYY, Y YYR, etc) in their set deviate from random in the way predicted by prior studies of wobble correlation; Y NNY and R NNR favored; Y NNR and R NNY disfavored.

Bulmer,[11] again concentrating on genes with weak codon bias, applied a subtle statistical test to the 3-1 doublet, based on the notion that if complementary sequences (e.g., NUR C and NGY A) are of equal occurrence in weakly expressed yeast and *E. coli* mRNAs, the implication is that the untranslated complementary strand is showing the same "context" tendencies as the translated one. In this case, Bulmer argues that nonrandomness would be more plausibly attributed to DNA structure or strand-symmetrical mutation pressure than to translation. In fact, complementary sequences are found to behave in a correlated fashion, and Bulmer[11] favors a sequence-specific mutational equilibrium as the source of 3-1 doublet bias.

IV. A SUMMARY OF STATISTICAL CONTEXT INVESTIGATIONS

Statistical study necessarily detects only structure that generalizes to many sequences, though more idiosyncratic interactions may also be important (see following).[6,8] Nevertheless, general patterns exist. The two studies that examined a broader window of sequences agree that the upstream 5' neighbor (a 3-1 doublet) and the downstream 3' triplet from a central codon are not randomly chosen in bacterial messages.[5,7]

Virtually all workers agree that codon boundaries (3-1 doublets) are biased.[3,6] However, a part of this variation is probably nontranslational; it is similar to doublet preference in nontranslated regions[7,9] and resembles the variation of the untranslated strand.[11]

However, we argue for a functional role for some 3-1 doublets for several reasons. Firstly, mutational pressure affects all sequences to some extent. Therefore, a finding that the 3-1 doublet responds to evolutionary forces that shape all DNA does not *exclude* selection of some translation-related doublet sequences. A parallel argument about mononucleotides may be illuminating; the composition of the wobble position closely follows the GC content of the DNA as a whole, though first and second codon positions do not.[12] Wobble and general GC content are linearly related over the full range of genomic GC content, and the correlation is even more regular than the correlation between 3-1 doublets just referred to.[7,9,11] Yet we think it would be precipitous to deduce that no translational role exists for wobble positions. Instead, with some freedom to specify the same function with more than one sequence, mRNA sequences can track the evolutionary forces on the DNA in which they are embedded without compromising an essential translational role.

Secondly, there are particular doublet sequences that are reproducibly significant outliers in the correlations previously described, as for example, avoidance of the 3-1 C-C doublet.[7,9] These outliers may account for some of the significant unexplained variance in complementary sequences.[11] Thirdly, controlled experimental tests[12-16] show that a 3-1 effect on translation is easily measured *in vivo* and *in vitro* (see following). Therefore, it would be surprising if some such sequences were not under selection. Fourthly, as seen below, the 3-1 doublet affects at least two translational properties, and selection for more than one effect may make its behavior less distinct in statistical studies.

Two studies[5,7] agree that the middle position of the downstream triplet differs from random expectation, and that the difference is more significant in weakly expressed genes. This difference has not attracted enough study; in particular, translational effects on this nucleotide need to be carefully distinguished from secondary doublet effects due to primary correlations with its flanking nucleotides. However, one observation suggesting further investigation of the 3-2 doublet is the existence of individual codons like GCC that show strong context bias at the middle position of the downstream codon, with much less effect at the flanking first and wobble positions.[5] In addition, the 3-2 doublet necessarily affects the choice of tRNA neighbors, which may be functionally significant (see following).

The least equivocal effect is the tendency to repeat a purine or pyrimidine at adjacent wobble positions in weakly expressed eubacterial genes. Varied selections of genes and several statistical techniques agree on the significance of this wobble correlation.[5,7-9] The unique triplet periodicity suggests a translational function. However, correlations rarely have a unique rationalization, and one can devise explanations in which wobble correlation has no relation to translation. Thus the detection, in controlled experiments,[13,14,17] of a translational effect due to neighboring wobbles is an essential supporting observation. The apparent absence of correlated 3-3 doublets from yeast and human mRNAs[7] is a useful reminder that any context effect may depend on a particular translation apparatus and/or set of functional constraints.

V. CLUES TO THE CONTEXT MECHANISM?

The summed nonrandomness 3′ of a fixed codon extends three nucleotides, then tails off sharply to levels characteristic of sampling error.[5] This suggests tRNA-tRNA interaction, because the 3′-adjacent triplet is the domain of action of the next tRNA to be selected. In addition, the partial attribution of this effect to wobble-wobble correlation strongly suggests an indirect effect like tRNA-tRNA contacts because the distance between wobble nucleotides makes direct interactions less probable. However, interaction through the ribosome cannot be ruled out. The finding that codon-codon correlations are local[8] is consistent with any short range interaction. However, the tendency of tRNAs to prefer neighbors preferentially as the 5′ or 3′ member of a pair specifically suggests tRNA-tRNA interactions unique to specific surfaces of the molecules.[8] These effects could be expressed either during the original selection of the tRNA or as it interacts during selection of its 3′ neighbor. Such interactions between the P-site and A-site tRNA appear to exist and have a functionally significant magnitude.[18]

Thus, while statistical data cannot prove a mechanism, they support tRNA-tRNA interaction. However, there is also likely to be a direct interaction hidden within the envelope of correlation attributable to tRNA-tRNA interaction; translational context effects of the 3′ neighbor alone have now been measured[15,16] (see following), and these clearly occur during tRNA selection, with no apparent role for the previous or next tRNA. We believe that context effects are both direct (by interaction between codons and a 3′ neighbor nucleotide) and indirect (through tRNA-tRNA interactions), and that this list of mechanisms will grow.

VI. STUDY OF TRANSLATION USING SUPPRESSION

Much that is known about translation was elucidated using suppressors (reviewed in Eggertsson and Söll[19]). In particular, the context effect was discovered by its effect on nonsense suppressors.[20] The virtues of context studies with suppressors, in contrast to statistical studies, is that they can quantitate the phenotype, and by design, can be explicit as to which tRNA and process is affected. Therefore, we briefly discuss measurement of suppression. Transfer RNA-mediated suppression requires selection of the suppressor by the ribosome. In nonsense suppression, a suppressor tRNA competes with one or both release factors for a stop codon. The percent successful insertion of a nonsense suppressor's amino acid rather than termination of the nascent peptide is called the efficiency of suppression. In missense suppression, the suppressor competes with a normal tRNA for its sense codon, and the percent of gene product in which it acts instead of the competing tRNA is termed its efficiency. Suppressor efficiencies, therefore, always indicate the outcome of the competition between two reactions, and context-dependent differences may be due to effects on either reaction or both.

This competition has consequences for the use of suppression to characterize translational variables. Suppression is a probability, and cannot exceed 1.0. Therefore, it is nonlinear, and this should be included in calculations, e.g., a tenfold effect on the rate of action of a weak suppressor will produce nearly a tenfold change in suppression efficiency. In contrast, a tenfold effect on a strong suppressor will have a barely measurable phenotype (see Reference 21). Further, in translational comparisons, context should be a controlled experimental variable. A change in suppression efficiency for mutationally altered tRNAs is attributable to an effect on tRNA quantity or action, whereas the suppression efficiency in different contexts combines effects which may need to be resolved.

In addition, suppression efficiency should be measured directly, by measuring the gene product. Assays depending on plating efficiency or growth rate are gloriously sensitive. Such assays are the indispensable mainstays of genetic selection and screening, because they are ideally suited to the essentially binary "does/doesn't work" decision that a geneticist wishes to make. However, for mechanistic purposes these measurements are treacherous; growth rate and plating efficiency may both be highly nonlinear with respect to a particular gene product, and frequently saturate at low efficiencies. For example, as cellular *trpA* product varies from nil to 6% of wild-type, growth rates in minimal medium go from nil to fully wild-type[22]. Growth accelerates slightly, if at all, at greater *trpA* levels. Therefore, growth rate both amplifies the differences between mutants in the range where it is sensitive, and compresses the differences between dissimilar mutations in the range where it is not sensitive.

More critically, suppression efficiency estimates the fraction of ribosomal transits in which the suppressor acts. The efficiency therefore determines whether suppression occurs with a rate of the same order as its competitor, an event of normal translation. Therefore, significant suppression efficiencies, and more particularly, context-dependent differences in suppression efficiency, comprise *prima facie* evidence that the phenomena under study are relevant to normal translation.

When a conclusion requires comparison of different messages, it is essential to measure mRNA level, or to control for effects on it. Many sites may be subject to polar effects and effects on mRNA stability discussed in Appendix I. For example, Stanssens et al.[23] show that premature transcriptional termination (polarity) can vary over, at least, a 170-fold range depending on translational initiation frequency. Even single nucleotide substitutions produce changes in mRNA concentration of magnitude similar to other phenomena of interest, including context effects.[15,24]

VII. CONTEXT EFFECTS ON tRNA-MEDIATED NONSENSE SUPPRESSION

Nonsense codons at 3′ purine contexts (e.g., UAG A) are better suppressed than are pyrimidine contexts (e.g., UAG U).[15,26] Stormo et al.[27] have reanalyzed the large body of data on context and suppressor efficiency generated by the above authors, using multiple regression to simultaneously fit all the data. The result is that the 3′neighbor of UAG should be a purine for efficient suppression, and UAG A ≈ UAG G > UAG C > UAG U. A second interesting result is that U at the second-nearest 3′ neighboring position is almost equally influential, and contributes almost additively to the effect of a 3′ purine. In light of statistical studies on the context of sense codons, there is another interesting result, not emphasized by Stormo et al.[27]. The wobble nucleotides of the flanking 5′ and 3′ codons also had a smaller, equal, significant effect on suppressor efficiency, equivalent to about one sixth the total free energy associated with context.

VIII. UAG SUPPRESSION REFLECTS 3′ NEIGHBOR EFFECTS ON BOTH tRNA AND RELEASE FACTOR

Pedersen and Curran[15] have estimated the effect of the 3′ neighbor, UAG N, on the action of release factor 1 (RF1) and an amber suppressor tRNA, Su7C33[21] *in vivo*. All data refer to a single site, *lacZ* codon 366, at which only the 3′ neighbor of UAG was varied. Alleles were controlled for differences that the context alterations have on message level and enzyme activity. Observed suppressor efficiencies were UAG A ≈ UAG G > UAG C > UAG U, matching the 3′ pattern familiar from other studies.

The rates of RF1 and tRNA action were separated by putting translation of the UAG codon in competition with a frameshift, using the frameshift mechanism of the *E. coli prfB* (RF2) gene.[28,29] Competition for the UAG at the shift point implies that the output of a reporter gene in the shifted frame is a function of the rate at which RF1 and the tRNA act to fix the original frame. For five codons, this *in vivo* assay correlated well with *in vitro* assays of tRNA selection velocity (k_{GTP}, see Appendix III). Thus, rate of selection of a ligand in the coding site is the quantity measured by the frameshift competition *in vivo*.[24,28] To resolve the effects of context, cells having RF1 alone and RF1 plus the tRNA simultaneously can be compared, after normalization for mRNA levels, to distinguish the release factor and the tRNA.

Calculated rates of aminoacyl-tRNA selection vary 5-fold for Su7C33 in the order UAG A > UAG G = UAG U > UAG C. Rates of RF1 action vary 2.6-fold in the order UAG U > UAG G > UAG C > UAG A. Therefore, the accustomed 3′ neighbor effect on amber suppression appears to be the sum of distinct tRNA and RF patterns.

The order of nucleotides of the RF1 context preference[15] suggests a protein site that hydrogen bonds to U or G, which have keto H bond acceptors and imino H bond donors in corresponding positions on their base-pairing faces.[30] RF1 also appears to have a superposed, but weaker preference for a pyrimidine rather than a purine.

In contrast, the tRNA pattern may reflect base stacking; that is, it qualitatively resembles a "dangling-end effect", in which an unpaired 3' nucleotide stabilizes an adjacent RNA helix: 3' A ≈ G > U > C.[31] When the data of Miller and Albertini[25] were grouped and averaged within 3' contexts, and relative tRNA selection velocity compared, such velocities for four different suppressors correlate strongly with 3' dangling-end free energy[15]. Ayer and Yarus[32] had suggested that the overall context effect matched the specificity of the dangling-end effect, and Stormo et al.[27] rationalized the stimulatory effect of the next-nearest-neighbor U by suggesting that its weak stacking decouples the nearest-neighbor 3' purine so it can better stack on the codon-anticodon complex. Further, this relation is quantitatively plausible. Dangling ends modify RNA duplex stability by up to 1.8 kcal/mol in solution,[31] or about 20-fold in binding or rate. Context modifies efficiency about 30-fold,[27] which is consistent with the free energy attributable to 3'-stacking of message nucleotides on the codon-anticodon helix. Such stacking not only supplies the physical basis for a 3-1 doublet effect on translation, but also explains why the effect appears statistically as a correlation specifically with the neighboring wobble nucleotide.[3,5,7]

There is a non-exclusive alternative to 3' stacking of a dangling end, raised by Miller and Albertini:[25] all natural tRNAs possess a U 5' of the anticodon that might form a base pair with A or G 3' to a codon. However, this notion has been tested by Ayer and Yarus[32] employing mutated amber suppressor tRNAs with all four nucleotides 5' of the anticodon. Comparing the 16 possible combinations, no evidence of base pairing was seen. For example, A is a similarly superior 3' codon context whether it is opposite U, C, G, or A in the tRNA. Thus the superiority of the purine 3' context most likely rests on its ability to facilitate tRNA selection, the reaction assayed here, by stacking on the codon in the codon-anticodon complex. Clearly, the detection of a direct 3' context action on tRNA selection, by any mechanism, suggests that 3'context influences both missense suppressors and normal sense tRNAs, as well as nonsense suppressors.

IX. A BIOCHEMICAL DEMONSTRATION OF A 3' NEIGHBOR EFFECT ON tRNA SELECTION

Dix and Thompson[33] have devised a system for translation *in vitro*, using purified components, suitable for transient kinetics. The message is a 52-mer transcript with a Shine-Dalgarno sequence, on which ribosomes are initiated using initiation factors and fMet-tRNA. In unpublished experiments, they have translated the codon UUU in the sequence ... AUC AUG UUU NAG AUC ..., where AUG is the initiation codon, and N is any nucleotide. By measuring the

TABLE 1
Rate of GTP Hydrolysis During Selection of Aminoacyl-tRNAs at UUU in Different 3' Contexts

mRNA	Aminoacyl-tRNA	k_{GTP}, $M^{-1}s^{-1}$
UUU U	Phe-tRNA^{Phe}	2.8×10^6
UUU C	Phe-tRNA^{Phe}	4.5×10^6
UUU A	Phe-tRNA^{Phe}	2.8×10^6
UUU G	Phe-tRNA^{Phe}	3.3×10^6
UUU U	Leu-tRNA$_2^{Leu}$	0.010×10^6
UUU C	Leu-tRNA$_2^{Leu}$	0.012×10^6
UUU A	Leu-tRNA$_2^{Leu}$	0.008×10^6
UUU G	Leu-tRNA$_2^{Leu}$	0.009×10^6

rate of [^{32}P] release from GTP in preformed EFT$_u$·GTP·aa-tRNA, they characterized the rate of tRNA selection, k_{GTP} (Appendix II; Table 1), at UUU in four 3' contexts.

These data (Table 1) demonstrate a 3' neighbor effect in a purified system *in vitro*. The 3' neighbor affects the selection of the aa-tRNA cognate to the codon in the coding site, at least over the range 2.8 to 4.5×10^6 M^{-1} s^{-1}. A non cognate tRNA, Leu-tRNA$_2^{Leu}$ (with a G:U mispair in the first position), is selected much more slowly, but is also affected by context to about the same extent (Table 1). While this 60% effect may seem excessively moderate, it is the same magnitude as the velocity difference between the alternative phenylalanine codons UUU and UUC,[34] a codon choice which is highly biased in real genes, and usually thought to be under selection.[35] The subsequent tRNA is not involved in the context effect. In fact, in the Dix-Thompson experiments, the subsequent aa-tRNA (the one complementary to the context nucleotides) is not even present in the reactions. Accordingly, these data strengthen the argument that the 3' neighbor context effect is a direct one, on the binding of the EFT$_u$·GTP·aa-tRNA ternary complex and subsequent hydrolysis of GTP (Appendix II).

By reasoning from other studies, one can be even more specific about the context-sensitive step. Thompson, Dix, and collaborators have shown that, once bound, cognate ternary complexes almost invariably hydrolyze their GTP.[36] That is, because their dissociation is so slow, they do not dissociate, but proceed to hydrolyze GTP with an invariant rate (see Appendix II). Therefore, since neither dissociation nor GTP hydrolysis can be responsible, it was the association of the ternary complex with the open coding or ribosomal A-site, in which the anticodon first contacts the codon, whose rate was probably altered by 3' neighbor context.

The pattern of 3' neighbor effects on tRNA selection velocity, C > G > U = A, is different from that observed for the codon UAG *in vivo*,[15] and does not suggest a stacking pattern. This raises the interesting possibility that there may be more than one 3' context effect on tRNA selection. In this connection, the stacking (dangling-end) effect in the sequence UAG N is expected to be larger than that in the sequence UUU N translated by Dix and Thompson.[31] It is conceivable that weakened stacking allows another effect to be observed.

X. CONTEXT BIASES AT NATURAL TERMINATORS

Kohli and Grosjean[37] first noted nonrandom sequences near terminators. They find different preferences in eukaryotes and prokaryotes, and even eukaryotic cells and their viruses differ at the same positions. Selection for termination efficiency therefore may differ among genes translated by the same ribosomes.

Brown et al.[38] have recently surveyed termination codons and their context biases for 862 natural E. coli terminators. UAA is by far the most common, used about 65% of the time. Next is UGA (29%) and then UAG (6.6%). These biases are more pronounced in highly expressed genes, where UAA is the only commonly used stop codon. Neighboring nucleotides are also used nonrandomly. The most significant bias is for a U immediately 3' to all three termination codons (43%). Brown et al.[38] propose that the termination codon usage and context biases suggests that termination signals have a decreasing order of efficiency: UAA U > UAA G > UAA A/C >...> UAG A/C. The superior efficiency of UAA U is supported by the observation that as expression level of genes increases, usage of both UAA and a 3' U increase together. Furthermore, *in vitro* mammalian RF action requires a nucleotide 3' to stop codons,[39] though only A was tested. In the same assay, *E. coli* RFs do not require a fourth nucleotide, suggesting that the 3' neighbor enhances, but may not be an essential participant in the reaction.

There are also nucleotide biases 5' to terminators. The biases differ among the termination codons; UAA, though it is decoded by both RFs, does not appear to be the superposition of effects at UAG and UGA. Choice of the last amino acid is biased toward the basic ones and glutamine and biased against residues with a large hydrophobic side chain. In addition, threonine is strongly avoided.[38]

XI. CONTEXT SENSITIVE ACTION BY RELEASE FACTORS

A context effect on RFs was demonstrated by Martin et al.,[40] who used the antisuppressor effect of cloned RF1 and RF2 genes to show that context affects their actions *in vivo*. Because both RFs act at UAA, overexpression of RF1 and RF2 will differ in their antisuppressor effect at different UAA codons if the RFs themselves have context response that is different. Comparison of 13 different UAA contexts showed just such differences; and we have reanalyzed these data in Appendix III to estimate that the relative rate of action of RF1 and RF2 varies at least sevenfold in different contexts. While the context elements that affect RFs partially overlap with those that affect the ochre suppressor tRNA, there are probably unique influences on the RFs also. These unique RF context sequences cannot be specified, but reanalysis (Appendix III) shows that significant RF context elements probably exist at sites more distant than the 5' and 3' neighbors of the UAA codon.

The velocity of action of RF1 *in vivo* is affected by the 3' context in the order UAG U > UAG G > UAG C > UAG A.[15] These data identify one context nucleotide for RF1, and suggest a direct type of mechanism for its action. The 3' neighbor may account for some of the context variation seen by Martin et al.,[40] but one cannot yet be sure that RF1 and RF2 would be *differentially* affected by the 3' neighbor, as required in those experiments. Remarkably, the RF1 3' nucleotide preference matches the natural order of preference at UAA terminators tallied by Brown et al.[38] In fact, the frequency of usage of these 3' contexts in natural terminators is precisely correlated with measured relative velocities of RF1 action.[15] These data suggest strongly that codons in natural terminators exist in a context selected for rapid RF1 action. Accordingly, this correlation also strengthens the argument that natural selection acts on codon contexts, by providing an explicit example in which a biochemical context phenotype and a consistent evolutionary response exist together.

XII. WOBBLE-WOBBLE CONTEXT INTERACTIONS ARE OBSERVABLE *IN VIVO*

Folley and Yarus[17] attempted to measure wobble-wobble interaction, suggested by previous studies of sequence bias (see previous review and Reference 5). To measure what might be a small effect, a homogeneous set of codons potentially translated by five wobble V-base (U 5-oxyacetate) tRNAs were used, and the context change was amplified by simultaneously modifying the context of eight codons (of 1043 in the gene). A ten-codon insertion was made in the 5' terminus of the *lacZ* gene, and the two sequences to be compared were related by permutation of the same ten codons to yield a pattern (separately repeated A and repeated G wobble nucleotides; the low pattern) found to be significantly favored by these particular tRNAs in weakly expressed genes, and a pattern (interspersed A and G wobbles; the high pattern) significantly favored by V-base tRNAs in highly expressed genes. Because the two sequences were related by permutation of codons (only four nucleotides are changed), they have the same codon composition and therefore are unaffected by possible differences between codons (codon choice). The primary experiment was also arranged so as to reproduce the same amino acid sequence in both high and low pattern mRNAs, so that β-galactosidase subunit assembly, stability, and specific activity cannot affect the result.

The low pattern gave 0.68 ± 0.03 and 0.86 ± 0.02 of the gene product of the high pattern message, when the oligonucleotides were inserted at two different sites. This result was unaffected by level of expression, as 0.64 ± 0.04 and 0.60 ± 0.04 of the high pattern product were observed for the first site in a repeated measurement with induced and uninduced cultures, respectively. There is no indication that these effects are due to different frequencies of initiation of translation; no structures are created near the ribosome binding site by computer analysis, and the effects of context are observable with three ribosome binding sites that differ in sequence. Further, the effect appears to depend on translational elongation, because it disappears at both affected sites when the context oligos are shifted out of frame by flanking single substitutions. Another indication of translational involvement is that the context effect disappears when a translational variable (ribosome density; see following) is manipulated. Furthermore, when translation is terminated just after the context insertions at the most effective site, and a normal reporter gene put downstream in a separate cistron (a transcriptional fusion), the output of the reporter is not affected. Thus, low pattern sequences do not generally destabilize their message, or autonomously cause polar effects. These authors conclude that ribosomal slowing during transit of the low pattern of wobbles creates the opportunity for a polar effect downstream, though in principle, ribosomal slowing might also produce an effect on mRNA decay at some downstream site. In fact, Ruteshouser and Richardson[41] subsequently reported rho-dependent transcription termination sites near the context sites of Folley and Yarus.[17]

Buckingham et al.[13,14] are in the midst of an analysis of the effect of the 5' and 3' neighboring codons on suppression at codon 234 in *E. coli trpA*. Codon 234 is flanked by serine codons, so the six serine codons UCN and AGY have been introduced on both sides of 234 without changing the amino acid sequence. This group of messages provides controlled comparisons bearing on the effect of neighboring wobble positions upstream and downstream (e.g., neighboring UCA vs UCG), on the effect of simultaneous changes in the nearest and next-nearest neighbors upstream and downstream (e.g., neighboring AGC vs UCC), and, most subtly, on the effect of wobble versus normal pairing by upstream versus downstream neighbors (e.g., neighboring AGU vs AGC, probably translated by the same tRNA). Relative growth rates have been published for codon 234 as UGA or AAA, characterizing nonsense and missense suppression, respectively, in each of the above respects. Suppression of AAA 234 with a glycine-inserting suppressor has a phenotype because lysine 234 inactivates *trpA*.

The results are simple to state, but provocative; *each* of the mentioned types of change alters the cellular growth rate under UGA nonsense suppression. Thus, in particular, both neighboring wobble positions apparently alter UGA nonsense suppression, even when the changes do not change the neighboring tRNA, and alter only base pairing at the wobble positions. For missense suppression, change in the 3' and 5' wobble nucleotide, and simultaneously in the two nearest 3' neighbors modifies growth rate; the other missense cases were not tested, as of this writing. As emphasized by Buckingham et al.,[13] these results could mean that these aspects of context are prejudicing the choice between two sense tRNAs (at the missense site). This means that different tRNAs translating the same codon might be used selectively in different contexts, a possibility implicit in the previous observation of context-specific missense suppression.[42] It will be of great interest to follow the quantitation of these experiments, which may allow the comparison of translational 3′ neighbor effects with the effects of wobble correlation, in a single system.

XIII. A MECHANISM FOR WOBBLE INTERACTION

The direct effect of a 3′ neighboring nucleotide on tRNA selection supplies a rationale for the 3-1 doublet effect, where the doublet is positioned at the 3′ side of the A site. A distinct, but related mechanism might apply to the 3-1 doublet positioned at the ribosomal P/A-site boundary, if it is found that this has an effect on tRNA selection.[14] The nucleotides of the A-site codon could be stacked on the P-site codon-anticodon helix before they are disturbed by formation of an mRNA kink due to the nascent tRNA-tRNA interface.[18] Yet the 3-3 doublet or wobble-wobble interaction seems unlikely to be mediated by stacking within the mRNA, and other direct interactions between adjoining wobble nucleotides seem even less likely than does stacking. Therefore, there is probably at least one more local context interaction, probably mediated by tRNA-tRNA contacts.

Smith and Yarus[18] point out that such contacts are likely between anticodon arms and CCA arms of A- and P-site tRNAs, which must approach each other during protein synthesis. They devised a means to measure tRNA-tRNA interactions at a test site in the N-terminus of *lacZ* by comparing test messages containing the sequence UAG UGA, and controls UGG UGA and UAG UGG. All these codons are translated by specific derivatives of tRNA^{Trp}, so tRNAs differ only at the sites of individual mutations. The measurement turns on the idea that if translation of the UGA (opal, A-site) codon is independent of the translation of the UAG (amber, P-site) codon, then the suppressor efficiencies in the control messages where the suppressors act in isolation can be multiplied to predict the efficiency when the two suppressors act side-by-side at UAG UGA. If the tRNAs interfere, their joint efficiency will be lower than predicted. In practice, two amber suppressor tRNAs are compared in the P site, because comparison of different tRNAs controls for nonspecific effects that affect all measurements.

Interference is observed for mutations in the anticodon loop of the P-site tRNA, but not the helix. These effects can be large, and are up to 170-fold when the conserved pyrimidines on the 5′ side of the anticodon loop of the P-site tRNA are replaced with purines. Smith and Yarus[18] interpret this as physical interference between the tRNAs during selection of the A-site tRNA, mediated by the larger purines. Modeling shows that the crystallographic structure of tRNA allows the 5′ side of the anticodon loop in the P site to closely approach the 3' side of the anticodon loop in the A site. In addition, normal tRNAs with various amino acid specificities whose anticodons had been altered to read UAG varied ninefold in apparent interference. This suggests that normal tRNAs in the P site vary about an order of magnitude in their effect on rate of selection of the tRNA 3' to them in the A site. This variation implies a plausible free energy difference (about 1.4 kcal/mol) for contacts over a large molecular surface between the tRNAs.

However, there is an alternative interpretation. Apparent interference effects could be instead stimulatory effects on the rate of RF2 action at the A-site codon. Such effects on RF2 seem *ad hoc,* requiring that the rate of RF2 be stimulated two orders of ten by 5' anticodon loop purines that do not occur naturally in tRNAs. In addition, tRNA-tRNA interactions are made likely by other experiments. Wittenberg, Dix, Thompson, and Uhlenbeck[43] have performed *in vitro* transient reaction studies with purine 33-substituted tRNAs in the P site, and shown interference with the A-site tRNA selection. In addition, Moazed and Noller[44] have shown that ribosomal A-site tRNA protections (characteristic of A-site occupation) do not appear if the P site is occupied with a tRNA purine-substituted at position 33. Thus, context effects on tRNA selection *in vitro* can be mediated by changes in the tRNA-tRNA interface. By hypothesis, this tRNA-tRNA interaction alters not only with the tRNAs themselves, but also is affected by changes in their pairing with the message.

XIV. tRNA MODIFICATION AFFECTS CONTEXT SENSITIVITY

Absence of posttranscriptional modifications in the tRNA anticodon loop can affect suppression efficiencies of nonsense suppressors[45,46] and can cause deattenuation of operons sensitive to the rates of translation of specific codons translated by modified tRNAs[47] (for more sources, see the review by Landick and Yanofsky[48]). Such modifications at purine 37 have been shown to increase stabilities of tRNA-tRNA complexes paired through complementary anticodons, presumably due to increased enthalpy change due to stacking of modified bases.[49-51] Interestingly, sensitivity of suppression to context can also be affected by modification.[52,53]

Björk and collaborators have shown that deficiency for ms^2io^6A adjacent to the anticodon (position 37) affects the context sensitivity of certain amber suppressors.[52,53] The most recent work examines suppressor efficiency at *S. typhimurium hisD2404,* with either C or A 3' to the amber codon. The A context variant is better suppressed than is the 3' C context allele, as expected. Each of three suppressors that normally carries the ms^2io^6A37 modification is more affected by 3' context *miaA1* cells, which are blocked in the first step of the modification pathway. The context-dependence of a fourth suppressor that is not modified is unaffected by *miaA*.

Nucleotide 37, the 3' neighbor of the anticodon, cannot interact directly with the message nucleotide 3' to the codon (they are separated by the codon:anticodon complex). How then does modification affect the response to 3' neighbor context? More quantitatively, how can the context effect and the lack of modification be more deleterious together than when assayed separately?

We interpret these measurements in terms of a scheme for translational selection of aa-tRNAs (see Appendix II). There is evidence suggesting that both 3' context and anticodon loop modification may alter the rate of dissociation EFT_u·GTP·aa-tRNA from the ribosome. 3' context modifies the binding of EFT_u·GTP·aa-tRNA to the ribosome.[16] It is natural to suppose that it may alter the dissociation, the reversal of the same reaction, also. Nucleotide 37 modification has been observed to alter the *in vitro* dissociation of tRNA from a message analogue in solution.[49] However, because a normal cognate tRNA is relatively immune to changes in its rate of dissociation, both context and modification affect such tRNAs only to the extent that they affect other reactions, for example, initial binding (k_1; see Appendix II).

However, when both poor context and lack of modification occur together, aa-tRNA selection may be forced to operate under conditions in which dissociation (k_{-1} and/or k_{-4}, Appendix II) becomes large enough to be significant to the translational phenotype. Thus, the apparent effect of both conditions together is much larger than when the consequences for suppression efficiency are measured for a single defect.

XV. A CONTEXT THAT DETERMINES THE MEANING OF THE UGA CODON

Remarkably, UGA in *E. coli* is a termination codon, yet, under anaerobic conditions can be translated as selenocysteine, depending on codon context. The selenocysteine tRNA,[54] tRNA^{Sec}, has numerous unique structural features required for UGA translation,[55] and an anticodon specific for UGA. tRNA^{Sec} is a serine accepting tRNA, but Ser-tRNA^{Sec} is subsequently converted to selenocysteinyl-tRNA^{Sec} through the action of *selA,* the selenocysteine synthase, and *selD,* which synthesizes the selenium donor.[56] Selenocysteine is incorporated during translation of internal in-phase UGA codons in a few proteins, notably formate dehydrogenase.[57] ^{75}Se labeling does not occur elsewhere, which testifies to the specificity of UGA translation by tRNA^{Sec}. Fusions of the critical *fdhF* sequences with *lacZ* have been used to show that the sequences upstream of the UGA are not crucial, but that 46 nucleotides downstream of the UGA in *fdhF* are required for full insertion of selenocysteine.[58] Within this required downstream region is a stable hairpin, whose structure and loop sequence is required for selenocysteine insertion at the UGA.[59]

Recognition of the hairpin is carried out by an unknown element, but a possibility is the product of the *selB* gene, which is homologous to EFT_u.[60] The *selB* protein binds GTP and sec-tRNA^{Sec}, but not ser-tRNA^{Ser}, or even ser-tRNA^{Sec}. In contrast, EFT$_u$ binds ser-tRNA^{Ser}, but ser- and sec-tRNA^{Sec} only weakly.[62] Thus, it is thought that *selB* is a specialized translation factor that delivers sec-tRNA^{Sec} to UGA in the correct context. This suggests that it may be *selB* that interacts with the essential sequences in the context hairpin, and accounts for the translation of the special UGA codons as selenocysteine. This context effect can even overwhelm a change in the codon itself. A tRNA^{Sec} anticodon mutant that creates a G:A mispair at the wobble position is nonetheless about 30% efficient in translation of the UGA in these genes.[55] Not all mispairs can be compensated; a tRNA^{Sec} mutation that creates a second position G:G mismatch has low translational efficiency.[55] If EFT$_u$ is also sensitive to message structure in somewhat the same way as is the *selB* protein, such sensitivity provides another indirect pathway for the context effect at normal codons.

Another aspect of the specificity of tRNA^{Sec} is that it does not translate normal UGAs; that is, it is a very inefficient opal suppressor. This prohibition is very robust; simultaneous mutations in the CCA stem, D arm, and long extra loop of tRNA^{Sec}, removing most of its special sequence features, are required to make it an efficient serine-inserting UGA suppressor.[62]

XVI. A DOWNSTREAM SEQUENCE DETERMINES THE MEANING OF THE UAG CODON

Expression of the Tobacco Mosaic Virus RNA replicase region involves ≈ 5% readthrough of an in-phase UAG codon. Skuzeski et al.[63] have studied the sequences around this UAG by fusing the UAG to β-glucuronidase, and transiently introducing these fused genes into tobacco protoplasts by electroporation. The results eliminate 5' sequences as essential; even if the initiation codon was moved adjacent to the UAG, readthrough was efficient. In contrast, the sequence two codons downstream of UAG was critical. Skuzeski et al.[63] assayed every single nucleotide change in the sequence UAG NNN NNN; for high frequency readthrough UAG CAR YYA was required. That is, the next codon could be either glutamine codon, but the next-nearest neighbor could be any of the four codons, YYA. Mutagenesis of the codon 3' of YYA had little effect. The readthrough mechanism also works when UAG is replaced with UAA or UGA. The critical events are probably translational. Expression of fused β-glucuronidase was high and equivalent when the sequences giving variable readthrough when in-frame were translated out-of-frame. The mechanism of this effect is not apparent, but some conceivable models are unlikely. For example, at the moment that readthrough is determined (UAG in the A site) it is improbable that the distal sequence YYA is interacting with the anticodon of a tRNA. The mechanism is probably indirect, for example, through factors or the ribosome. Product amino acid sequences may greatly clarify these events. As the authors point out, other plant RNA viruses conserve the critical sequence, and may use the same mechanism.

XVII. EFFECTS OF CONTEXT ON INITIATION

The nucleotides near start codons are not random (reviewed by Gold[64]). The most frequent second codons are AAA and GCU. The eight most highly expressed T7 genes all have GCU as the second codon[65] and, therefore, it is tempting to think that GCU at position 2 may contribute to high translational initiation rates.

Systematic surveys of second codon effects have been performed by Looman et al.[66] and Cantrell et al.[67] Looman et al. placed each codon at the second position of *lacZ*, and assayed immunoprecipitated protein as an index of expression. β-galactosidase expression varies 15-fold among the set of alleles with all possible sense codons at the second position. AAA as the second codon gave the highest activity, in agreement with its high use in natural start sites. However, no other simple explanation can fully explain the observed pattern, including codon usage. In fact, synonymous codons translated by the same tRNA were often quite different. Cantrell et al.[67] studied the effect of 16 second codons on the translation of a human IGFII gene, and also found an idiosyncratic order, that differs substantially from the order of Looman et al.[66] This suggests that the best second codon may vary from message to message. One reason for the differences may be that Cantrell et al.[67] used a system that gave extreme overexpression, and Looman et al. assayed normal levels of gene product. In addition, β-galactosidase output may be dependent on several factors in this system, which may combine to partially obscure trends. These factors may include variation in mRNA stability that can accompany mutations near the 5' end of *lacZ*,[68] and variable polar effects through dependence on translational initiation efficiency.[23] Further, De Smit and Van Duin[69] have shown convincingly that translational yield is inversely dependent on the strength of secondary structures that involve the initiation site in the MS2 coat gene and that, generally, natural initiation sites have low propensities for secondary structure. GCU, which is one of the two most frequently used second codons, gives only average β-galactosidase output in Looman et al..[66] However, we have noticed that GCU strengthens a potential secondary structure that includes the AUG start codon in their construct.

Stormo et al.,[27] also using *lacZ*, show that the identity of the triplet upstream of the initiation codon is important to β-galactosidase synthesis. No systematic patterns based on nucleotide identity were observed, though this study may also be subject to effects on mRNA synthesis and decay. Stormo et al.[27] suggest a novel context effect; that rates of correct initiation may be affected by nearby potential start codons in the wrong reading frames.

It is quite clear that sequences outside the canonical initiation sequences alter expression, and in that sense a context effect on initiation exists. However,

XVIII. CONTEXT EFFECTS ON MISSENSE ERRORS

It follows easily that if nucleotide context differentially affects tRNA selection, it will affect the mistaken substitution of one tRNA for another. There is evidence for context effects on missense error frequencies. For example, Carrier and Buckingham[70] found that UGU was more easily mistranslated as tryptophan *in vitro* when flanked by UUU than when flanked by GUG. These workers suggested that either base stacking, tRNA-tRNA interactions, or base pairing between 3' neighbor nucleotides and tRNA could influence error.

Differential error frequencies depending on position of the misincorporation error also suggest, among other possibilities, that context affects missense error frequencies *in vivo*. The basal frequency of misincorporation of cysteine at the sole arginine codon (CGU) in *rplL*[71] is about ten times greater than the average error frequency at the six CGU codons in the *hag* gene.[72] Misincorporation of cysteine at Phe codons is also more frequent in *rplL* than in *rplA*.[73] GAC-codon-GGC, or a related sequence, was suggested as a context for high translational error based on a number of observations of relatively high misincorporation frequencies in such contexts.[74] We do not know of any experimental test of this idea.

Precup et al.[75] show that either UUU or UUC (phenylalanine) at position 8 of *argI* can direct incorporation of leucine at frequencies near 50% during starvation of *relA* cells for phenylalanine. However, neither UUU nor UUC directs detectable leucine incorporation above background at position 3 under the same conditions.[75] Judging from the background in the published amino acid sequences, the difference in rate of incorporation is at least five- to ten fold between sites. The inability to detect leucine at position 3 was shown not due to an instability of the errant peptide, but could be due to frameshifting or dropoff.

XIX. CONTEXT AND FRAMESHIFTING

Because frameshifting involves base pairing between tRNA and the message in the wrong frame, frameshifting should appear strongly context-dependent. For example, peptide sequencing reveals that frameshifts may be caused by slippage of tRNA to a new site where stable base pairing can occur.[76,77] Translation of the T4 gene 60 protein apparently involves hopping of the P-site tRNA between the 46th and 47th amino acids, leaving 50 intervening nucleotides untranslated.[78] One possibility is that the P-site tRNA moves from one GGA codon, spanning a stimulatory secondary structure, to a second GGA codon downstream in virtually every ribosomal transit. Weiss et al.[76] have characterized a variety of shorter hops, in which it appears that similar five- and six-nucleotide hops over a stop codon occur with a frequency of 0.4 to 1.0%. Shorter hops, from the first to the second GUG in GUGUG, occur with apparent frequencies of 8 to 9% in wild-type *E. coli* cells, and these shorter jumps are facilitated by suppressor tRNAs with insertions in the anticodon loops.[77,79] These data testify to the ability of the P-site tRNA to explore alternative anticodon pairing possibilities, perhaps especially when a stop codon or another context feature allows a pause that gives time for unusual, slow activities.[28,80]

Frameshifts may also occur when aa-tRNA is selected during base pairing with an out-of-phase triplet. The sequence of the peptide produced by frameshifting at a slowly translated AAG (lysine) codon in the context GCC AAG CUU suggests that the first 'A' of AAG is skipped.[81] The shift is likely caused by translation of the overlapping AGC by a serine-inserting tRNA rather than slippage of the alanine tRNA from GCC to CCA, because such a slippage would result in mispairs at two out of three positions. Certain tRNAs can cause frameshifts by their acceptance while paired to the message in the −1 reading frame.[82] In those cases, adjacent nucleotides alter shift frequency, but apparently not by base pairing. The mechanism of these context effects remains obscure.

Context influences three- and four-base decoding by tRNAs that have an extra nucleotide in the anticodon loop. Many such tRNAs probably cause frameshifting by forcing the *next* translational cycle to occur out-of-phase.[83] These enlarged anticodon loops may occupy the P site in either of two conformations, corresponding to three- and four-base decoding. In a conformation in which the anticodon helix contains an extra nucleotide, this stacked structure probably occludes the first base of the next in-phase codon, which causes it to be skipped. A fourth base pair between suppressor and message is not necessary for four-base decoding,[83-85] but presumably by stabilizing the extended anticodon helix, it makes four-base decoding more likely than three-base decoding.[83] tRNAs with enlarged anticodon loops may also have an increased tendency to explore alternative pairings.[76,86]

The high-frequency frameshifting necessary for expression of certain genes occurs at homopolymeric (slippery) sites in contexts that create ribosomal pauses. In addition, context elements may catalyze tRNA-message slippage. In *E. coli prfB* (RF2), a +1 frameshift occurs by slippage of peptidyl-tRNA^{leu2} from CUU to the overlapping UUU in CUU UGA.[87] High-frequency frameshifting requires at least two context elements: (1) a run of purines upstream of the shift site,[28,29,76] and (2) the codon 3′ to CUU must be slowly translated to provide a ribosomal pause. Weiss et al.[29] provide strong evidence that the upstream run of purines contributes to frameshifting by interacting with the anti-Shine-Dalgarno sequence of 16 S rRNA. A G -> C mutation that disrupts frameshifting can be compensated by a complementary mutation in the anti-Shine-Dalgarno sequence. This putative interaction probably causes frameshifting by straining the message, which must be in contact with both the

P-site tRNA and the anti-Shine-Dalgarno sequence only three mRNA nucleotides apart. A tRNA-message slippage in the 3′ direction increases the spacing to four nucleotides and presumably relieves the strain. In addition, the Shine-Dalgarno-like interaction may contribute to a ribosomal pause.[28]

The requirement for a ribosomal pause at the codon following CUU is shown by observations that frameshifting is inhibited by rapid translation of that codon. For example, increasing concentrations of RF2 cause decreasing frameshifting *in vitro* by facilitating termination at the UGA 3′ to the shift-point CUU.[88] In addition, the *in vivo* frameshift frequencies of prfB-*lacZ* fusion alleles in which the UGA is replaced by other codons is decreased when the activities of cognate tRNAs are increased.[22,28]

It has been suggested that high levels of frameshifting require a stop codon 3′ to CUU. Stop codons 3′ to frameshift sites have also been observed to enhance frameshifting at other sites.[76,77] However, at least in the case of *prfB*, the high levels of frameshifting also occur when the UGA is replaced by codons that are read by rare tRNAs. For example, *prfB-lacZ* fusions that have CUA or CGG 3′ to the CUU, frameshift with frequencies similar to alleles with UGA or UAG.[15,24] In fact, within a set of 29 alleles that have sense codons in place of UGA, frameshift frequency is inversely related to codon usage.[24] Apparently, elevated levels of frameshifting require pauses, which can be provided by either rare sense codons or stop codons at the immediate 3′ position to the *prfB* frameshift.

Eukaryotic retroviral and *E. coli dnaX* frameshifts may also occur by pause-slippage mechanisms.[89] Comparisons of amino acid and nucleic acid sequences indicate that tRNA slippage occurs at homopolymeric shift sites (for recent reviews, see References 79 and 90). Downstream secondary structures may cause ribosomes to pause allowing time for tRNA slipping to occur with high frequencies.[89] Such a pause has been demonstrated for the *E. coli dnaX* frameshift site *in vitro*.[91] This site contains a purine heptamer just upstream of a putative strong secondary structure.[92-94] Peptide sequencing confirms that the heptamer is the site of tRNA-message rephasing.[92] Ribosomes pause in the vicinity of the heptamer *in vitro*.[91] Mutations that disrupt the slippery heptamer do not prevent ribosomal pausing,[91] but do block frameshifting.[91-94] Both pausing and frameshifting are reduced in a mutant in which the stem-loop structure is at least partially disrupted[91]. Thus, these frameshifts may be cases where nature has selected message contexts that inhibit normal translation thereby increasing the frequencies of frame slippage.

Even though *prfB* and the eukaryotic frameshifts both apparently occur by pause-slippage, there is an important distinction. *prfB* frameshifts when the A site is free, but the eukaryotic viral shifts occur when both A and P sites are filled with tRNA. *prfB* can therefore be regulated by the rate of translation of the A-site codon.[24,28,88] Two-site-slippage does not offer the same opportunity for regulation by rate of release factor or tRNA selection at a specific codon.

A final example of a context-dependent pause-slippage occurs when the rare arginine codons, AGG or AGA, appear in tandem in messages that are expressed at extremely high levels.[95] A second AGG in highly expressed messages becomes difficult to translate because the first AGG consumes the rare cognate tRNA^{Arg}.[96] As a result of the low rate of normal translation at the second AGG, slow events like rephasing or translation of an out-of-phase codon become likely. A pause contributes to the high-frequency frameshifting in this case, because frameshifting can be suppressed by overexpression of the normally rare arginine tRNA.[97]

XX. POSSIBLE CONSEQUENCES FOR THE CONTEXT STRUCTURE OF MESSAGES

Avoidance of frameshifts could constrain message contexts. It is well known that translational rates are variable among codons and perhaps contexts.[24,98-100] Put another way, pauses of some length will be frequent. Clearly, tens of pause sites within a gene with frameshift frequencies of a few percent would be intolerable. Measurements of the fractions of ribosomes that fail to complete peptide synthesis argue that the upper limit for loss of nascent chain, including those lost following frameshifting, is only one per several thousand codons in *lacZ*.[101,102] Frameshifting out of the normal into either alternative frame, or into the normal frame from –1, was found to occur with a similarly low frequency within segments of approximately 100 codons of *rpoC*.[86] Those results argue that the quotidian ribosomal pause is not frequently accompanied by frameshifting.

Are there contextual features that contribute to this observed low level of spontaneous frameshifting? For the purpose of this review, we searched 45 highly and 50 weakly expressed *E. coli* genes that contain 17,849 and 17,302 codons, respectively, to determine whether codon pairings that are plausible sites of rephasing are more or less frequent than expected.[103]

Codon pairs that resemble the frameshift site of *prfB* seem to be avoided. In highly expressed genes, codon pairs of the type YUU YNN are less than one third as common as expected. This is related to expression; such pairings are only slightly less common than expected in weakly expressed genes. In particular, sequences involving the phenylalanine codons (UUU/C) are highly nonrandom. The sequence UUU CNN is strongly avoided in both gene sets (31 expected, 4 observed in highly expressed genes; 97 expected, 65 observed in weakly expressed genes). UUU CNN might be especially prone to +1 slippage — a rephased tRNA^{Phe} might be more stably base paired in the new frame, having exchanged a G-U pair for a G-C pair at the wobble position. In contrast to the avoidance of UUU CNN, UUC CNN is unexpectedly frequent. We would not expect the latter site to be shifty because rephasing would generate an A-C pair at the second anticodon:mRNA position. It seems a reasonable hypothesis that when a CNN codon is preceded by a phenylalanine codon, UUC is preferred to minimize frameshifting.

Frameshifting in *dnaX* requires slippage of tandem tRNA^{Lys} from NNA AAA AAG onto overlapping AAAs in the −1 frame.[92] Such *dnaX*-like sites also seem to be avoided. In highly expressed genes, when the codon 3′ to the lysine codons AAA/G begins with A, the preferred lysine codon is AAG, which may be another mechanism for minimizing spontaneous frameshifting. In addition, highly expressed genes, both lysine codons (AAA/G) are preceded by A only about half as often as expected. This may decrease −1 frameshifting, particularly where the next codon may be slowly translated.

We have also looked for context constraints related to hopping. The demonstrated potential for high hopping frequencies[76,77] suggests that triplets potentially translated by the same tRNA close to each other in any phase would be underrepresented. However, this is not true for the normal phase (that is, for in-frame hops).[104] We studied this case by determining the distributions of codons read by the same tRNAs. With a few exceptions, such isospecific codons are distributed randomly out to a spacing of 25 codons (see Figure 1). Exceptions involve overly frequent repetition of neighboring codons read by alanine or glutamine tRNAs and aviodance of adjacent codons read by the major proline tRNA (CCA/G). All of those exceptions correspond to perturbed dipeptide frequencies and therefore might be attributed to protein structure. The only case not accompanied by an unusual dipeptide frequency is that of two codons, UCA/G, read by tRNA$_1^{Ser}$. These tend to repeat in weakly expressed genes; this departure is in the wrong direction to depress tRNA hopping. Further, in weakly expressed genes these serine codons strongly favor 3′ neighbors that have wobble purines.[5] This bias may include preference of UCA/G over the other serine codons, UCU/C and AGU/C, as 3′ neighbors. Thus, we find no support for the idea that in-phase tRNA hopping is controlled by avoiding juxtaposed codons read by the same tRNA. Hopping out-of-phase deserves separate examination.

XXI. CONTEXT AND EXPRESSION

How might effects on, for example, tRNA-tRNA interaction, alter gene expression when there is no overt translational error? The question is not rhetorical. Translation might be considered to be a pathway of hundreds of steps. If codon context modifies the rates of internal steps which are not rate limiting, context is necessarily irrelevant to steady-state velocity. However, we believe that this argument is fundamentally flawed, at least for eubacterial messages. Such mRNAs are not translated in the steady-state, and therefore single steps do not determine overall rates. As shown elsewhere,[105] in contrast to steady-state reactions, all reactions in a transient pathway can contribute to output. (It is clear from this argument that eukaryotic mRNAs, with their longer half-lives and more leisurely approach to steady-state, must be examined

The Translational Context Effect

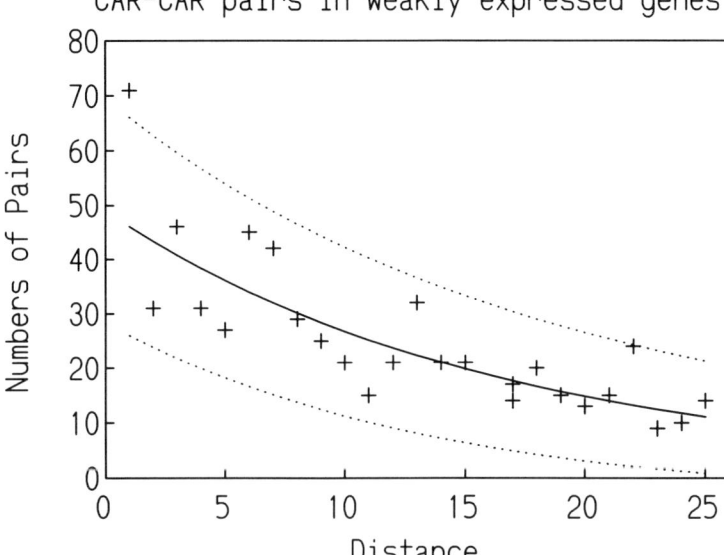

FIGURE 1. Number of CAR (CAA/G, glutamine, Q) codon pairs in weakly expressed *E. coli* genes as a function of the distance between them. Expected values are represented by the solid curve, which is flanked by dotted lines representing 1% binomial confidence limits. Crosses are observed values from our sequence database containing 17,302 codons; adjacent CAR codons are significantly more frequent than expected. Individual CAA and CAG codons are distributed similarly to the pooled CAR codons shown. There is also a less significant excess of CAR nearest neighbors in highly expressed genes.

Expected numbers of pairs (P_d) separated by distance d (d ≡ 1 for adjacent codons) were calculated as follows. The formula is complicated by the need to account for conditional probabilities during sampling from individual genes without replacement.

$$P_d = \sum_{i=1}^{N}(T_i - d)(Q_i/T_i)\{(Q_i - 1)/(T_i - d)\} \left\| \prod_{j=2}^{d}(T_i - Q_i - j + 2)/(T_i - j + 1) \right\|$$

N is the number of genes, Q the total number of CAR codons, T the total number of codons. The term in $\|$ vertical bars $\| \equiv 1$ for d = 1, otherwise having the values calculated from the contained expression.

separately.) In addition to this theoretical difficulty, the usual steady-state argument is also practically flawed, as it omits the most important sources for the context effect. More concretely, it is calculated in Appendix I, that P, the protein output from R_o bacterial transcripts can be written:

$$P \approx R_o \{t_d / t_s\} \exp(-T/t_{do}) \tag{1}$$

where the exponential is optional and sometimes relevant

t_d = the average time for mRNA decay during production of protein (closely related to the measured halflife, $t_{1/2}$)
t_s = the average spacing of ribosomes on the message in time units
T = the time for ribosomes to traverse the gene (the transit time)
t_{do} = the average time in which message decays, specifically during the first ribosomal transit.

Codon context (or other events during elongation) could conceivably alter every parameter in the equation above. In the following paragraphs, we roughly estimate the *maximal* size of these effects for single tracts with a poor context. These estimates may be crude but are based on an explicit set of assumptions (Appendix I) and can be improved as knowledge increases. Note that different effects must be of different magnitudes.

For the class of messages whose decay begins at the 5' end, near the ribosome binding site, context could alter expression:

1. via R_o, which may be reduced by polarity, modulated by ribosomal movement during transcription — we take our cue from the classical study on the gradient of polarity within a gene.[106] The observed effects of lacZ *amber* and *ochre* mutations on downstream gene expression span about 100-fold, rising sharply as distance between translational termination and the *lacZ-Y* boundary increases. We take from this an estimate of a maximal 20-fold effect on polarity, because the average gene is smaller than *lacZ* and gives less chance for adventitious termination. In addition, the highest values in Newton et al.[106] may be confounded with stability effects. These context effects may be most significant for sites near the 5' end of messages. Recall that these estimates are upper limits; the only experiment that may apply to this effect, using typical sense codons, gave 1.2 to 1.7-fold.[17]

2. via t_d, the average mRNA working lifetime, which may respond to the distribution of ribosomes — we estimate a *maximal* 15-fold range for $t_{1/2}$ on the basis of the half-lives of the translated and untranslated L11 message,[107] and thus a 15-fold range of expression (Equation 1).

3. via t_s, the temporal spacing of ribosomes, by changing clearance of the ribosome binding site — this effect should be mostly due to the first few codons. Ribosomal clearance will be most significant in messages that have ribosome binding sites that are rapidly filled (probably corresponding to highly expressed genes), because t_s is most affected by changes in ribosomal clearance time in this case (Appendix I). For the fastest-acting ribosome binding site having a size of 12 codons,[108] we estimate that if 2 codons have a transit time 10-fold longer than the others, this gives a maximal range of 2.5-fold in t_s and expression (Appendix I; text Equation 1).

4. via t_s, by ribosomal queuing in the 5' region, which modulates the frequency of initiation — this effect will be most significant for highly expressed genes, whose maximal rate of initiation is that which would pack the message with ribosomes. Ribosome binding sites, like messages as a whole, can be cleared of ribosomes at 12.5 codons per second.[99] If the fastest conceivable ribosome binding is assumed, clearance time would dominate the frequency of initiation. Sörensen et al.[99] have estimated ribosomal transit rates lower than 2.1 codons per second for slowly translated codons *in vivo*. This suggests a difference in expression of sixfold, for a very highly expressed gene with a maximally disruptive queue.

5. via t_s, by ribosomal dropoff mechanisms, including frameshifting — tentatively estimated as a twofold effect maximally, using the 50% loss of nascent chains due to frameshift at AGG AGG sequences as the extreme.[95] However, ribosome loss by any means may interact with polarity and halflife to give larger effects.

For messages whose decay begins at the 3' end, context potentially has the above effects, and in addition can work:

6. via t_s, because queuing anywhere in the mRNA alters the relevant average ribosomal spacing — we estimate a maximum of sixfold, as above, and the maximum is more plausible for this style of message.

7. via T/t_{do}, because the rate of ribosomal movement determines the transit time T, and via the protective effect of translocating ribosomes, may also be coupled to t_{do}, characterizing the initial decay of the message — if we assume the above 15-fold range for decay rates, and an average transit time of 30 seconds, this factor ranges from $e^{-0.17}$ to $e^{-2.5}$, corresponding to about 10-fold change in expression.

The problem of context seems to us not the conceivability of translational elongation effects on the output of genes (compare Liljenstrom and von Heijne[109]), but resolving the possibilities and putting authentic numbers to them.

Only one measurement has been attempted to measure context effects on the yield from a normal gene (a gene that does not require suppressor tRNA action[17]). Accordingly, this experiment presents an opportunity to investigate the mechanism of the regular context effect. Folley and Yarus[17] divide context mechanisms into three distinct classes with regard to their response to ribosome density or spacing within the message. Experimentally, messages are divided by their response to less or more active ribosome binding sites. Independent ribosomes engaged in dropoff (e.g., frameshift or missense) error should not respond to change in ribosome density. Ribosomal interference mechanisms (e.g., queuing) predict an increased effect as average ribosome density increases because they depend on interaction between ribosomes. Thirdly and uniquely, ribosome density-dependent mRNA stability, or polar effects, should have a diminished effect at high ribosome densities. The third result was observed at two context sites when several ribosome binding sites were introduced. Further, the context effect virtually disappeared when the most active ribosome binding site is substituted for the wild-type *lacZ* site. Thus, the context mechanism in this message appeared to be a polarity (via R_o) or stability effect (via t_o or t_{do}), abolished by densely packed ribosomes.

One puzzle has been that wobble-wobble and wobble-second position sequence bias seems more obvious in weakly expressed genes. One likely reason is that proposed by Gouy (1987); that weakly expressed genes use all codons and thus have more freedom to vary their contexts. However, there is an independent consideration. If the wobble-wobble context effect frequently has the property observed by Folley and Yarus,[17] then whatever its mechanism, it may also be obscured in highly expressed genes because the effect is minimized at high ribosome densities.

Skepticism is sometimes expressed (e.g., Shpaer[6] and Bulmer[11]) that low expression can be selected, as has been suggested for the prevalence of repeated wobbles in weakly expressed genes.[5] On one hand, there seems to be no doubt that translational signals are selected for expression. Ribosome binding sites of messages transcribed from highly expressed genes resemble a consensus for ribosome binding sites more than does the average message.[110] Even the more subtle choice between synonymous codons is selected, because synonyms are interchanged by evolution at slower rates in highly expressed genes, suggesting functional constraint.[35] Nevertheless, the question of selection for low expression is not settled. In particular, it is not settled by measurements that show that a sequences prevalent in weakly expressed genes can decrease expression.[17] Selection might exist for high expression only; determinants for expression of weakly expressed genes could simply drift into sufficient disarray to lower expression to the required level. However, the last sentence contains the key idea: correlated wobbles in weakly expressed genes are *not in disarray,* they are ordered beyond reasonable expectation. To explain this fact, the most plausible hypothesis seems that order is maintained by selection, and that contexts that both elevate and depress expression can be selected.

XXII. SUMMARY AND PROSPECTS

The sequence bias around codons in eubacteria seems to engage the upstream 5' neighbor and the downstream triplet of many codons.[5,7]

Of these statistically-defined positions, the 3' neighbor of a codon to be translated has been best explored biochemically. Some 3-1 doublets in mRNAs are likely to be selected because the 3' mRNA neighbor of a codon alters the rate of selection of the cognate tRNA (*in vivo:* Pedersen and Curran;[15] *in vitro:* Dix and Thompson,[16]), in some cases by stacking on the codon-anticodon helix.

The 3-1 doublet at its other position in the P-site/A-site boundary also has a distinct effect on the rate of tRNA selection, because it determines the identity of the two neighboring tRNAs. Relatively unperturbed tRNA-tRNA interfaces differ roughly 9-fold in the rate of tRNA selection they permit.[18] This may partially explain statistical nonrandomness at the 5′ neighbor of codons.[3,5,6] Nevertheless, there is no question that the junction nucleotides between codons are nonrandom in part because of the same nontranslational evolutionary forces that organize sequences in untranslated DNA.[7,9,11] However, statistical studies do not distinguish the 3-1 doublet 3′ to the A site from the same doublet at the P-site/A-site junction. These two positions are unlikely to show the same requirements, because their roles are mechanistically distinct. Therefore, superposition of two different selections may make the 3-1 context effect less apparent.

There is also a 3′ neighbor effect on peptide termination, acting on the velocity of RF1 action.[15] However, the effective context for RF1 and RF2 is likely to extend beyond the 5′ and 3′ neighbors (see Appendix III).[38,40]

The bias in adjacent wobble positions in messages transcribed from weakly expressed genes[5,7-9] is probably a translational one. Context affects the *relative* selection velocities of tRNAs;[42] therefore it affects the selection velocities of tRNAs. This effect may be traceable in part to variation in flanking wobbles.[13,14] Wobble interaction could be mediated by tRNA-tRNA contacts, and can alter the output of genes.[17]

With regard to the future of the general context effect on selection of tRNAs, one might wish for the extension of the above mechanistic information on tRNA selection to all nucleotides of the immediate context. However, the case for context effects on ribosomal progression will be incomplete without information about the rest of the ribosomal cycle (e.g., translocation). The question of how frequently and by what means such effects can be transduced into effects on gene expression particularly needs clarification.

Close inspection seems likely to reveal more examples like that of the RF2 gene, T4 gene 60, *selC* and UAG CAR YYA, where contexts have large, specific effects on a few sites. These four examples already implicate both prokaryotic and eukaryotic translation. Thus, there may be a general potential for novel indirect context effects due to mRNA interaction with the ribosome or with factors.

ACKNOWLEDGMENTS

Michael Yarus was supported by NIH Research grant GM30881 during the preparation of this review and James Curran was supported by NSF Research grant DMB-8904708. We thank Dan Dix and Bob Thompson for providing unpublished data. Dan O'Connell, Tony Palombella, and Dennis Schultz made useful comments on a draft manuscript. We also thank the W.M. Keck Foundation (Boulder, CO) for support of RNA science in Boulder.

APPENDIX I

MECHANISM OF CONTEXT ACTION

We believe that the context of sense codons affects expression because context has direct effects on translation, and also because translation interacts with both mRNA synthesis and decay. The latter two interactions may generate the largest context effects. For example, messages have a characteristic rate of decay, and therefore a (stochastically) limited time to perform. Therefore, an altered rate of translation can alter the amount of gene product derived from a message. To discuss these relationships, we will develop a simplified expression relating translational yields and mRNA properties (such modeling has been reviewed[111]). This calculation makes explicit and quantitative suggestions for the source of context effects, but also points to uncertainties that need elucidation.

MESSAGE DECAY

Ideas about message function have vague consequences except in the context of a particular model for mRNA birth and decay, so we first present a thumbnail review of this subject. The functional decay of bacterial mRNA is probably initiated by endonucleolytic cleavage at one or a few sites, then completed by the digestion of the resulting fragments to nucleotides by further endonucleolytic and 3' exonucleolytic action (reviewed by Belasco and Higgins[112]). Remarkably, the initial cleavages in a variety of messages are frequently near either the 3' or 5' ends of the corresponding cistrons, rather than internally. For example, *ompA* seems to be inactivated by endonucleolytic cleavages in the 5' untranslated region, close to the ribosome binding site.[113] Similarly, small changes in message structure in the 5' translated region of *lacZ* [68,114] and the insulin-like growth factor II gene[67] increase mRNA half-life. Message destabilization can result from the introduction of a sensitive structure anywhere, but its stabilization implies that rate-limiting structures for decay have been altered. Therefore, these data suggest that rate-determining structures are near the ribosome binding site. After the initial 5' cleavage inactivates the ribosome binding site, degradation may closely follow the exiting train of ribosomes.[115]

Conversely, Lambda *int*[116] and *rpsO*[117] (ribosomal protein S15) decay via exonucleolytic attack at their 3' ends. In both cases, there is evidence that the removal of a 3' mRNA secondary structure is the rate-limiting event. Just as 5' mRNA structure alters 5'-initiated decay, the addition of hairpins to the 3' ends of unstable mRNAs can stabilize them.[118] Individual mRNAs with both 3' and 5' susceptibility may exist, as *lacZ* is not notably affected by 3' structures until it is stabilized by 5' changes in the region of the ribosome binding site, apparently converting it into a message whose least stable region is now at the 3' end.[114] As one corollary of this focus on gene termini, operons like the *lac* operon can give rise to more or less gene-size pieces as early decay intermediates.[119,120] In the following we suggest a reason for this terminal tendency.

Ribosomes stabilize messages. During a kasugamycin-induced ribosomal runoff on *lacZ*,[115] or during translational repression of ribosomal protein L11 mRNA,[107] or distal to translational terminator mutations in the *trp* operon message,[121] mRNA is destabilized. It seems virtually certain that some ribosomal protection is attributable to straightforward steric occlusion of the critical decay targets discussed. It follows that ribosome density and distribution potentially alter message half-life. When amber and ochre mutations were introduced at different positions in bla,[122] the entire mRNA (whether traversed by translating ribosomes or not) was stabilized when normal translation was allowed to progress to codon 56, instead of being halted at codon 26. This bla example explicitly suggests that the rate of ribosomal progression to, or ribosomal density around, a specific, relatively short message site might alter functional mRNA halflife in some mRNAs, and thus the yield of protein from a fixed amount of message.

MESSAGE SYNTHESIS AND POLARITY

In addition to the half-life, ribosomes affect prokaryotic transcription by preventing the phenomenon called polarity, or premature transcription termination. For example, deletions and substitutions just upstream of the ribosome binding site of *lacZ* radically alter both the amount of β-galactosidase and full-size *lacZ* message, without substantial alterations in message half-life.[23] The effect on mRNA levels is removed by expression of lambda *N*, a transcriptional antiterminator, though expression differences still exist in the presence of *N*. Thus, these data suggest that weak resulting ribosomal binding sites, by lowering ribosome density, increase the probability that *lacZ* mRNA will be terminated before the end of the gene is reached. If alteration of the spacing between RNA polymerase and the first ribosome alters polarity, then it follows that internal mRNA sequences that differentially alter the relative velocity of ribosomes and RNA polymerase (e.g., codon contexts) can also alter the amount of completed message, and therefore may change gene expression. The combination of stability and polarity effects can make message quantity a very sensitive and idiosyncratic function of message sequence. While not usually discussed, the density of RNA polymerase pauses, and the velocity of transcription, is as significant to both polarity and stability as are slowly translated mRNA passages.

PROTEIN SYNTHESIS WITH 5' INITIATION OF mRNA DEGRADATION

Throughout these calculations, we assume that events on the message under discussion do not alter a much larger cellular steady-state. For example, the pool of preinitiation complexes and therefore initiation rates are not significantly perturbed by the fate of ribosomes on one species of mRNA. In the 5'-initiation-of-degradation model, ribosomes simply initiate at a ribosome binding site until the site decays. All ribosomes that leave the ribosome binding site are assumed to complete their protein products, because the rate limiting steps in decay happen around the ribosome binding site and are likely to propagate from there. The productivity of this type of message, once completed, depends on the number of ribosomes that use it:

$$P = R\,s\,t \tag{1}$$
$$dP/dt = R\,s,$$

where

P = the protein synthesized, and is the constant average spacing of ribosomes on the message in time; that is, the ribosomal frequency in $\min^{-1}$, or so many per minute
R = number or concentration of mRNAs
t = time

A frequent observation both *in vivo* and *in vitro* is that mRNA decay is first-order, or exponential, over a considerable range:

$$R/R_o = \exp(-k_d\,t) \tag{2}$$

where

k_d = the decay constant
R_o = the initial concentration of message

Combining,

$$dP/dt = R_o\,s\,\exp(-k_d\,t). \tag{3}$$

After integration over all times ($t = 0$ to $t = \infty$) to calculate total synthesis,

$$P = R_o\,s/k_d. \tag{4}$$

The average time to reaction **t**, for any first order or pseudo first-order reaction is $1/k$:

$$\mathbf{t} = 1/k \tag{5}$$

using a bold font for averaged times, **t**.

The average time of reaction **t**, where the "reaction" is functional mRNA decay in time t_d, is closely related to the more usual functional mRNA half-life $t_{1/2}$:

$$t_{1/2} = \ln 2 / k_d = 0.693 / k_d = 0.693\, t_d.$$

Thus a useful form of Equation 4 based on use of **t** recognizes that s can be a first-order rate constant for the initiation of protein synthesis:

$$P = R_o\, t_d/t_s \qquad (6)$$

where

t_d = the average time to decay
t_s = the average time spacing between translocating ribosomes, or average time to initiate synthesis.

Thus, the primary result is that the yield of protein from a 5'-decaying message is determined by the ratio of the average time to decay to that for initiation. If message lifetime is extended by a factor of 2, we expect twice as much protein; if the average time to decay is 20 times t_s, we expect $P = 20\, R_o$ or a yield of 20 molecules of protein for every molecule of message, and so on.

Though t_s in Equation 6 is an average with spacing constant in time, we will make arguments for other conditions where t_s is a function of time, even though this is not mathematically fastidious. However, for the ribosomal distributions likely to exist on real messages, interpretation of t_s as the arithmetic mean seems likely to be sufficiently accurate.

The utility of an equation employing times instead of rate constants is that the average times for complex events are the sums of the properly averaged times for successive individual stages: t_s is the sum of the average times for binding of a ribosome t_b and subsequent clearance of (exit from) the ribosome binding site t_c, for example:

$$t_s = t_b + t_c, \qquad (7)$$
$$P = R_o\, t_d/\{t_b + t_c\}.$$

It is relatively easy to reason about the average time required for defined events, and therefore relatively easy to use Equations 6 or 7 to reason about the effects of molecular changes on P, the translational yield of protein from R_o molecules of mRNA.

In particular, Equations 6 and 7 show that the 5' decaying message presents little opportunity for the direct effects of context (or other elongational events).

The yield is directly affected by translational elongation only via t_c, during clearance of the ribosome binding site. Thus, the choice and ordering of the 5' codons may have a special importance, but most of the message usually has no direct translational effect (compare Bergmann and Lodish[123]). In addition, if decay is initiated at the ribosome binding site, no ribosomal movement is required to reach the critical decay targets, so effects of translational elongation rate on t_d are also probably minimized. However, ribosomal movement can alter P by influencing R_o via polarity. Thus the 5' decaying messages as a class may be expected to frequently manifest context effects through polarity, and such a case has been described for *lacZ*.[17]

There is, conceivably, a distinct class of t_s effects resulting from hindrance of the ribosome or from ribosomal accidents. Average ribosomal frequency (spacing; t_s, Equation 6) could be determined during transit of a message rather than at initiation if ribosomes queue at a passage of slowly translated codons. *Queues* have been detected[124] at internal message features in eukaryotic mRNAs being translated in vitro. In the 5'-degraded message, a queue might extend back to the ribosome binding site, so that the interval between ribosomes t_s would primarily be determined by the mean time required to pass the slow codons (modeled by Harley et al.[125]). By hypothesis this time to pass the slow codons is greater than the time to initiate (or no queue will form), thereby proportionately (Equation 6) reducing the output from the message. If the queue does not extend to the ribosome binding site before the message decays, it has no t_s effect on the output of 5'-decay-initiating mRNAs. Therefore, somewhat as for clearance of the ribosome binding site, (t_c, Equation 7), the sequences near the 5' end of the translated region are of greatest concern. t_s may be even more dramatically increased than by queues if ribosomes drop off the message, or frameshift and terminate at certain codons or contexts with significant probability. There is considerable evidence for the existence of such abortive translation.[101,102,126]

PROTEIN SYNTHESIS WITH 3' INITIATION OF mRNA DEGRADATION

In this model, rapid functional inactivation follows the penetration of a 3' mRNA structure by exonucleolysis, or follows the removal of such a 3' mRNA structure by an endonucleolytic cut 5' proximal to it. The simplest assumption consistent with this picture is that degradation is rarely initiated during synthesis of the message, because the major RNAse-sensitive target is created at or near the terminator. Because the first ribosome is expected to closely follow the RNA polymerase, this model is perhaps unexpectedly similar to the case of 5' degradation, where the message does not ordinarily decay until the message has made some protein. The observable tendency of the most sensitive targets for mRNA decay to occur at cistron boundaries may be accounted for in this way. Decay targets near or just outside either the 5' ribosome binding site or the 3' terminus give the ribosomes and the machinery for mRNA degradation virtually simultaneous access to the RNAse-sensitive parts of the message, at times when some product is assured. Therefore futile rounds of transcription in which the nascent mRNA decays before any, or little, protein product can be completed are minimized (compare Kennell[115]).

For analysis of 3′-initiated decay, we use the same sort of deterministic model of translation used previously; ribosomes initiate with a frequency s, and each traverses the structural gene in time T, the transit time. Contribution to P, the number of molecules of protein, now requires that ribosomes arrive at the end of the gene. P has a discontinuous form that can be broken into three eras.

From t = 0 to T, there is no product because the first ribosome is still in transit (era I).

$$P_I = 0,$$
$$dP_I/dt = 0.$$

In a short interval surrounding T, 1 mole of product is made per mole of mRNA, R (era II):

$$P_{II} = \int dP(t)/dt \, dt = R.$$

After T, the transit time (era III), protein is made linearly with frequency s, as ribosomes arrive at the end of the gene:

$$P_{III} = R\,s\,(t - T),$$
$$dP_{III}/dt = R\,s.$$

No product is made in era I. If the mRNA does not decay during the first ribosomal transit, integration over the brief era II gives R_o (= initial mRNA) molecules of protein. During era III, the message decays exponentially:

$$\int dP = \int_0^\infty R_o\,s\,\exp(-k_d\,t)\,dt = R_o\,s/k_d = R_o\,t_d/t_s \quad (8)$$

$$P = P_I + P_{II} + P_{III} = 0 + R_o + R_o\,t_d/t_s \quad (9)$$

Because t_d (mean time to decay) is usually much larger than t_s (mean frequency of ribosomes; most messages yield many proteins):

$$P \approx R_o\,t_d/t_s \quad (10)$$
$$P \approx R_o\,t_d/\{t_b + t_c\}.$$

Equation 10 is virtually identical to Equations 6 and 7 for the 5′-decaying message, as anticipated above. In particular, the effect of polarity on R_o and the effect of the 5′ codons and context on t_s provide routes for codon context to alter P, the output from a message. The 3′-initiating messages are even more sensitive to queues, because queues can increase t_s, the interval between ribosomes at the end of the gene, even if the queue does not reach back to the ribosome binding site. However, there is also a new sensitivity to t_d, because ribosomes arrive at the signal for mRNA decay after traversing the entire message. Thus the time required to traverse all sections of the message is reflected in the mean spacing between the ribosome and RNA polymerase. Because the 3′ decay target may be vulnerable until protected by the arrival of the leading ribosome, delayed ribosomes could have a substantial effect on survival of the message during the first transit (before any product). For this and other reasons, it is of interest to modify the above model to allow decay of the message during the first transit, era I, at rate k_{do}:

$$P \approx R_o \{t_d / t_s\} \exp(-k_{do} T)$$
$$P \approx R_o \{t_d / t_s\} \exp(-T/t_{do}) \tag{11}$$

where the exponential term represents the reduction due to decay during the initial transit. As the second relation shows, this cost exponentially increases with the ratio of the transit time to the average time to decay. If the transit time increases 5% without change in decay time, the output from the gene decreases 5%, because small changes in an exponent have a linear effect. However, such effects could be amplified because opposite effects on T and t_{do} are likely. When T increases 5%, t_{do} could simultaneously decrease because the lag increases the exposure of the decay target, perhaps by a large factor. Thus T/t_{do} could increase disproportionately, with a corresponding increase in the cost of the first transit, $\exp(-T/t_{do})$.

SUMMARY

Equation 11 summarizes the output from a fixed amount of message for the above mRNA life cycles if it is understood that k_{do} is small (t_{do} large) if there is no decay of the message during initial ribosomal transit. We have found the five parameters in Equation 11 a useful checklist. Change in gene expression signifies an alteration in one of the five, or alternatively, suggests a new effect not embodied in these models.

APPENDIX II

THE MECHANISM OF aa-tRNA SELECTION

The minimum kinetic mechanism for the selection of aa-tRNA and the formation of a peptide bond has been defined by transient kinetic studies in vitro, using purified reactants.[127,128] As for any enzyme, such transient reactions are needed to resolve and study individual steps in the overall reaction. Because we have often found that scheme useful, and have made deductions based on it at several points in this review, a brief treatment of aa-tRNA

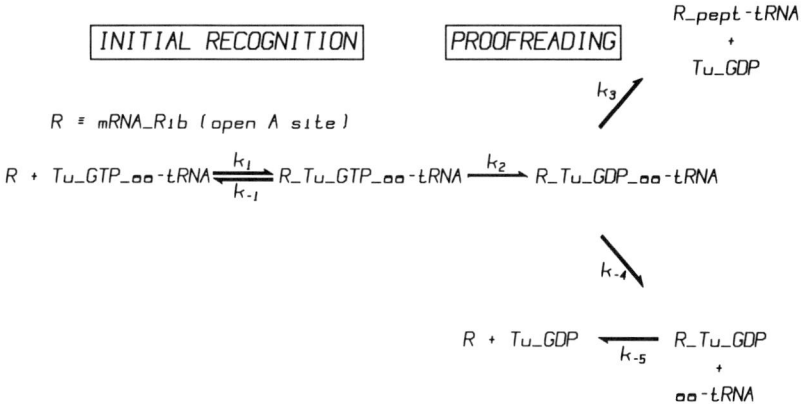

FIGURE 2. A minimal kinetic scheme for the selection of aa-tRNA and the formation of a peptide bond.[128] See text for explanation.

selection and GTP hydrolysis is appended here. Appendix II Figure 2, drawn after Thompson,[128] summarizes the process.

k_1 is the initial binding of the $EFT_u \cdot GTP \cdot aa$-tRNA ternary complex to the open A site of the mRNA•ribosome (R); this complex can dissociate via k_{-1} before GTP hydrolysis. Dissociation is the fate of the vast majority of complexes with mismatched anticodons. The aa-tRNA becomes committed to the rest of the cycle if GTP is hydrolyzed with rate constant k_2. At this point, initial discrimination is completed; cognate codon-anticodon complexes will be stable enough to have lasted until GTP hydrolysis. k_2 seems to be roughly constant for all aa-tRNA; noncognates or mistakes will therefore be rejected only because they have large k_{-1} and dissociate before GTP is hydrolyzed. However, residual mistakes in which the codon and anticodon are mismatched can now be subjected to a second level of discrimination (proofreading). This discrimination is based on the relative rates of $EFT_u \cdot GDP$ and aa-tRNA dissociation. $EFT_u \cdot GDP$ dissociates via k_3, approximately constant for all complexes; when $EFT_u \cdot GDP$ departs, its inhibition of peptidyl transferase is released, and a peptide bond is rapidly formed. After k_3, an aa-tRNA has passed both levels of discrimination, and contributed its amino acid to the nascent peptide. However, an erroneous aa-tRNA·mRNA complex stands a good chance of dissociating before $EFT_u \cdot GDP$ via k_{-4}, and therefore does not transfer its amino acid to the growing chain. Thereafter, the ribosome is recycled for another try at aa-tRNA selection when $EFT_u \cdot GDP$ dissociates via k_{-5}. See the original references for a more complete discussion.

In this review, we have usually discussed only original discrimination, k_1, k_{-1}, and k_2, through GTP hydrolysis. At this point, the bulk of tRNA selection is completed. The apparent rate constant for the hydrolysis of GTP, starting with free ternary complex and mRNA-instructed ribosomes, is k_{GTP}:

$$v_{GTP} = k_1 k_2/(k_{-1} + k_2) \; EFT_u \cdot GTP \cdot aa\text{-}tRNA \; R \qquad (1)$$
$$k_{GTP} = k_1 k_2/(k_{-1} + k_2).$$

The situation is simplified for a cognate complex, because its dissociation is extraordinarily slow:[128]

$$k_{-1}^{cognate} \ll k_2, \qquad (2)$$

so

$$k_{GTP}^{cognate} \approx k_1. \qquad (3)$$

We have referred to this finding, (Equation 3), that the rate of acceptance of a cognate aa-tRNA is determined by its initial binding, in this review. As one consequence, cognates are unaffected by substantial changes in k_{-1}. This should not be read to mean that nothing changes $k_{-1}^{cognate}$; instead, such changes normally do not have a phenotype. In contrast, a noncognate (or near-cognate, in the language of Thompson et al.) dissociates quickly (has a large k_{-1}). Therefore an error, and by extension perhaps a tRNA damaged by mutation, or a tRNA in a poor context, tends to have greater k_{-1}, a smaller k_{GTP}, and to be affected by changes in k_{-1}. That is, less-than-optimal translation has a phenotype that depends on the rate of ternary complex dissociation. A similar discussion, with similar conclusions, can be conducted for the other aa-tRNA dissociation during proofreading, k_{-4}.

APPENDIX III

THE DATA OF MARTIN ET AL.[40] AND THE CONTEXT EFFECT ON RELEASE FACTORS

UAA termination codons are translated by both RF1 and RF2. Martin et al.[40] measured action of each RF at 13 UAA nonsense sites in *lacIZ* fusions. These workers report decreases in suppression efficiency (antisuppression) by 1.5 to 3.4-fold in cells that carry an ochre (UAA) suppressor tRNA and overexpress either RF1 or RF2. Intuitively, antisuppression at a given UAA site should indicate the relative roles of the two RFs in termination at that site. If, for example, RF1 were the most active factor, then overexpression of RF1 (but not RF2) would antisuppress by a large factor. Comparisons of effects among sites with different contexts should, then, reveal differential context effects on the RFs.

In the following, we reanalyze the data of Martin et al. to more sensitively reveal the relative effect of message context on RF1 and RF2 acting at UAA. The development is based on the ideas originated in Yarus et al.,[21] and extended in Smith and Yarus[129] and Curran and Yarus.[28] The velocity, v, of virtually any enzyme-like reaction may be expressed as:

$$v = (k_{cat}/K_m) \, e \, S, \qquad (1)$$

where

e and S = free enzyme and substrate concentrations
  k_{cat} = the catalytic (turnover) constant
  K_m = the Michaelis (saturation) constant for the reaction

Equation 1 is very robust, being true for any concentration of the single substrate S, and also for any concentration of any number of alternative substrates competing for a single active site. If the ribosomal coding site is the active site in question, and subscripts $_{tRNA}$, $_{RF1}$, and $_{RF2}$ indicate quantities pertinent to an aa-tRNA suppressor, RF1, and RF2, respectively:

$$\begin{aligned} E &= v_{tRNA}/(v_{tRNA} + v_{RF1} + v_{RF2}) \\ &= (k_{cat}/K_m)_{tRNA} \, tRNA / \{(k_{cat}/K_m)_{tRNA} \, tRNA + (k_{cat}/K_m)_{RF1} RF1 \\ &\quad + (k_{cat}/K_m)_{RF2} \, RF2\} \end{aligned} \qquad (2)$$

where

e is the measured suppressor efficiency. $(k_{cat}/K_m)_{tRNA}$ tRNA, and the other terms like it, are pseudo-first order rate constants characterizing the rate of the relevant tRNA and factor reactions per unit concentration of ribosomal coding site.

The k_{cat}s, and the K_ms are potentially unique for each UAA locus tested. However, the normal concentrations tRNA, RF1, and RF2 seem likely to be constant throughout these experiments. In fact, "tRNA" is the effective concentration of $EFT_u \cdot GTP \cdot aa\text{-}tRNA$, a refinement that is not necessary in this calculation (compare Appendix II). We symbolize the pseudo-first order rate constants as "R" in the interest of compactness:

$$R_{tRNA} \equiv (k_{cat}/K_m)_{tRNA} \, tRNA, \quad R_{RF1} \equiv (k_{cat}/K_m)_{RF1} \, RF1,$$
$$\text{and } R_{RF2} \equiv (k_{cat}/K_m)_{RF2} \, RF2,$$

For each locus,

$$E = R_{tRNA}/(R_{tRNA} + R_{RF1} + R_{RF2}). \qquad (2a)$$

In the Martin et al.[40] experiments RF1 and RF2 are overexpressed from plasmids, so let O_{RF1} and O_{RF2} be the fold overexpression:

$$\begin{aligned} E_{RF1} &= R_{tRNA}/(R_{tRNA} + R_{RF1} \cdot O_{RF1} + R_{RF2}), \\ 1/E_{RF1} &= 1 + R_{RF1} \cdot O_{RF1}/R_{tRNA} + R_{RF2}/R_{tRNA} \\ 1/E_{RF2} &= 1 + R_{RF1}/R_{tRNA} + R_{RF2} \cdot O_{RF2}/R_{tRNA} \end{aligned} \qquad (3)$$

where

E_{RF1} = the suppressor efficiency when RF1 is overexpressed, and similarly for RF2.

So at each of the 13 UAAs where E, E_{RF1}, and E_{RF2} have been measured:

$$1/E_{RF1} - 1/E = (O_{RF1} - 1) R_{RF1}/R_{tRNA},$$
$$(1/E_{RF2} - 1/E) = (O_{RF2} - 1) R_{RF2}/R_{tRNA}, \qquad (4)$$

$$(1/E_{RF1} - 1/E)/(1/E_{RF2} - 1/E) = (E/E_{RF1} - 1)/(E/E_{RF2} - 1) \qquad (5)$$

$$= \{(O_{RF1} - 1)/(O_{RF2} - 1)\} \{R_{RF1}/R_{RF2}\}. \qquad (5a)$$

Equation 5a is the desired result: it shows that the E/E_{RF1} and E/E_{RF2} terms measured and tabulated by Martin et al.[40] as the "antisuppression" are related to a constant determined by the fold overexpression (left-hand term in curly brackets), times the ratio of the rates of action of RF1 and RF2 (right-hand curly brackets). This relation has several notable qualities:

1. The left-hand side is all experimental quantities, measured for a single UAA context.
2. The overexpressions are not known, save that they are five- to eight-fold.[40] However, Equation 5a shows that values for overexpression are not needed; calculation of Equation 5 provides a quantity proportional to the relative rates of action of RF1 and RF2 because overexpression is presumably constant for all loci.
3. The rate constant for action of the suppressor tRNA cancels because it is constant at a single locus, and 5a is proportional solely to the desired ratio of RF1 to RF2 rates. The analysis is independent of the usual competition between suppressor tRNA and RF.

In Appendix III Figure 3, the relative rates for RF1 and RF2 at 13 UAA contexts are compared. As in Figure 1 of Martin et al.,[40] we have distributed the relative RF rates versus suppressor efficiency to see if there is any relation between the context effect on tRNA action, and the relative context effect on RF action:

1. RFs respond to context. Figure 3 shows a relation between the tRNA and relative RF context effects, in complete agreement with Martin et al.[40] However, relative context dependence appears about threefold as great as in the original calculations, because in the original plot relative RF rates are confounded with constants. At contexts where the tRNA works faster (greater k_{cat}/K_m), RF1 tends to be faster than RF2. Over the contexts surveyed, the relative RF rate constants apparently

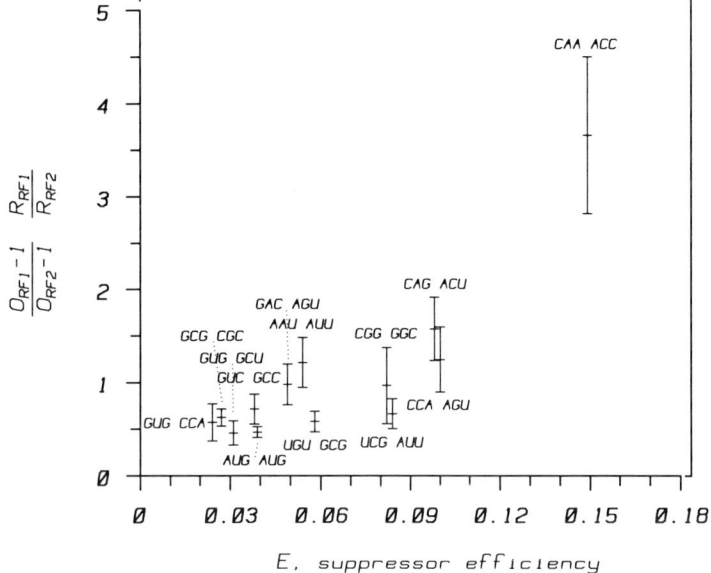

FIGURE 3. Relative rates of release factor action (RF1 rate/RF2 rate) versus UAA suppressor tRNA efficiency in the same context. The data of Martin et al.[40] have been replotted according to Appendix III, Equation 5. The central tick on each bar represents the point, the flanking bars show the standard error. Errors are propagated to relative RF velocity, $\{(O_{RF1} - 1)/(O_{RF2} - 1)\} \{R_{RF1}/R_{RF2}\}$, from the standard errors of antisuppression listed by Martin et al., using standard means for arithmetic operations. The legend associated with each bar is the UAA context, 5' neighboring codon to the left and 3' neighboring codon to the right.

   span sevenfold, though the exact range is dependent on one uncertain point at the highest suppressor efficiency.
2. Suppression efficiency and relative RF rates vary together to some extent, suggesting that context effects on the protein RFs and suppressor tRNA partially overlap. However, the relation does not seem linear or even very regular, and there is no reason to expect that it will be, because the RFs and tRNA are unlikely to interact with mRNA in exactly the same way.
3. Associated with each point-with-error-bar is the context, with the 5' codon on the left and 3' codon on the right. It is likely that the relative RF context effect, and therefore the context effect on RF1 and/or RF2, extends beyond the nucleotides 5' and 3' of the UAA. There are three pairs of cases in which the 5' and 3' neighboring nucleotides are reproduced, but the ratio of RF1 and RF2 rates is significantly different.
4. Finally, because overexpression is in the range five- to eightfold,[40] the overexpression term (first term in Equation 5a) varies only from 4/7 to 7/4, implying that the relative rates of action of RF1 and RF2 (second term in Equation 5a) at UAA at a typical context *in vivo* are similar.

REFERENCES

1. **Andersson, S. G. E. and Kurland, C. G.,** Codon preferences in freeliving microorganisms, *Microbiol. Rev.,* 54, 198, 1990.
2. **Buckingham, R. H.,** Codon context, *Experientia,* 46, 1126, 1990.
3. **Lipman, D. J. and Wilbur, W. J.,** Contextual constraints on synonymous codon choice, *J. Mol. Biol.,* 163, 363, 1983.
4. **Nussinov, R.,** The universal dinucleotide asymmetry rules in DNA and the amino acid codon choice, *J. Mol. Evol.,* 17, 237, 1981.
5. **Yarus, M. and Folley, L.S.,** Sense codons are found in specific contexts, *J. Mol. Biol.,* 182, 529, 1985.
6. **Shpaer, E. G.,** Constraints on codon context in *E. coli* genes, their possible role in modulating the efficiency of translation, *J. Mol. Biol.,* 188, 555, 1986.
7. **Gouy, M.,** Codon contexts in enterobacterial and coliphage genes, *J. Mol. Evol.,* 4, 426, 1987.
8. **Gutman, G. A. and Hatfield, G. W.,** Nonrandom utilization of codon pairs in *Escherichia coli, Proc. Natl. Acad. Sci. U.S.A.,* 86, 3699, 1989.
9. **Hanai, R. and Wada, A.,** Novel third-letter bias in *E. coli* codons revealed by rigorous treatment of coding constraints, *J. Mol. Biol.,* 207, 655, 1989.
10. **Shields, D. C. and Sharp, P. M.,** Synonymous codon usage in *B. subtilis* reflects both translational selection and mutational biases, *Nucleic Acids Res.,* 15, 8023, 1987.
11. **Bulmer, M.,** The effect of context on synonymous codon usage in genes with low codon usage bias, *Nucleic Acids Res.,* 18, 2869, 1990.
12. **Muto, A. and Osawa, S.,** The guanine and cytosine content of genomic DNA and bacterial evolution, *Proc. Natl. Acad. Sci. U.S.A.,* 84, 166, 1987.
13. **Buckingham, R. H., Sörenson, P., Pagel, F. T., Hijazi, K. A., Mims, B. H., Brechemier-Baey, D., and Murgola, E. J.,** Third position changes in codons 5′ and 3′ to UGA affect suppression *in vivo, Biochem. Biophys. Acta,* 1050, 259, 1990a.
14. **Buckingham, R. H., Murgola, E. J., Sörenson, P., Pagel, F. T., Hijazi, K. A., Mims, B. H., Figuroa, N., Brechemier-Baey, D., and Coppin-Raynal, E.,** Effects of codon context on the suppression of nonsense and missense mutations in the *trpA* gene of *E. coli.,* in *The Ribosome,* Hill, W., Dahlberg, A., Garrett, R.A., Moore, P. B., Schlessinger, D., and Warner, J. R., Eds., American Society for Microbiology, Washington, D.C., 1990b, 541.
15. **Pedersen, W. T. and Curran, J. F.,** Effects of the nucleotide 3′ to an amber codon on ribosomal selection rates of suppressor tRNA and release factor 1, *J. Mol. Biol.,* 219, 231, 1991.
16. **Dix, D. B. and Thompson, R. C.,** personal communication, 1991.
17. **Folley, L. S. and Yarus, M.,** Codon contexts from weakly expressed genes reduce expression *in vivo, J. Mol. Biol.,* 209, 359, 1989.
18. **Smith, D. and Yarus, M.,** tRNA-tRNA interactions within cellular ribosomes, *Proc. Natl. Acad. Sci. U.S.A.,* 86, 4397, 1989.
19. **Eggertsson, G. and Söll, D.,** Transfer RNA-mediated suppression of termination codons in *E. coli, Microbiol. Rev.,* 52, 354, 1988.
20. **Salser, W.,** The influence of the reading context upon the suppression of nonsense codons, *Mol. Gen. Genet.,* 105, 125, 1969.
21. **Yarus, M., Cline, S. W., Wier, P., Breeden, L., and Thompson, R. C.,** Actions of the anticodon arm in translation: on the phenotypes of RNA mutants, *J. Mol. Biol.,* 192, 235, 1986.
22. **Yarus, M.,** unpublished data.
23. **Stanssens, P., Remaut, E., and Fiers, W.,** Inefficient translation initiation causes premature transcription termination in the *lacZ* gene, *Cell,* 44, 711, 1986.
24. **Curran, J. F. and Yarus, M.,** Rates of aminoacyl-tRNA selection at 29 sense codons *in vivo, J. Mol. Biol.,* 209, 65, 1989.

25. Miller, J. H. and Albertini, A. M., Effects of surrounding sequence on the suppression of nonsense codons, *J. Mol. Biol.,* 164, 59, 1983.
26. Bossi, L., Context effects: translation of UAG codon by suppressor tRNA is affected by the sequence following UAG in the message, *J. Mol. Biol.,* 164, 73, 1983.
27. Stormo, G. D., Schneider, T. D., and Gold, L., Quantitative analysis of the functional relationship between nucleotide sequence and functional activity, *Nucleic Acids Res.,* 14, 6661, 1986.
28. Curran, J. F. and Yarus, M., Use of tRNA suppressors to probe regulation of *Escherichia coli* release factor 2, *J. Mol. Biol.,* 203, 75, 1988.
29. Weiss, R. B., Dunn, D. M., Dahlberg, A. E., Atkins, J. F., and Gesteland, R. F., Reading frame switch caused by base-pair formation between the 3' end of 16S rRNA and the mRNA during elongation of protein synthesis in *Escherichia coli, EMBO J.,* 7, 1503, 1988.
30. Saenger, W., *Principles of Nucleic Acid Structure,* Springer-Verlag, New York, 1984.
31. Freier, S. M., Kierzek, R., Jaeger, J. A., Sigimoto, N., Caruthers, M. H., Neilson, T., and Turner, D. H., Improved free energy parameters for predictions of RNA duplex stability, *Proc. Natl. Acad. Sci. U.S.A.,* 83, 9373, 1986.
32. Ayer, D. and Yarus, M., The context effect does not require a fourth base pair, *Science,* 231, 393, 1986.
33. Dix, D. B. and Thompson, R. C., Codon choice and gene expression: synonymous codons differ in translational accuracy, *Proc. Natl. Acad. Sci. U.S.A.,* 86, 6888, 1989.
34. Thomas, L. K., Dix, D. B., and Thompson, R. C., Codon choice and gene expression: synonymous codons differ in their ability to direct aa-tRNA binding to ribosomes *in vitro, Proc. Natl. Acad. Sci., U.S.A.,* 85, 4242, 1988.
35. Sharp, P. M. and Li, W.-H., The rate of synonymous substitution in enterobacterial genes is inversely related to codon usage bias, *Mol. Biol. Evol.,* 4, 222, 1987.
36. Eccleston, J. F., Dix, D. B., and Thompson, R. C., The rate of cleavage of GTP on the binding of Phe-tRNA·Elongation Factor Tu·GTP to poly(U)-programmed ribosomes of *Escherchia coli, J. Biol. Chem.,* 260, 16237, 1985.
37. Kohli, J. and Grosjean, H., Usage of the three termination codons: compilation and analysis of the known eukaryotic and prokaryotic translation termination sequences, *Mol. Gen. Genet.,* 182, 430, 1981.
38. Brown, C. M., Stockwell, P. A., Trotman, C. N. A., and Tate, W. P., The signal for termination of protein synthesis in prokaryotes, *Nucleic Acids Res.,* 18, 2079, 1990.
39. Beaudet, A. L. and Caskey, C. T., Mammalian peptide chain termination. II. Codon specificity and GTPase activity of release factor, *Proc. Natl. Acad. Sci. U.S.A.,* 68, 619, 1971.
40. Martin, R., Weiner, M., and Gallant, J., Effects of release factor context at UAA codons in *Escherichia coli, J. Bacteriol.,* 170, 4714, 1988.
41. Ruteshouser, E. C. and Richardson, J. P., Identification and characterization of transcription termination sites in the *E. coli lacZ* gene, *J. Mol. Biol.,* 208, 23, 1989.
42. Murgola, E. J., Pagel, F. T., and Hijazi, K. A., Codon context effects in missense suppression, *J. Mol. Biol.,* 175, 19, 1984.
43. Wittenberg, W., Dix, D. B., Thompson, R. C., and Uhlenbeck, O., personal communication, 1991.
44. Moazed, D. and Noller, H. F., personal communication, 1991.
45. Colby, D. S., Schedl, P., and Guthrie, C., A functional requirement for modification of the wobble nucleotide in the anticodon of a T4 suppressor tRNA, *Cell,* 9, 449, 1976.
46. Hagervall, T. and Björk, G., Undermodification in the first position of the anticodon of *supG*-tRNA reduces translational efficiency, *Mol. Gen. Genet.,* 196, 194, 1984.
47. Yanofsky, C. and Soll, L., Mutations affecting tRNA^{Trp} and its charging and their effect on transcription termination at the attenuator of the tryptophan operon, *J. Mol. Biol.,* 113, 663, 1977.

48. **Landick, R. and Yanofsky, C.**, Transcription attenuation, in *Escherichia coli* and *Salmonella typhimurium: Cellular and Molecular Biology,* Ingraham, J. L., Low, K. B., Magasanik, B., Schaechter, M., and Umbarger, H. E., Eds., American Society for Microbiology, Washington, D.C., 1987, 1276.
49. **Grosjean, H., Söll, D., and Crothers, D.**, Studies of the complexes between transfer RNAs with complementary anticodons. I. Origins of enhanced affinity between complementary triplets, *J. Mol. Biol.*, 103, 491, 1976.
50. **Weissenbach, J. and Grosjean, H.**, Effect of threonylcarbamyl modification (t^6A) in Yeast tRNA$_{III}^{Arg}$ on codon-anticodon interactions, *Eur. J. Bochem.*, 116, 207, 1981.
51. **Labuda, D., Striker, G., Grosjean, H., and Porschke, D.**, Mechanism of codon recognition by transfer RNA studied with oligonucleotides larger than triplets, *Nucleic Acids Res.*, 13, 3667, 1985.
52. **Bouadloun, F., Srichaiyo, T., Isaksson, L. A., and Björk, G.**, Influence of modification next to the anticodon in tRNA on codon context sensitivity of translational suppression and accuracy, *J. Bacteriol.*, 166, 1022, 1986.
53. **Ericson, J. U. and Björk, G. R.**, tRNA anticodons with the modified nucleoside 2-methylthio-N^6-(4-hydroxyisopentenyl)adenosine (ms^2io^6A) distinguish between base 3' of the codon, *J. Mol. Biol.*, 218, 509, 1991.
54. **Leinfelder, W., Zehelein, E., Mandrand-Berthelot, M.-A., and Böck, A.**, Gene for a novel tRNA species that accepts L-serine and cotranslationally inserts selenocysteine, *Nature (London)*, 331, 723, 1988.
55. **Baron, C., Heider, J., and Böck, A.**, Mutagenesis of *selC,* the gene for the selenocysteine-inserting tRNA-species in *E. coli:* effects on *in vivo* function, *Nucleic Acids Res.*, 18, 6761, 1990.
56. **Forchhammer, K. and Böck, A.**, Selenocysteine synthase from *E. coli, J. Biol. Chem.*, 266, 6324, 1991.
57. **Zinoni, F., Birkmann, A., Stadtman, T. C., and Böck, A.**, Nucleotide sequence and expression of the selenocysteine-containing polypeptide of formate dehydrogenase from *E. coli, Proc. Natl. Acad. Sci. U.S.A.*, 83, 4650, 1986.
58. **Zinoni, F., Heider, J., and Böck, A.**, Features of the formate dehydrogenase mRNA necessary for decoding of the UGA codon as selenocysteine, *Proc. Natl. Acad. Sci. U.S.A.*, 87, 4660, 1990.
59. **Heider, J., Zinoni, F., and Böck, A.**, personal communication, 1991.
60. **Forchhammer, K., Leinfelder, W., and Böck, A.**, Identification of a novel translation factor necessary for the incorporation of selenocysteine into protein, *Nature (London)*, 342, 453, 1989.
61. **Förster, C., Ott, G., Forchhammer, K., and Sprinzl, M.**, Interaction of a selenocysteine-incorporating tRNA with elongation factor T_u from *E. coli, Nucleic Acids Res.*, 18, 487, 1990.
62. **Li, W.-q. and Yarus, M.**, Bar to normal UGA translation by the selenocysteine tRNA, *J. Mol. Biol.*, 223, 9, 1992.
63. **Skuzeski, J. M., Nichols, L. M., Gesteland, R. F., and Atkins, J. F.**, The signal for a leaky UAG stop codon in several plant viruses includes the two downstream codons, *J. Mol. Biol.*, 218, 365, 1991.
64. **Gold, L.**, Posttranscriptional regulatory mechanisms in *E. coli, Ann. Rev. Biochem.*, 57, 199–223, 1988.
65. **Dunn, J. J. and Studier, F. W.**, Complete nucleotide sequence of bacteriophage T7 DNA and the locations of T7 genetic elements, *J. Mol. Biol.*, 166, 477, 1983.
66. **Looman, A. C. , Bodlaender, Constock, L. J., Eaton, D., Jhurani, P., DeBoer, H. A., and van Knippenberg, P. H.**, Influence of the codon following the AUG initiation codon on the expression of a modified *lacZ* gene in *Escherichia coli, EMBO J.*, 6, 2489, 1987.
67. **Cantrell, A. S., Burgett, S. G., Cook, J. A., Smith, M. C., and Hsiung, H. M.**, Effects of second-codon mutations on expression of the insulin-like growth factor-II-encoding gene in *E. coli, Gene,* 98, 217, 1991.

68. Petersen, C., The functional stability of the *lacZ* transcript is sensitive towards sequence alterations immediately downstream of the ribosome binding site, *Mol. Gen. Genet.*, 209, 179, 1987.
69. DeSmit, M. H. and Van Duin, J., Control of prokaryotic translational initiation by mRNA secondary structure, *Prog. Nucleic Acid Res. Mol. Biol.*, 38, 1, 1990.
70. Carrier, M. J. and Buckingham, R. H., An effect of codon context on the mistranslation of UGU codons *in vitro*, *J. Mol. Biol.*, 175, 29, 1984.
71. Bouadloun, F., Donner, D., and Kurland, C. G., Codon-specific missense errors *in vivo*, *EMBO J.*, 2, 1351, 1983.
72. Edelmann, P. and Gallant, J., Mistranslation in *E. coli*, *Cell*, 10, 131, 1977.
73. Laughrea, M., Latulippe, J., Filion, A.-M., and Boulet, L., Mistranslation in twelve *Escherichia coli* ribosomal proteins. Cysteine misincorporation at neutral amino acid residues other than tryptophan, *Eur. J. Biochem.*, 169, 59, 1987.
74. Laughrea, M., GAC codon GGC: a very favorable context for translation errors?, *FEBS Lett.*, 195, 185, 1986.
75. Precup, J., Ulrich, A. K., Roopnarine, O., and Parker, J., Context specific misreading of phenylalanine codons, *Mol. Gen. Genet.*, 218, 397, 1989.
76. Weiss, R. B., Dunn, D. M., Atkins, J. F., and Gesteland, R. F., Slippery runs, shifty stops, backward steps and forward hops: -2, -1, $+1$, $+2$ and $+6$ ribosomal frameshifting, *Cold Spring Harbor Symp. Quant. Biol.*, 52, 687, 1987.
77. O'Connor, M., Gesteland, R. F., and Atkins, J. F., tRNA hopping: enhancement by an expanded anticodon, *EMBO J.*, 8, 4315, 1989.
78. Huang, W. M., Ao, S.-Z., Casjens, S., Orlandi, R., Zeikus, R., Weiss, R., Winge, D., and Fang, M., A persistent untranslated sequence within bacteriophage T4 DNA topoisomerase gene 60, *Science*, 239, 1005, 1988.
79. Atkins, J. F., Weiss, R. B., and Gesteland, R. F., Ribosome gymnastics — degree of difficulty 9.5, style 10.0, *Cell*, 62, 413, 1990.
80. Yarus, M. and Thompson, R. C., Precision in protein biosynthesis, in *Gene Expression in Prokaryotes*, Beckwith, J., Davies, J., and Gallant, J., Eds., Cold Spring Harbor Laboratory Press, Cold Spring Harbor, NY, 1983, 23.
81. Weiss, R., Lindsley, D., Falahee, B., and Gallant, J., On the mechanism of ribosomal frameshifting at hungry codons, *J. Mol. Biol.*, 203, 403, 1988.
82. Dayhuff, T. J., Atkins, J. F., and Gesteland, R. F., Characterization of ribosomal frameshift events by protein sequencing, *J. Biol. Chem.*, 261, 7491, 1986.
83. Curran, J. F. and Yarus, M., Reading frame selection and tRNA anticodon loop stacking, *Science*, 238, 1545, 1987.
84. Gaber, R. F. and Culbertson, M. R., Codon recognition during frameshift suppression in *Saccharomyces cerevisiae*, *Mol. Cell Biol.*, 4, 2052, 1984.
85. Bossi, L. and Smith, D. M., Suppressor *sufJ*: a novel type of tRNA mutant that induces translational frameshifting, *Proc. Natl. Acad. Sci. U.S.A.*, 81, 6105, 1984.
86. Weiss, R. B., Dunn, D. M., Atkins, J. F., and Gesteland, R. F., Ribosomal frameshifting from -2 to $+50$ nucleotides, *Prog. Nucl. Acids Res. Mol. Biol.*, 39, 159, 1990.
87. Craigen, W. J., Cook, R. G., Tate, W. P., and Caskey, C. T., Bacterial peptide chain release factors: conserved primary structure and possible frameshift regulation of release factor 2, *Proc. Natl. Acad. Sci. U.S.A.*, 82, 3616, 1985.
88. Craigen, W. J. and Caskey, C. T., Expression of the peptide chain release factor 2 requires high efficiency frameshifting, *Nature (London)*, 322, 273, 1986.
89. Jacks, T., Madhani, H. D., Masiarz, F. R., and Varmus, H. E., Signals for ribosomal frameshifting in the Rous Sarcoma Virus *gag-pol* region, *Cell*, 57, 447, 1988.
90. Hatfield, D. and Oroszlan, S. The *where, what* and *how* of ribosomal frameshifting in retroviral protein synthesis, *TIBS*, 15, 186, 1990.
91. Tsuchihashi, Z., Translational frameshifting in the *Escherichia coli* dnaX gene *in vitro*, *Nucleic Acids Res.*, 19, 2457, 1991.

92. **Tsuchihashi, Z. and Kornberg, A.**, Translational frameshifting generates the τ subunit of DNA polymerase holoenzyme, *Proc. Natl. Acad. Sci., U.S.A.*, 87, 2516, 1990.
93. **Blinkowa, A. L. and Walker, J. W.**, Programmed ribosomal frameshifting generates the *Escherichia coli* DNA polymerase III τ subunit reading frame, *Nucleic Acids Res.*, 18, 1725, 1990.
94. **Flower, A. M. and McHenry, C. S.**, The τ subunit of DNA polymerase III holoenzyme of *Escherichia coli* is produced by ribosomal frameshifting, *Proc. Natl. Acad. Sci. U.S.A.*, 87, 3713, 1990.
95. **Spanjaard, R.A. and van Duin, J.**, Translation of the sequence AGG-AGG yields 50% ribosomal frameshift, *Proc. Natl. Acad. Sci. U.S.A.*, 85, 7967, 1988.
96. **Varenne, S. and Lazdunsky, C.**, Effect of unfavorable codons on the maximum rate of gene expression by an heterologous organism, *J. Theor. Biol.*, 120, 99, 1986.
97. **Spanjaard, R. A., Chen, K., Walker, J. R., and van Duin, J.**, Frameshift suppression at tandem AGA and AGG codons by cloned tRNA genes: assigning a codon to *argU* tRNA and T4 tRNA^{Arg}, *Nucleic Acids Res.*, 18, 5031, 1990.
98. **Varenne, S., Buc, J., Lloubes, R., and Lazdunski, C.**, Translation is a non-uniform process. Effect of tRNA availability on the rate of elongation of nascent polypeptide chains, *J. Mol. Biol.*, 180, 549, 1984.
99. **Sörensen, M. A., Kurland, C. G., and Pedersen, S.**, Codon usage determines translation rate in *Escherichia coli, J. Mol. Biol.*, 207, 365, 1989.
100. **Pedersen, S.**, *Escherichia coli* ribosomes translate *in vivo* with variable rate, *EMBO J.*, 3, 2895, 1984.
101. **Manley, J. L.**, Synthesis and degradation of termination and premature termination fragments of β-galactosidase *in vitro* and *in vivo, J. Mol. Biol.*, 125, 407, 1978.
102. **Jørgensen, F. and Kurland, C. G.**, Processivity errors of gene expression in *E. coli, J. Mol. Biol.*, 215, 511, 1990.
103. **Schwartz, R., Curran, J. F., and Yarus, M.**, unpublished data, 1991.
104. **Curran, J. F. and Yarus, M.**, unpublished data, 1991.
105. **Yarus, M.**, Accurate biochemistry, G proteins, and translation, *TIBS*, 17, 130, 1992.
106. **Newton, W. A., Beckwith, J. R., Zipser, D., and Brenner, S.**, Nonsense mutants and polarity in the *lac* operon of *E. coli, J. Mol. Biol.*, 14, 290, 1965.
107. **Cole, J. R. and Nomura, M.**, Changes in the half-life of ribosomal protein messenger RNA cause by translational repression, *J. Mol. Biol.*, 188, 383, 1986.
108. **Gold, L., Pribnow, D., Schneider, T., Shinedling, S. Singer, B., and Stormo, G.**, Translational initiation in prokaryotes, *Ann. Rev. Microbiol.*, 35, 365, 1981.
109. **Liljenstrom, H. and von Heijne, G.**, Translation rate modification by preferential codon usage: intragenic position effects, *J. Theor. Biol.*, 124, 43, 1987.
110. **Stormo, G. D., Schneider, T., Gold, L., and Ehrenfeucht, A.**, Use of the 'Perceptron' algorithm to distinguish translational initiation sites in *E. coli, Nucleic Acids Res.*, 10, 2997, 1982.
111. **von Heijne, G., Blomberg, C., and Liljenström, H.**, Theoretical modeling of protein synthesis, *J. Theor. Biol.*, 125, 1, 1987.
112. **Belasco, J. G. and Higgins, C. F.**, Mechanisms of RNA decay in bacteria: a perspective, *Gene*, 72, 15, 1988.
113. **Melefors, Ö. and von Gabain, A.**, Site-specific endonucleolytic cleavages and the regualtion of stability of *E. coli ompA* mRNA, *Cell*, 52, 893, 1988.
114. **Petersen, C.**, Multiple determinants of functional mRNA stability: sequence alterations at either end of the *lacZ* gene affect the rate of mRNA inactivation, *J. Bacteriol.*, 173, 2167, 1991.
115. **Kennell, D. E.**, The instability of messenger RNA in bacteria, in *Maximizing gene expression*, Reznikoff, W. and Gold, L., Eds., Butterworth, Stoneham, MA, 1986, 101.
116. **Plunkett, G. and Echols, H.**, Retroregulation of the bacteriophage lambda *int* gene: limited secondary degradation of the RNAse-III-processed transcript, *J. Bact.*, 171, 588, 1989.

117. **Régnier, P. and Hajnsdorf, E.,** Decay of mRNA encodong ribosomal protein S15 of *E. coli* is initiated by an RNAse-E dependent endonucleolytic cleavage that removes the 3′ stabilizing stem and loop structure, *J. Mol. Biol.,* 217, 283, 1991.
118. **Newbury, S. F., Smith, N. H., Robinson, E. C., Hiles, I. D., and Higgins, C. F.,** Stabilization of translationally active mRNA by prokaryotic REP sequences, *Cell,* 48, 297, 1987.
119. **Lim, L. W. and Kennell, D.,** Models for decay of *E. coli lac* messenger RNA and evidence for inactivating cleavages between its messages, *J. Mol. Biol.,* 135, 369, 1979.
120. **Murakawa, G. J., Kwan, C., Yamashita, J., and Nierlich, D. P.,** Transcription and decay of the *lac* messenger: role of an intergenic terminator, *J. Bact.,* 173, 28, 1991.
121. **Morse, D. E. and Yanofsky, C.,** Polarity and degradation of mRNA, *Nature (London),* 224, 329, 1969.
122. **Nilsson, G., Belasco, J. G., Cohen, S. N., and von Gabain, A.,** Effect of premature termination of translation on mRNA stability depends on the site of ribosome release, *Proc. Natl. Acad. Sci. U.S.A.,* 84, 4890, 1987.
123. **Bergmann, J. E. and Lodish, H. L.,** A kinetic model of protein synthesis, *J. Biol. Chem.,* 254, 11927, 1979.
124. **Wolin, S. L. and Walter, P.,** Ribosome pausing and stacking during translation of a eukaryotic mRNA, *EMBO J.,* 7, 3559, 1988.
125. **Harley, C. B., Pollard, J. W., Stanners, C. P., and Goldstein, S.,** Model for messenger translation during amino acid starvation applied to the calculation of protein synthetic error rates, *J. Biol. Chem.,* 256, 10786, 1981.
126. **Menninger, J. R.,** Peptidyl transfer RNA dissociates during protein synthesis from ribosomes of *Escherichia coli, J. Biol. Chem.,* 251, 3392, 1976.
127. **Thompson, R. C., Dix, D. B., and Karim, A. R.,** The reaction of ribosomes with elongation factor Tu-GTP complexes, *J. Biol. Chem.,* 261, 4868, 1986.
128. **Thompson, R. C.,** EF-Tu provides an internal kinetic standard for translational accuracy, *TIBS,* 13, 91, 1988.
129. **Smith, D. and Yarus, M.,** Transfer RNA structure and coding specificity. I. Evidence that a D-arm mutation reduces tRNA dissociation from the ribosome, *J. Mol. Biol.,* 206, 489, 1989.

Chapter 12

CODON DISCRIMINATION IN TRANSLATION

Ulf Lagerkvist

TABLE OF CONTENTS

I.	Introduction	368
II.	The Classic Principles of Codon Reading	368
III.	Unconventional Methods in the Cell's Normal Repertoire of Codon Readings	369
IV.	Mechanistic Considerations	372
V.	Influence of Structures Outside the Anticodon on Codon Discrimination	373
	A. The Role of the Nucleotide in Position 32	373
VI.	Influence of Structures Outside the Anticodon on Suppressor Efficiency	375
VII.	The Evolution of Codon Discrimination	376
VIII.	Concluding Remarks	377
References		378

I. INTRODUCTION

The canonical principles of codon reading as laid down in the wobble rules, undoubtedly predict in a very useful way the probability of a certain anticodon being able to read a particular codon in the elongation of the peptide chain during protein synthesis. However, the numerous exceptions to the rules that have by now been observed both *in vivo* and *in vitro* raise the question of the generality of the rules, and the validity of the structural arguments offered in support of them. Furthermore, it is by now obvious that the principles governing codon-anticodon recognition in the formation of the initiation complex are different from those of elongation. Not only does initiation involve the 30S subunit and not the whole ribosome, but the initiator tRNAs are also recognized as such, not on the basis of their anticodon, but rather because of other structural elements in the tRNA molecule. In addition to this, the initiation codons AUG, UUG, and GUG are all read by the same anticodon (CAU) indicating that the reading of the first nucleotide of the codon is relaxed and not according to the Watson-Crick pairing rules. This is in contrast to elongation in which the reading of the third codon position involves wobble interactions not permitted in the reading of the first two codon nucleotides. It is also worth noting that the exceptions to the canonical codon reading principles observed in elongation mainly concern the third codon nucleotide.

In what follows, the discussion will focus on the classic codon reading scheme in elongation with particular reference to the instances of aberrant reading that have been observed here. The emphasis of the discussion will be on the conceptual aspects, and we will not attempt to make an inventory of all the data in the literature pertaining to these questions.

II. THE CLASSIC PRINCIPLES OF CODON READING

The genetic code can be thought of as being made up of 16 boxes, each containing 4 codons. All codons in a box have the same nucleotides in the first two positions, the variation between the codons being in the third nucleotide. In half of the boxes the codons are divided between amino acids or between amino acids and stopwords, we will refer to such boxes as "split". When reading the codons of split boxes, it is obviously necessary to be able to distinguish between the nucleotides in the third codon position in order to avoid translational errors. A minimum requirement is that one must always discriminate between purines and pyrimidines and in two cases, the AUA/AUG and UGA/UGG codons, it is also necessary to make a distinction between A and G.

Half of the codons in the genetic code are contained in the so-called family boxes where all four codons denote the same amino acid (Figure 1). In such codon families, it makes no difference, as far as translational fidelity is con-

U U U Phe	U C U Ser	U A U Tyr	U G U Cys
U U C "	U C C "	U A C "	U G C "
U U A Leu	U C A "	U A A Stop	U G A Stop
U U G "	U C G "	U A G "	U G G Trp

C U U Leu	C C U Pro	C A U His	C G U Arg
C U C "	C C C "	C A C "	C G C "
C U A "	C C A "	C A A Gln	C G A "
C U G "	C C G "	C A G "	C G G "

A U U Ile	A C U Thr	A A U Asn	A G U Ser
A U C "	A C C "	A A C "	A G C "
A U A "	A C A "	A A A Lys	A G A Arg
A U G Met	A C G "	A A G "	A G G "

G U U Val	G C U Ala	G A U Asp	G G U Gly
G U C "	G C C "	G A C "	G G C "
G U A "	G C A "	G A A Glu	G G A "
G U G "	G C G "	G A G "	G G G "

FIGURE 1. The genetic code. Family codon boxes are indicated by bold lines.

cerned, how the third codon nucleotide is read, since the first two are sufficient to specify the amino acid.

In his wobble hypothesis, Crick[1] summarized the rules that presumably govern the reading of the codons by the tRNA anticodons. The canonical principles of codon reading allow only Watson-Crick-type base pairs between the anticodon and the first two codon nucleotides, while the wobble nucleotide of the anticodon is allowed a wider scope in its interaction with the third codon nucleotide. Nevertheless, there are a number of pairings that are not permitted even in the wobble position: U·U, U·C, C·U, C·C, C·A, I·G, G·A, and G·G. Furthermore, the principles demand that an anticodon must form stable base pairs with all three nucleotides of a codon in order to read it.

III. UNCONVENTIONAL METHODS IN THE CELL'S NORMAL REPERTOIRE OF CODON READINGS

Even if the great majority of the data concerning codon reading *in vivo* that have accumulated over the years indicate that codons are read in the cell according to the wobble rules there has, nevertheless, been a steady trickling of information suggesting that this is not always the case, and that there may be exceptions to the rules. In fact, a perusal of the literature shows that there is by now evidence for codon readings *in vivo* involving many of the forbidden base pairs. These exceptions are of two different types: one that involves unconventional readings with discrimination between the nucleotides in the third codon position,[2-11] albeit using other principles than those laid down in the wobble rules, and another that comprises readings without discrimination in this position.[12-14]

Unconventional codon readings of the first type are always solitary events, observed against a predominant background of conventional readings. For instance, there are a considerable number of reports of codon readings involving the forbidden base pair C·A.[2-7] Nevertheless, in all the organisms and organelles where this has been observed, an A in the third position is in the vast majority of cases read according to the wobble rules. Thus, it is not a question of the cell or organelle in question having established a new, generally adopted principle of codon reading that it adheres to in all applicable cases. One could rather think of unconventional readings with modified discrimination in the third codon position as having been caused by an isolated genetic event, the mutation of a single tRNA, to give it somewhat different reading properties.

Unconventional codon readings of the second category, however, are entirely different. Here the four codons of a family box are read without discriminating between the nucleotides in the third codon position. The analysis of mitochondrial tRNA genes and their gene products in the laboratories of Sanger,[12] Tzagoloff,[13] and RajBhandary[14] has established beyond doubt that in both mammalian and yeast mitochondria the family boxes are read by only one tRNA each. In all these cases, the tRNA gene has a T in the position corresponding to the wobble nucleotide and whenever the primary structure of the tRNA itself is known the wobble nucleotide has turned out to be an unsubstituted U.[15] The only exception to this rule is the tRNA that reads the arginine family box in yeast, which has A in this position.[15] Thus, in unconventional codon readings of the second type a new general principle has been established as illustrated by the fact that in yeast and mammalian mitochondria all family codons are read without discrimination in the third codon position.

It has been suggested that the mitochondrion is of prokaryotic origin and one may consequently ask if there are prokaryotes today that rely on unconventional codon reading without discrimination in the third position in order to read family boxes using a minimum number of tRNAs? Kilpatrick and Walker[16] have reported that *Mycoplasma mycoides* sp. *capri* contains only one glycine tRNA and that this tRNA has an unsubstituted U in the wobble position. Nothing is known about codon usage in this Mycoplasma, but in the very closely related *M. capricolum* the glycine codon GGU is used frequently. Unless there is an absolute bias against GGU and GGC in *M. mycoides*, its single glycine tRNA must be able to read these codons.

In order to elucidate further the possibility of such reading in *M. mycoides* we have initiated an inventory of the tRNA genes and their gene products in this organism. We have so far cloned and sequenced one isolated gene for arginine tRNA, a cluster of nine genes for the arginine, proline, alanine, methionine, isoleucine, serine, formyl methionine, aspartate, and phenylalanine tRNAs[17] and another cluster containing the genes for the asparagine, glutamate, valine, and threonine tRNAs.[18]

Thus, we have sequence information for a total of 14 tRNA genes in *M. mycoides*, 6 of which code for tRNAs that read family codons. In each of these latter cases there is only one gene per family box and the gene sequence contains a T in the position corresponding to the wobble nucleotide, with the exception of the arginine tRNA gene that has an A in this position. This is reminiscent of the situation in the yeast mitochondrion.[13] The resemblance is further strengthened by our finding that the mycoplasma tRNAs that read the family box codons for alanine, proline and valine all have an unsubstituted U in the wobble position,[18] as has the glycine tRNA sequenced by Kilpatrick and Walker.[16] The purification of these tRNAs yielded only one isoacceptor of each species consistent with our finding of only one gene each for alanine, proline, and valine tRNAs. However, the resemblance to the mitochondrial situation is probably not complete since our sequencing of the mycoplasma tRNAs has indicated that there are two isoacceptor tRNAs for threonine.[19]

Working with the closely related species *M. capricolum*, Andachi et al.[20] have arrived at similar conclusions based on extensive sequencing of tRNA genes and their gene products. It would seem, therefore, that in these small genome mycoplasmas the family codon boxes are read by only one tRNA each with the exception of the threonine box. These organisms apparently use an unconventional method without discriminating between the nucleotides in the third codon position. This would represent the first case so far encountered of a free-living organism using this type of unconventional codon reading.

A similar, albeit somewhat modified scenario was observed when the chloroplast genomes of liverwort[21] and tobacco leaves[22] were sequenced. The results showed that in the latter case half of the codon family boxes were read by only one tRNA each, while in liverwort chloroplasts three of the eight family boxes were read in this way.

To summarize, because they have been forced to jettison tRNA genes as part of a general genomic squeeze, mitochondria (with the exception of plant mitochondria) as well as chloroplasts and the small genome mycoplasmas have resorted to a simplified translational machinery where family codon boxes are read without discrimination between the nucleotides found in the third codon position. In mitochondria, all eight family boxes are read in this way while in the mycoplasmas there is one exception in that the threonine box is read by two tRNAs. In chloroplasts, the situation is even more complicated since only three (liverwort) or four (tobacco) of the family boxes are read without discrimination while the rest are read according to the classic principles.

IV. MECHANISTIC CONSIDERATIONS

Two mechanisms have been advanced to explain the undiscriminating codon reading already described. The fact that almost all the tRNAs that read without discrimination have an unsubstituted U in the wobble position prompted the suggestion that in the pertinent organelles and organisms this nucleotide can form stable base pairs with all four standard nucleotides.[14] We will refer to this tentative mechanism as "U reads all". One problem with this neat and straight forward explanation is that in some instances the wobble nucleotide in the undiscriminating tRNA is A or I instead of U. One would therefore have to assume that, for instance, the mispair A·A is also a stable base pair. On the other hand, the fact that U in the wobble position of tRNAs reading codons in the split boxes always seems to be modified might be an argument in favor of the "U reads all" assumption. Obviously, if an unsubstituted U had a tendency to read undiscriminatingly, it should be avoided in tRNAs reading split boxes.

The alternative explanation, that in undiscriminating reading the "forbidden" codon-anticodon interactions can lead to the formation of a peptide bond in spite of the fact that the anticodon cannot form a stable base pair with the third codon nucleotide, we will refer to as reading by "two-out-of-three". This concept was first put forward in order to explain the results of experiments using an *in vitro* protein synthesizing system from *E. coli* programmed with the phage message MS2 RNA. We observed a type of unconventional reading whereby a codon may be read by an anticodon that cannot, according to the wobble rules,[1] form a stable base pair with the third codon nucleotide.[23-26] Furthermore, it was found that as a group, the family codons, which were represented by the alanine, glycine, and valine codons, were considerably more amenable to this type of unconventional reading (that seemingly ignored the wobble restrictions) than were the nonfamily codons, which were represented by the lysine and glutamine codons.

What, then, might be the mechanism of this phenomenon? We have suggested that codons may be read according to the two-out-of-three principle that relies mainly on the Watson-Crick base pairs formed with the first two codon positions, while the mispaired nucleotides in the third codon and anticodon wobble positions make a comparatively small contribution to the total stability of the reading interaction. This hypothesis also attempts to explain why family codons are more amenable to such reading than codons in the split boxes, based on certain structural properties of the two codon categories.[27,28]

The two-out-of-three hypothesis does not imply that family codons *in vivo* are normally read by two-out-of-three even in situations where all the prerequisites for conventional codon readings are at hand. It only requires that, in such circumstances, the family codons should be read by two-out-of-three with a frequency that is not negligible, i.e., we assume it to be higher than the translational error frequency. Even on this limited assumption, such reading would become a potential source of translational error and therefore also an important restriction in the evolution of the genetic code.

The two-out-of-three hypothesis has some attractive features. For instance, there is strong evidence that this mode of codon reading can sustain effective protein synthesis *in vitro*. There is also no need to assume that mispairs like

U·U and U·C are as stable as regular base pairs. Finally, the hypothesis provides an explanation of why two-out-of-three reading does not take place on codons where it would cause translational errors. However, this hypothesis does not *per se* explain why almost all tRNAs that read family codons without discrimination contain U in the wobble position. A possible interpretation might be that the U·U pair has a marginal stability that would increase the probability of an effective reading when it occurs in the interaction with the third codon position. The result could perhaps be described as a reinforced two-out-of-three reading. On this assumption, U might be considered the best choice, if one had to select the most effective base in the wobble position of an anticodon that must read all codons in a family.[28] In the same vein, U in the wobble position of tRNAs that read codons in split boxes should always be substituted to guard against translational errors by reinforced two-out-of-three reading.

V. INFLUENCE OF STRUCTURES OUTSIDE THE ANTICODON ON CODON DISCRIMINATION

We do not know yet what mechanism operates in unconventional codon reading without discrimination in the third codon position, whether it is of the two-out-of-three or the "U reads all" type. However, regardless of what the answer to this question may be, it is pertinent to ask what structural features, other than the anticodon itself, might influence the ability of a tRNA to read a certain codon. An obvious answer of course is that the ribosomes of mitochondria, for instance, might have unusual properties that could stabilize an unconventional codon-anticodon interaction. Another possibility would be that the structural context, which the tRNA molecule provides for the anticodon, could affect its reading ability. This suggestion is consistent with our finding that the mycoplasma $tRNA_{UCC}^{Gly}$ showed efficient codon reading without discrimination between the nucleotides in the third position when it was tested in an *in vitro* protein synthesizing system derived from *E. coli*.[26]

A. THE ROLE OF THE NUCLEOTIDE IN POSITION 32

What structural elements are important for the ability of a tRNA to discriminate between the nucleotides that can occupy the third codon position? The $tRNA_1^{Gly}$ (anticodon CCC) from *E. coli* can be regarded as being at the other end of the spectrum compared to the mycoplasma $tRNA_{UCC}^{Gly}$ in the sense that it discriminates very intensely between the glycine codons in our *in vitro* system, as required by the wobble rules.[26] We have used site-directed mutagenesis to change the nucleotide C in the wobble position of $tRNA_1^{Gly}$ (anticodon CCC) to U. The mutant tRNA was tested for its ability to read glycine codons in an *in vitro* protein-synthesizing system programmed with the phage message MS2 RNA that had been modified by site-directed mutagenesis so as to make it possible to monitor conveniently the reading of all four glycine codons. The results showed that, while the tRNA$_1^{Gly}$ (anticodon UCC) and the mycoplasma tRNA$_{UCC}^{Gly}$ were equally efficient in the reading of the codon GGA, the mycoplasma tRNA^{Gly} was far more efficient than the tRNA$_1^{Gly}$ (anticodon UCC) in the reading of the codons GGU and GGC. Thus, the anticodon UCC, when present in the structural context of the tRNA$_1^{Gly}$ molecule, behaved as predicted by the wobble rules, while in the structural context of the mycoplasma tRNA^{Gly} it read without discrimination between the nucleotides in the third codon position, in violation of the wobble restrictions.[29]

To determine the influence of structures in the anticodon loop on the ability of the UCC anticodon to discriminate between the glycine codons we have used a tRNA construct based on tRNA$_1^{Gly}$ which we called tRNA$_1^{Gly}$-C32. In this construct, where the anticodon UCC is present in a structural context derived entirely from tRNA$_1^{Gly}$, except that the U in position 32 has been replaced by a C, the anticodon could not discriminate between the glycine codons, i.e., it behaved as in mycoplasma tRNA^{Gly}.[30] It may of course be argued that the tRNA constructs tested have varying modification patterns and that this is the reason for their different reading properties. However, although a definite settling of this question must await the outcome of experiments with completely unmodified tRNAs transcribed *in vitro,* we feel that this explanation is unlikely.

Instead, we think the results strongly suggest that the nature of the nucleotide in position 32 of the anticodon loop has a decisive influence on the reading properties of the anticodon UCC. When this position is occupied by a U it confers on the anticodon the ability to discriminate between the glycine codons, while with a C in the same position the ability to discriminate is lost. In this context, it is interesting to note that all tRNAs in *M. capricolum* and *M. mycoides* that can read each of the four codons in a family codon box, have a C in position 32, the only exception being tRNA^{Pro} that has a U in this position.[15]

One may then ask by what mechanism the nucleotide C, when present in position 32 instead of U, could deprive the anticodon UCC of its ability to discriminate between the glycine codons. One obvious possibility would be that a C in this position affects the conformation of the tRNA and in particular the anticodon loop so as to alter the reading properties of the anticodon. While such a mechanism is certainly possible there is no structural evidence to suggest how this could be brought about. It would be very interesting to determine and compare the structures of the tRNA constructs tRNA$_1^{Gly}$ (anticodon UCC) and tRNA$_1^{Gly}$-C32, preferably by X-ray crystallographic analysis.

An alternative explanation would be that the ability to discriminate between the nucleotides in the third position of the glycine codons is mainly a question of kinetics and has to do with the sitting time of the tRNA on the ribosome. The presence of a C instead of a U in position 32 might make for a more stable interaction, for instance, with the 16S RNA thus increasing the sitting time.

Finally, it is appropriate to consider if the present results have any bearing on the question of how undiscriminating reading is achieved. Is it by a two-out-of-three or a "U reads all" mechanism? Obviously, if the U to C shift in position 32 influences the reading properties of the anticodon by way of a change in conformation, both mechanisms are equally possible. However, if it is a question of an increased sitting time on the ribosome, when the tRNA has a C in position 32, this kinetic explanation would be more consistent with a two-out-of-three mechanism. The idea that sitting time is the crucial issue where discrimination is concerned, would in itself argue that undiscriminating reading of the glycine codons GGU and GGC by the anticodon UCC cannot involve good base pairs between U in the anticodon wobble position and U or C in the third codon position. If these mispairs were stable there should be no need for a prolonged sitting time.

To illustrate this point let us consider the following simple model. In conventional codon reading the formation of three good base pairs between codon and anticodon triggers a number of unspecific interactions between tRNA and ribosome which ensure a sitting time that is long enough to allow the formation of a peptide bond. In two-out-of-three reading the sitting time could either be considerably reduced compared to conventional reading or else, because of a lack of precision in the positioning of the tRNA anticodon, a prolonged sitting time compared to normal might be required to form the peptide bond. In either case a longer sitting time, caused by a structural change in the tRNA outside the anticodon, would increase the probability of a peptide bond being formed. On the other hand, it would have no effect on conventional readings like the reading of GGA by the anticodon UCC where presumably the sitting time is already optimal.

VI. INFLUENCE OF STRUCTURE OUTSIDE THE ANTICODON ON SUPPRESSOR EFFICIENCY

The UGA suppressor $tRNA^{Trp}$ (anticodon CCA) originally described by Hirsch[7] is the classic example of a mutation outside the anticodon, a G to A24 transition in the D-arm, that influences the ability of the tRNA to discriminate. This suppressor tRNA recognizes the stop codon UGA in spite of the mispair C·A formed with the third codon nucleotide. Smith and Yarus[31] have investigated the effect of the A24 mutation on a number of tRNA constructs based on the Hirsch mutant in terms of their suppressor efficiency. They suggest that the A24 mutation reduces the tRNA dissociation from the ribosome which would selectively enhance noncognate pairings, while cognate pairings would not be affected. In this hypothesis, the Hirsch suppressor and the $tRNA_1^{Gly}$-C32 construct discussed earlier, regardless of the great difference in the position of the mutation, might operate according to the same general kinetic model. Another case in point here is the report by Baumann et al.[32] that the conversion of 2-

thiocytidine to cytidine in position 32 of the anticodon loop of a tRNA^{Arg} from *E. coli* affects its ability to suppress frameshifting in a protein-synthesizing *in vitro* system programmed with MS2-RNA.

Finally, an interesting hypothesis should be mentioned here although it does not directly pertain to the codon discrimination problem. Yarus and co-workers[33] have provided evidence for what they call an extended anticodon. They find that the efficiency of a nonsense suppressor tRNA is dependent on the relation of the anticodon, in particular, the so called "cardinal" nucleotide in the 3' position on the anticodon, to the structure of the rest of the anticodon loop and stem. Thus, the extended anticodon hypothesis deals with the influence of the anticodon structural context on the efficiency of a codon reading that is entirely according to the wobble rules. The problem that we are primarily interested in, discrimination between the nucleotides in the third codon position, is different from nonsense suppression, and there is *a priori* no reason why it should be influenced by the same structures outside the anticodon. Furthermore, the basic concept of the extended anticodon hypothesis of a possible incompatibility between the cardinal nucleotide of a mutated anticodon and the unchanged anticodon loop and stem, does not apply in our case, since we are concerned with situations where, for instance, a glycine anticodon functions in a glycine tRNA molecule. Nevertheless, there is at least a formal similarity in the sense that a nucleotide outside the anticodon has a decisive influence on the reading properties of the anticodon.

VII. THE EVOLUTION OF CODON DISCRIMINATION

In the vast majority of organisms today, where we have any knowledge of their translational machinery, all codon boxes would seem to be read with discrimination between the nucleotides in the third codon position. However, this was not necessarily the case in the very primitive progenitor of cells when the genetic code first emerged. The following model for the evolution of codon discrimination illustrates this point.[28]

We will assume that, at some stage in its evolution, the code had reached a degree of sophistication where four distinct bases were employed to code for an assortment of amino acids more limited than that found in proteins today. However, the anticodons could not distinguish among the nucleotides occupying the third codon position; although the code was structurally a three-letter code, it was operationally a two-letter code, in terms of the number of letters per codon actually read with discrimination. Thus, all codons in the primitive code must have belonged to codon families. Setting aside one of the 16 families for the stop words, this leaves a maximum of 15 amino acids that could be unambiguously coded. At a more advanced stage of development, the cell would have perfected its translational machinery so that it could read the third codon nucleotide of some codons without ambiguity. Consequently, new amino acids could be accommodated in the code. It then became necessary to make a decision as to which groups of four codons could be divided up between two different amino acids without compromising translational fidelity. As a result of this selection, codons with a high probability of being read without discrimination in the third position would remain as family codons, and only those with a low probability would be used as nonfamily codons. Thus, it is possible to understand some of the structural features of the present code in terms of an evolution aimed at minimizing translational errors resulting from the reading of codons without third position discrimination.

One may then ask what structural developments made an improved codon discrimination possible. An obvious candidate would be the ribosome, the center piece of modern protein synthesis. However, the results discussed tend to focus our attention on the tRNA molecule and its properties. If we accept the idea of an RNA-dominated prebiotic world, the task of the primitive tRNAs at that time presumably included also the functions of the modern aminoacyl-tRNA synthetases, since there were no protein enzymes. This preeminence of the tRNA called for great versatility which must at the same time have seriously reduced the possibility of obtaining great precision in such a variety of functions. With the advent of protein molecules this would change drastically. The synthetase function could be taken over by protein enzymes leaving the tRNA free to develop a structure that allowed greater sophistication in codon reading, including the discrimination between the nucleotides in the third codon position. With this development we stand on the threshold of the biotic world.

VIII. CONCLUDING REMARKS

In the classic scheme of codon reading the emphasis is on the anticodon. It is this structure in the tRNA molecule that recognizes the codon and it is the property of the anticodon alone that determines whether it will be able to read a particular codon or not. Furthermore, we have had a tendency to perceive the canonical principles laid down in the wobble hypothesis, as cast-iron rules that allow no exceptions because of the structural constraints that presumably govern codon-anticodon interactions. With the realization that there are a number of instances where codons are read in violation of the wobble restrictions, and that in some organisms and organelles new general principles of codon reading have been established, the universality of the wobble rules as well as the validity of the structural arguments proposed to support them, should be reexamined. As for the supremacy of the anticodon in the selection of the correct codon, it would seem to be in doubt based on some of the results discussed. Instead, there is renewed interest in the interplay between the anticodon and the rest of the tRNA molecule.

REFERENCES

1. **Crick, F. H. C.**, Codon-anticodon pairing: the wobble hypothesis, *J. Mol. Biol.*, 19, 548, 1966.
2. **Hsu Chen, C.-C., Cleaves, G. R., and Dubin, D. T.**, A major lysine tRNA with a CUU anticodon in insect mitochondria, *Nucleic Acids Res.*, 11, 8659, 1983.
3. **Gupta, R.**, Halobacterium volcanii tRNAs. Identification of 41 tRNAs covering all amino acids, and the sequences of 33 class 1 tRNAs, *J. Biol. Chem.*, 259, 9461, 1984.
4. **Fukuda, K. and Abelson, J.**, DNA sequence of a T4 transfer RNA gene cluster, *J. Mol. Biol.*, 139, 377, 1980.
5. **Kashdan, M. A. and Dudock, B. S.**, The gene for a spinach chloroplast isoleucine tRNA has a methionine anticodon, *J. Biol. Chem.*, 257, 11191, 1982.
6. **Kuchino, Y., Watanabe, S., Harada, F., and Nishimura, S.**, Primary structure of AUA-specific isoleucine transfer ribonucleic acid from *E. coli*, *Biochemistry*, 19, 2085, 1980.
7. **Hirsch, D.**, Tryptophan transfer RNA as the UGA suppressor, *J. Mol. Biol.*, 58, 439, 1971.
8. **Clary, D. O. and Wolstenholm, D. R.**, A cluster of six tRNA genes in *Drosophila* mitochondrial DNA that includes a gene for an unusual tRNA$^{Ser}_{AGY}$, *Nucleic Acids Res.*, 12, 2367, 1984.
9. **Bibb, M. J., Van Etten, R. A., Wright, C. T., Walberg, M. W., and Clayton, D. A.**, Sequence and gene organization of mouse mitochondrial DNA, *Cell*, 26, 167, 1981.
10. **Beier, H., Barciszewska, M., Krupp, G., Mitnacht, R., and Gross, J. J.**, UAG read through during TMV RNA translation: isolation and sequence of two tRNAs^{Tyr} with suppressor activity from tobacco plants, *EMBO J.*, 351, 1984.
11. **Bienz, M. and Kubli, E.**, Wild-type tRNA$^{Tyr}_G$ reads the TMV RNA stop codon, but Q base-modified tRNA$^{Tyr}_Q$ does not, *Nature (London)*, 294, 188, 1981.
12. **Barell, B. G., Anderson, S., Banker, A. T., De Bruijn, M. H. L., Chen, E., Coulson, A. R., Drouin, J., Eperon, I. C., Nierlich, D. P., Roe, B. A., Sanger, F., Schreier, P. H., Smith, A. J. H., Staden, R., and Young, I. G.**, Different pattern of codon recognition by mammalian mitochondrial tRNAs, *Proc. Natl. Acad. Sci. U.S.A.*, 77, 3164, 1980.
13. **Bonitz, S. G., Berlani, R., Coruzzi, G., Li, M., Macino, G., Nobrega, F. G., Nobrega, M. P., Thalenfeld, B. E., and Tzagoloff, A.**, Codon recognition rules in yeast mitochondria, *Proc. Natl. Acad. Sci. U.S.A.*, 77, 3167, 1980.
14. **Heckman, J. E., Sarnoff, J., Alzner-Deweerd, B., Yin, S., and RajBhandary, U. L.**, Novel features in genetic code and codon reading patterns in *Neurospora crassa* mitochondria based on sequences of six mitochondrial tRNAs, *Proc. Natl. Acad. Sci. U.S.A.*, 77, 3159, 1989.
15. **Sprinzl, M., Hartman, T., Weber J., Blank, J., and Zeidel, R.**, Compilation of tRNA sequences and sequences of tRNA genes, *Nucleic Acid. Res.*, 17, (Seq. Suppl.), 1989
16. **Kilpatrick, M. W. and Walker, R. T.**, The nucleotide sequence of glycine tRNA from *Mycoplasma mycoides* sp. *capri*, *Nucleic Acids Res.*, 8, 2783, 1980.
17. **Samuelsson, T., Elias, P., Lustig, F., and Guindy, Y. S.**, Cloning and nucleotide sequence analysis of transfer RNA genes from *Mycoplasma mycoides*, *Biochem. J.*, 232, 223, 1985.
18. **Samuelsson, T. Guindy, Y. S., Lustig, F., Borén, T., and Lagerkvist, U.**, Apparent lack of discrimination in the reading of certain codons in *Mycoplasma mycoides*, *Proc. Natl. Acad. Sci. U.S.A.*, 84, 3166, 1987.
19. **Guindy, Y. S., Samuelsson, T., and Johansen, T.-I.**, Unconventional codon reading by *Mycoplasma mycoides* tRNA as revealed by partial sequence analysis, *Biochem. J.*, 258, 869, 1989.
20. **Andachi, Y., Yamao, F., Muto, A., and Osawa, S.**, Codon recognition patterns as deduced from sequences of the complete set of transfer RNA species in *Mycoplasma capricolum*, *J. Mol. Biol.*, 209, 37, 1989.

21. Ohyama, K., Fukuzawa, H., Kohchi, T., Shirai, H., Sano, T., Sano, S., Umesono, K., Shiki, Y., Takeuchi, M., Chang, Z., Aota, S.-I., Inokuchi, H., and Ozeki, H., Chloroplast gene organization deduced from complete sequence of liverwort *Marchantia polymorpha* DNA, *Nature (London)*, 322, 572, 1986.
22. Shinozaki, K., Ohme, M., Tanaka, M., Wakasugi, T., Hayashida, N., Matsubayashi, T., Zaita, N., Chunwongse, J., Obokata, J., Yamaguchi-Shinozaki, K., Ohto, C., Torazawa, K., Meng, B. Y., Sugita, M., Deno, H., Kamogashira, T., Yamada K., Kusuda, J.,Takaiwa, F., Kato, A., Tohdoh, N., Shimada, H., and Sugiura, M., The complete nucleotide sequence of the tobacco chloroplast genome: its gene organization and expression, *EMBO J.*, 5, 2043, 1986.
23. Mitra, S. K., Lustig, F., Åkesson, B., Axberg, T., Elias, P., and Lagerkvist, U., Relative efficiency of anticodons in reading the valine codons during protein synthesis *in vitro, J. Biol. Chem.*, 254, 6397, 1979.
24. Samuelsson, T., Elias, P., Lustig, F., Axberg, T., Fölsch, G., Åkesson, B., and Lagerkvist, U., Aberrations of the classic codon reading scheme during protein synthesis *in vitro, J. Biol. Chem.*, 255, 4583, 1980.
25. Lustig, F., Elias, P., Axberg, T., Samuelsson, T., Tittawella, I., and Lagerkvist, U., Codon reading and translational error. Reading of the glutamine and lysine codons during protein synthesis *in vitro, J. Biol. Chem.*, 256, 2635, 1981.
26. Samuelsson, T., Axberg, T., Borén, T., and Lagerkvist, U., Unconventional reading of the glycine codons, *J. Biol. Chem.*, 258, 13178, 1983.
27. Lagerkvist, U., 'Two out of three': an alternative method for codon reading, *Proc. Natl. Acad. Sci. U.S.A.*, 75, 1759, 1978.
28. Lagerkvist, U., Unorthodox codon reading and the evolution of the genetic code, *Cell*, 23, 305, 1981.
29. Lustig, F., Borén, T., Guindy, Y. S., Elias, P., Samuelsson, T., Gehrke, C. W., Kuo, K. C., and Lagerkvist, U., Codon discrimination and anticodon structural context, *Proc. Natl. Acad. Sci. U.S.A.*, 86, 6873, 1989.
30. Lustig, F., Borén, T., and Lagerkvist, U., The nucleotide in position 32 of the anticodon loop determines the ability of the anticodon UCC to discriminate between the glycine codons, unpublished results.
31. Smith, D. and Yarus, M., Transfer RNA structure and coding specificity. I. Evidence that a D-arm mutation reduces tRNA dissociation from the ribosome, *J.Mol. Biol.*, 206, 489, 1989.
32. Baumann U., Fischer, W., and Sprinzl, M., Analysis of modification-dependent structural alterations in the anticodon loop of *E. coli* tRNA^{Arg} and their effects on the translation of MS2 RNA, *Eur. J. Biochem.*, 152, 645, 1985.
33. Yarus, M., Translational efficiency of transfer RNAs: uses of an extended anticodon, *Science*, 218, 646, 1982.

Chapter 13

UNIVERSAL RULE OF TA/CG DEFICIENCY AND TG/CT EXCESS

Susumu Ohno

TABLE OF CONTENTS

I. Introduction 382

II. The Codon Assignments: *A Priori* Reason or a Frozen Accident ... 382

III. Codon Assignment Changes in Evolution 383

IV. The Combination of the Universal Codon Assignment and TA/CG Deficiency Rule Yields the Average Amino Acid Composition Typical of Most Proteins 384

V. The Antiquity of the Universal TA/CG Deficiency Rule 389

VI. The Notion of Codons Preference as an Illusion 392

VII. Base Trimers are not the Unit of Coding Sequences 394

VIII. Summary 395

References 396

I. INTRODUCTION

In his very revealing book, *Theory of Self-Replicating Machines,* von Neumann, the conceptual father of computers, sets forth three essential components that enable a machine to replicate itself.[1] The first is a set of instructions that can be transmitted more or less immutably from one generation to the next. The second is a set of machineries that carry out the above noted instructions, and the third is a container that sequesters the innards from the harsh outside world. In modern computer terms, "software" fulfills the first requirement while "hardware" carries out the second function. As stated in Rudolph Virchow's immortal dictum, *Ominis Cellula a Cellula,* the unit of life on earth is a cell. In the cell, the first task is assigned to nucleic acids, the second to proteins, and the third to plasma membranes. Therefore, the emergence of the first cell, and hence, life itself on earth was contingent upon the successful linkage between the base sequences of self-replicating nucleic acids and the amino acid sequences of the corresponding proteins, so that the amino acid sequences of individual proteins, also became immutable. It follows, then, that codon assignments were the key to the once and only spontaneous generation of life that occurred more than one billion years ago.

II. THE CODON ASSIGNMENTS: *A PRIORI* REASON OR A FROZEN ACCIDENT

Because of the great antiquity of the event, there is no ready answer to the crucial question of whether or not primitive tRNAs were already involved in the initial linkage between the base sequences of self-replicating nucleic acids (RNA rather than DNA) and the amino acid sequences of proteins. Nevertheless, it needs to be pointed out that one plausible digital deciphering mechanism which would have directly translated the double-stranded base sequences of nucleic acids to the amino acid sequences of proteins was proposed in 1954 by Gamov, who also suggested the big bang theory as the origin of the universe.[2] In any case, the amino acid compositions of proteins are greatly influenced by the base compositions of nucleic acids encoding them as we shall shortly see. Thus, it would have been very helpful to know the average base compositions of self-replicating RNAs that were present in the prebiotic world. Unfortunately, here, too, one cannot be too certain. In accordance with current fashion, if life originated in the vicinity of a geothermal vent in the ocean, prebiotic RNAs had to be rather GC-rich so that they could have maintained a double-stranded state at least part of the time. The resulting GC-richness would have also helped to minimize a high copying error rate of prebiotic RNA replication in the presence of Zn^{++} ion without enzymatic help, for the error rate is thought to be of the order of 10^{-2}/base pair/replication.[3] At this high error rate, prebiotic RNA can hardly be thought of as an immutable set of instructions. If, on the other hand, life evolved in a part of the primitive earth with a more moderate environment, disassociation of complementary RNA strands as a necessary prelude to their replication would have to depend primarily upon midday heat generated by the faint, early sun. Under the circumstances, AU-richness would have been forced upon prebiotic RNA. Indeed, there are a number of unanswerable questions and not a single readily available answer as to the set of prebiotic circumstances that culminated in the once and only spontaneous generation of life more than one billion years ago. Even if there were primitive tRNAs that required no enzymatic help in being charged with particular amino acids, their mere presence did not assure constant codon assignment in the prebiotic world. Under the extremely high replication error rate of the order of 10^{-2}, the anticodons of tRNAs would have undergone rapid, successive changes. Viewed in this light, there should have been no *a priori* reason for the assignment of a set of particular base trimers as codons to a specific amino acid.

III. CODON ASSIGNMENT CHANGES IN EVOLUTION

Indeed, it is now apparent that the so-called universal coding system was not as ubiquitous as originally thought. Although deviations from the universal genetic code were first found in the mitochondrial genomes of higher metazoans, codon changes that occurred in the chromosomal genomes of certain unicellular eukaryotes were probably more pertinent in evolution. For reasons that are not at all clear, the genomes of unicellular eukaryotes are characterized by their high AT-content, which are 75% or greater. At the lower end of this extremely AT-rich scale, such unicellular eukaryotes as malarial protozoan parasites still cling to the universal coding system. Inevitably, they suffer from deficiencies of Ala and Gly, encodable only by codons beginning with GC and GG, as these two amino acids rank second and third in the order of abundance within the average amino acid composition of proteins.[4] In proteins of those unicellular eukaryotes which still cling to the universal coding system, we find that sites which are normally occupied by Gly and Ala are replaced by Asn which is coded mostly by AAU, a codon with neither G nor C (e.g., see Reference 5).

When the genomic AT-content increased beyond 75%, however, the incidence of the three potential chain-terminators, TAA, TAG, and TGA, must have reached unbearably high levels as base trimers within coding sequences. Accordingly, in ciliates such as *Tetrahymena,* some of the classical glutamine tRNA species have undergone anticodon changes to recognize UAA and UAG, and thus, these two former chain-terminators became Gln codons.[6,7]

As originally proposed by Lynn Margulis, the ancestry of mitochondria found in all eukaryotes should be sought in an early prokaryote where it is thought to have existed as a parasite before coming into a symbiotic relationship with one of the first eukaryotes. Further evolution of eukaryotes to more complex forms was generally accompanied by progressive reductions in the genome sizes of their mitochondria. The original mitochondrial genomes must have been comparable to that of early prokaryotes with the genome approaching 10^6 base pairs encoding 1000 or so different proteins based upon the universal coding system. Curiously, the mitochondrial genomes approaching the original ones noted above are found in higher plants with the genome sizes ranging from 2.5×10^5 to 2.5×10^6 base pairs. This large size of higher plant mitochondrial genomes, however, is not likely to reflect a continuous retention of the original size. Rather, it is likely to represent a secondary increase, for the mitochondrial genome sizes of more primitive organisms such as *Saccharomyces cerevisiae* (a representative of unicellular eukaryotes) is already reduced to 78 to 68×10^3 base pairs, which is about the same genome size found in mitochondria of fungi such as *Neurospora crassa*. Further reduction to roughly one fifth the above size has been found in mitochondria of insects, enchinoderms, and mammals.

A minimum of 32 different species of tRNAs are required if translation is to depend upon the universal coding system, and smaller mitochondrial genomes cannot readily accommodate genes for that many different species of tRNAs. Hence, progressive deviation from the universal coding system observed in mitochondrial genomes can be seen as an accommodation to make do with only 27 or so species of tRNAs. In mitochondria of *S. cerevisiae* and *N. crassa,* one of the three universal chain terminators, UGA, is a Trp codeword, along with the traditional UGG codon. Thus, both UGA and UGG are translated by the same species of tRNA^{Trp} which contains a U*CA anticodon (where U* denotes an unknown modification of U) instead of the traditional CCA or CmCA anticodon which recognizes only UGG. There are other codon changes which may be class-specific to genus-specific. For example, in the mitochondrial genomes of echinoderms, the traditional Arg codons, AGA and AGG, code for serine.[8] It should be recalled that in the universal coding system, six codons are assigned to Ser. In addition to the four codons beginning with UC, there are AGC and AGU. As these unusual codon usages are discussed further in other chapters of this book (e.g., see Chapter 7), I shall only use these unusual codon assignments as evidence against an *a priori* relationship between sets of base trimers and the amino acids they encode.

IV. THE COMBINATION OF THE UNIVERSAL CODON ASSIGNMENT AND TA/CG DEFICIENCY RULE YIELDS THE AVERAGE AMINO ACID COMPOSITION TYPICAL OF MOST PROTEINS

Fred Hoyle's argument for the extraterrestrial origin of life is as follows: Assuming an average length of 100 amino acid residues to be the prerequisite for proteins to acquire functional competence of any kind, the initial set of proteins in the prebiotic world had to be 20^{100} different kinds. Inasmuch as only a very small fraction of the above noted astronomical variety would be functionally competent, the escape from this random chaos would have taken far longer than four or five billion years. Hence, life must have originated on an older planet before the formation of the earth, and a passenger from this older planet must have seeded earth to start life here.[9]

Indeed, if the initial codon assignments were extremely even handed, giving equal opportunity to each of the 20 amino acid residues to be represented in more or less equal numbers in most proteins, Hoyle's random chaos would surely have precluded the emergence of life roughly 1.5 billion years ago. Thus, we owe our lives to the extremely biased codon assignments that characterize the universal coding system, for it is this bias that provided the initial escape from the quagmire of randomness. Met and Trp, which were assigned only one codon each, are destined forever to remain minor components in all proteins. On the other hand, it would seem that the three residues with the assignment of six codons each were destined to be major components in nearly all proteins. At this point, however, one readily realizes that an exclusive obedience to the universal codon assignment also would have precluded the emergence of life, for the bias is also prominent in the assignment of codons to the charged residues. While the two acidic residues, Glu and Asp, are endowed with only two codons each, a total of eight codons are assigned to the two basic residues, Lys and Arg, with the assignment of six codons to Arg being the cause of this twofold disparity. Furthermore, if His is included as a basic residue, the disparity in the codon assignment of acidic and basic residues becomes four to ten. The utility of the resulting proteins with very strong positive charges would be very limited, their only attribute being the ability to interact with nucleic acids. They would seem to be quite inadequate either as cytosolic enzymes or as cytoskeletal components.

The fact is that most proteins extracted from all organisms have nearly neutral isoelectric points, albeit slightly below 7.0. Indeed, in an electrophoretic field buffered to nearly neutral pH, most proteins readily move toward the anode, thus revealing their slightly negative charges. The only consistent exceptions to this observation are provided by histones, other nuclear proteins and those nuclear gene encoded enzymes that are targeted for mitochondria which move toward the cathode.

At the end of 1988, there were 18,383 entries in the Genbank Database. At this point, a glance at the average protein amino acid composition deduced from all the entries would be most instructive.[4] The most abundant amino acid is Leu with six codons, constituting 9.2% of the total number of residues, followed by Ala, Gly, Ser, and Val in decreasing order. These five residues with either four or six codons comprise 38.1% of the total. At the other extreme, the five least abundant amino acid residues in decreasing order are Tyr, His, Met, Cys, and Trp, which are coded by only one or two codons. These five amino acids together comprise only 10.3% of the total number of residues (Figure 1). The percent abundances among the four charged residues are: (No. 6) Glu, 6.2%; (No. 8) Lys, 5.9%; (No. 9) Arg, 5.4%; and (No. 10) Asp, 5.3%.

No. 1 LEU 9.2% 6 COD.	No.2 ALA 7.8% 4 COD.	No.3 GLY 7.4% 4 COD.	No. 4 SER 7.2% 6 COD.
No.5 VAL 6.5% 4 COD.	No.6 GLU 6.2% 2 COD.	No.7 THR 6.1% 4 COD.	No.8 LYS 5.9% 2 COD.
No. 9 ARG 5.4% 6 COD.	No.10 ASP 5.3% 2 COD.	No.11 PRO 5.2% 4 COD.	No.12 ILE 5.1% 3 COD.
No.13 ASN 4.2% 2 COD.	No.14 GLN 4.1% 2 COD.	No.15 PHE 3.7% 2 COD.	No.16 TYR 3.1% 2 COD.
No.17 HIS 2.3% 2 COD.	No.18 MET 2.2% 1 COD.	No.19 CYS 1.9% 2 COD.	No.20 TRP 1.0% 1 COD.

$$\frac{LYS + ARG}{GLU + ASP} = \frac{5.9\% + 5.4\%}{6.2\% + 5.3\%} = 0.98 \qquad \frac{TYR}{PHE} = \frac{3.1\%}{3.7\%} = 0.84$$

FIGURE 1. Average amino acid composition deduced from 18,383 entries in GENBANK database.[4] Each residue is accompanied by the number of codons assigned to it. Four charged residues are underlined by solid bars. While the top four residues together comprise 31.6% of the total, the bottom four residues account for only 6.4%. Other pertinent points are made in the text.

Indeed, the average protein is ever so slightly negatively charged, for the combination of Glu and Asp outnumbers that of Lys and Arg by 1.0 to 0.98 (Figure 1). Why is Arg with six codons outnumbered not only by Glu, but also by Lys, both of which have two codons? The answer is found in the universal grammatical rule of coding sequences, which entails a deficiency in TA/CG base pairs as we shall see shortly.[10,11] Provided that the average amino acid composition shown in Figure 1 is nearly ideal for the versatile and functional competence of proteins, this grammatical rule had to be present from the beginning of life.

Inasmuch as coding sequences reside in one or the other of two complementary DNA strands, the A = T and G = C rule does not apply, since some coding sequences are purine-rich while others are pyrimidine-rich. Nevertheless, one finds that as long as each of the four bases comprising coding sequences remain within the range of 25% (± 5%), the amino acid compositions of proteins encoded by them remain reasonably close to the average compositions shown in Figure 1. Only those coding sequences with notably lopsided base compositions encode proteins of rather unusual amino acid compositions, but even here, the unusual compositions involve only several of the 20 residues rather than all 20. As shown at the top of Figure 2, the 1599 base pair coding sequence for the chicken c-*src* tyrosine kinase (the cellular counterpart of Rous sarcoma virus v-*src* oncogene) is extremely GC-rich (61.5%) and T-poor (16.7%).[12]

At first glance, it would seem that this GC-richness in its coding sequence should have greatly favored Arg to become a prominent residue in the c-*src* protein, since four of the six Arg codons start with CG. However, it is not the unusual prominence of Arg, but an elevation in the prominence of Glu and Thr, accompanied by the decline of Ala and Ile which makes this protein highly unusual. The nucleotide dimer analysis of the c-*src* coding sequence shown near the top of Figure 2 provides insight into the reason for the unusual amino acid composition of the corresponding protein. The dimeric composition of the coding sequence reflects not only the universal grammatical rule, but also manifests a few unusual features in its own right. As usual, the two dimers, TA and CG, are grossly underrepresented. The TA deficiency is compensated by a proportional excess of TG among the four dimers starting with T, and by an

Universal Rule of TA/CG Deficiency and TG/CT Excess 387

```
      C: 502 (31.4%)      G: 481 (30.1%)      A: 349 (21.8%)      T: 267 (16.7%)

   T G: 114 (80)     G T:  66 (80)      A T: 45 (58)      T A: 31 (58)
      1.43              0.82               0.78              0.53

   C A: 133 (109)    A C: 116 (109)     
      1.22              1.06

   C T: 109 (84)     T C:  80 (84)      G C: 138 (151)    C G: 87 (151)
      1.30              0.95               0.91              0.58
```

1) 49 LEU		2) 43 GLU		3) 41 THR		4) 40 GLY	
TRIMER	CODON	TRIMER	CODON	TRIMER	CODON	TRIMER	CODON
49 C T G	32	42 G A G	35	38 A C C	18	45 G G C	17
24 C T C	13	20 G A A	8	25 A C T	9	41 G G G	12
18 T T G	3			30 A C A	7	43 G G A	8
17 C T A	1			23 A C G	7	21 G G T	3
19 C T T	0						
1 T T A	0						

5) 37 SER		6) 35 ALA		7) 34 ARG		8) 33 PRO	
TRIMER	CODON	TRIMER	CODON	TRIMER	CODON	TRIMER	CODON
40 A G C	13	48 G C C	17	36 C G G	16	54 C C C	23
26 T C C	12	35 G C T	8	20 C G C	10	42 C C A	5
18 T C G	7	39 G C A	7	30 A G G	4	30 C C G	3
22 T C A	2	16 G C G	3	14 C G T	2	35 C C T	2
14 T C T	2			30 A G A	1		
18 A G T	1			17 C G A	1		

9A) 30 LYS		9B) 30 VAL		11A) 23 ASP		11B) 23 TYR	
TRIMER	CODON	TRIMER	CODON	TRIMER	CODON	TRIMER	CODON
34 A A G	27	28 G T G	17	35 G A C	19	23 T A C	19
9 A A A	3	22 G T C	10	15 G A T	4	5 T A T	4
		8 G T T	3				
		8 G T A	0				

13) 21 PHE		14) 20 GLN		15) 18 ASN		16) 16 ILE	
TRIMER	CODON	TRIMER	CODON	TRIMER	CODON	TRIMER	CODON
19 T T C	15	42 C A G	18	22 A A C	17	15 A T C	13
8 T T T	6	35 C A A	2	5 A A T	1	6 A T T	2
						5 A T A	1

17) 12 MET		18) 10 HIS		19A) 9 CYS		19B) 9 TRP	
TRIMER	CODON	TRIMER	CODON	TRIMER	CODON	TRIMER	CODON
19 A T G	12	36 C A C	6	31 T G C	9	48 T G G	9
		20 C A T	4	13 T G T	0		

```
                    TERMINATORS
              TRIMER        TERM.
              22 T G A      0
               3 T A A      0
               0 T A G      0
```

FIGURE 2. The analysis of the extremely CG-rich, T-poor 533-codon long chicken c-*src* coding sequence. Actual numbers and percentages in parentheses of the four bases present in the c-*src* coding sequences are shown of the figure. Two overabundant (above 25%) bases are underlined by solid bars, while two underrepresented bases (below 25%) are underlined by open bars. Immediately below the base composition, the incidences of five pairs of mirror-image base dimers are shown. The observed number is accompanied by an expected number in parentheses and the observed/expected ratio is shown below each base dimer. Three persistently excessive dimers are underlined by solid bars, while two invariably deficient dimers are underlined by open bars. Their respective mirror-image dimers are underlined by shaded bars. Shown below the base dimers are the 64 base trimers grouped into 20 sets of potential amino acid encoding codons and a set of potential chain terminators. with regard to each base trimer, the number shown on its left is its incidence as a base trimer, while the number shown on its right is that used as a codon.

excess of CA among the four dimers ending in A. In the case of the CG deficiency, a compensatory excess affects CT among the four dimers starting with C, while a TG excess also compensates the CG deficiency among the four dimers ending in G. These excesses and deficiencies in dimers comprise the parts of the universal rule that govern the construction of most, if not all, of the coding sequences.[10,11] As to the unusual features mentioned previously, it should be noted that GT, which is the reciprocal of the always excessive TG, is as deficient as AT, which is the reciprocal of the always deficient TA. These two reciprocals of the two always deficient dimers, GT and AT, are represented at nearly the expected numbers in most coding sequences. One immediately sees the consequences of an AT deficiency in the decrease of Ile, encoded by three codons starting with AT. Ile, ranking 12th and comprising 5.1% of the total amino acid residues in the average composition, has fallen to 16th in ranking, comprising only 3% of the total. Since the deficiency of GT was not as drastic as that of AT, the decline of Val, encoded by four codons starting with GT, was not as prominent as that of Ile in the c-*src* protein. The average rank and percent are 5th and 6.5%, respectively, for Val among the proteins shown in Figure 1, which dropped to 9th and 5.6% in c-*src*. Thus, the consequences of a universal CG deficiency is very evident as shown in Figure 2. Two C's and one G can be arranged in three ways to yield an Ala codon, GCC, an Arg codon, CGC, and a Pro codon, CCG. The c-*src* coding sequence contains 48 GCC, 36 CGC, and 30 CCG base trimers. A scarcity of second and third base trimers compared to the first is a consequence of the CG deficiency. Because of the universal CG deficiency which affects four of the six Arg codons, Arg did not increase much from the extreme CG-richness of the c-*src* coding sequence. Although the frequency of Arg is greater than that of Lys, the number of positively charged Arg and Lys residues is still less than the negatively charged Glu and Asp residues by 64 to 70. Thus, the c-*src* protein still remains slightly acidic.

As to the far reaching effects of TA deficiency, it should first be noted that the 1599 Nucleotide c-*src* coding sequence contains only one TTA, and not a single TAG trimer. All four base trimers ending in TA are very scarce. Accordingly, GTA, ATA, and CTA, which code for Val, Ile, and Leu, respectively, as well as TTA, which also codes for Leu, make only negligible contributions. The use of TAA and TAG as two of the three chain terminators by the universal coding system was also fortuitous, if TA a deficiency was already present at the beginning of life. A scarcity in these TA-rich termination codons would have endowed many prebiotic nucleic acids with long, open reading frames. The third chain terminator, TGA, on the other hand, occurs rather frequently, as shown at the bottom of Figure 2, due to the TG excess aspect of the universal rule. In this respect, Juke's view that TGA was originally a Trp codon and not a chain terminator finds a favorable audience.[13] As already noted, the metazoan mitochondrial genome appears to have reverted to its original state, since both TGA and TGG function as Trp codons.

Both Leu and Ser are encoded by six codons each. However, the universal TA/CG deficiency more acutely affects Leu codons than Ser codons. The TA deficiency, as already noted, affects two of the six Leu codons, CTA and TTA. Only one of the six Ser codons, TCG, is affected by the CG deficiency. Yet, in the average amino acid composition of a typical protein (see Figure 1), Leu substantially outnumbers Ala, the second most prominent amino acid, while Ser ranks only fourth. In fact, it takes a rather extremely TC-rich sequence to encode a protein in which Ser becomes the most prominent residue rather than Leu. The usual prominence of Leu is due more to the presence of one codon, CTG, than to the assignment of six codons. Since this base trimer is an amalgam of two of the three always excessive dimers, CT and TG, CTG is usually the most numerous of the 64 base trimers in most sequences. Thus, CTG encodes as much as two thirds of the Leu residues. Figure 2 shows that the 49 CTG trimers are outnumbered only by 54 CCC trimers in spite of the extreme T-poorness in the c-*src* coding sequence. Not surprisingly, 32 of the 42 Leu residues are encoded by CTG.

The fact that even a very lopsided base composition of c-*src* coding sequence managed to cause but a few deviations from the average amino acid composition shown in Figure 1, attests to the compatibility of the average amino acid composition with the functional versatility of protein sequences. This is no surprise, when one realizes that peptide domains comprising a protein are endowed with only three basic conformational alternatives: either an α-helix, a β-pleated sheet, or failing these two, a random coil configuration.

Four of the first five most numerous residues in average amino acid compositions, Leu, Ala, Gly, and Val are hydrophobic residues. Being hydrophobic, they are conducive to the formation of both α-helical and β-pleated sheet structures. Furthermore, Leu and Ala are two of the eight residues with S values greater than 1.0.[14] The order of abundance in the average amino acid composition of these eight residues is No. 1, Leu; No. 2, Ala; No. 6, Glu; No. 9, Arg; No. 12, Ile; No. 15, Phe; No. 18, Met; and No. 20, Trp. Of these eight, Leu and Ala, being more numerous make the most significant contributions to α-helix formation. On the other hand, Ser, if located in the midst of hydrophobic residues, is very conducive to the formation of a β-pleated sheet structure. The same role is also played by the usually less numerous Thr.

All in all, it would appear that the universal codon assignment, indeed, was an extremely fortunate accident, provided that the universal rule of TA/CG deficiency was also present from the very beginning of life.

V. THE ANTIQUITY OF THE UNIVERSAL TA/CG DEFICIENCY RULE

CG deficiency is generally attributed to CG methylation; methylated CG being converted to TG or CA.[15] As CG methylation is extremely widespread, affecting prokaryotes and eukaryotes alike, the universality of CG deficiency, indeed, appears explainable by CG methylation. Although there have been sporadic secondary losses of methylation, as in dipteran insects, they may be overlooked as minor evolutionary diversions. Nevertheless, in the prebiotic and peribiotic world of RNAs, it can be speculated that methylation could not possibly have played any role. Thus, it became of interest to examine the RNA genomes of certain viruses not endowed with reverse transcriptase to determine the efficacy of the CG deficiency rule.

Judging from its wide host range (e.g., man and duck), influenza virus must have lived as an independent entity for a considerable period. Moreover, since it lacks reverse transcriptase, its RNA genome is not likely to have experienced even a single conversion to DNA throughout its existence. Thus, the single, contiguous coding sequence of influenza A virus Aichi/2/68 encoding two hemagglutinins with a combined total of 550 residues is analyzed in Figure 3.[16] This figure shows that the universal rule applies equally to this RNA coding sequence, as well as to any other DNA coding sequence in spite of its independence from DNA for a considerable period of time. For the sake of uniformity, this RNA sequence is treated here as a DNA sequence using T in place of U. The extent of TA and CG deficiencies observed in Figure 3 is almost identical with those found in Figure 2. Furthermore, deficiencies in the corresponding base trimers are once again compensated primarily by excesses of the same three base dimers, TG, CA, and CT. It can be taken that these observations are rather persuasive evidence that the universal rule was already present in RNAs of the prebiotic and peribiotic world.

The influenza virus coding sequence also has a very lopsided base composition which is extremely A-rich and moderately C-poor. Despite this extreme imbalance in the base composition of its coding sequence, deviations from the average amino acid composition of Figure 1 again involve only a small number of residues in the influenza virus hemagglutinins. Only four residues are noticeably affected. The extreme A-richness in its coding sequence elevated Ile, which is ranked 12th in the composite protein analysis shown in Figure 1, to the No. 1 position (Figure 3). Asn, which is ranked 13th in Figure 1, also came into prominence sharing the No. 3 position with Leu (Figure 3). A milder degree of C-poorness only caused a decline of Ala from its customary No. 2 position (Figure 1) to 12th position in the influenza virus hemagglutinins. As shown in Figure 3, the decline of Pro was less marked. The number of each residue actually present in the influenza virus hemagglutinins is accompanied by a number in parenthesis in Figure 3, which is the expected number for that residue, if this protein consisted of an average amino acid composition. It should be noted that the actual numbers of Ser, Asp, Glu, Lys, Arg, Gln, Phe, Tyr, and His residues are either identical, or nearly identical, with the numbers shown in parentheses. Furthermore, noticeable changes affecting Leu, Val, Ala, and Gly residues are functionally inconsequential ones, since Ile is equivalent to Leu as well as Val, and, in proteins encoded by extremely AT-rich sequences, Asn readily replaces Ala or Gly, as already noted. Figures 2 and 3, indeed, appear to indicate

Universal Rule of TA/CG Deficiency and TG/CT Excess 391

<u>A: 533 (32.3%)</u>   <u>G: 394 (23.9%)</u>   <u>T: 370 (22.4%)</u>   <u>C: 353 (21.4%)</u>

<u>T G</u>: 124 (96)   <u>G T</u>: 80 (96)   <u>A T</u>: 128 (127)   <u>T A</u>: 74 (127)
 1.29   1.00   0.58

<u>C A</u>: 151 (111)   <u>A C</u>: 114 (111)
 1.36   1.02

<u>C T</u>: 97 (83)   <u>T C</u>: 91 (83)   <u>G C</u>: 89 (84)   <u>C G</u>: 40 (84)
 1.17   1.10   1.06   0.47

1) <u>45 ILE</u> (28)	2) <u>44 GLY</u> (40)	3A) <u>42 ASN</u> (23)	3B) <u>42 LEU</u> (51)	5) <u>39 THR</u> (34)
CODON	CODON	CODON	CODON	CODON
A T C 19	G G A 14	A A T 22	C T G 17	A C T 16
A T A 14	G G T 12	A A C 20	C T T 9	A C A 14
A T T 12	G G G 12		C T A 7	A C C 5
	G G C 6		T T G 5	A C G 4
			C T C 3	
			T T A 1	

6) <u>38 SER</u> (39)	7) <u>31 ASP</u> (29)	8) <u>30 GLU</u> (34)	9) <u>29 LYS</u> (32)	10) <u>29 ARG</u> (30)
CODON	CODON	CODON	CODON	CODON
A G C 14	G A C 19	G A A 16	A A A 19	A G G 12
T C A 9	G A T 12	G A G 14	A A G 10	A G A 10
T C T 6				C G G 3
T C C 4				C G C 2
A G T 3				C G A 1
T C G 2				C G T 0

11) <u>28 VAL</u> (36)	12) <u>26 ALA</u> (43)	13) <u>23 GLN</u> (23)	14A) <u>20 PRO</u> (28)	14B) <u>20 PHE</u> (21)
CODON	CODON	CODON	CODON	CODON
G T T 10	G C A 11	C A A 16	C C A 7	T T C 11
G T A 8	G C T 8	C A G 7	C C T 6	T T T 9
G T G 6	G C C 5		C C C 4	
G T C 4	G C G 2		C C G 3	

16A) <u>17 TYR</u> (18)	16B) <u>17 CYS</u> (11)	18) <u>12 TRP</u> (5)	19) <u>11 HIS</u> (13)	20) <u>8 MET</u> (12)
CODON	CODON	CODON	CODON	CODON
T A T 9	T G C 12	T G G 12	C A T 8	A T G 8
T A C 8	T G T 5		C A C 3	

FIGURE 3. Analysis of the extremely A-rich, T,C-poor coding sequence of influenza A virus hemagglutinin I and II. The analysis was carried out essentially as given in Figure 2. Although influenza A virus is an RNA virus its coding sequence is treated here as though it is c-DNA. Incidences of 64 base trimers are not shown because the points are already made in Figure 2. Instead, the observed number of each residue is accompanied by a numbers in parentheses which are the expected number in hemagglutinin if it was of the average amino acid composition shown in Figure 1.

that the universal codon assignment and the equally universal TA/CG deficiency rule minimize deviations from the average amino acid composition of proteins encoded by coding sequences with very unbalanced base compositions.

Thus far, only two groups of genomes have been found which violate the universal TA/CG deficiency rule. One is the extremely AT-rich genome of some of the unicellular eukaryotes such as *Tetrahymena*. It should be recalled that this group also deviates from the universal codon assignment. The second is the metazoan mitochondrial genome which also deviates from the universal codon assignment.

VI. THE NOTION OF CODON PREFERENCE AS AN ILLUSION

Four codons are assigned to each of the five amino acids, Ala, Gly, Val, Thr, and Pro. The third base in these codons is redundant, since the combination of the first two bases determine the assignment; e.g., GCC, GCG, GCA, and GCT encode Ala. The same applies to four of the six codons assigned to each of the three amino acids, Leu, Ser, and Arg. The redundancy in the third position of codons applies to 8 of the 20 amino acids. Since four synonymous codons assigned to each of these eight amino acids are never utilized with equal frequency, the notion of codon preference has evolved. Few other subjects in molecular biology have received as much attention as the notion of codon preference. Its underlying theme is that since two or more different tRNA species designed to recognize a set of four synonymous codons are produced in different amounts, a particular one of the four codons to which a major species of tRNA is assigned come to be favored by natural selection, for the sake of improving the translational efficiency of coding sequences. There is one glaring weakness with this idea. Since the relative amounts of synonymous tRNA species can more easily be adjusted by, for example, varying the number of gene copies of each, it would be far more sensible to improve the overall translational efficiency by adjusting tRNA levels to codon usage than vice versa. It would be a folly to adjust codon usages of tens of thousands of individual coding sequences one by one, while keeping relative production rates of synonymous tRNA species fixed.[17]

One codon which enjoys universally frequent usage among the six codons for Leu is CTG. This frequent usage is seen in *E. coli* as well as in man.[18] Indeed, 32 of the 49 Leu residues in the chicken c-*src* are encoded by CTG (Figure 2), and so are 17 of the 42 residues in the influenza A virus hemagglutinins (Figure 3). Nevertheless, there is no need to invoke natural selection to explain the frequent use of CTG, since the preponderance of CTG as a Leu codon is a reflection of the universal TA/CG deficiency rule which can be restated as the CT/TG/CA compensatory excess rule. Being an amalgam of two excessive CT and TG dimers, CTG as a base trimer is invariably overrepresented, as already noted. In Figure 2, each base trimer is accompanied by two numbers. The number to its left is the total number of times it is used as a base trimer in the chicken c-*src* coding sequence, while the number to its right denotes the number of times it is used as a codon. Of the six base trimers that are potential Leu codons, CTG outnumbers the other five as a base trimer, and outdistances the next most numerous Leu trimer, CTC, by 49 to 24. Hence, CTG is utilized most frequently as a Leu codon. With regard to the 17 other amino acid residues with two or more codons, the base trimer that occurs the most numerous times is also used most frequently as a synonymous codon. Not a single exception in this observation occurs in the data shown in Figure 2. Thus, it is clear from Figure 2 that codon usages are largely determined by base trimer frequencies which, in turn, are solely determined by the combination of two factors: (1) the universal TA/CG deficiency rule and (2) the base composition of a given coding sequence. In the c-*src* coding sequence, which is extremely C-rich (31.4%), GCC outnumbers the other three base trimers that are potential Ala codons, whereas GCG, due to the CG-deficiency part of the universal rule, is the least numerous. Accordingly, of the 35 Ala residues, 17 are encoded by GCC and only 3 by GCG. The C-richness of this coding sequence was further reflected in the preponderant use of ACC, GGC, AGC, CCC, GAC, TAC, TTC, AAC, CAC, and TGC as Thr, Gly, Ser, Pro, Asp, Tyr, Phe, Asn, His, and Cys codons, respectively.

The universal TA/CG deficiency (or CT/TG/CA excess) rule is virtually ubiquitous, even though the base compositions of coding sequences may be very different. This is readily apparent in the codon usage found in the coding sequences for the influenza A virus hemagglutinins (see Figure 3) which are very different from those found for c-*src* (see Figure 2). The only constant in these two genomes is the preponderant use of CTG among the six Leu codons which is dictated by the universal rule. The rule also dictates that codons ending in TA, as well as in CG, should fail to gain in prominence as is observed in Figure 2. In all other codons, the effect of the extreme A-richness (32.3%) is very evident. Accordingly, codons having A as the third base, such as GGA, GAA, AAA, GCA, CAA, CCA, became preponderant codons for Gly, Glu, Lys, Ala, Gln, and Pro.

Inasmuch as the genomes of complex organisms such as vertebrates contain coding sequences of very diverse base compositions, there can be no universally preferred codon for any of the amino acid residues with the single exception of Leu, where the universal rule dictates CTG as the preponderant codon. As already noted, the metazoan mitochondrial genome does not violate the universal rule. Thus, in this case, we find that the preponderant Leu codon can be CTA, provided the coding sequence is very A-rich and G-poor. Such an exception attests to the power of the universal rule in keeping CTG as the preponderant Leu codon, thereby giving the usual prominence to Leu.

VII. BASE TRIMERS ARE NOT THE UNIT OF CODING SEQUENCES

As shown in Figure 2, 32 of the 49 CTG base trimers in the c-*src* coding sequence are utilized as a Leu codon with 65% indeed being a very large fraction. In sharp contrast, only 9 of the 48 TGG base trimers encode Trp. Thus, the bottom rank of Trp in the c-*src* protein is due neither to the single codon assignment, nor to the scarcity of TGG base trimers, but rather to the poor utilization rate (19%) of TGG base trimers as its codon. In fact, the abundance of Leu encoded by CTG and the scarcity of Trp encodable only by TGG are two sides of the same coin. Forty-two percent of the 49 CTG base trimers, which is the same percentage of the 48 TGG base trimers occur as 20 CTGG base tetramers. Of those, 14 are translated in the first reading frame to yield Leu, while the second reading frame is utilized only four times to encode Trp. The remaining 2 CTGG are translated in the third reading frame to produce a pair of Ala-Gly dipeptides. The similar preferential translation in the first reading frame of 15 CTGC tetramers is the major cause of the poor ranking of Cys with Trp, since only 2 of the 15 CTGC tetramers are translated in the second reading frame to yield Cys. The preferential translation of the above two base tetramers in the first reading frame, which results in an enrichment of Leu at the expense of Trp and Cys, is an almost universal practice seen in most coding sequences. However, different preferences are shown by individual coding sequences in translation of other sets of base tetramers. One set of examples follows.

As shown in Figure 2, 35 of the 42 GAG trimers (83%) in the chicken c-*src* coding sequence are utilized as Glu codons. This preferential utilization of GAG as a Glu codon is, in fact, the major cause in the rise of Glu to second position in the order of abundance. Victims of this preference are two potential Arg codons, AGA and AGG. It should be noted that only 4 of the 30 AGG (13%) and 1 of the 30 AGA trimers (3%) serve as Arg codons. This poor utilization of the these two potential Arg codons keeps Arg ranked 7th, despite the extreme CG-richness of this coding sequence which mitigates the effect of CG deficiency on the other 4 codons for Arg. Forty-two GAG and 30 AGG trimers were united 12 times to form the base tetramer, GAGG, while the remaining 30 GAG and 30 AGA trimers united 14 times to form another base tetramer GAGA. Both of these tetramers are preferentially translated in the first reading frame to yield 16 Glu residues, while translation in the second reading frame yields only 4 Arg residues. Figure 2 also shows that Arg is a victim of Lys as well, since 10 AAGG and 9 AAGA base tetramers are again preferentially translated in the first reading frame to yield 15 Lys residues, but not a single Arg residue.

The situation is quite different in the coding sequence for the influenza A virus hemagglutinins shown in Figure 3. Here, the 29 Arg codons are nearly equivalent to 30 Glu codons, and AGG and AGA are the major Arg coding elements. Although not shown in Figure 3, there are 33 GAA and 26 GAG trimers in the influenza virus coding sequence, and 16 and 14 of them, respectively, are utilized as Glu codons. Thus, the GAG trimer, which is utilized 54% of the time as a Glu codon in the influenza virus coding sequence, is considerably less than that of the 83% utilization rate of this trimer in the c-*src* coding sequence shown in Figure 2. In sharp contrast, 12 of the 21 AGG (57%) and 10 of the 29 AGA (35%) base trimers in the influenza virus hemaglutinins function as Arg codons. The 26 GAG and 21 AGG trimers are united 8 times to form GAGG base tetramers, while the remaining 18 GAG and 29 AGA trimers are united 10 times to form GAGA tetramers (Figure 4). It should be noted that in the translation of these two tetramers, equal opportunities occur in the first and second reading frames, and thus, GAGG and GAGA together yield 8 Glu and 8 Arg codons. Also, Arg is no longer a victim of Lys, since the translation of 4 AAGG base tetramers yields 4 Arg and no Lys codons.

VIII. SUMMARY

Aberrant codon usages typically seen in unicellular eukaryotes with AT-rich genomes, as well as in mitochondria with extremely small genomes, appear to indicate that there has never been *a priori* coupling between two parts of tRNA molecules, i.e., the upper domain, recognized by an amino acid acylating enzyme and a lower domain containing an anticodon. However, this is not universally true as is observed in some of the tRNAs for the 20 amino acids in which the anticodon has an essential role in determining the attachment of the correct amino acid (e.g., see tRNA^{Met} in Chapter 11). In any case, the universal codon assignment must be viewed as a fortunate frozen accident. Were it not for the universal CT/TG/CA excess-CG/TA deficiency grammatical rule, the very uneven codon assignments for the 20 amino acid residues should have yielded basic proteins rich in Leu, Arg and Ser with each being assigned 6 codons. Thanks to the CG-deficiency part of the universal rule, Arg is not a prominent amino acid in most proteins. The isoelectric point of the average protein is slightly below 7.0. Because of the CT/TG excess part of the universal rule, CTG is usually the most numerous base trimer in coding base sequences. The prominence of Leu residues in most proteins is due not so much to its assignment of six codons, but rather to this abundance of CTG.

The codon assignments and the universal rule are invariant. Hence, amino acid compositions of proteins and codon usages for individual residues are primarily determined by the base compositions of coding sequences, since base compositions determine the relative abundances of the 64 base trimers. The shortest repeating unit in coding sequences, however, are not base trimers, but would appear to be tetramers. Thus, the fraction of base trimers to be used as codons is determined by a reading frame choice in translation of a set of repeating base tetramers in which a given trimer resides.

REFERENCES

1. **von Neumann, J.**, *Theory of Self-Reproducing Automata*, Burks, A. W., Ed., University of Illinois, Press, Urbana, IL, 1966.
2. **Gamow, G.**, Possible relation between deoxyribonucleic acid and protein structures, *Nature (London)*, 173, 318, 1954.
3. **Bridson, P.K. and Orgel, L.E.**, Catalysis of accurate poly(C) directed synthesis of 3'-5'-linked oligoguanytes by Zn^{+2}, *J. Mol. Biol.*, 144, 567, 1980.
4. **Seto, Y.**, Formation of proteins on the primitive earth. Evidence for the oligoglycine hypothesis, *Viva Origino*, 17, 153, 1989.

5. **Robson, K. J. H., Hall, J. R. S., Jennings, M. W., Harris, T. J. R., Marsh., K., Newbold, C. I., Tate, V. E., and Weatheral, D. J.**, A highly conserved amino-acid sequence in thrombospondin, properidin and in proteins from sporozoites and blood stages of a human malaria parasite, *Nature (London)*, 335, 79, 1988.
6. **Hanyu, N., Kuchino, Y., Nishimura, S., and Beier, H.**, Dramatic events in ciliate evolution: alteration of UAA and UAG termination codons to glutamine codons due to anticodon mutations in two Tetrahymena tRNAs^{Gln}, *EMBO J.*, 5, 1307, 1986.
7. **Kuchino, Y., Hanyu, N., Tashiro, F., and Nishimura, S.**, Tetrahymena thermophilia glutamine tRNA and its gene that corresponds to U A A termination codon, *Proc. Natl. Acad. Sci. U.S.A.*, 82, 4758, 1985.
8. **Himeno, H., Masaki, H., Kawai, T., Ohta, T., Kumagai, I. Miura, K., and Watanabe, K.**, Unusual genetic codes and a novel gene structure for tRNA$^{Ser}_{AGY}$ in starfish mitochondrial DNA, *Gene*, 56, 219, 1987.
9. **Hoyle, F.**, *Ten Faces of the Universe*, Freeman Press, London, 1979.
10. **Ohno, S.**, Universal rule for coding sequence construction: TA/CG- deficiency-TG/CT-excess, *Proc. Natl. Acad. Sci. U.S.A.*, 85, 9630, 1988.
11. **Yomo, T. and Ohno, S.**, Concordant evolution of coding and noncoding regions of DNA made possible by the universal rule of TA/CG deficiency-TG/CT-excess, *Proc. Natl. Acad. Sci. U.S.A.*, 86, 8452, 1989.
12. **Takeya, T. and Hanafusa, H.**, Structure and sequence of the cellular gene homologous to the RSV *src* gene and the mechanism for generating the transforming virus, *Cell*, 32, 881, 1983.
13. **Jukes, T. H.**, *Molecules and Evolution*, Columbia University Press, New York, 1966.
14. **Vasquez, M., Pincus, M. R., and Sheraga, H. A.**, Helix-coil transition theory including long-range electrostatic interactions: application to globular proteins, *Biopolymers*, 26, 351, 1987.
15. **Bird, A. P.**, CpG islands as gene markers in the vertebrate nucleus, *Trends Genet.*, 3, 342, 1987.
16. **Verhoeyen, M., Fang, R., Jou, W. M., Devos, R., Huylebroeck D., Saman, E., and Fiers, W.**, Antigenic drift between the haemagglutinin of the Hong Kong influenza strains A/Aichi/2/68 and A/Victoria/3/75, *Nature (London)*, 286, 771, 1980.
17. **Ohno, S.**, Codon preference is but an illusion created by the construction principle of coding sequences, *Proc. Natl. Acad. Sci. U.S.A.*, 85, 4378, 1988.
18. **Grantham R., Gautier, C., Gouy, M., Jacob, M., and Mercier, R.**, Codon catalog usage is a genome strategy modulated for gene expressivity, *Nucleic Acids Res.*, 9, 543, 1981.
19. **Ohno, S.**, Many peptide fragments of alien antigens are homologous with host proteins, thus channelizing T cell responses, *Proc. Natl. Acad. Sci. U.S.A.*, 88, 3065, 1991.
20. **Ohno, S.**, To be or not to be a responder in T cell responses: ubiquitous oligopeptides in all proteins, *Immuno-gen.*, in press, 1991.

Chapter 14

SELECTIVE USE OF TERMINATION CODONS AND VARIATIONS IN CODON CHOICE

Paul M. Sharp, Conal J. Burgess,
Andrew T. Lloyd, and Kevin J. Mitchell

TABLE OF CONTENTS

I. Introduction 398

II. *Escherichia coli* 398
    A. Sense Codon Usage 398
    B. Stop Codon Usage 399
    C. Evolution at Stop Codons 406

III. Gram-Positive Bacteria 407

IV. Eukaryotes 409
    A. *Saccharomyces cerevisiae* 409
    B. Human 409
    C. Other Eukaryotes 419
        1. Mouse 419
        2. *Drosophila melanogaster* 419
        3. Maize 421
        4. Fungi 421

V. Conclusions 422

References 422

I. INTRODUCTION

With the determination of the first gene sequences, it was noticed that alternative synonymous codons are not used in equal frequencies. Grantham and colleagues pioneered the compilation of codon usage data for many genes,[1,2] allowing the description of codon usage frequency patterns, and the use of multivariate statistics to elucidate variations in these patterns. As the gene sequence database has expanded, it has become clear that synonymous codon usage varies both among genes from one genome, and between species.[1-4] More recently, a firm theoretical basis for these observed patterns has been established, in terms of the effects of natural selection, biased mutation, and genetic drift.[5-7] Most of this work has focused on sense codons, and rather little attention has been paid to termination (or "stop") codons.

The primary reason why stop codon frequencies have not been investigated is that there is much less data — there is typically only one termination codon per gene. However, cursory examination of large compilations of data from diverse species of prokaryotes and eukaryotes indicates that the three stop codons are used to differing extents (Table 1). The relative frequencies of the three stop codons vary considerably between species. In many species, the most commonly used termination codon is UAA, but in several others it is UGA. UAG is rare in some genomes, but common in others. [Note that, for simplicity, we will discuss throughout codons containing U (rather than T), when referring to DNA or mRNA.]

Unequal frequencies of stop codons could simply result from mutational biases. To take an extreme example, the genomic G+C content of the slime mold *Dictyostelium discoideum* is around 22%, which presumably reflects a strong mutational bias towards A+T-richness, and 96% of genes examined in this species terminate in UAA (Table 1). In contrast, in the G+C-rich prokaryote *Micrococcus luteus* (genomic G+C content around 74%) only 10% of the genes so far investigated end with UAA. That UGA is used much more often than UAG in *M. luteus* (and in many other species, see Table 1) may reflect the fact that mutational biases can be influenced by neighboring bases.[14] However, we shall see that stop codon frequencies vary among different groups of genes from the same genome, and that at least in certain species, the unequal usage of stop codons appears to have a selective basis.[15]

To understand stop codon frequencies, it is necessary to examine them in the context of sense codon usage. As sense codon usage varies between species, we will take a "taxonomic" approach; necessarily, we will concentrate on those organisms in which sense codon usage is best understood.

II. *ESCHERICHIA COLI*

A. SENSE CODON USAGE

Escherichia coli was the first species in which codon usage was investigated in detail, and remains the species in which codon usage patterns are best understood. Codon usage varies among genes in a manner related to expression

TABLE 1
Stop Codon Usage in Different Species

Species	UAA N[a] (RSCU)[b]	UGA N (RSCU)	UAG N (RSCU)	Ref.
Escherichia coli	687 (1.99)	285 (0.82)	66 (0.19)	
Bacillus subtilis	158 (1.94)	55 (0.68)	31 (0.38)	8
Mycoplasma capricolum	16 (2.00)	—	8 (1.00)	
Micrococcus luteus	2 (0.32)	17 (2.68)	2 (0.32)	9
Saccharomyces cerevisiae	288 (1.56)	155 (0.84)	111 (0.60)	10
Aspergillus nidulans	19 (1.27)	7 (0.46)	19 (1.27)	11
Schizosaccharomyces pombe	57 (2.22)	5 (0.19)	15 (0.58)	
Dictyostelium discoideium	49 (2.88)	1 (0.06)	1 (0.06)	12
Zea mays	13 (0.50)	33 (1.27)	32 (1.23)	
Caenorhabditis elegans	34 (2.08)	3 (0.18)	12 (0.73)	
Drosophila melanogaster	163 (1.58)	68 (0.66)	79 (0.76)	
Homo sapiens	322 (0.90)	513 (1.44)	237 (0.66)	
Mus musculus	227 (0.89)	396 (1.54)	146 (0.57)	

Note: References (in most-righthand column) are to recent codon usage compilations. In other cases, species-specific collections of gene sequences were taken from the GenBank/EMBL DNA sequence data library using the ACNUC retrieval system,[13] and screened for duplicate entries (which were excluded).

[a] Number of genes ending with this stop codon.
[b] Relative synonymous codon usage. RSCU is calculated as N/E, where N is the observed number of codons, and E is the number expected if all synonymous codons are used equally. For example, in *E. coli,* if the 3 stop codons were used with equal frequency, the expected number of each would be E = 346, and so the RSCU for UAA = 687/346 = 1.99.

level.[16,17] Highly expressed genes exhibit a strong "preference" for a subset of codons (one, or sometimes two, per amino acid),[18] which are those translated most efficiently and/or accurately. When more than one isoacceptor is used for an amino acid, the codons recognized by the more abundant tRNA are more frequently used.[16] When a tRNA recognizes more than one codon, the codon best complementing the anticodon is most frequently used. Ideas about the actual mechanism of selection, i.e., how differences in tRNA abundances and codon-anticodon interactions can lead to fitness differences, have recently been reviewed.[19] Genes expressed in low amounts (or lowly expressed genes) in *E.coli* have a more uniform pattern of codon usage, apparently influenced primarily by mutational biases.[20]

The strength of codon usage bias in an *E.coli* gene can be estimated by using the Codon Adaptation Index, or CAI.[21] "Relative adaptedness" (w) values have been assigned to each of the sense codons (excluding those encoding Trp and

Met), according to their frequency of use in a group of very highly expressed *E.coli* genes. w is obtained by dividing the codon usage value for any codon by the value for the most abundant, and presumably optimal, codon for that amino acid. For example, in these genes the two Phe codons UUU and UUC occur 47 and 159 times, respectively, and are therefore assigned w values of 0.296 (= 47/159) and 1.0 (see reference 21 for the full table of w values). The CAI value for a gene is calculated as the average (using the *geometric* mean) of the w values corresponding to each of the codons in that gene. For example, the *gutB* gene, which is 124 codons long, has the sequence:

$$(AUG) \cdot ACC \cdot GUU \cdot AUU \cdot UAU \cdot CAG \cdot ACC \cdot AUC.......$$

The AUG codon is in brackets because Met and Trp codons are excluded from the calculation. The CAI for *gutB* is given by:

$$(w_{ACC} \times w_{GUU} \times w_{AUU} \times w_{UAU} \times w_{CAG} \times w_{ACC} \times w_{AUC} ......)^{1/L}$$

where

L is total number of sense codons minus Met and Trp.

The calculation is:

$$(1.0 \times 1.0 \times 0.466 \times 0.239 \times 1.0 \times 1.0 \times 1.0 ......)^{1/121} = 0.34$$

The CAI has a maximum possible value of 1.0, if only optimal codons are used in a gene (i.e., all of the w values are 1.0). In an *E. coli* gene with random codon usage, a CAI value of 0.17 would be expected. Among the 1038 genes collated in Table 1, the values range from 0.08 to 0.85.

In Table 2, codon usage values are presented for the top and bottom 10% of these *E. coli* genes, when ranked according to CAI. This form of presentation reveals the extremes of the heterogeneity of codon usage among genes. The top 10% are very highly expressed genes, with codon usage values not dissimilar to those used[21] in assigning w values. The optimal *E. coli* codons are easily distinguished as those codons used more often than expected in this highly expressed group of genes. There are 23 codons with relative synonymous codon usage (RSCU) values significantly greater than 1.0. In fact, the low group of genes probably give the best indication of the expected codon usage due to mutational biases, since it is expected that translational selective constraints on those genes are very weak.[20]

B. STOP CODON USAGE

In the large sample of *E. coli* genes considered here, approximately $2/3$ end with the ochre codon, UAA, while only 1 in 16 end with the amber codon UAG (Table 1). However, as with synonymous sense codons, the usage of alternative

TABLE 2
Codon Usage in Prokaryotes

		Escherichia coli				Bacillus subtilis				Mycoplasma capricolum		Micrococcus luteus	
		High		Low		High		Low					
		N[a]	RSCU[b]	N	RSCU	N	RSCU	N	RSCU	N	RSCU	N	RSCU
Phe	UUU	230	0.43	604	1.41	35	0.59	253	1.41	117	1.72	1	0.01
	UUC	828	1.57	251	0.59	83	1.41	106	0.59	19	0.28	194	1.99
Leu	UUA	44	0.10	573	1.63	84	1.53	119	0.94	254	4.92	0	0.00
	UUG	70	0.16	322	0.92	29	0.53	137	1.08	8	0.15	5	0.06
Leu	CUU	107	0.25	384	1.09	161	2.94	158	1.25	17	0.33	2	0.02
	CUC	166	0.38	181	0.52	6	0.11	88	0.70	0	0.00	214	2.37
	CUA	10	0.02	174	0.50	31	0.57	30	0.24	30	0.58	0	0.00
	CUG	2201	5.08	472	1.34	18	0.33	226	1.79	1	0.02	320	3.55
Ile	AUU	459	0.69	688	1.41	101	1.15	272	1.50	273	2.36	1	0.01
	AUC	1528	2.30	313	0.64	161	1.84	188	1.04	38	0.33	323	2.99
	AUA	3	0.00	463	0.95	1	0.01	83	0.46	36	0.31	0	0.00
Met	AUG	919	—	464	—	102	—	214	—	93	—	127	—
Val	GUU	1269	1.85	436	1.43	177	1.91	89	0.77	196	2.50	2	0.01
	GUC	248	0.36	230	0.75	24	0.26	144	1.25	3	0.04	260	1.81
	GUA	654	0.96	278	0.91	135	1.46	88	0.76	101	1.29	0	0.00
	GUG	568	0.83	279	0.91	34	0.37	140	1.21	14	0.18	313	2.18

TABLE 2 (continued)
Codon Usage in Prokaryotes

| | | Escherichia coli | | | | Bacillus subtilis | | | | Mycoplasma capricolum | | Micrococcus luteus | |
| | | High | | Low | | High | | Low | | | | | |
		N[a]	RSCU[b]	N	RSCU	N	RSCU	N	RSCU	N	RSCU	N	RSCU
Ser	UCU	593	2.32	303	1.17	130	3.16	79	1.03	56	1.45	5	0.10
	UCC	476	1.86	159	0.62	3	0.07	91	1.19	0	0.00	183	3.61
	UCA	43	0.17	385	1.49	62	1.51	88	1.15	119	3.08	2	0.04
	UCG	59	0.23	158	0.61	0	0.00	61	0.80	2	0.05	83	1.64
Pro	CCU	114	0.38	210	1.10	69	2.00	62	0.98	46	1.38	10	0.14
	CCC	9	0.03	165	0.86	0	0.00	33	0.52	0	0.00	105	1.43
	CCA	161	0.53	253	1.32	49	1.42	40	0.63	84	2.53	1	0.01
	CCG	922	3.06	139	0.72	20	0.58	118	1.87	3	0.09	178	2.42
Thr	ACU	619	1.37	333	1.11	113	1.81	32	0.36	144	2.42	5	0.05
	ACC	1004	2.22	262	0.87	3	0.05	67	0.76	4	0.07	265	2.53
	ACA	52	0.12	410	1.36	115	1.84	139	1.58	90	1.51	2	0.02
	ACG	133	0.29	198	0.66	19	0.30	114	1.30	0	0.00	147	1.40
Ala	GCU	1157	1.39	351	1.05	210	2.10	111	0.80	153	2.32	10	0.07
	GCC	380	0.46	278	0.83	11	0.11	127	0.91	2	0.03	347	2.47
	GCA	848	1.02	442	1.32	116	1.16	169	1.22	106	1.61	16	0.11
	GCG	943	1.13	272	0.81	63	0.63	149	1.07	3	0.05	188	1.34

AA	Codon	N	F	N	F	N	F	N	F	N	F	N	F
Tyr	UAU	246	0.56	556	1.50	17	0.50	170	1.47	73	1.57	2	0.03
Tyr	UAC	633	1.44	183	0.50	51	1.50	61	0.53	20	0.43	153	1.97
ter	UAA	88	2.54	53	1.53	24	2.88	14	1.68	16	2.00	2	0.29
ter	UAG	2	0.06	14	0.40	1	0.12	4	0.48	8	1.00	2	0.29
His	CAU	126	0.43	334	1.35	29	0.97	106	1.49	34	1.39	7	0.12
His	CAC	460	1.57	159	0.65	31	1.03	36	0.51	15	0.61	113	1.88
Gln	CAA	191	0.32	351	0.91	160	1.74	109	0.92	151	1.97	0	0.00
Gln	CAG	1021	1.68	420	1.09	24	0.26	128	1.03	2	0.03	236	2.00
Asn	AAU	141	0.22	809	1.41	49	0.47	147	1.23	170	1.57	9	0.09
Asn	AAC	1139	1.78	337	0.59	159	1.53	92	0.77	47	0.43	182	1.91
Lys	AAA	1811	1.61	808	1.42	288	1.77	285	1.33	448	1.85	2	0.01
Lys	AAG	435	0.39	329	0.58	38	0.23	152	0.70	36	0.15	303	1.99
Asp	GAU	763	0.77	667	1.47	124	1.10	171	1.26	121	1.79	18	0.11
Asp	GAC	1221	1.23	242	0.53	102	0.90	100	0.74	14	0.21	310	1.89
Glu	GAA	1993	1.58	667	1.29	258	1.58	249	1.21	225	1.89	2	0.01
Glu	GAG	535	0.42	366	0.71	69	0.42	162	0.79	13	0.11	450	1.99
Cys	UGU	92	0.79	180	1.20	9	0.90	22	0.98	11	1.38	1	0.08
Cys	UGC	140	1.21	119	0.80	11	1.10	23	1.02	5	0.63	23	1.92
ter	UGA	14	0.40	37	1.07	0	0.00	7	0.84	25	1.85[c]	17	2.43
Trp	UGG	244	—	296	—	10	—	73	—	2	0.15	26	—
Arg	CGU	1277	4.11	293	1.42	122	3.44	43	0.97	51	1.49	58	0.70
Arg	CGC	563	1.81	220	1.06	69	1.94	44	0.99	3	0.09	324	3.93
Arg	CGA	8	0.03	164	0.79	2	0.06	32	0.72	3	0.09	4	0.05
Arg	CGG	7	0.02	158	0.76	0	0.00	49	1.11	0	0.00	100	1.21
Ser	AGU	34	0.13	300	1.16	16	0.39	47	0.61	44	1.14	2	0.04
Ser	AGC	328	1.28	244	0.95	36	0.87	93	1.22	11	0.28	29	0.57

TABLE 2 (continued)
Codon Usage in Prokaryotes

| | | Escherichia coli | | | | Bacillus subtilis | | | | Mycoplasma capricolum | | Micrococcus luteus | |
| | | High | | Low | | High | | Low | | | | | |
		N[a]	RSCU[b]	N	RSCU	N	RSCU	N	RSCU	N	RSCU	N	RSCU
Arg	AGA	8	0.03	260	1.26	20	0.56	67	1.51	147	4.30	0	0.00
	AGG	1	0.00	145	0.70	0	0.00	31	0.70	1	0.03	9	0.11
Gly	GGU	1543	2.19	366	1.16	164	1.74	77	0.66	139	1.79	54	0.42
	GGC	1195	1.70	258	0.82	101	1.07	144	1.23	0	0.00	395	3.04
	GGA	25	0.04	364	1.16	108	1.14	139	1.19	159	2.05	11	0.08
	GGG	53	0.08	269	0.86	5	0.05	108	0.92	13	0.17	59	0.45

Note: In *E. coli* and *B. subtilis* codon usage varies among genes according to expression level, and values for the top and bottom 10% of genes are presented: High (and Low) indicate the genes with highest (and lowest) codon usage bias (and expression level).

[a] Number of codons.
[b] Relative synonymous codon usage. RSCU is calculated as N/E, where N is the observed number of codons, and E is the number expected if all synonymous codons are used equally. For example, in *M. capricolum*, if the 4 Ala codons were used with equal frequency, the expected number of each would be E = 246/4 = 66, and so the RSCU for GCA is 106/66 = 1.61.
[c] UGA is translated as Trp in *M. capricolum*.

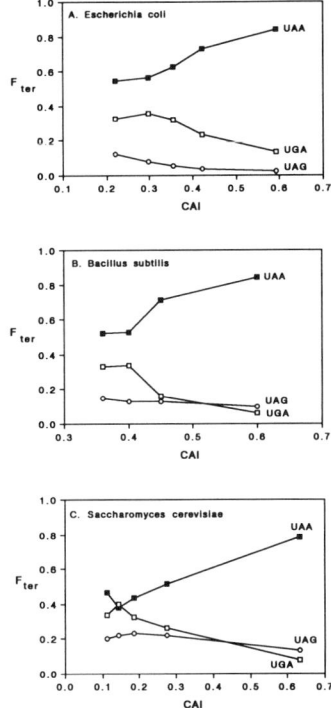

FIGURE 1. Frequency of use of termination codons in genes of different expression levels. (A) 1038 *Escherichia coli* genes, divided into 5 classes on the basis of codon usage bias measured by CAI, the codon adaptation index (see text). (B) 244 *Bacillus subtilis* genes, divided into 5 classes on the basis of codon usage bias measured by CAI. (C) 553 *Saccharomyces cerevisiae* genes, divided into 5 classes on the basis of codon usage bias, measured by CAI. In each panel the frequency of each stop codon (F_{ter}) in each group of genes is plotted against the mean CAI for that group.

stop codons varies with gene expression level.[15,22] This can be seen by grouping genes according to their sense codon bias. Stop codon frequencies are clearly different in the two extreme groups of genes in Table 2. When the 1038 genes are divided into five groups based on their CAI values, a clear pattern is seen: the stop codon frequencies change gradually with the strength of sense codon bias (Figure 1A).

The basis of stop codon preference is not yet clear. The three stop codons are recognized by two different (though probably ancestrally related) proteins, release factors RF1 and RF2.[23] RF1 recognizes UAA and UAG, while RF2 recognizes UAA and UGA. Thus, whatever the relative abundance of RF1 and RF2, UAA will have the most cognate release factor available. The excess use of UAA, which is most exaggerated in highly expressed genes, then seems to parallel the correlation of codon usage with tRNA abundance. The actual mechanism of stop codon recognition by the release factors is unknown.

It has also been noted that the context of termination codons in *E. coli* is nonrandom.[22] Highly significant departures from random nucleotide frequencies were found in the region of five bases preceding the stop codon, and at the site immediately following the stop codon. This is perhaps not unexpected, since the activity of both RF1 and RF2 varies according to the surrounding nucleotides.[24] The bias at sites 5' to the termination codon could be at least partly due to amino acid composition at the carboxyl terminus of proteins. However, it was also observed that the nature of the bias differed depending on the stop codon, suggesting that the cause is at the DNA or mRNA level rather than at the protein level. The bias did not vary with sense codon bias.

The strongest context effect relates to the nucleotide after the stop codon,[22] where there is an excess of U. Most interestingly, this bias was most pronounced in genes with high sense codon usage bias (again assessed by the CAI), where there is an extreme deficit of A and C. This led Brown et al.[22] to speculate that the signal for efficient termination of protein synthesis may be a tetranucleotide, rather than a trinucleotide codon.

C. EVOLUTION AT STOP CODONS

Sense codon usage in closely related species is usually very similar, and this is true for *E. coli* and *Salmonella typhimurium*. The similarity does not merely reflect recent common ancestry, since silent sites in genes in these species have accumulated an average of approximately one substitution per site during their divergence.[25] Rather, tRNA populations in the two species are similar,[26] and so the selection pressures on codon usage are likely to be the same. The *S. typhimurium* genome is slightly more G+C-rich than *E. coli*, and silent sites in some genes reflect this,[27] but the overall patterns of codon usage have not diverged significantly.

Termination codons should be free to accept certain mutations without impairment of their translational meaning ("stop"); these are synonymous substitutions in the same sense as mutations between alternative codons for the same amino acid. However, given that the likelihood of simultaneous mutations at both the second and third codon positions is small, synonymous mutations in stop codons are limited to two particular classes of transitions:

$$UGA \longleftrightarrow UAA \longleftrightarrow UAG$$

Among 96 pairs of homologous genes which can be compared between *E. coli* and *S. typhimurium*, 78 have the same stop codon (Table 3). The overall termination codon frequencies in this sample are not significantly different from those for *E. coli* in Table 1. Of 14 genes with high codon usage bias (here taken as CAI >0.5), all end with UAA in *E. coli*, and only 1 is different (ending in UGA) in *S. typhimurium*. Strong codon usage bias constrains the rate of nucleotide substitution at silent sites in *E. coli* and *S. typhimurium* genes,[25,26,28] and clearly there is a constraint on the stop codon also. Among the genes with

TABLE 3
Stop Codon Divergence Between *Escherichia coli* and *S. typhimurium*

		S. typhimurium		
		UGA	UAA	UAG
	UGA	18 (0)	8 (0)	0 (0)
E. coli	UAA	9 (1)	59 (13)	0 (0)
	UAG	0 (0)	1 (0)	1 (0)

Note: Values are the number of homologous gene pairs with the combination of stop codons in the margins. For example, in 9 gene pairs, the stop codon is UAA in *E. coli*, but UGA in *S. typhimurium*. Values in brackets refer to genes with high codon usage bias (CAI > 0.50).

lower codon usage bias, and excluding the small number of UAG codons, there are 16 UAA <-> UGA transitions among 80 gene pairs. After statistical correction for possible superimposed multiple mutations,[29] the rate of transitions becomes 0.27 transitions per stop codon second position. If we count each second position in a UAA or UGA codon as 1/3 of a silent site (because only 1 in 3 mutations are silent),[30] then the rate of divergence at this site is 0.80 substitutions per site. This rate is not dissimilar to that at silent sites in sense codons in the same genes compared between these two species, and suggests that stop codons in lowly expressed genes are not constrained to any greater extent than sense codons.

III. GRAM-POSITIVE BACTERIA

After *E. coli*, the gram-positive bacterium *Bacillus subtilis* is genetically the best understood prokaryote. *B. subtilis* is quite distantly related to *E. coli*, being placed in a separate phylum.[31] Codon usage in *B. subtilis* resembles that in *E. coli* insofar as highly expressed genes prefer a subset of translationally "optimal" codons, and the strength of bias towards these codons is correlated with gene expression level.[8,32] However, there are several ways in which codon usage differs between *B. subtilis* and *E. coli*. Some of these features can be seen in Table 2. For certain amino acids (e.g., Leu, Pro, or Gln) the codons which are optimal in the two species are different; many of the optimal codons in *B. subtilis* end in A or U. Also, the overall degree of codon usage bias is lower in *B. subtilis* than in *E. coli*, as seen by the lower RSCU values for the optimal codons in the high group. Under the selection-mutation balance hypothesis of codon usage,[5,6] it is easy to suggest a basis for these similarities and differences. First, it would be anticipated that the identities of the optimal codons in each species echo differences in their tRNA populations, although comparatively little is known about tRNAs in *B. subtilis*. Second, the lower bias in *B. subtilis* genes might suggest that selective differences between synonymous codons in this species are either smaller, or less effective, possibly because of its life history and effective population size. *B. subtilis* has a lower genomic G+C content than *E. coli,* presumably as a result of a mutational bias towards A+T-richness. Under such a bias, it would be predicted that A+T-rich codons should become optimal.[7]

Stop codon usage in *B. subtilis* closely parallels that in *E. coli*.[15,22] Overall, the relative frequencies of the three codons are quite similar in the two species (Table 1), and the frequencies vary with sense codon bias (and thus gene expression level) in a very similar manner (contrast Figure 1B and 1A). That is, highly expressed genes almost invariably terminate with UAA, while in genes expressed at lower levels the frequencies of UAG and particularly UGA are increased. Furthermore, there is an excess of U bases at the site 3′ to the stop codon, which is more pronounced in highly expressed genes.[22] Thus, despite the great evolutionary divergence of *E. coli* and *B. subtilis,* signals for the termination of protein synthesis in the two species appear to be very similar. This is in contrast to features at the 5′ end of genes.[8]

Codon usage in two other gram-positive organisms, *M. capricolum*[33] and *M. luteus,*[9] has received attention because of the extreme base compositions of these genomes. Genomic base composition should reflect the underlying mutational bias within a genetic system,[34] and the effect on codon usage can be dramatic. The *M. capricolum* genome is very A+T-rich (G+C content around 25%), and the vast majority of codons end in A or U (Table 2). Theoretical considerations suggest that strong mutational bias would eventually lead to a reassignment of optimal codons.[7] In *M. capricolum,* this base compositional bias has also led to a deviation from the standard genetic code: Trp, which must normally be encoded by UGG, is also encoded by the erstwhile stop codon UGA.[35] Of the two remaining stop codons, UAA is used twice as often as UAG. This bias is not as extreme as might be expected from consideration of other codon groups since, for example, AAA is used 12 times more often than AAG to encode Lys, and the bias among Glu and Gln codons is even stronger (Table 2). This is even more surprising because the majority of the genes examined are highly expressed. Thus, while the number of *M. capricolum* genes considered is as yet small, there is a hint here that the termination signal may differ between *M. capricolum* and *B. subtilis,* even though they are comparatively closely related.[31]

By way of contrast, *M. luteus* is extremely G+C-rich (genomic G+C around 74%), and this is evident in both sense and stop codon usage (Table 2). Again, the genes which have been sequenced in this species largely belong to the highly expressed category.[9] However, the codon which usually terminates highly expressed genes in the other prokaryotes considered here, UAA, is rarely used in *M. luteus*. Either the mutational bias towards G+C-richness has overwhelmed any selection for UAA, or in some way analogous to sense codon usage,[7] the mutation pattern has brought about a change in the termination signal, for example, by reshaping the recognition properties of release factors.

Examination of overall stop codon usage in a range of prokaryotes with different base compositions indicates that there is a strong general correlation between the frequency of UAA usage and genomic A+T-richness.[22] There is also a tendency for the ratio of UGA:UAG to increase as G+C content increases. In the case of the most A+T-rich species, *M. capricolum,* this ratio is, in a sense, zero, since UGA is not used as a stop codon.

IV. EUKARYOTES

A. SACCHAROMYCES CEREVISIAE

Among eukaryotes, the budding yeast *S. cerevisiae* has the best understood codon usage. From some of the first yeast gene sequences, it was recognized that highly expressed genes utilize a very small subset of codons.[36] These codons are different from those used in highly expressed *E. coli* genes, but are those best recognized by the most abundant yeast tRNAs.[37] Subsequent investigations of much larger data sets have confirmed these observations, and revealed that lowly expressed genes have less biased codon usage.[10,38] The pattern of codon usage in highly expressed genes in *S. cerevisiae* (Table 4) is, if anything, even more biased than in *E. coli.* The pattern of codon usage in lowly expressed genes is consistent with (nearest-neighbor-dependent) mutational biases being the most important influence.[39] Thus, yeast rivals *E. coli* as a clear exemplar for the selection-mutation balance theory of codon usage.[5,40]

The pattern of termination codon usage in *S. cerevisiae* is remarkably similar to that in *E. coli* and *B. subtilis* (Figure 1C). In highly expressed genes, there is a strong preference for UAA. In lowly expressed genes, the frequencies of three stop codons are more nearly uniform, but the rank order of usage is UAA > UGA > UAG.

S. cerevisiae appears to differ from *E. coli* with respect to the nucleotide following the stop codon. In yeast genes as a whole, stop codons are most often followed by A (rather than U).[41] This is true irrespective of which stop codon is used. However, and perhaps more interestingly, in highly expressed genes the most common nucleotide after UAA is G; the frequency of G 3′ to UAA is highly correlated with CAI (Figure 2). Thus, the optimal signal for translation termination in *S. cerevisiae* appears to be UAAG.

B. HUMAN

Codon usage in the human genome differs quite markedly from the prokaryotes and unicellular eukaryotes previously discussed. Codon usage in some human genes is very biased, and codon usage varies enormously among human genes. However, no relationship has yet been established between the particular pattern of codon usage and the level of expression of the gene. Rather, the largest source of variation among genes is in base composition, or more precisely the G+C content at silent sites in codons.[3,4] G+C content at third positions of codons varies among individual genes from under 30 to over 90%;

TABLE 4
Codon Usage in Fungi

		Saccharomyces cerevisiae				Aspergillus nidulans				Schizosaccharomyces pombe			
		High		Low		High		Low		High		Low	
		N[a]	RSCU[b]	N	RSCU	N	RSCU	N	RSCU	N	RSCU	N	RSCU
Phe	UUU	96	0.32	527	1.23	2	0.05	40	0.91	22	0.44	168	1.44
	UUC	495	1.68	332	0.77	78	1.95	48	1.09	77	1.56	65	0.56
Leu	UUA	157	0.73	383	1.26	0	0.00	35	0.68	14	0.37	168	1.83
	UUG	1063	4.94	403	1.33	7	0.37	43	0.83	93	2.44	112	1.22
Leu	CUU	15	0.07	261	0.86	26	1.36	62	1.20	91	2.38	136	1.48
	CUC	0	0.00	178	0.59	62	3.23	62	1.20	30	0.79	43	0.47
	CUA	52	0.24	263	0.87	1	0.05	47	0.91	0	0.00	51	0.55
	CUG	5	0.02	330	1.09	19	0.99	62	1.20	1	0.03	42	0.46
Ile	AUU	511	1.53	439	1.08	28	0.79	40	1.00	116	1.97	165	1.63
	AUC	486	1.46	307	0.76	78	2.19	48	1.20	61	1.03	57	0.56
	AUA	4	0.01	470	1.16	1	0.03	32	0.80	0	0.00	82	0.81
Met	AUG	318	—	436	—	58	—	76	—	70	—	86	—
Val	GUU	709	2.18	291	1.14	52	1.52	36	1.14	121	2.14	104	1.79
	GUC	578	1.78	184	0.72	84	2.45	42	1.33	100	1.77	45	0.77
	GUA	3	0.01	287	1.12	0	0.00	21	0.67	3	0.05	51	0.88
	GUG	10	0.03	262	1.02	1	0.03	27	0.86	2	0.04	33	0.57

Ser	UCU	568	3.39	381	1.28	30	1.35	82	1.26	87	3.63	157	1.97
	UCC	369	2.20	231	0.78	82	3.70	69	1.06	41	1.71	45	0.57
	UCA	23	0.14	412	1.39	2	0.09	75	1.15	4	0.17	105	1.32
	UCG	4	0.02	246	0.83	6	0.27	65	0.99	1	0.04	45	0.57
Pro	CCU	54	0.34	228	1.07	26	1.49	77	1.12	70	2.31	73	1.60
	CCC	4	0.03	165	0.78	43	2.46	49	0.71	50	1.65	15	0.33
	CCA	582	3.64	293	1.38	1	0.06	81	1.18	1	0.03	67	1.47
	CCG	0	0.00	163	0.77	0	0.00	68	0.99	0	0.00	27	0.59
Thr	ACU	509	2.05	302	1.09	24	0.97	62	1.05	99	2.32	90	1.42
	ACC	467	1.88	214	0.77	69	2.79	67	1.14	69	1.61	41	0.65
	ACA	15	0.06	351	1.27	4	0.16	57	0.97	3	0.07	81	1.28
	ACG	2	0.01	242	0.87	2	0.08	50	0.85	0	0.00	41	0.65
Ala	GCU	1171	3.05	266	1.07	66	1.71	67	1.08	179	2.68	107	1.71
	GCC	349	0.91	203	0.82	80	2.08	73	1.17	76	1.14	39	0.62
	GCA	17	0.04	315	1.27	2	0.05	63	1.01	11	0.16	83	1.32
	GCG	1	0.00	208	0.84	6	0.16	46	0.74	1	0.01	22	0.35
Tyr	UAU	39	0.18	352	1.10	4	0.15	34	0.76	31	0.73	138	1.48
	UAC	389	1.82	289	0.90	51	1.85	55	1.24	54	1.27	48	0.52
ter	UAA	49	2.67	23	1.25	3	1.80	2	1.20	10	3.00	7	2.33
ter	UAG	2	0.11	10	0.55	1	0.60	2	1.20	0	0.00	1	0.33
His	CAU	68	0.44	242	1.22	2	0.13	64	1.04	25	0.91	87	1.47
	CAC	239	1.56	154	0.78	29	1.87	59	0.96	30	1.09	31	0.53
Gln	CAA	509	1.98	467	1.16	6	0.17	81	0.92	82	1.84	129	1.45
	CAG	5	0.02	336	0.84	65	1.83	96	1.08	7	0.16	49	0.55

TABLE 4
Codon Usage in Fungi

		Saccharomyces cerevisiae				Aspergillus nidulans				Schizosaccharomyces pombe			
		High		Low		High		Low		High		Low	
		N[a]	RSCU[b]	N	RSCU	N	RSCU	N	RSCU	N	RSCU	N	RSCU
Asn	AAU	58	0.18	704	1.20	6	0.16	61	0.82	24	0.51	180	1.33
	AAC	597	1.82	474	0.80	67	1.84	87	1.18	70	1.49	90	0.67
Lys	AAA	188	0.28	734	1.17	3	0.07	39	0.76	27	0.30	215	1.39
	AAG	1175	1.72	524	0.83	85	1.93	63	1.24	152	1.70	95	0.61
Asp	GAU	376	0.82	625	1.23	26	0.63	89	0.92	85	1.19	164	1.44
	GAC	546	1.18	391	0.77	57	1.37	104	1.08	58	0.81	63	0.56
Glu	GAA	1035	1.95	657	1.18	17	0.29	81	1.01	90	0.96	193	1.30
	GAG	25	0.05	461	0.82	99	1.71	79	0.99	98	1.04	103	0.70
Cys	UGU	118	1.87	134	1.07	4	0.40	15	0.70	11	0.61	62	1.32
	UGC	8	0.13	116	0.93	16	1.60	28	1.30	25	1.39	32	0.68
ter	UGA	4	0.22	22	1.20	1	0.60	1	0.60	0	0.00	1	0.33
Trp	UGG	138	—	251	—	21	—	50	—	31	—	49	—
Arg	CGU	97	0.72	126	0.77	42	2.90	40	1.06	120	5.26	63	1.57
	CGC	0	0.00	97	0.59	40	2.76	53	1.41	13	0.57	25	0.62
	CGA	1	0.01	127	0.78	2	0.14	45	1.19	0	0.00	42	1.05
	CGG	0	0.00	80	0.49	1	0.07	32	0.85	0	0.00	15	0.37
Ser	AGU	20	0.12	276	0.93	0	0.00	35	0.54	3	0.13	82	1.03
	AGC	21	0.13	236	0.79	13	0.59	66	1.01	8	0.33	43	0.54

		N[a]	RSCU[b]	N	RSCU	N	RSCU	N	RSCU	N	RSCU	N	RSCU
Arg	AGA	713	5.26	341	2.09	1	0.07	28	0.74	4	0.18	60	1.49
	AGG	2	0.01	208	1.27	1	0.07	28	0.74	0	0.00	36	0.90
Gly	GGU	1228	3.89	264	1.16	111	2.74	31	0.69	208	3.25	70	1.34
	GGC	29	0.09	227	0.99	43	1.06	54	1.20	31	0.48	33	0.63
	GGA	4	0.01	284	1.24	8	0.20	44	0.98	17	0.27	81	1.55
	GGG	3	0.01	138	0.60	0	0.00	51	1.13	0	0.00	25	0.48

Note: Codon usage varies among genes in each of these species and two extreme groups of genes are presented for each species to illustrate this variation (for the definition of groups see Table 2 and text).

[a] Number of codons.
[b] Relative synonymous codon usage (see Table 2).

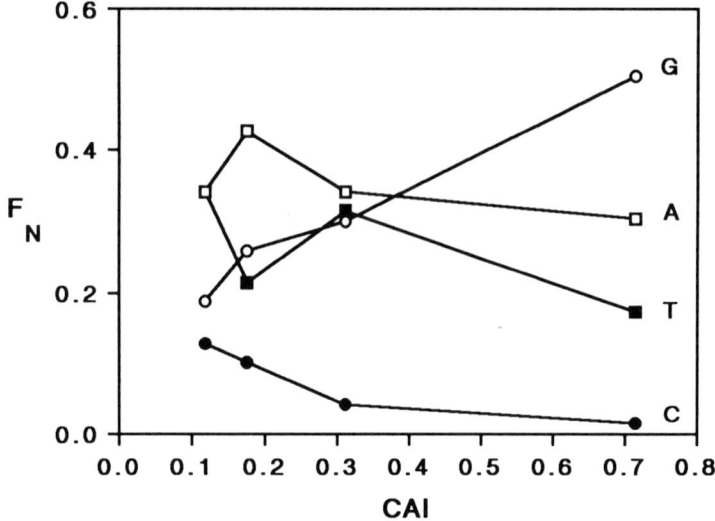

FIGURE 2. Frequency of nucleotides (F_N) at the site 3' to UAA in *S. cerevisiae* genes, divided into four classes on the basis of codon usage bias, measured by CAI.

the average values for the A+T-rich and G+C-rich groups in Table 5 are 34 and 85%, respectively. The base composition in a gene seems to reflect its chromosomal location, insofar as G+C content at silent sites within codons is highly correlated with G+C content in introns and flanking sequences of the same genes.[42,43]

Before an extensive examination of human gene sequence data had begun, Bernardi and colleagues[44] had concluded from density gradient centrifugation that the human genome comprises a mosaic of long regions of differing base compositions. These regions, termed "isochores", may be 200–1,000 kb in length, and within an isochore base composition is thought to be quite homogeneous. Subsequent analyses of sequence data seem to confirm this view of the human genome.[42,45,46]

Bernardi and colleagues speculated on the possible evolutionary advantages of the isochore structure, and argued that base composition across the entire genome is maintained at particular levels by natural selection (see, for example, Reference 47). However, no strong evidence for this exists, and counter arguments have been made by several groups. For example, it has been suggested that it is much simpler to envisage isochores as having an origin in mutational biases which vary around the genome.[48-50] A variety of origins for the mutational biases have been proposed. We have suggested that an isochore may correspond to an array of continuous regions of a chromosome which replicate simultaneously.[50] Several factors suggest that mammalian codon usage patterns are largely determined by mutational biases. First, selection on codon usage is unlikely to be effective in humans, because the expected fitness

TABLE 5
Codon Usage in Animals

		Drosophila				Human				Mouse			
		Low		High		A+T		G+C		A+T		G+C	
		N[a]	RSCU[b]	N	RSCU	N	RSCU	N	RSCU	N	RSCU	N	RSCU
Phe	UUU	354	1.17	29	0.16	1357	1.25	259	0.35	610	1.14	178	0.46
	UUC	250	0.83	329	1.84	810	0.75	1222	1.65	462	0.86	590	1.54
Leu	UUA	231	1.03	3	0.02	754	0.99	24	0.04	312	0.84	23	0.07
	UUG	273	1.21	61	0.48	846	1.11	191	0.29	393	1.06	147	0.45
Leu	CUU	257	1.14	34	0.27	991	1.29	151	0.23	434	1.17	131	0.40
	CUC	144	0.64	111	0.88	571	0.75	957	1.48	303	0.82	428	1.32
	CUA	170	0.75	20	0.16	460	0.60	119	0.18	273	0.74	78	0.24
	CUG	276	1.23	530	4.19	968	1.26	2445	3.77	511	1.38	1144	3.52
Ile	AUU	335	1.35	93	0.52	1542	1.55	249	0.51	759	1.35	162	0.54
	AUC	184	0.74	434	2.42	730	0.73	1174	2.40	527	0.94	701	2.35
	AUA	227	0.91	11	0.06	711	0.71	43	0.09	396	0.71	33	0.11
Met	AUG	269	—	250	—	1290	—	831	—	620	—	500	—
Val	GUU	277	1.17	81	0.49	1204	1.33	92	0.16	531	1.17	59	0.20
	GUC	189	0.80	237	1.42	611	0.68	632	1.09	388	0.86	303	1.04
	GUA	192	0.81	17	0.10	745	0.82	71	0.12	309	0.68	45	0.15
	GUG	288	1.22	333	1.99	1057	1.17	1529	2.63	586	1.29	754	2.60
Ser	UCU	243	1.02	49	0.48	1001	1.52	195	0.45	606	1.64	214	0.73
	UCC	241	1.02	314	3.08	549	0.83	804	1.86	314	0.85	483	1.64
	UCA	224	0.94	9	0.09	897	1.36	167	0.39	515	1.39	122	0.41
	UCG	195	0.82	112	1.10	86	0.13	363	0.84	53	0.14	188	0.64

TABLE 5 (continued)
Codon Usage in Animals

		Drosophila					Human					Mouse			
		Low		High		A+T		G+C		A+T		G+C			
		N[a]	RSCU[b]	N	RSCU	N	RSCU	N	RSCU	N	RSCU	N	RSCU		
Pro	CCU	219	0.98	27	0.28	1113	1.54	347	0.57	571	1.56	241	0.80		
	CCC	185	0.83	261	2.69	506	0.70	1188	1.95	255	0.70	512	1.71		
	CCA	299	1.34	48	0.49	1140	1.58	328	0.54	551	1.51	173	0.58		
	CCG	191	0.85	52	0.54	131	0.18	576	0.94	83	0.23	273	0.91		
Thr	ACU	221	0.96	59	0.39	988	1.36	189	0.38	649	1.41	183	0.60		
	ACC	231	1.01	474	3.11	614	0.85	1099	2.22	405	0.88	633	2.08		
	ACA	275	1.20	23	0.15	1151	1.59	241	0.49	698	1.51	177	0.58		
	ACG	192	0.84	54	0.35	144	0.20	452	0.91	95	0.21	222	0.73		
Ala	GCU	337	1.19	151	0.76	1368	1.63	415	0.52	602	1.65	407	0.92		
	GCC	306	1.08	573	2.88	712	0.85	1861	2.32	326	0.89	860	1.94		
	GCA	302	1.07	23	0.12	1138	1.35	285	0.35	473	1.29	167	0.38		
	GCG	184	0.65	50	0.25	147	0.17	653	0.81	63	0.17	338	0.76		
Tyr	UAU	234	1.12	43	0.26	1147	1.25	199	0.35	680	1.29	162	0.47		
	UAC	184	0.88	285	1.74	692	0.75	934	1.65	371	0.71	531	1.53		
ter	UAA	17	1.64	23	2.23	53	1.47	20	0.56	34	1.32	23	0.91		
ter	UAG	4	0.39	6	0.58	14	0.39	35	0.98	9	0.35	16	0.63		
His	CAU	157	0.92	33	0.32	792	1.27	126	0.31	508	1.22	132	0.48		
	CAC	186	1.08	173	1.68	453	0.73	690	1.69	327	0.78	413	1.52		

Selective Use of Termination Codons and Variations

AA	Codon	n	ratio	n	ratio	n	ratio	n	ratio	n	ratio	n	ratio
Gln	CAA	357	0.87	46	0.22	1123	0.93	122	0.16	532	0.90	130	0.24
	CAG	460	1.13	378	1.78	1299	1.07	1427	1.84	653	1.10	943	1.76
Asn	AAU	475	1.21	47	0.22	1691	1.26	191	0.31	977	1.31	160	0.41
	AAC	307	0.79	389	1.78	991	0.74	1023	1.68	518	0.69	624	1.59
Lys	AAA	471	1.07	44	0.13	2127	1.17	254	0.31	1168	1.12	274	0.39
	AAG	407	0.93	624	1.87	1511	0.83	1391	1.69	905	0.87	1121	1.61
Asp	GAU	500	1.25	223	0.72	1965	1.36	351	0.36	944	1.22	269	0.56
	GAC	297	0.75	396	1.28	932	0.64	1571	1.63	597	0.77	697	1.44
Glu	GAA	515	1.15	60	0.15	2763	1.38	320	0.27	1411	1.24	309	0.39
	GAG	383	0.85	739	1.85	1252	0.62	2081	1.73	872	0.76	1278	1.61
Cys	UGU	128	0.90	10	0.11	865	1.23	220	0.45	602	1.32	147	0.60
	UGC	156	1.10	170	1.89	539	0.77	762	1.55	307	0.67	339	1.39
ter	UGA	10	0.97	2	0.19	41	1.14	52	1.46	34	1.32	37	1.46
Trp	UGG	151	—	99	—	677	—	473	—	424	—	193	—
Arg	CGU	116	0.87	136	1.84	317	0.74	152	0.38	143	0.62	154	0.61
	CGC	137	1.03	244	3.30	207	0.48	951	2.41	113	0.49	655	2.61
	CGA	148	1.11	16	0.22	385	0.90	149	0.38	147	0.64	119	0.47
	CGG	93	0.70	14	0.19	214	0.50	625	1.58	79	0.34	273	1.09
Ser	AGU	263	1.11	10	0.10	794	1.21	142	0.33	417	1.13	136	0.46
	AGC	257	1.08	117	1.15	617	0.94	916	2.12	316	0.85	622	2.11
Arg	AGA	181	1.36	3	0.04	968	2.26	130	0.33	606	2.64	84	0.33
	AGG	122	0.92	31	0.42	479	1.12	363	0.92	290	1.26	219	0.87

TABLE 5 (continued)
Codon Usage in Animals

		Drosophila				Human					Mouse			
		Low		High		A+T		G+C			A+T		G+C	
		N[a]	RSCU[b]	N	RSCU	N	RSCU	N	RSCU		N	RSCU	N	RSCU
Gly	GGU	270	1.10	197	1.13	1089	1.07	266	0.39		368	0.78	269	0.60
	GGC	267	1.08	347	2.00	792	0.78	1498	2.18		431	0.91	915	2.03
	GGA	340	1.38	144	0.83	1625	1.60	242	0.35		797	1.68	240	0.53
	GGG	108	0.44	7	0.04	556	0.55	744	1.08		297	0.63	375	0.83

Note: Codon usage varies among genes in each of these species and two extreme groups of genes are presented for each species to illustrate this variation (for the definition of groups see Table 2 and text). Unlike the situation in the other species presented, codon usage bias is *not* correlated with gene expression level in differences between alternative synonymous codons are probably too small, given the small effective population size in mammals.[51] Second, rates of nucleotide substitution at silent sites in mammalian genes are consistent with their neutrality.[50] Third, patterns of point mutations in primate sequences appear to vary with chromosome localization.[52] Finally, patterns of codon usage in human genes have the characteristics expected if they are determined by mutational biases.[53]

In mammals, there appears to be just one release factor,[23] which must recognize all three stop codons, although it might have differing affinities for each of them. However, if sense codon usage in humans is determined by local mutational biases, it is possible that stop codon usage would also reflect this. The relative usage of the different stop codons does indeed differ between the A+T-rich and G+C-rich genes (Table 5). The changes occur gradually, correlated with GC3s (GC3s is the G+C content at silent third codon positions [Figure 3B]). The decrease in usage of UAA as G+C-richness increases is particularly striking. Thus, the likelihood that a human gene terminates with UAA as opposed to UAG or UGA will vary with chromosomal location, and seems to depend on the local chromosomal base composition. There is a consistent, approximately twofold, excess of UGA over UAG at all levels of G+C content (Figure 3B). This might be due to the interaction between the release factor and the termination codon, but it should be noted that the UpA dinucleotide occurs generally less often than expected in mammalian DNA.[54] This may reflect nearest-neighbor-dependent mutational biases,[14] or some form of selection on DNA structure.

C. OTHER EUKARYOTES
1. Mouse

As already remarked, closely-related species usually have similar patterns of codon usage. On average, at the DNA level, primates and rodents[55] have diverged somewhat less than *E. coli* and *S. typhimurium*,[25] and so codon usage in human and mouse genes would be expected to be similar. Density gradient centrifugation analyses suggest that the murine genome has an isochore structure similar to the human genome, but lacking the more extremely G+C-rich and A+T-rich components.[56] Sequence data support this: the extreme groups of genes from the mouse genome have less biased codon usage.[57] This is illustrated by the average G+C content at silent third positions of genes in the top and bottom 20% (contrast Figure 3C with 3B), and by the RSCU values of individual codons in the top and bottom 10% (Table 5). The average GC3s values for these 2 groups of genes are 80% and 37%. Otherwise, the relationship of stop codon usage to sense codon bias in the mouse genome (Figure 3C) is similar to that in the human genome.

2. Drosophila melanogaster

From the species surveyed so far it is tempting, in summarizing codon usage patterns, to make a distinction between prokaryotes and unicellular eukaryotes,

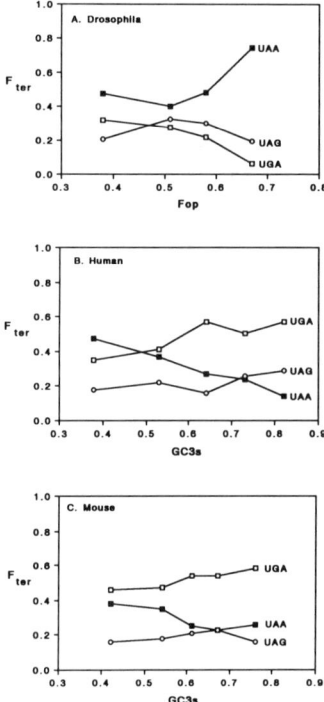

FIGURE 3. Frequency of use of termination codons in animal genes. (A) 310 *Drosophila melanogaster* genes, divided into 4 classes on the basis of codon usage bias, measured by the frequency of optimal codons, F_{op} (see text). (B) 1072 human genes, divided into 5 classes on the basis of GC3s, the G+C content at silent third codon positions (see text). (C) 769 mouse genes, divided into 5 classes on the basis of GC3s, the G+C content at silent third codon positions. In each panel, the frequency of each stop codon (F_{ter}) in each group of genes is plotted against the mean codon bias measure (F_{op} or GC3s) for that group.

on the one hand, and multicellular eukaryotes on the other.[3] However, an analysis[58] of codon usage in *D. melanogaster* led us to suggest that codon usage in this fruit fly more closely resembles that of *E. coli* and yeast, than it does that in the human genome. While G+C content at silent sites does vary among *Drosophila* genes, it is not correlated with base composition in flanking or intronic sequences. Instead, multivariate analyses revealed that the main trend among genes is in the extent of usage of a particular subset of codons. Genes with a high frequency of these codons are at one end of the trend, while codon usage is more uniform at the other extreme (Table 5). While "expression level" is a complicated issue in a multicellular organism with a complex life cycle, nevertheless, where relative expression levels could be compared between members of gene families, it was found that the frequencies of these (presumably optimal) codons were higher in the more highly expressed genes.[58] Therefore, the frequency of these "optimal" codons in a gene, F_{op} (calculated

Selective Use of Termination Codons and Variations 421

as the number of occurrences of these optimal codons, divided by the total number of codons excluding Met, Trp, and stop codons) is used as the species-specific measure of codon usage bias.

Stop codon usage varies with sense codon usage in *D. melanogaster* (Figure 3A). As in *E. coli* and yeast, in the genes with the highest codon usage bias (here measured by F_{op}), it is UAA that is strongly preferred. It should be noted that, because many of the codons identified as optimal are C- or G-ending, these genes with high F_{op} values also have the highest GC3s values, but feature the more A+T-rich termination codons. Thus, consideration of termination codon frequencies seems to strengthen the interpretation of the sense codon usage data in *D. melanogaster*, i.e., that codon usage in this species is influenced by translational selection. One relevant difference between *Drosophila* and mammals is the somewhat larger longterm effective population size in the former, which may have allowed the small selective differences between codons to overcome random genetic drift.[51]

3. Maize

Codon usage in plants has been described,[59] but not as yet elucidated. Certain monocotyledenous angiosperms, the grasses, appear to have isochores reminiscent of the genomic compartments found in mammals,[60,61] and this variation in G+C content dominates codon usage in the species from which the largest number of genes have been sequenced, maize (*Zea mays*). In maize, both UAG and UGA are used more often than UAA (Table 1). The usage of these codons varies with sense codon usage in an interesting way. When genes are ranked according to G+C content at silent sites (in the same way that human and mouse genes were categorized previously), UAA is found at a similar (low) frequency irrespective of sense codon G+C content. However, in the G+C-rich genes UGA predominates, while in the A+T-rich genes UAG predominates. The group of A+T-rich genes is comprised largely of members of the zein multigene family, but even after excluding the zein genes the stop codon frequencies vary with sense codon G+C content. Perhaps the current data sets are too small, and not enough is yet known about the causes of codon bias in plant genes, but it will be interesting to follow up this observation.

4. Fungi

Codon usage has been investigated in some species of fungi other than *S. cerevisiae*. *Aspergillus nidulans* is a filamentous fungus classified in the ascomycetes with *S. cerevisiae*. Optimal *A. nidulans* codons have been identified, from their higher frequency in highly expressed genes.[11] Many of these optimal codons are different from those in yeast (Table 4), indicating that the codon "dialects" of the two species have diverged quite rapidly. The overall frequencies of stop codons in these two species (Table 1) differ significantly, but this may be due to the particular genes in the rather small sample from *A. nidulans*. More interestingly, when the *A. nidulans* genes are ranked according to codon usage bias, 15 of the top 18 genes end with UAA, indicating that as in yeast UAA appears to be favorable in highly expressed genes.

The fission yeast *Schizosaccharomyces pombe* is only very distantly related to *S. cerevisiae*. Again, codon usage varies among genes in a manner correlated with gene expression level.[4] The codons preferred in highly expressed *S. pombe* genes differ from those in *S. cerevisiae* for some amino acids (e.g., Pro, Arg), but not others (Table 4). Note that these highly expressed genes *all* terminate with UAA.

V. CONCLUSIONS

In parallel with sense codon usage, the frequencies of the alternative stop codons vary among species, and the likelihood that a gene will terminate with a particular stop codon varies among genes within the same genome. Mutational biases clearly influence stop codon frequencies. However, in species where sense codon usage is strongly influenced by translational selection, stop codons also appear to be influenced. In most species, UAA appears to be the optimal termination codon, though the exact basis of this superiority is yet to be understood. Thus, the frequencies of the three stop codons in groups of similar genes will depend on a balance between mutational biases and selective differences, if the latter are strong enough to overcome genetic drift.

ACKNOWLEDGMENTS

This is a publication from the Irish National Centre for Bioinformatics. This work was supported by EOLAS grant SC/91/603. ATL was supported by bursary 900027 from the EC BRIDGE program.

REFERENCES

1. **Grantham, R., Gautier, C., Gouy, M., Mercier, R., and Pave, A.,** Codon catalog usage and the genome hypothesis, *Nucleic Acids Res.,* 8, r49, 1980.
2. **Grantham, R., Gautier, C., Gouy, M., Jacobzone, M., and Mercier, R.,** Codon catalog usage is a genome strategy modulated for gene expression, *Nucleic Acids Res.,* 9, r43, 1981.
3. **Ikemura, T.,** Codon usage and tRNA content in unicellular and multicellular organisms, *Mol. Biol. Evol.,* 2, 13, 1985.
4. **Sharp, P. M., Cowe, E., Higgins, D. G., Shields, D. C., Wolfe, K. H., and Wright, F.,** Codon usage in *Escherichia coli, Bacillus subtilis, Saccharomyces cerevisiae, Schizosaccharomyces pombe, Drosophila melanogaster* and *Homo sapiens;* a review of the considerable within-species diversity, *Nucleic Acids Res.,* 16, 8207, 1988.
5. **Sharp, P. M. and Li, W.-H.,** An evolutionary perspective on synonymous codon usage in unicellular organisms, *J. Mol. Evol.,* 24, 28, 1986.
6. **Bulmer, M.,** The selection-mutation-drift theory of synonymous codon usage, *Genetics,* 129, 897, 1991.

7. **Shields, D. C.,** Switches in species-specific codon preferences: the influence of mutation biases, *J. Mol. Evol.,* 31, 71, 1990.
8. **Sharp, P. M., Higgins, D. G., Shields, D. C., Devine, K. M., and Hoch, J. A.,** *Bacillus subtilis* gene sequences, in *Genetics and Biotechnology of Bacilli,* Vol. 3, Zukowski, M. M., Ganesan, A. T., and Hoch, J. A., Eds., Academic Press, San Diego, 1990, 89.
9. **Ohama, T., Muto, A., and Osawa, S.,** Role of GC-biased mutation pressure on synonymous codon choice in *Micrococcus luteus,* a bacterium with a high genomic GC-content, *Nucleic Acids Res.,* 18, 1565, 1990.
10. **Sharp, P. M. and Cowe, E.,** Codon usage in *Saccharomyces cerevisiae, Yeast,* 7, 657, 1991.
11. **Lloyd, A. T. and Sharp, P. M.,** Codon usage in *Aspergillus nidulans, Mol. Gen. Genet.,* 230, 288, 1991.
12. **Sharp, P. M. and Devine, K. M.,** Codon usage and gene expression level in *Dictyostelium discoideum:* highly expressed genes do 'prefer' optimal codons, *Nucleic Acids Res.,* 17, 5029, 1989.
13. **Gouy, M., Gautier, C., Attimonelli, M., Lanave, C., and di Paola, G.,** ACNUC — a portable retrieval system for nucleic acid sequence databases: logical and physical designs and usage, *CABIOS,* 1, 167, 1985.
14. **Bulmer, M.,** Neighboring base effects on substitution rates in pseudogenes, *Mol. Biol. Evol.,* 3, 322, 1986.
15. **Sharp, P. M. and Bulmer, M.,** Selective differences among translation termination codons, *Gene,* 63, 141, 1988.
16. **Ikemura, T.,** Correlation between the abundance of *Escherichia coli* transfer RNAs and the occurrence of the respective codons in its protein genes: a proposal for a synonymous codon choice that is optimal for the *E. coli* translational system, *J. Mol. Biol.,* 151, 389, 1981.
17. **Gouy, M. and Gautier, C.,** Codon usage in bacteria: correlation with gene expressivity, *Nucleic Acids Res.,* 10, 7055, 1982.
18. **Post, L. E., Strycharz, G. D., Nomura, M., Lewis, H., and Dennis, P. P.,** Nucleotide sequence of the ribosomal protein gene cluster adjacent to the gene for RNA polymerase subunit beta in *Escherichia coli, Proc. Natl. Acad. Sci. U.S.A.,* 76, 1697, 1979.
19. **Andersson, S. G. E. and Kurland, C. G.,** Codon preferences in free-living microorganisms. *Microbiol. Rev.,* 54, 198, 1990.
20. **Sharp, P. M. and Li, W.-H.,** Codon usage in regulatory genes in *Escherichia coli* does not reflect selection for "rare" codons, *Nucleic Acids Res.,* 14, 7737, 1986.
21. **Sharp, P. M. and Li, W.-H.,** The Codon Adaptation Index — a measure of directional synonymous codon usage bias, and its potential applications, *Nucleic Acids Res.,* 15, 1281, 1987.
22. **Brown, C. M., Stockwell, P. A., Trotman, C. N. A., and Tate, W. P.,** The signal for the termination of protein synthesis in prokaryotes, *Nucleic Acids Res.,* 18, 2079, 1990.
23. **Craigen, W. J., Lee, C. C., and Caskey, C. T.,** Recent advances in peptide chain termination, *Mol. Microbiol.,* 4, 861, 1990.
24. **Martin, R., Weiner, M., and Gallant, J.,** Effects of release factor context at UAA codons in *Escherichia coli, J. Bacteriol.,* 170, 4714, 1988.
25. **Sharp, P. M.,** Determinants of DNA sequence divergence between *Escherichia coli* and *Salmonella typhimurium:* codon usage, map position and concerted evolution, *J. Mol. Evol.,* 33, 23, 1991.
26. **Ikemura, T.,** Codon usage, tRNA content, and rate of synonymous substitution, in *Population Genetics and Molecular Evolution,* Ohta, T. and Aoki, K., Eds., Springer-Verlag, Berlin, 1985, 385.
27. **Riley, M. and Krawiec, S.,** Genome organization, in *Escherichia coli and Salmonella typhimurium,* Neidhardt, F. C., Ingraham, J. L., Low, K. B., Magasanik, B., Schaechter, M., and Umbarger, H. E., Eds., American Society for Microbiology, Washington DC, 1987, 967.

28. **Sharp, P. M. and Li, W.-H.**, The rate of synonymous substitution in enterobacterial genes is inversely related to codon usage bias, *Mol. Biol. Evol.*, 4, 222, 1987.
29. **Bulmer, M., Wolfe, K. H., and Sharp, P. M.**, Synonymous nucleotide substitution rates in mammalian genes: implications for the molecular clock and the relationships of mammalian orders, *Proc. Natl. Acad. Sci. U.S.A.*, 88, 5974, 1991.
30. **Li, W.-H., Wu, C.-I., and Luo, C.-C.**, A new method for estimating synonymous and nonsynonymous rates of nucleotide substitution considering the relative likelihood of nucleotide and codon changes, *Mol. Biol. Evol.*, 2, 150, 1985.
31. **Woese, C. R.**, Bacterial evolution, *Microbiol. Rev.*, 51, 221, 1987.
32. **Shields, D. C. and Sharp, P. M.**, Codon usage in *Bacillus subtilis* reflects both translational selection and mutational biases, *Nucleic Acids Res.*, 15, 8023, 1987.
33. **Andachi, Y., Yamao, F., Muto, A., and Osawa, S.**, Codon recognition patterns as deduced from sequences of the complete set of transfer RNA species in *Mycoplasma capricolum*, *J. Mol. Biol.*, 209, 37, 1989.
34. **Sueoka, N.**, Variation and heterogeneity of base composition of deoxyribonucleic acids: a compilation of old and new data, *J. Mol. Biol.*, 3, 31, 1962.
35. **Yamao, F., Muto, A., Kawauchi, Y., Iwami, M., Iwagami, S., Azumi, Y., and Osawa, S.**, UGA is read as tryptophan in *Mycoplasma capricolum*, *Proc. Natl. Acad. Sci. U.S.A.*, 82, 2306, 1985.
36. **Bennetzen, J. L. and Hall, B. D.**, Codon selection in yeast, *J. Biol. Chem.*, 257, 3026, 1982.
37. **Ikemura, T.**, Correlation between the abundance of yeast transfer RNAs and the occurrence of the respective codons in protein genes, *J. Mol. Biol.*, 158, 573, 1982.
38. **Sharp, P. M., Tuohy T. M. F., and Mosurski, K. R.**, Codon usage in yeast: cluster analysis clearly differentiates between highly and lowly expressed genes, *Nucleic Acids Res.*, 14, 5125, 1986.
39. **Bulmer, M.**, The effect of context on synonymous codon usage in genes with low codon usage bias, *Nucleic Acids Res.*, 18, 2869, 1990.
40. **Bulmer, M.**, Are codon usage patterns in unicellular organisms determined by selection-mutation balance?, *J. Evol. Biol.*, 1, 15, 1988.
41. **Cavener, D. R. and Ray, S. C.**, Eukaryotic start and stop translation sites, *Nucleic Acids Res.*, 19, 3185, 1991.
42. **Ikemura, T. and Aota S.-I.**, Global variation in G+C content along vertebrate genome DNA, *J. Mol. Biol.*, 203, 1, 1988.
43. **Aota, S.-I. and Ikemura, T.**, Diversity in G+C content at the third position of codons in vertebrate genes and its cause, *Nucleic Acids Res.*, 14, 6345, 1986.
44. **Bernardi, G., Olofsson, B., Filipski, J., Zerial, M., Salinas, J., Cuny, G., Meunier-Rotival, M., and Rodier, F.**, The mosaic genome of warm-blooded vertebrates, *Science*, 228, 953, 1985.
45. **Mouchiroud, D., Fichant, G., and Bernardi, G.**, Compositional compartmentalization and gene composition in the genomes of vertebrates, *J. Mol. Evol.*, 26, 198, 1987.
46. **Filipski, J., Salinas, J., and Rodier, F.**, Two distinct compositional classes of vertebrate gene-bearing DNA stretches, their structures and possible evolutionary origins, *DNA*, 6, 109, 1987.
47. **Bernardi, G. and Bernardi, G.**, Compositional constraints and genome evolution, *J. Mol. Evol.*, 24, 1, 1986.
48. **Filipski, J.**, Why the rate of silent codon substitutions is variable within a vertebrate's genome, *J. Theor. Biol.*, 134, 159, 1988.
49. **Sueoka, N.**, Directional mutation pressure and neutral molecular evolution, *Proc. Natl. Acad. Sci. U.S.A.*, 85, 2653, 1988.
50. **Wolfe, K. H., Sharp, P. M., and Li, W.-H.**, Mutation rates differ among regions of the mammalian genome, *Nature (London)*, 337, 283, 1989.
51. **Sharp, P. M.**, Evolution at 'silent' sites in DNA, in *Evolution and Animal Breeding; Reviews on Molecular and Quantitative Approaches in Honour of Alan Robertson*, Hill, W. G. and Mackay, T. F. C., Eds., C.A.B. International, Wallingford, U.K., 1989, 24.

52. **Filipski, J., Salinas, J., and Rodier, F.,** Chromosome localization-dependent compositional bias of point mutations in *Alu* repetitive sequences, *J. Mol. Biol.,* 206, 563, 1989.
53. **Eyre-Walker, A.,** An analysis of codon usage in mammals: selection or mutation bias? *J. Mol. Evol.,* 33, 442, 1991.
54. **Nussinov, R.,** Doublet frequencies in evolutionary distinct groups, *Nucleic Acids Res.,* 12, 1749, 1984.
55. **Li, W.-H., Tanimura, M., and Sharp, P. M.,** An evaluation of the molecular clock hypothesis using mammalian DNA sequences, *J. Mol. Evol.,* 25, 330, 1987.
56. **Salinas, J., Zerial, M., Filipski, J., and Bernardi, G.,** Gene distribution and nucleotide sequence organization in the mouse genome, *Eur. J. Biochem.,* 160, 469, 1986.
57. **Mouchiroud, D. and Gautier, C.,** Codon usage changes and sequence dissimilarity between human and rat, *J. Mol. Evol.,* 31, 81, 1990.
58. **Shields, D. C., Sharp, P. M., Higgins, D. G., and Wright, F.,** "Silent" sites in *Drosophila* genes are not neutral: evidence of selection among synonymous codons, *Mol. Biol. Evol.,* 5, 704, 1988.
59. **Murray, E. E., Lotzer, J., and Eberle, M.,** Codon usage in plant genes, *Nucleic Acids Res.,* 17, 477, 1989.
60. **Salinas, J., Matassi, G., Montero, L. M., and Bernardi, G.,** Compositional compartmentalization and compositional patterns in the nuclear genomes of plants, *Nucleic Acids Res.,* 16, 4269, 1988.
61. **Matassi, G., Montero, L. M., Salinas, J., and Bernardi, G.,** The isochore organization and the compositional distribution of homologous coding sequences in the nuclear genome of plants, *Nucleic Acids Res.,* 17, 5273, 1989.

INDEX

A

Abnormal proteins, 242
ac^{4}C34, 31, 36, 40, 55
Acidic residues, codon assignment of, 385
acp^{3}U48, 35
Affinity chromatography, 147
Alphaviruses, 212
Alternative readings, 192
Amber anticodon, 288
Amber codons, 281
Amber suppressors, efficiency of, 53
Amber suppressor tRNAs, 300
Amino acid
  limitation, 240
  sequences, 382
  starvation, 201, 226, 238
Aminoacylation, tRNA discrimination in, 279–318
  anticodon and acceptor arm as major recognition sites in tRNAs, 283
  E. coli glutamine tRNA-glutamine synthetase complex, 299–301
  recognition of other tRNA domains, 301–308
    E. coli arginine tRNA, 303–304
    phenylalanine tRNAs, 304–306
    tRNAs containing large variable arm, 301–303
    yeast aspartate tRNA, 306–308
  role of acceptor stem and discriminator base at position 73, 293–298
    alanine tRNAs, 293–296
    E. coli histidine tRNA, 297
    E. coli tRNAs, 297–298
  role of anticodon, 284–293
    E. coli methionine tRNA, 289–291
    in vitro studies, 284–285
    in vivo studies, 285–289
    valine tRNAs, 292–293
  role of modified bases, 308–309
    E. coli isoleucine tRNAs, 308
    recognition by yeast arginine tRNA synthetase, 308–309
  tRNA identity assays, 280–283
Aminoacyl-tRNA (anticodon):codon
  adaptation, in higher eukaryotes, 113–123
  adaptation of tRNA content to requirements of protein synthesis, 114–115

aminoacyl-tRNA (anticodon):codon compositions, 117–120
aminoacyl-tRNA:protein amino acid compositions, 115–117
possible mechanisms for determining cellular amounts of isoacceptors, 120–121
Aminoacyl-tRNA ligases, 36
Aminoacyl-tRNA synthetases, 281, 283
Aminoglycoside antibiotics, 197
AMV, see Avian myeloblastosis virus
Animals, codon usage in, 415–418
Anticodon base requirements, 299
Anticodon-codon complex, 51
Anticodon:codon interaction, 59, 65, 158
Anticodon-codon pairing, 133
Anticodon nucleotides, 289
Anticodon swap experiments, 290
Antideterminants, 28
Antisuppression, 356
Aspergillus nidulans, 421
ASV, see Avian sarcoma virus
Attenuation, 163
A+T rich genome, 408
AT-rich sequences, 390, 391
Autogenous repression, 106
Avian myeloblastosis virus (AMV), 5, 14
Avian sarcoma virus (ASV), 6

B

Bacillus subtilis, 35, 407
Basal-level errors, 239
Basal-level frequency, 205
Base pair, 328
Base sequences, 382
Basic residues, codon assignment of, 385
Binding protein, 147
Binding specificity, degree of, 11
Bombyx mori, 114, 117, 143
Bovine tRNA, 9–10

C

CA, see Capsid
Caenorhabditis elegans, 272, 399
Calf lens, internal cortex of, 115
Capsid (CA), 14
5-Carboxymethylaminomethyluridine, 133
Casein, 121

CAT, see Chloramphenicol
    acetyltransferase
Cattle, selenocysteine tRNA in, 272
Cell
  aminoacyl-tRNA in, 114
  extracts, 148
  growth rate of, 175
  wall biosynthesis, 24
Cellular organelles, aminoacyl-tRNAs in,
    114
Chickens, selenocysteine tRNA in, 272
Chloramphenicol acetyltransferase (CAT),
    163
Chlorophyll, synthesis of, 24
Chloroplast genetic code, 126
Chloroplast tRNAs, 126
Chromosomal banding, 88
cis-acting DNA sequence motifs, 152
Cm34, 40
cmnm^{5}s^{2}U34, 63
cmnm^{5}Um, 63
cmo^{5}U, 62
cmo^{5}U34, 40
cmo^{5}U34/mcmo5U34, 40
Coding sequences, base compositions of,
    393, 395
Codon
  assignments, 118, 382, 385
  capture, 132, 206
  choices, 89, 102
  context, 53, 65, 158, 206
  dialect, 89
  families, 62, 376
  G+C%, 88
  misreading, 193
  pair bias, 183
  preference, 392
  reading, canonical principles of, 368
  reassignment, 131, 132
  recognition, 118
  usage, 88, 160, 202, 227, 392, 398, 400
Codon Adaptation Index, 399
Codon pair utilization bias, in bacteria,
    yeast, and mammals, 157–189
  biases in codon pair utilization, 164–173
    codon pair bias in E. coli, 164–167
    codon pair bias in yeast, human, rat,
      and mouse, 167–171
    nonadjacent codon pair bias, 171–173
  codon context and translational
    efficiency, 158–164
    influence of nonsense suppression by
      bases adjacent to codon triplet,
      159–160
    nonsense versus missense suppression,
      158–159

translational efficiency of frequently
  and infrequently used codons, 162–
  164
tRNA-tRNA interactions on surface of
  translating ribosome, 160–161
functional significance of codon pair
  utilization patterns, 173–184
  codon pairs and translational step-time,
    173–175
  interactions between nascent polypep-
    tide chains and cellular factors,
    177–178
  messenger RNA stability, 179–180
  metabolic growth rate control in E.
    coli, 180–182
  practical applications of codon pair
    rules, 182–184
    identification of protein-coding
      regions, 183–184
    optimizing genetically engineered
      systems, 182–183
  protein folding, 175–177
  protein stability, 178
  regulation of protein expression, 178–
    179
Codon usage, adaptation of tRNA
    population to in cellular organelles,
    125–140
  anticodon content and codon recognition
    mechanism, 132–135
    codon recognition in chloroplasts, 134
    codon recognition in mammalian
      mitochondria, 135
    codon recognition in yeast, 133–134
  evolution of genetic code, 131–132
  genetic code, 130
  levels of isoaccepting tRNAs and codon
    usage in chloroplasts and mitochon-
    dria, 135–137
  mitochondrial tRNAs of nematodes, 137
  number of mitochondrial or chloroplast
    tRNAs in different organisms, 126–
    130
  tRNA species in *Mycoplasma
    capricolum*, 138
Codon usage, correlation between tRNA
    content and, in microorganisms,
    87–111
  cellular tRNA contents of E. coli, S.
    typhimurium, and S. cerevisiae, 89–
    95
  codon choices in genes of unicellular
    organisms and distinction from
    higher eukaryotes, 107–108
    distinction from higher eukaryotes,
    107–108

Index

unicellular organisms, 107
codon usage in genes of *E. coli, S. typhimurium,* and *S. cerevisiae,* 89
codons optimal for organism's translation system, 98–100
  frequency of optimal codon usage, 99–100
  optimal codons, 98–99
constraints on codon choices, 100–104
correlation between codon usage and tRNA content, 95–98
  frequency of isoaccepting tRNA usage, 95–96
  frequency of tRNA usage, 95
  preference among multiple codons recognized by single tRNA, 97–98
mechanisms for correlation between codon usage and tRNA content, 104–107
  evolutionary viewpoints, 106–107
  molecular mechanisms, 104–106
Collagen mRNA, 117
Competition, 326, 358
Complementary DNA strands, 386
Complementary RNA strands, 383
Context effect, 97, 406
Context-specific effects, 159
Correlation coefficient, 115, 118

D

Darwinian selection, 108
Deacylated tRNAs, 120
Decay targets, 352
DHFR, see Dihydrofolate reductase
DHU arm-replacement loop, 137
Dictyostelium discoideum, 398
Differential tRNA gene expression, in eukaryotes, 141–155
  developmental regulation of tRNA gene transcription, 145–148
  differential expression of tRNA gene families, 150–151
  species-specific tRNA gene transcription, 148–150
  tissue-specific tRNA usage and gene expression, 142–145
Differentiation, erythroid, 144
Dihydrofolate reductase (DHFR), 281, 282
Dinucleotide bias, 170
Discrimination, unconventional readings with, 370
Discriminator nucleotide, role of, 306
Dispensable tRNAs, 93
dnaX, 60
Double frameshift, 226, 228

Drop-off, 218, 239, 242, 243
Drop-off frequency, 219
Drosophila, selenocysteine tRNA in, 272

E

Editing mechanism, 58
Effective population size, 419
EF-G, see Elongation factor G
EF-Tu, see Elongation factor Tu
Elongation factor G (EF-G), 24, 220
Elongation factor Tu (EF-Tu), 24, 240
Equilibrium binding conditions, 11
Error coupling, 221, 240
Error frequency, 227
Error-prone codons, 237
Erythroid differentiation, 120
Escherichia coli, 26, 31, 89, 96, 158, 398
Eukaryotes, evolution of, 383
Eukaryotic discrimination, 309
Extended anticodon, 376

F

Family boxes, 368
Family codons, 372
Fibroin, 116, 117
Fitness change, level of, 106
Fold overexpression, 358
Forbidden base pairs, 369–370
Formate dehydrogenase, 270
Frameshift, 219, 242, 327, 341
  frequency, 227
  mutations, 220, 221
  products, 193, 220, 225
  suppression, 222
  window, 224
Friend leukemia cells, 120
Fungi, codon usage in, 410–413

G

G+C content, 409, 421
G+C% mosaic structures, 88, 108
GC-rich anticodons, 41
G+C-rich gene, 88
G+C rich genome, 408
G-terminated codon, 97
Gene
  development of, 145
  expressivity, 99
  number, 93
Gene expression, 342
  change in, 354
  development-specific, 142, 145

differential, 150
differentiation-specific, 142
species-specific, 142, 148
tissue-specific, 142
Genetic code, variations in reading, 191–267
  control of error, 236–241
    balanced cell, 240–241
    codon usage and codon context, 236–237
    proofreading and editing, 237–239
  error level in cells, 241–244
  errors involving termination, 207–219
    leaky termination codons, 209–210
    native tRNA, 215–217
    premature termination and ribosome pausing, 217–219
    required readthrough, 210–215
    UGA as alternative sense codon, 217
  misreading sense codons, 196–206
    misacylation, 197–199
    missense alternatives, 206
    missense misreading, 199–205
  ribosomal frameshifts, 220–236
    frameshift errors, 222–228
    required programmed frameshifts, 228–235
    shift to terminate, 235–236
  tolerance of error, 244–245
  types of errors and alternatives in codon reading, 193–196
Genetic codes, anticodon content in, 127–128
Genetic origin of mitochondrial tRNAs, 126, 129, 130
Genome G+C%, 102
Genomic base composition, 408
Globin mRNA, 52
Glutamine tRNA, 58–59, 383
Glutathione peroxidase, 270
Glycine codons, 373
Gm34, 40

H

Heat shock response, 243
Heme, synthesis of, 24
Hemoglobin B, 115
Hemoglobin C, 115
hisT, 35, 49, 58, 59, 65, 160
HIV-1, see Human immunodeficiency virus, type 1
HTLV-1, see Human T-cell lymphocyte virus, type 1
Human genome, 409
Human immunodeficiency virus, type 1 (HIV-1), 2, 7, 8, 12–16
Humans
  reticulocytes of, 117
  selenocysteine tRNA in, 272
Human T-cell lymphocyte virus, type 1 (HTLV-1), 2
Hydrophobic residues, 389
Hyperaccurate ribosomes, 238
Hypermodified nucleoside, 53, 60
Hypermodified nucleoside Q, 32

I

I34, 40
i^6A37, 33, 38, 39, 46, 47
ICRs, see Internal control regions
Influenza A virus hemagglutinins, 393, 394
Influenza virus coding sequence, 390
Initiation assay, 282
Internal control regions (ICRs), 142
Internal promoter elements, 148
In vitro protein-synthesizing system, 373
Iodothyronine (type I) deiodinase, 270
Isochores, 414, 421
Isoelectric points, 385
Isoleucine, 32
Isoleucine tRNA, 115

K

k^2C34, 31, 36, 56
Knuckle, 14

L

L34, 31, 56
Large variable arms, 303
Leucine tRNA, 115
Liverwort, chloroplast genomes of, 371
Long terminal repeat (LTR) sequences, 2, 3
LTR, see Long terminal repeat sequences
Lysidine, 31, 56, 132
Lysine, 287

M

m^2A37, 39
m^6A37, 39
MA, see Matrix proteins
Mammalian mitochondria, 370
Mammary gland, lactoproteins of, 114
Matrix (MA) proteins, 14
m^5C, 45
m^5C34, 40, 45

mcm^{5}s^{2}U, 38, 64
mcm^{5}s^{2}U34, 45
mcmo^{5}U, 62
Metal-binding residues, 14
Metazoan mitochondrial genome, 392, 393
Methionine tRNA, 115
Mevalonic acid, starvation for, 33
m^{1}G, 59
m^{1}G37, 33, 39, 59
m^{2}G10, 30
m$^2_2$G26, 31
m^{7}G46, 35
m^{1}I37, 39
miaA, 26, 38, 56, 57
miaB, 26, 47
miaC, 26
miaE, 26
Micrococcus luteus, 408
Microorganisms, aminoacyl-tRNAs in, 114
Minus strand, 4
Misacylation, 195, 240
Misincorporations, 199
Mismatched anticodons, 355
Mispaired nucleotides, 372
Misreading frequency, 200, 204
Missense error, 63, 193, 242
Missense misreading, 196, 205
Mitochondrial genetic code, 126
Mitochondrial tRNAs, 126
M-MulV, see Moloney murine leukemia virus
MMTV, see Mouse mammary tumor virus
mnm^{5}Se^{2}U, 64
mnm^{5}Se^{2}U34, 45
mnm^{5}s^{2}U, 63
mnm^{5}s^{2}U34, 32, 36, 39, 42, 44, 45, 52, 63
mnm^{5}U, 63
Moloney murine leukemia virus (M-MulV), 4, 6, 14, 16
mo^{5}U, 62
mo^{5}U34, 40
Mouse mammary tumor virus (MMTV), 6
mRNA
  cognate codon usage within, 115
  -instructed ribosomes, 355
  level, 326
  passages, slowly translated, 349
  stability, 175
ms^{2}i^{6}A37, 27, 32, 33, 46, 47, 53, 56–58
ms^{2}io^{6}A37, 26, 36, 46, 47, 53, 58
m^{5}U54, 35, 37, 49, 50
Mutational bias, 408
Mutational biases, 398, 414
Mycoplasma
  capricolum, 408
  mycoides sp. *capri*, 370

N

Nascent peptide, 355
Nascent polypeptide interactions, regulation of, 175
Natural selection, 398
NC, see Nucleocapsid proteins
ncm^{5}U, 64
nm^{5}s^{2}U34, 63
Nonfamily codons, 372
Nonrandom codon contexts, 321
Nonrandomness, 325
Nonsense codons, 158, 327
Nonsense suppression, 158, 207, 208
Nonsense suppressor tRNAs, 158
Nonsense suppressors, 325
Nonuniversal codes, 215
Nucleic acids, self-replicating, 382
Nucleocapsid (NC) proteins, 14, 15, 17
Nucleotide dimer analysis, 386

O

Opal suppressor tRNA, 304
Optimal codons, 420
Origin of life, extraterrestrial, 384
Overlapping reading, 222, 223

P

Paused ribosome, 235
Pause sites, 179
PBS, see Primer binding site
Peptidyl-tRNA hydrolase, 218
Phage message MS2 RNA, 373
Phenylalanine, 338
Phosphoseryl-tRNA, 271
Polar effects, 326
Polarity, 326
Polypeptide chain release factors, 158
Polypeptide elongation rates, 162
Position 64, 36, 37
Positions 38, 39, and 40, 35, 39, 49
PR, see Protease
Pre-B cells, endoplasmic reticulum of, 177
Prebiotic RNA replication, 382
Premature termination, 195
Primary recognition determinants, 297
Primer binding site (PBS), 4, 16
Primitive code, 376
Processivity errors, 219, 227
Programmed frameshifting, 234
Programmed frameshifts, 221, 223
Programmed translational alternatives, 236
Programming signal, 230

Prolactin, 120
Promoter elements
  extragenic, 142
  internal control regions, 142
  intragenic, 142
  tRNA genes, 142
Promoter regions, genes
  5S rRNA
    A box, 142
    C box, 142
  tRNA
    A box, 142
    B box, 142
    ICRs, 142
Proofreading, 105, 237, 355
Protease (PR), 14
Protein
  alanine insertion into, 295
  amino acid composition of, 114
  folding, 175
  production levels, 106
  stability, cellular components of, 175
  synthesis translational mechanisms, 17
Proteolysis, 182
Proteolytic degradation, 24
Pseudo-first order rate constants, 357
Pseudoknot, 230, 231
pyrBI operons, 163
pyrE operons, 163

Q

Q34, 32, 36, 40, 43, 44, 52, 54, 55, 60, 61, 65
queA, 68
Queues, 352

R

Rabbit
  globin, α and β chains of, 115
  reticulocyte tRNA population, 115
  selenocysteine tRNA in, 272
  tRNA, 11
Random codon usage, 400
Random coil configuration, 389
Random genetic drift, 421
Rare codons, 223
RAV, see Rous associated virus
Readthrough, 195, 207, 208, 336
  proteins, 211, 212, 214
  regulated, 215

relA, 222
relA mutants, 201, 238
Relative adaptedness values, 399
Relative synonymous codon usage
  (RSCU), 399, 400, 419
Release factor, 159, 207, 215, 225, 232, 359, 405
Release factor 1 (RF1), 159, 327
Replication accessory factor, 17
Replication error, 383
Restrictive phenotyping, 197
Restrictive *rpsL* allele, 214, 217
Restrictive *rpsL* mutants, 235
Restrictive *rpsL* mutation, 209
Reticulocytes, hemoglobin in, 114
Retroviral replication, 2
Retroviral-type zinc finger, 14
Retroviruses, 213, 216, 228
Retroviruses, tRNA in molecular biology of, 1–21
  historical evidence for tRNA involvement, 5–7
  HIV RT and tRNA$_3^{Lys}$ interaction, 8–13
  retroviral nucleocapsid protein acts as replication-accessory factor, 13–17
  retroviruses and their life cycle, 2–5
  tRNA in HIV life cycle, 7–8
Reverse transcriptase (RT), 2, 4, 6, 8, 60
RF1, see Release factor 1
Ribosomal ambiguity, 197
Ribosomal binding assay, 271
Ribosomal coding site, 357
Ribosomal distributions, 351
Ribosomal frameshifting, 5
Ribosomal pause sites, 219
Ribosomal transit, 353
Ribosome pauses, 218, 239
2′-Ribosylated guanosine, 37
RNA
  adaptor, 137
  -dominated prebiotic world, 377
  -DNA synthesis, priming of reverse, 24
  editing, 272
  polymerase, 180, 181, 349, 354
  polymerase II, 152
  polymerase III, 142, 152
  self-replicating, 382
  viruses, 211, 232
Rous associated virus (RAV), 6
Rous sarcoma virus (RSV), 2, 5, 14, 16, 17
rpsL, 163, 222
RSCU, see Relative synonymous codon usage

Index

RSV, see Rous sarcoma virus
RT, see Reverse transcriptase

S

s^5C32, 43
Saccharomyces cerevisiae, 30, 89, 96, 384
Salmonella typhimurium, 26, 35, 89, 96, 160, 406
s^2C32, 31, 43
Schizosaccharomyces pombe, 422
selA1, 45
Selenium, use of, 270
Selenocysteine, 217, 269–278, 335
  selenocysteine biosynthesis and incorporation of selenocysteine into protein, 272–273
  selenocysteyl-tRNA in nature, 273–275
  selenocysteyl-tRNA[Ser]Sec, 271–272
  UGA specifies selenocysteine, 270–271
Selenoprotein P, 270
Selenoprotein P RNA, 273
Sense-codon-containing mutant, 212
Sense codons, misreading of, 54
Sequence bias, 321
Sericin, 116
Serine-containing mutant, 212
Sheep reticulocytes, 115
Shifty stops, 225, 232
Silkworm moth, silk gland of, 114, 117
Simplified translational machinery, 371
Simultaneous mutations, likelihood of, 406
Simultaneous slippage, 229–231
SIN, see Sindbis virus
sin3, 45, 64
sin4, 64
Sindbis virus (SIN), 212
Single-stranded DNA, 16
7SK RNA gene, 152
Slippery codons, 229, 235
Slippery sequences, 229–231, 235
Slippery sites, 339
s^2m^5U54, 36, 50
Specificity constant, 284
Split boxes, 372, 368
SP6 RNA polymerase, 10
Stalled ribosomes, 235
Stem-loops, 230, 233
Stop codons, 398
Stop codon usage, 419
Strand-displacement activity, 5
Streptomycin, 238
Stringent response, 237–239
s^2U34, 63
s^4U8, 42
Suppression assay, 282
Suppressor efficiency, 42, 44, 293, 326, 357
Suppressor minus, 215
Suppressor tRNAs, 158
Synonymous codons, 88
Synonymous substitution, rate of, 107
Synonymous substitutions, 406
Synthetic peptide, 15

T

t^6A37, 33, 39, 41, 48, 52
TA/CG deficiency, universal rule of TG/CT excess and, 381–396
  antiquity of universal TA/CG deficiency rule, 389–392
  average amino acid composition of proteins yielded by universal codon assignment and TA/CG deficiency rule, 384–389
  codon assignment changes in evolution, 383–384
  codon assignments, 382–383
  notion of codon preference as illusion, 392–393
  unit of coding sequences, 393–395
Tandem codons, 53
TATA-like motifs, 152
TATA motifs, 152
Termination codon, 207, 213
Termination codons, variations in codon choice and selective use of, 397–425
  Escherichia coli, 398–407
    evolution at stop codons, 406–407
    sense codon usage, 398–400
    stop codon usage, 400–406
  eukaryotes, 409–422
    Drosophila melanogaster, 419–421
    fungi, 421–422
    human, 409–419
    maize, 421
    mouse, 419
    Saccharomyces cerevisiae, 409
  Gram-positive bacteria, 407–409
Ternary complex, 329
Ternary complex dissociation, 356
TFIIIA, see Transcription factor IIIA
TGA, 270

tgt, 43, 64
Tissues, aminoacyl-tRNAs in, 114
Tissue-specific gene expression, 151
Tissue-specific regulation, 144
Tobacco leaves, chloroplast genomes of, 371
Tobacco rattle virus, 212
trans-acting factor, 144
Transcription
 activation, 143
 differential, 143
 factor
  IIIA (TFIIIA), 13, 14, 142
  τ, 142
  TFIIIA, 142
  TFIIIB, 142
  TFIIIC, 142
  TFIIIC1, 142
  TFIIIC2, 142
  TFIIIR, 142
 modulatory element, 149
Transcriptional pause sites, 179
Transferase protein, 228
Transfer RNA, silkgland-specific, 144
Transframe protein, 229
Translation, 384
 efficiency, 99, 100
 in-frame suppression during, 5
 termination, 207, 208
Translation, codon discrimination in, 367–379
 classic principles of codon reading, 368–369
 evolution of codon discrimination, 376–377
 influence of structures outside anticodon on codon discrimination, 373–375
 role of nucleotide in position 32, 373–375
 influence of structures outside anticodon on suppressor efficiency, 375–376
 mechanistic considerations, 371–373
 unconventional methods in cell's normal repertoire of codon readings, 369–371
Translational context effect, 319–365
 biochemical demonstration of 3' neighbor effect on tRNA selection, 328–330
 clues to context mechanism, 325
 context biases at natural terminators, 330
 context effect on release factors, 356–359
 context effects on missense errors, 338
 context effects on tRNA-mediated nonsense suppression, 327

context and expression, 342–346
context and frameshifting, 338–341
context sensitive action by release factors, 330–331
context sensitivity affected by tRNA modification, 334–335
determination of UAG codon meaning by downstream sequence, 336–337
determination of UGA codon meaning, 335–336
effects of context on initiation, 337–338
existence of context constraints, 321–323
mechanism of aa-tRNA selection, 354–356
mechanisms of context action, 320–321, 348
mechanism for wobble interaction, 333–334
message decay, 348–349
message synthesis and polarity, 349
3' neighbor effects on tRNA and release factor by UAG suppression, 327–328
possible consequences for context structure of messages, 341–342
protein synthesis with 3' initiation of mRNA degradation, 352–354
protein synthesis with 5' initiation of mRNA degradation, 349–352
study of translation using suppression, 325–326
summary of statistical context investigations, 323–325
wobble-wobble context interactions observable *in vivo*, 331–333
Translational efficiency, 158, 392
Translational elongation rate, 352
Translational fidelity, 376
Translational initiation, 178, 179
Translational step-times, 158
Translational yield, 351
trmA5, 50
trmC1, 52, 63
trmC2, 52, 63
trmD3, 59
tRNA
 abundance, 162, 399
 allogenes, 150
 availability, 96
 gene family, 150
 genes, 93, 142
  Bombyx mori, 143
  DNA sequences, 151
  Drosophila, 148
  human, 150

Index **435**

oocyte-specific, 146
Saccharomyces cerevisiae, 149
silkgland-specific, 143
somatic-specific, 146
tyrosine, 146
Xenopus, 145, 146
hopping, 225, 226, 233
import, see Genetic origin of mitochondrial tRNAs
isoacceptors, 88
in *Mycoplasma capricolum,* 138
oligonucleotide, 10
population, functional adaptation of, 114
populations, 151
primer, 17
pseudogenes, 151
selection, 338
slippage, 223, 224, 229
posttranscriptional modifications in, 10, 11
recognition elements of, 283
role of anticodon in recognition of, 286–287
sitting time of, 374
specialization of, 114
tRNA interactions, role of modified nucleosides in, 23–85
with aminoacyl tRNA ligases, 27–36
general modification deficient tRNAs, 27–28
role of modified nucleosides in specific positions, 28–36
position 8, 30
position 10, 30–31
position 26, 31
position 32, 31
position 34, 31–32
position 35, 32
position 37, 32–35
position 46, 35
position 48, 35
position 54, 35–36
position 55, 36
positions 38, 39, and 40, 35
anticodon:codon interactions, 39–65
codon choice, 62–65
codon context sensitivity, 51–54
position 34, 52
position 37, 52–54
efficiency of translation, 41–51
position 32, 42–43
position 34, 43–45
position 35, 45–46
position 37, 46–49
position 54, 49–51
positions 38, 39, and 40, 49
fidelity, 54–62
frameshifting errors, 59–62
missense errors, 54–59
not apparently involved in translation, 37–39
with modifying enzymes, 26–27
with translation factors, 36–37
tRNA(m^{2}g10)methyltransferase, 30
tRNA(m^{5}U54)methyltransferase, 51
tRNA(m^{7}G46)methyltransferase, 35
tRNA-tRNA interaction, 41, 334
Trp codons, 388
Turnip yellow mosaic mosaic virus (TYMV), 292
TV-replacement loop, 137
21st naturally occurring amino acid, 270
Two-dimensional polyacrylamide gel electrophoresis, 89
Two-out-of-three reading, 372
Two out of three recognition mechanism, 134, 135, 138
TYMV, see Turnip yellow mosaic virus
Tyrosine tRNA, 45

U

UAA termination, 356
Unassigned codon, 207, 219
Unconventional codon readings, 370
Unicellular eukaryotes, 383, 395
Unicellular organisms, codon dialect for, 89
Universal coding system, 383, 384
Universal codon assignment, 385, 391, 392
Universal genetic code, deviations from, 130
U reads all, 372
Uridines, 62
U6 snRNA gene, 152

V

Variable arm, recognition determinants in, 303
Vitamin B12, synthesis of, 24

W

Watson-Crick-type base pair, 40, 369
Wobble, 132–135, 138
base pairs, 295
nucleoside, 64
rules, 368, 377

V-base tRNAs, 331
-wobble interaction, 331

X

Xenopus, selenocysteine tRNA in, 272
xm^{5}s^{2}U34, 62
xo^{5}U34, 40, 62

Y

Y base, 33, 52
Ψ35, 32, 36, 45, 46, 55
Ψ55, 27, 35, 36
Ψ+ factor, 208, 225
Yeast mitochondria, 370
yW37, 36, 41, 48, 60

Z

Zinc finger, 13
Zinc-binding domain, 14, 15